Medicinal Applications of Coordination Chemistry

Medicinal Applications of Coordination Chemistry

Chris Jones
School of Chemistry, University of Birmingham, UK
and School of Chemistry, The University of Manchester, UK

John Thornback
Mass Tag Technologies Ltd, Oxford, UK

RSC Publishing

Cover image reproduced by permission of Mark Woods from *Dalton Trans.*, 2005, 3829–3837.

ISBN: 978-0-85404-596-9

A catalogue record for this book is available from the British Library

Reprinted 2009

Published by The Royal Society of Chemistry,
Thomas Graham House, Science Park, Milton Road,
Cambridge CB4 0WF, UK

Registered Charity Number 207890

For further information see our web site at www.rsc.org

Typeset by Macmillan India Ltd, Bangalore, India
Printed by Henry Lings Ltd, Dorchester, Dorset, UK

Preface

During the later part of the 20th century, pharmaceutical compounds containing metals began to play an increasingly important rôle in medicine. In particular, the discovery of platinum anticancer drugs and the appearance of organ specific diagnostic-imaging agents containing technetium were prominent developments that have stimulated further interest in so-called 'metallopharmaceuticals'. Today, the metallopharmaceuticals industry has a global market measured in billions of pounds and has well-established applications in both diagnostic and therapeutic medicine.

The aim of this book is to review the diagnostic and therapeutic applications of selected metallic elements and consider the chemistry underlying the formulation of metallopharmaceuticals. In order to do this it is necessary to assume some basic understanding of the chemical behaviour of metals and, in particular, their 'coordination chemistry'. A very brief introduction to coordination chemistry is also presented for the benefit of those unfamiliar with this aspect of chemistry. The level of presentation of this section is designed for a general but scientifically literate audience, some of whom may have only a rudimentary knowledge of chemistry. It is hoped the text will be helpful to pharmacists as well as to clinicians and medical researchers using metal containing drugs in a clinical environment. However, it is also hoped that the book might also serve as an introductory text of interest to chemistry students wishing to learn about an aspect of their subject within an applied context, rather than just as a subject in its own right.

The discussion of medicinal applications focuses on the coordination chemistry of what are known as the 'transitional elements', particularly vanadium, manganese, technetium, ruthenium rhodium, platinum, gold and gadolinium. Compounds containing metals of this type are the most amenable to manipulation through chemical design so that coordination chemistry plays a significant rôle in the development of metallopharmaceuticals involving these metals. The so-called 'main group' metals (*e.g.* gallium, indium, tin, lead, bismuth) are, for the most part, not included though some of their applications are mentioned. These elements offer less scope for the application of coordination chemistry in product development as their compounds typically break down quickly *in vivo* transferring the metal to biological binding agents and a fate largely beyond the chemist's direct control.

The first introductory chapter aims to set the scene with a brief historical review and a consideration of some general aspects of the use of chemical substances in medicine. The second chapter provides a concise description of the basic chemical principles required for understanding the particular properties of transitional metal elements and which are exploited in medical applications. The chapter assumes relatively little chemical knowledge and those with prior University level chemical training may wish to skip most or all of this material. However, it is hoped that it will help those without a strong inorganic chemistry background to appreciate the origins and nature of the particular properties of metals that are so useful in biology and in medical applications. It could serve as a concise introduction to coordination chemistry within the context of applications in medicine for students on some chemistry courses.

Subsequent chapters examine the utilisation of the metallic elements in various medical contexts divided into chapters on Diagnosis and Therapy. In addition to describing what is happening in a clinical environment, these chapters seek to consider past and current research into metallodrugs and the lessons this provides for the future. Particularly important and well-established applications include diagnostic imaging through the use of radioactive or paramagnetic compounds and the treatment of cancer with metallodrugs based primarily on platinum compounds. The final chapter considers the design of new metallopharmaceuticals and some examples of the relationship between structure and activity. Serendipity is still important in the development of metallopharmaceuticals but, as more information emerges about the mechanism of action of metallopharmaceuticals, rational design is playing an increasingly important part.

In addition to established applications in medicine, the chemistry of metal compounds offers new opportunities for the diagnosis or treatment of disease. This potential for the future development of metallopharmaceutical agents has been recognised by major US medical research funding agencies and, in 2000, a conference entitled 'Metals in Medicine – Targets, Diagnostics and Therapeutics' was held on the National Institutes of Health (NIH) campus at Bethesda, Maryland. The meeting identified applications of metal compounds in magnetic resonance imaging, radiology and radiation therapy as expanding areas of activity. Research into therapeutic applications of metal complexes was thought to be relatively under-developed. One area identified as particularly ripe for research growth was metal metabolism, including studies of the mechanisms of metal homeostasis and the roles of metals in the regulation of cell function and cell–cell interaction. Finally, a need was identified for a set of basic principles to guide the development of new metallopharmaceuticals. At a chemical level at least, coordination chemistry provides a basis for the rational design of metal complexes. This will need to be combined with medicinal experience and an expanding knowledge of bioinorganic chemistry if such a set of guiding principles is to be developed. It is hoped this text will help stimulate wider interest in the potential of metal containing pharmaceuticals, encourage readers to explore more advanced texts and contribute to realising the opportunities metallopharmaceuticals present.

Acknowledgements

We are most grateful to Lee Cronin for looking over and commenting on the section reviewing platinum anticancer agents. In addition CJJ is indebted to his long-suffering wife Judy for all her help and support during the preparation of the manuscript.

Disclaimer

This text is for general educational purposes only and must not be relied upon in any way to guide clinical decisions. The authors and publishers make no claim that the applications and dosages mentioned herein would be suitable, or appropriate, for any particular individual. The manufacturers product information and medical literature applying to clinical usage must be consulted for each individual case. Even in the absence of a specific statement the use of general descriptive names, registered names, trademarks *etc.*, does not imply exemption of such names from the relevant regulations, or that they are free or available for general use.

Contents

Abbreviations

AO	atomic orbital
ATP	adenosine triphosphate
BBB	blood brain barrier
BNCT	boron neutron capture therapy
CAT	computer aided tomography
CEA	carcinoembryonic surface antigen
CFSE	crystal field stabilisation energy
CFT	crystal field theory
CN	coordination number
CSAP	monocapped square antiprism
CT	computed tomography
DAT	dopamine transporter
DDP	diaminedichloroplatinum(+2)
DELFIA	dissociation enhanced lanthanide enhanced fluorescence immuno-assay
DNA	deoxyribonucleic acid
DSA	digital subtraction angiography
FDA	US Food and Drug Administration
GI	gastrointestinal
GSH	glutathione
HPLC	high pressure liquid chromatography
HSA	human serum albumin
LGO	ligand group orbital
LMCT	ligand to metal charge transfer
Me	methyl
MLCT	metal to ligand charge transfer
MO	molecular orbital
MRI	magnetic resonance imaging
NMG^+	*N*-methylglucammonium
NMR	nuclear magnetic resonance
NOS	nitric oxide synthase
OER	oxygen enhancement ratio
PE	pairing energy
PET	positron emission tomography
Ph	phenyl
rCBF	regional cerebral blood flow

RNA	ribonucleic acid
SBM	Solomon Bloembergen Morgan
SER	sensitiser enhancement ratio
SOD	superoxide dismutase
SPECT	single photon emission computed tomography
STZ	streptozotocin
$T_{1/2}$	is used to denote radioactive half life
$t_{1/2}$	is used to denote a reaction half time
Tf	transferrin
TRFIA	time resolved fluorescence immunoassay
TTP	tricapped trigonal prism

Ligand Abbreviations

acacH		Chapter 4 in **133a**
boptaH$_5$		Chapter 3 **3b**
bpy	2,2′-bipyridyl	Chapter 3 **13**
cdtaH$_4$		Chapter 4 **12**
chrysi		Chapter 4 in **115**
cyclam		Chapter 2 in Figure 36
cyclen		Chapter 2 in Figure 36
dach		1,2-diaminocyclohexane
dfoH$_3$	DFO	Chapter 4 **7**
dfohopoH$_4$		Chapter 4 **14**
dipicH$_2$		Chapter 4 in **129a**
dmgH$_2$	MeC(=NOH)C(=NOH)Me	dimethylglyoxime
dmpaH$_2$		Chapter 4 **6**
dmpe	$Me_2PCH_2CH_2PMe_2$	Chapter 3 in Scheme 4
dmpsH$_3$		Chapter 4 **5**
dmsaH$_4$	DMSA	Chapter 4 **4**
dmso		dimethylsulphoxide
do3a-butrolH$_3$		Chapter 3 **5c**
dota-tppH$_4$		Chapter 3 **8**
dotaH$_4$		Chapter 3 **5a**
dotmpH$_8$		Chapter 4 **149**
dpaH$_2$		Chapter 4 **8**
dpdpH$_4$		Chapter 3 in **9**
dppe	$Ph_2PCH_2CH_2PPh_2$	Chapter 4 in **116**
dtpa-bmaH$_3$		Chapter 3 **3d**
dtpa-bmeaH$_3$		Chapter 3 **3e**
dtpa-eobH$_5$		Chapter 3 **3c**
dtpaH$_5$	DTPA	Chapter 3 **3a**

ECD		Chapter 3 in **52**
$eddaH_2$		Chapter 4 in **134**
edt^-	diethyldithiocarbamate	Chapter 3 **15**
$edtaH_4$	EDTA	Chapter 3 **4**
$edtmpH_8$		Chapter 4 **147**
$egtaH_4$		Chapter 3 **12**
emaH		Chapter 4 in **127b**
en	1,2-diaminoethane	Chapter 4 in **30**
$hbedH_4$		Chapter 4 **21**
$hedpH_4$	HEDP	Chapter 4 **148**
$hmdpH_4$	HMDP	Chapter 3 **45c**
HMPAO		Chapter 3 in **69**
hp-do3aH_3		Chapter 3 **5b**
HYNIC	hydrazinonicotinic acid	Chapter 3 **39**
$idaH_2$	IDA, iminodiacetic acid	Chapter 4 **10**
ImH		imidazole
ImMe		N-methylimidazole
$lihopoH_4$		Chapter 4 **13**
maH	maltol	Chapter 4 in **127a**
$mdpH_4$	MDP	Chapter 3 **45b**
ms325H_6	MS-325	Chapter 3 **7**
$ntaH_3$		Chapter 4 **11**
$pdtaH_4$		Chapter 3 **2**
phen	1,10-phenanthroline	Chapter 3 **14**
PnAO		Chapter 3 in **68**
$pypH_4$	PYP	Chapter 3 **45a**
$salenH_2$		Chapter 4 in scheme 13
$tetaH_2$	TETA, 2,2,2-tet	Chapter 4 **9**
$tthaH_6$		Chapter 3 **11**
tu		thiourea

CHAPTER 1

Introduction

1.1 Metals in Medicine – A Historical Perspective

The use of particular metals, or their compounds, in medicinal preparations can be traced back for thousands of years. Copper sulfate and alums were among the many substances used by the ancient Egyptians to prepare potions, possibly because they had a sterilising effect on the concoction produced. In Arabia and China gold preparations appear to have been used by physicians as long ago as 2500 BC and, more recently, mercury was used to treat syphilis during the European epidemic of the late 15th and early 16th centuries. Aqueous suspensions of gold flakes known as Goldschlager or Geldwasser have also been used in medicinal preparations, although there is no proven medical value in the ingestion of metallic gold. Koch's observation of the bactericidal action of gold cyanide in 1890 offered a more scientific basis for the use of gold in pharmacy. However, gold compounds were subsequently found to be ineffective in the treatment of pulmonary tuberculosis. A more successful application followed in the 1930s when gold drugs were used to treat rheumatoid arthritis. In this case double-blind studies showed them to be effective for many, though not all, patients. Earlier, in 1909, Erlich had introduced the arsenic compound Salvarsan for treating syphilis. This was followed by another arsenic compound, mapharsen, and in 1921 bismuth compounds were also introduced and used in combination with mapharsen to treat syphilis. These pharmaceuticals, particularly those involving arsenic, could have severe side effects and no doubt this contributed to a common perception that metals are generally toxic and not well suited to use in pharmaceuticals.

In the second half of the 20th century, two elements in particular played a large part in arousing much greater interest in the medicinal use of metal compounds; one of these was platinum and the other technetium. In Michigan State University in 1964, Barnett Rosenberg was investigating the effect of electric fields on the growth of bacteria and made a quite serendipitous discovery that some platinum compounds could inhibit cell division. This observation led on to the development of the platinum compound cisplatin, which was approved by the US Food and Drug Administration (FDA) in 1978 for use in the treatment of ovarian and testicular cancer. Cisplatin had been known for over 100 years previously but its medicinal potential remained unrecognised until Rosenberg's investigations. Since then other second generation platinum drugs have followed, including compounds suitable for oral administration. The discovery of cisplatin has also stimulated research into a variety of other metal compounds with tumorocidal properties and the potential to become clinically useful anticancer drugs.

Beyond these innovations in therapy, the man-made element technetium began to make an important contribution to diagnostic medicine during the later part of the 20th century. Technetium was first identified by C. Perrier and E. Segrè in 1937, being found in molybdenum targets after bombardment with deuterium nuclei in a cyclotron. All forms of technetium are radioactive and one form in particular has nuclear properties which make it particularly suitable for use in diagnostic medicine. This form emits γ-rays which, when originating from a source within the body, pass through living tissues and can be detected externally. This allows an image to be created of the distribution of the technetium γ-ray source within the body. Fortuitously, technetium also has a rich and versatile chemistry allowing it to be incorporated into many different kinds of compound. This allows the use of chemical compounds with affinities for different specific organs or tissue types to selectively transport technetium to specific locations in the body. In this way images of diseased or damaged regions can be obtained without the need for invasive surgical examinations. Other radioactive elements can also be used in non-invasive diagnostic imaging procedures but technetium has become pre-eminent in this application.

The use of radioactive elements as components of drugs suitable for use in therapeutic medicine offers a rather greater technical challenge than that posed by diagnostic imaging applications. However, recently the chemistry of metals has even begun to bear fruit in this difficult arena. It is over 100 years since Paul Erlich envisioned the development of a 'magic bullet', a dye carrying a toxic heavy metal which would target disease causing agents while leaving healthy tissue unharmed. He had developed the technique of staining tissue types with dyes (1877–1890), shown that a dye could kill trypanosomes infecting blood (1907) and prepared the arsenic compound Salvarsan to kill syphilis spirochetes (1909). The 'magic bullet' idea was visionary extension of these developments but the means to properly realise it did not exist at that time. Radioactivity was still a newly discovered phenomenon at the end of the 19th century, although its potential for use in therapeutic medicine was recognised around 1911. Radium preparations were used to treat various ailments including tumors, *e.g.* when inserted in vials for cervical cancer treatment at the Holt Radium Institute in Manchester. It was almost half a century later (1953) before Korngold and Pressman showed that antibodies labelled with radioactive iodine could be localised in tumors in rats. The use of radioactive emissions to kill tumors, rather than using a chemical toxin, offered the advantage that the pharmaceutical did not have to be internalised by the cell to exert its toxic effect. However, the necessary tumor specific antibodies were difficult to obtain in any quantity, restricting the clinical viability of the approach. It was further quarter of a century before Köhler and Milstein (1975) fused B-cells producing antigen specific antibody with myeloma cells to form hybrid cells expressing antibodies specific to a single target. Even then the development of immortalised monoclonal antibody cell lines did not address the problems of loading sufficient antibody onto the target tissue within an acceptable timescale, and with sufficiently rapid clearance from non-target tissue. Further developments in immunology, including the ability to manipulate fragment antibodies to obtain improved rates of uptake, were necessary for the 'nuclear magic bullet' approach to become viable. Unsurprisingly the first approvals for the clinical use of radiolabelled antibodies, or their fragments, were for diagnostic imaging purposes involving relatively low radiation doses to tissues. However, finally in 2002 Zevalin® (Ibritumomab) received FDA approval for treating types of B-cell non-Hodgkin's lymphoma. This compound contains a monoclonal antibody labelled with radioactive yttrium and heralds the clinical application of what is known as radioimmunotherapy using a radioactive metallic element. It seems that, for one type of disease at least, Erlich's 'magic bullet' had finally arrived.

Current medical practice has access to a variety of metal containing pharmaceuticals. In addition to the continued use of gold drugs to treat rheumatoid arthritis, lithium is now used to treat depression, platinum to treat certain types of cancer, bismuth to treat stomach ulcers, vanadium to treat some cases of diabetes, iron to treat anaemia, iron compounds to control blood pressure, cobalt in vitamin B_{12} to treat pernicious anaemia and certain radioactive metals to alleviate the pain of bone cancer. Beyond these therapeutic uses metals have also become important in diagnostic medicine, particularly diagnostic imaging applications. In addition to technetium, radioactive forms of thallium, gallium and indium are also used routinely for diagnostic imaging purposes. Another important diagnostic imaging technique, developed more recently, uses what is known as magnetic resonance to produce images of internal organs by examining the water content of the tissues involved. Metals with magnetic properties, particularly gadolinium, are finding use as a means of enhancing some of the images produced by this method. In addition to these examples various other metal compounds, still in a preclinical research and development phase, show promise for clinical use in therapeutic and diagnostic applications.

1.2 Metals and Human Biochemistry

The special chemical properties of metals have long been exploited by biological systems and various metals are essential for human health. However, despite their biological importance, metals nonetheless constitute a rather small proportion of living organisms. The human body is mostly water and an elemental analysis of a typical individual (Figure 1) reveals that hydrogen, oxygen, carbon and nitrogen together account for just over 99% of the atoms present. Calcium and phosphorus make up much of the remainder being the

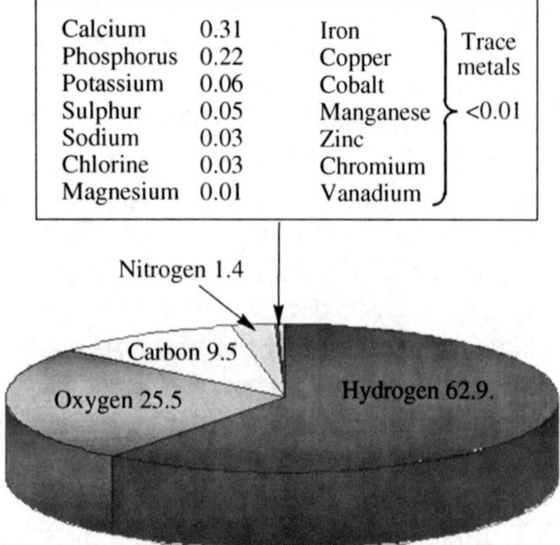

Approximate elemental composition as percentages of the total number of atoms.

Figure 1 *The elemental composition of a typical human*

primary constituents of bone, together with oxygen and some hydrogen. A particular characteristic of metals in aqueous media, contributing to their biological importance, is that they carry a positive charge. This distinguishes them from most organic species which, even when uncharged, tend to be more negative in character often attracting positively charged species in solution.

The metals present in the human body can be crudely divided into two types, bulk metals and trace metals. These two types share some chemical similarities but the trace metals introduce additional chemical characteristics beyond those displayed by the bulk metals sodium, potassium, magnesium and calcium. The bulk metals are present in relatively larger amounts compared to the trace metals (Figure 1) and have various biological rôles. Apart from its structural function in bone calcium, together with sodium and potassium, is involved in signal transmission through nerves. Calcium is also important in triggering actions, *e.g.* in muscles, while magnesium is found in some phosphohydrolase and phosphotransferase enzymes. Sodium and calcium are important extracellular bulk metals while potassium and magnesium are the major cellular metals. As a group, the bulk metals share broadly similar chemical properties but differ in the sizes and charges of the forms in which they exist in aqueous media. These differences are reflected in the properties of the metals so that sodium, for example, can be discriminated from potassium, which is larger in size, and from calcium which has a higher charge. The bulk metals are chemically promiscuous and generally do not form long-term associations with any particular entity, such as a protein. Rather they will usually move between hosts being quite rapidly exchanged. Compared to the trace metals their chemical reactivity is more limited and they remain in the same positively charged state throughout their time in the body.

In contrast with bulk metals, the trace metals show more varied chemical reactivity and, with the exception of zinc, can change the magnitude of their positive charge under the right conditions. The trace metals are usually found incorporated in a biological host entity such as a protein with which they will have a comparatively long-term relationship. Taken together the trace metals iron, zinc, copper, nickel, chromium, manganese, molybdenum, cobalt and vanadium barely constitute 0.01% by mass of the human body (Table 1). Despite this they are essential elements and any deficiency can have serious adverse consequences for health. Trace metals are involved in a wide variety of biochemical processes including the transport and storage of dioxygen, the dissolution of carbon dioxide in blood, the extraction of energy from food-stuffs, the breakdown of proteins, the removal of harmful species such as superoxide and many other chemical transformations. They can also have structural functions such as controlling the folding of proteins. While sharing some characteristics with the bulk metals as a result of their positive charges, the trace metals also possess additional chemical capabilities enabling them to participate in a wider range of chemical processes. For example, zinc is present in the enzyme carbonic anhydrase to catalyse the conversion of gaseous

Table 1 *Trace metals in humans*

Metal	*Mass*[a] *(mg)*	*Daily intake*[b] *(mg day^{-1})*	*Examples of some biological rôles*
Iron	4200	12 (male) 15 (female)	Dioxygen storage and transport, cytchromes enzymes – oxidases, reductases, hydrogenases
Zinc	2300	15	Structural control in proteins, enzymes involved in the chemical addition of water, alcohol dehydrogenase
Copper	72	2	Dioxygen storage and transport, electron transfer proteins
Nickel	15	–	Enzymes – urease, hydrogenases
Chromium	14	0.05–0.2	May be essential in mammalian glucose metabolism
Manganese	12	2	Enzymes – phosphatase, mitochondrial superoxide dismutase
Molybdenum	5	0.075–0.25	Enzymes – reductases, hydroxylases, nitrogenases
Cobalt	3	3(as vitamin B_{12})	Enzymes – as vitamin B_{12} coenzyme
Vanadium	0.11	–	Enzymes – nitrogenases, haloperoxidases

[a] Approximate amount in 70 kg average adult human
[b] Recommended adult daily intake requirement.

carbon dioxide to dissolved carbonic acid, while iron is present in hemoglobin and myoglobin to reversibly bind dioxygen. Iron is also found in systems, which add oxygen to organic substrates or transfer energy between biological components in metabolic energy transfer chains. Cobalt, in the form of vitamin B_{12} coenzyme, is used to carry out chemical transformations associated with structural changes in carbon-compounds, and copper is present in some superoxide dismutase enzymes to convert superoxide to dioxygen and peroxide. In each case the metal brings its own distinctive properties to the task and these are exploited by the biochemical system in which the metal participates.

Living organisms require efficient systems for managing metals. The bulk metals tend to be quite freely mobile but selective means of transporting them across cell membranes are necessary to maintain a delicate electrolyte balance and for nerve cells to function efficiently. Trace metals may be transported and stored by proteins. In particular iron is transported by transferrin and stored in ferritin while copper can be transported and stored in ceruloplasmin. Metals are also stored in the proteins, *e.g.* cytochromes or enzymes, within which they express their biological function. Some clinical applications of the chemistry of metals are concerned with the management of metals in the body. Metal containing pharmaceutical formulations can be used to treat deficiencies of a particular metal, *e.g.* iron for anaemia or cobalt as vitamin B_{12} for pernicious anaemia. Conversely the chemistry of metals may be exploited in treating problems arising from an excess of a metal in tissues, *e.g.* copper in Wilson disease or iron overload in patients receiving repeated blood transfusions. One approach to the treatment of iron overload exploits a natural microbial iron-sequestering agent. Since iron in the environment is generally found in a rather insoluble and inaccessible form, some micro-organisms excrete agents known as siderophores which strongly bind iron and facilitate its absorption into the micro-organism. Agents of this type can be adapted to bind iron in the body and promote its excretion providing an example of the application of bioinorganic chemical knowledge in a clinical application.

Beyond clinical applications which relate to the natural utilisation of trace metals, others exploit quite unnatural and non-physiological features of metal chemistry. As examples the toxicity of platinum complexes is managed and targeted in treatments involving anticancer drugs such as cisplatin while, in diagnostic medicine, the differing biodistributions of various technetium complexes can be exploited in imaging applications.

1.3 Metallopharmaceuticals

1.3.1 General Requirements

Metals may not seem an obvious choice as components of pharmaceuticals and it is a common perception that metal compounds are toxic, unstable and generally not well suited for pharmaceutical applications. Certainly in some historic uses of metal compounds, the treatment may have been as dangerous as the disease and early beneficial applications exploited the microbiocidal properties of metals such as mercury, arsenic and bismuth. Pharmaceuticals are more usually expected to be organic carbon based compounds, *e.g.* aspirin or penicillin, these being relatively unreactive chemically, often uncharged and amenable to structural variation to optimise their medicinal properties. However, over the past 30 years the particular chemical reactivities of metals, their magnetic and nuclear properties and the structural variety of their compounds, have become important in a variety of medical applications. Although not exactly Erlich's 'magic bullet', the organ specific uptake of technetium radiopharmaceuticals and the highly specific nature of the binding of cisplatin to DNA demonstrate the potential of this class of compound in specific medicinal applications.

In order to be useful in medicine, chemical compounds need to meet a variety of criteria. The most obvious requirement is that they must exhibit a medically beneficial effect with minimal toxic side effects. The relative benefit of a drug compared to its toxicity can be expressed in terms of its therapeutic index. This can be defined as the ratio of the dose required to kill 50% of test animals (LD_{50}) to the dose required to produce an effective therapeutic response in 50% of test animals (ED_{50}), *i.e.* Therapeutic Index = LD_{50}/ED_{50}. A high therapeutic index is clearly desirable, indicating that a relatively large dose is required to produce a toxic effect compared to that required to produce the therapeutic effect. However, because LD_{50} values cannot be obtained using human subjects, the therapeutic index can only have limited value for comparing different test compounds as it cannot be based on wholly human data.

As the dose of a compound with some medically beneficial effect is increased, and its *in vivo* concentration in the subject rises, it may be expected that the therapeutic benefit will increase. However, at some point this benefit will start to be negated by toxic effects until the dosage becomes so high as to be positively detrimental to the patient (Figure 2). Thus there will be a particular concentration range, associated with a particular dosage regime, which maintains the desired therapeutic effect. (A similar argument could be applied to foodstuffs. As an example carrots, a source of vitamin A, will produce a beneficial effect in normal dietary amounts. However, excessive consumption can lead to chronic or acute toxic effects, particularly in the liver where vitamin A accumulates. Those with liver damage, *e.g.* from over consumption of alcohol, are especially at risk). Typically the concentration of a drug will rise after administration then fall again as the compound is processed in the body and excreted. The timescale over which this process takes place, and any tendency for the drug to accumulate, will affect the dose regime necessary to maintain concentrations within the optimal therapeutic range.

A pharmaceutically useful compound must have an appropriate bioavailability and biodistribution, allowing it to pass through barriers such as cell membranes which lie between the outside world and the site of action (Figure 3). The rates of absorption, distribution, metabolism and elimination of a drug determine its pharmokinetics. Parenteral drugs enter the bloodstream directly, typically by intravenous injection, but oral drugs must be absorbed

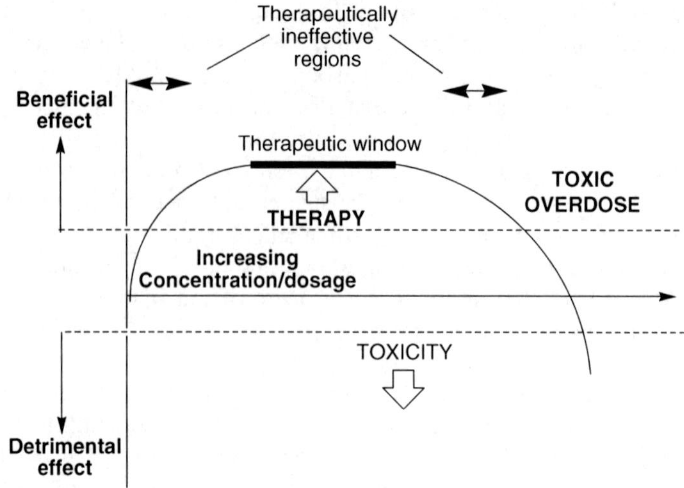

Figure 2 *The effect of increasing pharmaceutical dosage, or concentration in vivo, on benefit to the patient. Initially the beneficial effect increases with increasing concentration but at high doses, toxic effects predominate. Dosage regimes need to be adjusted to keep concentrations within the therapeutic window*

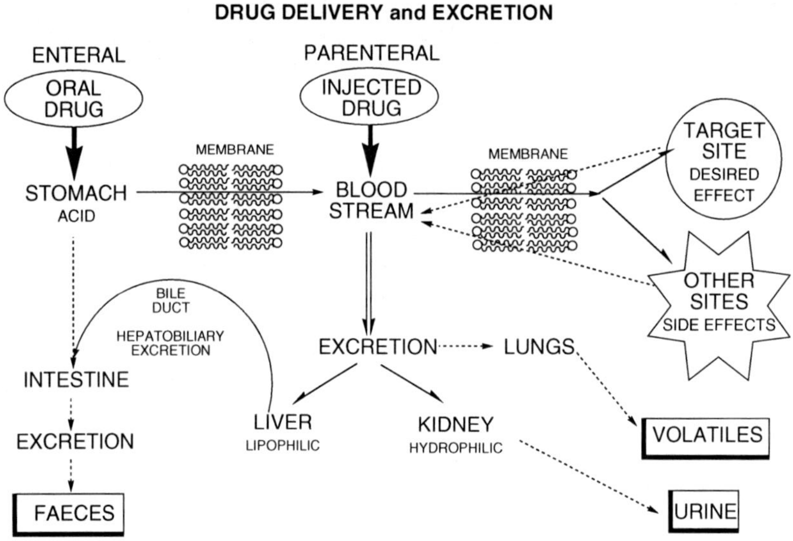

DRUG DELIVERY and EXCRETION
ENTERAL
ORAL DRUG
PARENTERAL
INJECTED DRUG
MEMBRANE
STOMACH ACID
BLOOD STREAM
MEMBRANE
TARGET SITE DESIRED EFFECT
OTHER SITES SIDE EFFECTS
BILE DUCT
HEPATOBILIARY EXCRETION
INTESTINE
EXCRETION
LUNGS
LIVER LIPOPHILIC
KIDNEY HYDROPHILIC
VOLATILES
EXCRETION
FAECES
URINE

Figure 3 *A diagram summarising the intake, distribution and excretion routes of a pharmaceutical*

through the gut (Figure 3). Consequently, oral drugs must be able to withstand the acidic conditions in the stomach and pass through the gut wall to enter the bloodstream in a suitable form. This presents further challenges in the design of oral drugs compared to parenteral drugs. Since uncharged compounds are generally absorbed through membranes more readily than charged compounds, neutral compounds, or those which become neutral under acidic conditions, tend to be more promising candidates for the formulation of oral drugs.

To be effective a drug will need to be sufficiently stable *in vivo* that it is not degraded by biological processes before having had chance to exert its effect. It will also need to reside in the body long enough to exert its effect but not so long as to irreversibly accumulate to the extent that unacceptable toxic effects start to arise. Sometimes drugs are administered in a form which is not the therapeutically active agent. Rather they are precursors which will be converted to the active form by metabolic processes *in vivo*. Such compounds are known as prodrugs and must show suitable pharmokinetics before and after their conversion to the active form.

The primary means of distributing a drug around the body will be *via* the bloodstream. Here pharmaceutical compounds can encounter a variety of challenges. Firstly, the compound will need to retain sufficient solubility in the aqueous saline environment of the blood. Secondly, interactions with the proteins and other species present in blood need to be considered. This is particularly important for metal containing drugs since proteins may compete to bind the metal and so influence its biodistribution and properties. In some cases this may be a desirable effect, in others it may not. As an example, radioactive gallium is easily converted to an insoluble form in water and so is injected as its citrate which is more stable. In the blood the iron transport protein transferrin competes with citrate for the gallium and transfers it to other sites where it is deposited. In other cases binding to serum proteins in the blood may lead to an unwanted distribution of the metal, so it may be necessary to inject it in a form resistant to competitive binding by serum proteins. A further challenge is provided by natural metabolic processes for affecting the breakdown of chemicals in the body. In the case of metal containing drugs, this process can sometimes release the metal and allow it to be converted to a form which may accumulate in the body. This issue needs to be considered in the design and formulation of a metallopharmaceutical agent. The elimination and excretion of a drug and its metabolic products can usually take place through the liver or kidneys. Water soluble, hydrophilic, compounds are generally excreted in urine *via* the kidneys (Figure 3). Here the process of glomerular filtration eliminates simple salts and small molecules while tubular secretion deals with larger molecules such as proteins, some of which may be readsorbed. Fatty, lipophilic, compounds tend to be processed through the liver and bile duct (hepatobiliary system) passing into the gut for readsorption or excretion in faeces.

1.3.2 Structure-Activity Relationships

Understanding the relationship between the molecular structure, and properties, of a chemical compound and its biological activity can provide an important tool to aid the design of new pharmaceuticals. Unfortunately, the complicated nature of the interactions between chemical compounds and biological systems does not make it easy to devise accurate quantitative structure-activity relationships. However, by collecting data on the behaviour of a large number of closely related compounds it may sometimes be possible to develop a general understanding of the importance of different structural features in the molecule. In some cases more quantitative correlations can be developed between a parameter describing a particular property of the compound and a particular biological effect.

One example of a parameter which has been used in this way is provided by a partition coefficient describing the distribution of a compound between water and a selected oily liquid which does not mix with water. The partition coefficient provides a numerical value reflecting the relative preference of a compound for oily organic regions, such as membranes, compared to aqueous media, such as blood serum. The numerical value may be correlated with a chosen pharmacological parameter such as the minimum concentration required to induce a particular physiological response. Data from a large number of experiments can be used to define the relationship between the partition coefficient and the physiological response. Once the form of this relationship has been established, the chemical structure of new trial compounds can be designed so as to optimise the partition coefficient without modifying other structural features essential for biological activity. In this way the search for new active compounds can be focused on those most likely to prove effective.

Usually drugs will need to interact with a site in the body which has a specific chemical structure so that the size and shape of a molecule are important design features. The presence of specific chemical groups at particular locations in the molecule can also be important as can the polarisability of parts of the molecule. Modern computational methods offer a powerful tool for modelling the properties of compounds and their compatibility with potential binding sites. Together with structure-activity relationships, such computer modelling can facilitate the design of new active compounds offering significant economies in the costly process of drug development. This approach is now well established for organic compounds but the inclusion of metals creates some additional complications when it comes to the design of new metallopharmaceuticals.

Where a metal is fully contained within an organic host, modification of the host structure to optimise biological distribution and activity might follow precedents set with non-metal containing drugs. However, often the properties or reactivity of metal will be an important feature of a metallopharmaceutical. In such cases the host structure containing the metal must not be modified in a way which might impair the ability of the metal to perform its function. Thus, although modelling and structure-activity relationships can be applied to metallopharmaceuticals, it is necessary to introduce additional considerations relating to the rôle of the metal and the nature of its interaction with the host structure in which it is contained. Some examples of the importance of structure in determining the efficacy of specific metallopharmaceuticals can be found in Chapters 3 and 4.

1.3.3 Clinical Trials

The evaluation of new compounds for clinical use involves several stages. There will be a discovery or design phase in which a serendipitous discovery, or a new example of a class of compound known to, or thought likely to, have medicinal properties, is evaluated. If sufficiently promising results are obtained, and acceptable toxicity levels established, the drug may proceed into clinical trials. These can be divided into four phases as follows:

PHASE I	Small groups of healthy volunteers receive the drug so that assessments can be made of its absorption, biodistribution, pharmokinetics, accumulation, side effects and dosage. Initially the trial will start with smaller doses than needed for therapeutic effects to check for adverse reactions. If the drug is shown to be safe in humans the trial may proceed to Phase II.
PHASE II	Small groups of patients suffering from the ailment to be treated receive the drug to establish its effectiveness. The optimum dosage regime and any adverse reactions are assessed.
PHASE III	A large number of patients with the ailment are evaluated in double blind trials involving the new drug, comparison treatments and placebos to establish the efficacy of the new drug and obtain safety data to support applications for its licensing and approval for clinical use.
PHASE IV	Monitoring of patients treated with the drug continues following approval and throughout general clinical use in order to further optimise procedures and check for previously undetected side effects.

This whole process can take many years. As an example US FDA approval for the clinical use of cisplatin was granted in 1978, some 14 years after the serendipitous discovery that platinum compounds suppressed cell division and 7 years after the start of Phase I clinical trials.

1.4 The Special Properties of Metals

1.4.1 Comparison with Organic and Biological Compounds

In aqueous solutions metals differ from organic compounds in a number of important respects. In organic compounds carbon–carbon bonds are generally rather stable and provide the robust skeleton, which gives parts of biological molecules their shape and stability. Where there is a need for carbon containing units, such as amino acids in proteins, to be connected and disconnected, carbon–nitrogen or carbon–oxygen bonds are usually used to form the linkages. Although bonds of this type are stable, compared to carbon–carbon bonds they can be more easily broken or formed through the addition or elimination of water. This can be done under biochemical control allowing proteins, for example, to be broken down and reassembled with comparative ease. Metals are often quite reactive towards changing the atoms they are bonded to in aqueous media. However, rates of reaction can vary substantially between different metals and between different compounds of the same metal. This offers a wide range of chemical behaviour. Sometimes it is useful to have metal compounds, which are inert and do not readily convert to other forms. Sometimes it is important for the metal compound to be sufficiently reactive to allow the metal to be exchanged and incorporated into specific biological systems. Controlling the reactivity of the metal compound is one way of controlling which biological system can have access to the metal.

The interactions between charged metal atoms and other species in solution are somewhat different from those involved in the interactions between organic biological compounds. Two biological compounds may enter into quite complicated interactions through mutual recognition and binding between several complementary sites in the two different species (Figure 4). The interactions between enzymes and their substrates or the complementary strands of DNA in a DNA duplex offer examples of this type of behaviour. The individual linkages may not be particularly strong compared to conventional chemical bonds, but can act in concert to give quite strong overall binding interactions. Unlike organic compounds, metals typically show what is known as 'Lewis acid' behaviour, that is they bind to atoms such as oxygen in water or nitrogen in ammonia. This allows them to spontaneously assemble groups of atoms around them to form a 'metal complex'. They can also bind to certain oxygen, nitrogen or sulfur atoms in proteins and other biochemical species. In this way they can influence the structures of organic substrates to which they become attached (Figure 5). They can also form reactive centres within biological species to which they have become bound. Metals bound within enzymes may act to

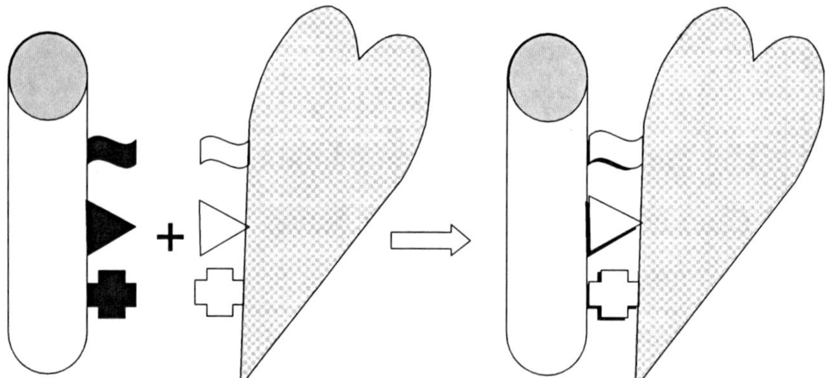

Figure 4 *A schematic representation of two biological molecules with complementary binding sites recognising each other and combining to form a "complex". Examples might be an enzyme and its substrate or the two strands of a DNA double helix*

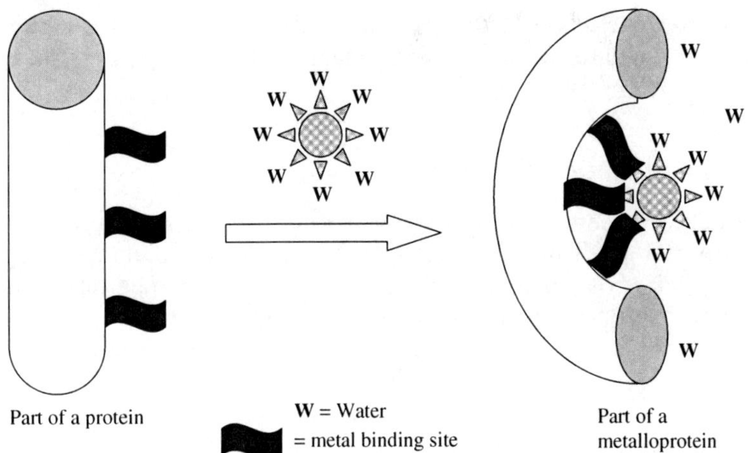

Figure 5 *A schematic representation of an aquated metal ion interacting with part of a biological substrate, such as a protein, to release some water and make the protein fold. This may just have a purely structural effect, or the metal may become a reactive centre within the biological molecule*

polarise a substrate to promote a particular reaction. An example is provided by zinc in carboxypeptidase acting as a centre catalysing protein degradation through the addition of water. Some metals can also change their charge and so act as electron transfer agents. This behaviour is found with iron in the cytochromes involved in mitochondrial electron transfer chains. Copper is also used as an electron carrier, for example in catalysing the conversion of superoxide to dioxygen and peroxide. Metals can also bind and activate small molecules such as dioxygen or nitric oxide in a much simpler and more reversible manner than would be possible using organic molecules. The use of iron centres to transport dioxygen in hemoglobin provides one obvious example.

Beyond the particular chemical features which metals bring to living systems and pharmaceuticals, they offer other attributes important for medical applications. Metals can form stable magnetic materials of a type becoming important in Magnetic Resonance Imaging applications. Organic compounds with magnetic properties tend to be very reactive and unstable *in vivo*. In addition the magnetism they can create is very limited compared to metals such as iron, manganese or gadolinium. Stabilising suitable forms of magnetic metals for use in magnetic resonance imaging applications presents a particular chemical challenge in (i) maintaining suitable magnetic behaviour (ii) controlling their biodistribution and (iii) preventing unwanted accumulation of the metal, for example in the liver. Radioactivity is another phenomenon with significant medical applications and a number of metallic elements are available in radioactive forms suitable for clinical use. In order to exploit this radioactivity it is again important to incorporate the metal in a compound with biodistribution and pharmokinetic behaviour suitable for the clinical application. This requires careful selection of the form in which the metal is administered and this, in turn, requires a good understanding of the chemistry of the metals involved.

1.4.2 Coordination Chemistry

The metallic elements used in medicine have characteristic features crucial to their particular application and not shared with organic compounds. In order to understand why a particular metal should be chosen for a particular medical application it is necessary to appreciate the properties and chemistry of the metals concerned. The branch of chemistry most closely concerned with the behaviour of metals under conditions relevant to living systems is known as coordination chemistry. This encompasses the chemistry of metals in aqueous media and their interactions with materials such as those encountered within living organisms or used in the formulation of metallopharmaceuticals. The origins of coordination chemistry as a distinct branch of chemistry date back to the beginning of the 20th century and are marked by the award of a Nobel prize to Alfred Werner in 1913. Among other achievements, Werner established the structure of the compound now know as cisplatin and used in cancer therapy.

To understand why metal atoms form compounds with particular structures and reactivities, it is necessary to have some appreciation of atomic structure and of chemical bonding models which go beyond those applicable to lighter elements such as carbon. The Crystal Field Theory proposed by Bethe in 1929 offered a starting point, although this was not developed into a generally used bonding model for coordination compounds until the 1950s. More elaborate theoretical models of bonding in metal compounds have since been developed but the Crystal Field Theory model continues to offer a relatively simple and accessible insight into the chemistry and magnetic properties of many metals. Chapter 2 attempts to provide a concise introduction to coordination chemistry. Those with prior experience of the chemistry of metals may wish to skip part or all of this chapter. Those with a limited chemistry background will hopefully find it a useful introduction to coordination chemistry, which underpins the subsequent chapters on diagnosis and therapy, and a platform for further reading on the chemistry of metals.

CHAPTER 2

The Chemistry of Metals in a Nutshell

2.1 The Structure of Materials

2.1.1 States of Matter

Our bodies, and the things which surround them in daily life, are made up from a variety of different materials. At the obvious level we can see that some are solids, some are liquids and some are gases. However, some apparently solid materials, *e.g.* human flesh, may turn out to be rather more complicated when examined in detail. In fact five different states of matter may be defined, all of which may be encountered in the context of medicine.

Solid – applies to rigid materials in which the location of the component particles remains essentially fixed over time for example in bones, teeth, *etc*. In crystalline materials the component particles are arranged in arrays with regular ordering in 3-dimensions.

Liquid crystalline – a fluid state in which the component particles can undergo relocation, and exchange positions, but some degree of long-range order persists in one or two dimensions. An example is provided by cell membranes in which large groups of rod shaped particles pack with their long axes aligned in a common direction. Compounds exhibiting this behaviour typically undergo a phase change from solid to liquid crystalline at their melting point, forming one or more liquid crystalline phases on heating before being ultimately converted to an isotropic liquid.

Liquid – a relatively high density, incompressible fluid state in which the component particles undergo continuous relocation with no long-range ordering, exemplified by water at physiological temperature.

Gas – a low-density compressible state in which the component particles undergo rapid more or less random relocation exemplified by air in the lungs. Gases such as carbon dioxide, oxygen and nitric oxide can be present in body tissues or fluids in dissolved form although they are gaseous in their normal state. Under some conditions of temperature and pressure a gas may enter a 'super critical fluid' state in which the sharp distinction between the high-density liquid and a low-density gas states disappears. Instead a fluid forms which can vary continuously in density without showing a distinct phase change from liquid to gaseous at a particular boiling point. Under a pressure of 55 bar at room temperature carbon dioxide, for example, forms a super critical fluid.

Plasma – a gaseous state containing separate charged atoms (ions) and electrons. This is usually associated with very high temperatures and plasma may be present during high power laser ablation procedures.

2.1.2 Chemical Elements

A material may be a single chemical compound,[1] such as common salt (sodium chloride, NaCl) or glucose ($C_6H_{12}O_6$), or it may be a composite mixture of different compounds. A compound is constructed from arrays of atoms, each atom corresponding with a particular chemical element.[2] At the time of writing over 112 different chemical elements have been identified, although some of the heavier elements are too unstable to exist long enough to be important in the natural world and can only be studied inside specialised laboratory equipment. The known elements are given different symbols for convenience (Table 1) and these are used in chemical formulae (*e.g.* $C_6H_{12}O_6$ for glucose), which denote the composition of materials. The elements may be arranged in a Periodic Table in which elements with related chemical properties are grouped together.[3] The modern Periodic Table (Table 2) directly reflects the structures of atoms of the elements and may be divided into blocks (Figure 1). The labels given to these blocks, *viz.* s, p, d and f, relate to the detailed atomic structures of the elements in the block, as explained later in Section 2.2.1. In effect these labels refer to the different outermost layers, or subshells, of electrons which surround the atomic nucleus.

Starting from hydrogen, with only one proton in the nucleus and an outer electron in an s subshell, the elements are constructed by adding further electrons to fill successive subshells and balance the positive charge on the nucleus. An s subshell has a capacity of 2 electrons giving the possible electron configurations s^1 and s^2 depending on whether one or two electrons are present. A p subshell has a capacity of 6 electrons giving possible electron configurations from p^1 to p^6, a d subshell has a capacity of 10 electrons giving possible electron configurations from d^1 to d^{10} and, finally, an f subshell has a capacity

[1] The term chemical compound relates to a single substance of well-defined and invariant chemical composition.

[2] A chemical element is a substance which cannot be divided further into components by chemical methods. An atom is the smallest unit of a chemical element, which retains the intrinsic properties of that element.

[3] Although the concept goes back to the ancient Greeks, it was Dalton who, in his Atomic Theory of 1803, identified atoms of the chemical elements as the fundamental building blocks of matter at the chemical level. This fundamental insight was followed by the development of the Periodic Table between 1860 and 1870 by Newlands, Meyer and Mendeleev, providing a basis for classifying elements and rationalising the similarities and differences in their chemical behaviour. Only in the early part of the 20th century was the structure of the atom revealed as containing a nucleus made up of positively charged protons and uncharged neutrons, surrounded at a distance by negatively charged electrons.

Table 1 *The symbols of the elements in order of atomic number*

1.	H	Hydrogen	39.	Y	Yttrium	77.	Ir	Iridium
2.	He	Helium	40.	Zr	Zirconium	77.	Pt	Platinum
3.	Li	Lithium	41.	Nb	Niobium	79.	Au	Gold
4.	Be	Beryllium	42.	Mo	Molybdenum	80.	Hg	Mercury
5.	B	Boron	43.	Tc	Technetium	81.	Tl	Thallium
6.	C	Carbon	44.	Ru	Ruthenium	82.	Pb	Lead
7.	N	Nitrogen	45.	Rh	Rhodium	83.	Bi	Bismuth
8.	O	Oxygen	46.	Pd	Palladium	84.	Po	Polonium
9.	F	Fluorine	47.	Ag	Silver	85.	At	Astatine
10.	Ne	Neon	48.	Cd	Cadmium	86.	Rn	Radon
11.	Na	Sodium	49.	In	Indium	87.	Fr	Francium
12.	Mg	Magnesium	50.	Sn	Tin	88.	Ra	Radium
13.	Al	Aluminium	51.	Sb	Antimony	89.	Ac	Actinium
14.	Si	Silicon	52.	Te	Tellurium	90.	Th	Thorium
15.	P	Phosphorus	53.	I	Iodine	91.	Pa	Protactinium
16.	S	Sulfur	54.	Xe	Xenon	92.	U	Uranium
17.	Cl	Chlorine	55.	Cs	Cesium	93.	Np	Neptunium
18.	Ar	Argon	56.	Ba	Barium	94.	Pu	Plutonium
19.	K	Potassium	57.	La	Lanthanum	95.	Am	Americium
20.	Ca	Calcium	58.	Ce	Cerium	96.	Cm	Curium
21.	Sc	Scandium	59.	Pr	Praesodymium	97.	Bk	Berkelium
22.	Ti	Titanium	60.	Nd	Neodymium	98.	Cf	Californium
23.	V	Vanadium	61.	Pm	Promethium	99.	Es	Einsteinium
24.	Cr	Chromium	62.	Sm	Samarium	100.	Fm	Fermium
25.	Mn	Manganese	63.	Eu	Europium	101.	Md	Mendelevium
26.	Fe	Iron	64.	Gd	Gadolinium	102.	No	Nobelium
27.	Co	Cobalt	65.	Tb	Terbium	103.	Lr	Lawrencium
28.	Ni	Nickel	66.	Dy	Dysprosium			
29.	Cu	Copper	67.	Ho	Holmium			
30.	Zn	Zinc	68.	Er	Erbium			
31.	Ga	Gallium	69.	Tm	Thulium			
32.	Ge	Germanium	70.	Yb	Ytterbium			
33.	As	Arsenic	71.	Lu	Lutetium			
34.	Se	Selenium	72.	Hf	Hafnium			
35.	Br	Bromine	73.	Ta	Tantalum			
36.	Kr	Krypton	74.	W	Tungsten			
37.	Rb	Rubidium	75.	Re	Rhenium			
38.	Sr	Strontium	76.	Os	Osmium			

Notes: Generic symbols An denotes an actinide (Ac to Lr), E usually denotes a non-metallic element, Ln a lanthanide (La to Lu), M a metal, R denotes a hydrocarbyl group *e.g.* CH_3, C_2H_5 and X and Z are general symbols for a variety of groups.

of 14 electrons giving f^1 to f^{14}. In going from hydrogen ($1s^1$) to helium ($1s^2$) the first shell is filled. Next the second shell comprising the 2s and 2p subshells is filled on going from lithium to fluorine and the $1s^2$ electrons now represent a 'core' electron shell. Thereafter the third shell, comprising the 3s and 3p subshells is filled with $1s^2$, $2s^2$, $2p^6$ constituting the core electrons. Because of the ways in which the electrons in the different subshells can interact, the 3d subshell does not become occupied until after the 4s shell has been filled. Similarly the 4d and 5d subshells start to fill after the respective 5s and 6s

Table 2 *The periodic table emphasising the transition series d-block and inner transition series f-block elements*[a]

IUPAC[b]	1	2	3	4	5	6	7	8	9	10	11	12	13	14	15	16	17	18
	H	He																
	Li	Be											B	C	N	O	F	Ne
	Na	Mg											Al	Si	P	S	Cl	Ar
4s 3d 4p	K	Ca	**Sc 21**	**Ti 22**	**V 23**	**Cr 24**	**Mn 25**	**Fe 26**	**Co 27**	**Ni 28**	**Cu 29**	**Zn 30**	Ga	Ge	As	Se	Br	Kr
5s 4d 5p	Rb	Sr	**Y 39**	**Zr 40**	**Nb 41**	**Mo 42**	**Tc 43**	**Ru 44**	**Rh 45**	**Pd 46**	**Ag 47**	**Cd 48**	In	Sn	Sb	Te	I	Xe
6s 5d 4f 6p	Cs	Ba	**La 57**	**Hf 72**	**Ta 73**	**W 74**	**Re 75**	**Os 76**	**Ir 77**	**Pt 78**	**Au 79**	**Hg 80**	Tl	Pb	Bi	Po	As	Rn
7s 6d 5f 7p	Fr	Ra	**Ac 89**	Rf 104 (Unq)	Db 105 (Unp)	Sg 106 (Unh)	Bh 107 (Uns)	Hs 108 (Uno)	Mt 109 (Une)	110 (Uun)	111 (Uuu)	112 (Uub)						

Ce 58	**Pr 59**	**Nd 60**	**Pm 61**	**Sm 62**	**Eu 63**	**Gd 64**	**Tb 65**	**Dy 66**	**Ho 67**	**Er 68**	**Tm 69**	**Yb 70**	**Lu 71**
Th 90	**Pa 91**	**U 92**	**Np 93**	**Pu 94**	**Am 95**	**Cm 96**	**Bk 97**	**Cf 98**	**Es 99**	**Fm 100**	**Md 101**	**No 102**	**Lr 103**

[a] Atomic numbers Z are shown beneath the element symbol for the d- and f- block elements, International Union of Pure and Applied Chemistry (IUPAC) symbols are shown in parentheses for the fourth row of the d-block.
[b] This numbering system for groups has been approved by the IUPAC and is the one used throughout this text.

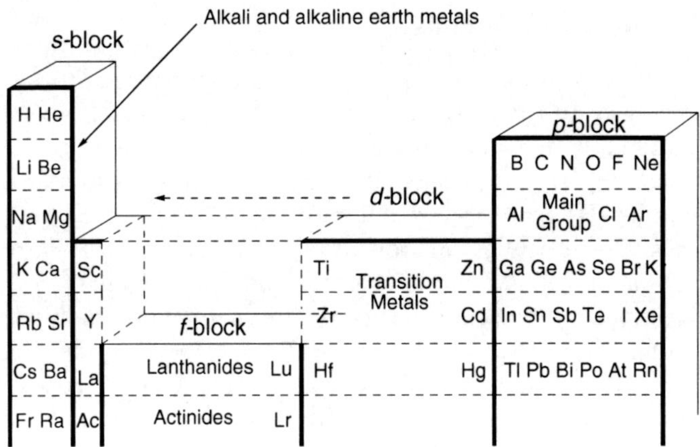

Figure 1 *Blocks within the Periodic Table*

subshells are filled. Again because of the details of the electronic structures of the atoms, the 4f subshell starts to fill after La and 5f after Ac. The noble gases He to Rn represent 'closed shell' configurations of electrons after which a new electron shell is started with an s subshell to give a new line in the Periodic Table. The electron configuration of the preceding noble gas being the core electron configuration. In the Periodic Table the vertical columns of atoms are known as groups and the horizontal rows of atoms as periods. Because they have similar numbers of electrons in their outermost layers, elements in the same group show some chemical similarities one to another.

The metallic elements occupy a large part of the Periodic Table and are normally characterised by being lustrous in appearance when pure, being malleable and ductile and by showing electrical conductivity. On the left side of the Periodic Table the reactive alkali metals (Na to Fr) and alkaline earth metals (Be to Ra), together with H and He form the s-block. On the right side are the p-block elements (B, C, N, O, F, Ne down, respectively to Tl, Pb, Bi, Po, As, Rn) among which the elements below the line from B to At show metallic properties. Between these two blocks are two series of transitional elements, which are all metallic in character. The f-block elements comprise the lanthanide series from La to Lu and the actinide series from Ac to Lr. To the right of the f-block lie the d-block elements also known as the transition metals. There are three rows of transition metals, the first from Sc to Zn, the second from Y to Cd and the third from La to Hg excluding the lanthanides.[4]

[4]Strictly speaking, the term transition element applies to an element with a partly filled d or f subshell and so excludes those with only filled or empty d or f subshells. However, it is convenient to include copper, silver and gold with the transition metals since, although their neutral atoms have d^{10} electron configurations, these elements commonly form compounds in which they have partly filled d subshells. Similar arguments apply to ytterbium and nobelium. It makes little sense to exclude them from the transitional lanthanide or actinide element series since both form stable trications with incomplete f subshell electron configurations. Despite the fact that their atoms, and their stable dications, have filled d subshell electron configurations, zinc, cadmium and mercury are also conveniently considered together with the transition metals.

2.1.3 Chemical Compounds

It is the nature of the atoms in a compound, and the particular ways in which they are arranged and can interact, which determine the properties of the compound and ultimately of the materials in which it is contained. As a simple example water (H_2O), constructed by joining an oxygen atom (O) to two hydrogen atoms (H) to form a water molecule (H_2O),[5] is a liquid at room temperature and an essential component of the human body [Figure 2(a)].

In contrast, under similar conditions common salt (sodium chloride, NaCl), is a solid composed of a cubic array of alternating positively charged, or cationic, sodium ions (Na^+) and negatively charged chloride anions (Cl^-) [Figure 2(b)]. Sodium chloride is a brittle solid which dissolves in water to form a liquid solution containing independent freely moving Na^+ and Cl^- ions, each surrounded by water molecules [Figure 2(c)]. Both Na^+ and Cl^- ions are essential components of much of the fluid in our bodies. In order to understand why water and sodium chloride are such different materials and have such different properties, it is first necessary to understand the structures of the atoms from which they are made. It is this atomic structure which determines the chemical properties of individual elements and so the nature of the compounds they form. Sodium easily loses an electron to form the sodium cation Na^+ while chlorine readily gains an electron to form the chloride anion Cl^-. Together these two ions form a solid salt containing alternating cations and anions. This arrangement is known as ionic bonding since, in forming sodium chloride, each sodium atom gives up an electron to a chlorine atom forming Na^+Cl^- in which the two oppositely charged ions are held together by electrostatic forces of attraction. Like chlorine, oxygen also readily adds electrons but elements like hydrogen or carbon would, if their outer valence shell of electrons were ionised, form small positively charged ions (H^+ or C^{4+}). These ions are harder to form than Na^+ and, being smaller, have a much higher charge to radius ratio. Thus they are far more polarising in character than Na^+ and so can distort the electron distribution in an anion such as Cl^- or O^{2-} leading to a situation in which one or more electrons are not fully transferred from one element to another but are rather shared between the two atoms. This leads to the formation of molecules such as water, H_2O or carbon dioxide, CO_2 in which the atoms are held together by covalent bonds.[6]

Although the interactions between adjacent atoms within these molecules may be strong, the interactions between the molecules themselves are weak compared to the interactions between Na^+ and Cl^- ions of NaCl or other ionic solids. Thus NaCl forms a high melting solid because a lot of heat energy is required to disrupt the lattice of Na^+ and Cl^- ions in solid NaCl. In contrast CO_2 is a gas at room temperature and water a liquid, since relatively little

[5] A molecule is the smallest unit of a compound which can exist in a free state, for water it is a single H_2O molecule and for sodium chloride it would be a single NaCl unit.

[6] A covalent bond is formed when atoms 'share' electrons so that a substantial energy input is required to separate the atoms. Atoms tend to stick together slightly even without any bond formation but in such cases little energy is needed to separate them.

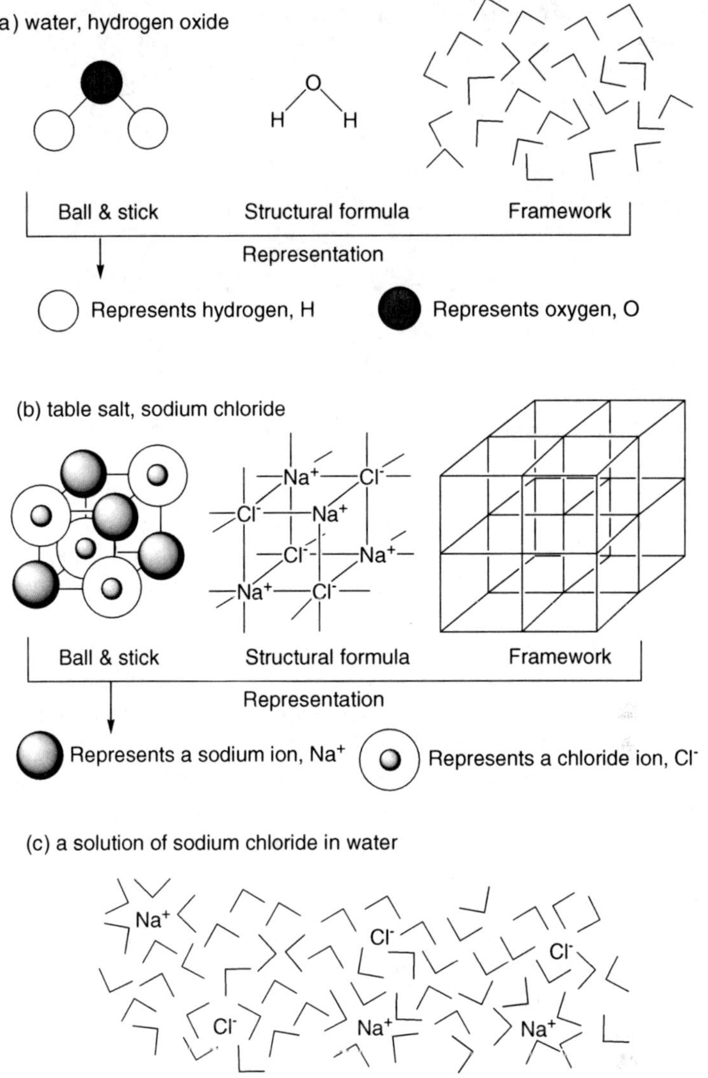

Figure 2 *Diagrams showing ways of representing (a) water, (b) part of a sodium chloride crystal lattice and (c) a solution of sodium chloride in water*

energy is needed to separate the CO_2 or H_2O molecules. Only by reducing the temperature (to 0 °C in the case of water and −79 °C in the case of carbon dioxide) and thereby the heat energy available to separate the component H_2O or CO_2 molecules, can these compounds be obtained as solids.

In practice bonds between atoms are not usually purely ionic or purely covalent in character but somewhere in between. Extremes are represented by the dipolar almost fully ionic K^+ Cl^- bond in potassium chloride and the non-polar covalent C–C bond in ethane. $H_3C–CH_3$. In $F_3C–CH_3$ the C–C bond is still covalent but the fluorine atoms attract electrons so as to polarise the bond and impart some polar character so that the situation might be represented by $F_3C^{\delta+}$–$^{\delta-}CH_3$ (δ+ and δ– represent fractional charges). Similarly in water the H–O bonds are polarised and might be represented $H^{\delta+}$–$^{\delta-}O^{\delta-}$–$^{\delta+}H$. In fact there is sufficient ionic character that a small proportion of the molecules in water may break up or ionise to give H^+ and ^-OH ions which can interact with other water molecules

$$H^+—^{\delta-}OH_2{}^{\delta+}—^-OH$$

The tendency of an atom of an element to attract electrons to itself is measured in its electronegativity, given the symbol χ (Table 3). Thus the most electronegative elements such as fluorine or oxygen have a strong tendency to attract electrons and form anions (*e.g.* F^- and O^{2-}) and the largest χ values. The least electronegative, or more electropositive, elements tend to lose electrons and form cations (*e.g.* Na^+ and Ca^{2+}) and have the smaller χ values. When atoms of similar electronegativity combine less polar covalent bonds are formed. Where there are large differences in electronegativity, more polar more ionic bonds arise. Generally, metals are electropositive elements and their chemistry is dominated by the tendency to lose electrons. However, it is worth noting that some metals such as platinum have electronegativities, which allow for significant covalent contributions to their bonding with elements such as carbon.

Trends in properties such as electronegativity become apparent when elements are arranged in the Periodic Table. Accordingly metals may be divided into different types according to their general chemical properties and their position in the Periodic Table. Perhaps the simplest chemical features are associated with the s-block metals which exist in aqueous solution as positively charged ions with closed shell electron configurations as described in Section 2.2.3. Among these Na^+, K^+, Mg^{2+} and Ca^{2+} are important in biology and Li^+ is used in the treatment of manic depression. Radioactive strontium is also sometimes used in radiation therapy to alleviate pain in bone cancer patients. Soluble barium compounds are highly toxic but the very insoluble salt $BaSO_4$ is used as a contrast agent in X-ray imaging. The toxic Ba^{2+} is rendered unavailable through the extremely low solubility of its sulfate salt and is excreted without toxic effect.

The p-block metals may lose their outermost, valence, electrons to form cations having closed shells of electrons, *e.g.* Al^{3+} and Ga^{3+} which have the same electronic structure as argon and krypton, respectively. However, they may also form compounds containing cations in which the outer s subshell remains filled and only electrons in the p subshell are removed by ionisation, as in Tl^+ for example. In the main the metallic p-block elements are relatively unimportant in biology but radioactive forms of metals such as gallium, thallium and indium have been widely used in diagnostic medical imaging.

The d-block metals often form cations with a partially occupied outer d subshell and the lighter examples from manganese to zinc play an important rôle in biology. Heavier elements in this block, such as technetium or platinum, have important medical applications, although they have no known role in biology.

Table 3 *The electronegativites of the elements*[a]

1	2	3	4	5	6	7	8	9	10	11	12	13	14	15	16	17	18
H	He																
2.2																	
Li	Be											B	C	N	O	F	Ne
1.0	1.6											2.0	2.6	3.0	3.4	4,0	
Na	Mg											Al	Si	P	S	Cl	Ar
0.9	1.3											1.6	1.9	2.2	2.6	3.2	
K	Ca	**Sc**	**Ti**	**V**	**Cr**	**Mn**	**Fe**	**Co**	**Ni**	**Cu**	**Zn**	Ga	Ge	As	Se	Br	Kr
0.8	1.0	**1.4**	**1.5**	**1.6**	**1.7**	**1.6**	**1.8**	**1.9**	**1.9**	**1.9**	**1.7**	1.8	2.0	2.2	2.6	3.0	3.0
Rb	Sr	**Y**	**Zr**	**Nb**	**Mo**	**Tc**	**Ru**	**Rh**	**Pd**	**Ag**	**Cd**	In	Sn	Sb	Te	I	Xe
0.8	1.0	**1,2**	**1.3**	**1.6**	**2.2**	**1.9**	**2.2**	**2.3**	**2.2**	**1.9**	**1.7**	1.8	2.0	2.1	2.1	2.7	2.6
Cs	Ba	**La**	**Hf**	**Ta**	**W**	**Re**	**Os**	**Ir**	**Pt**	**Au**	**Hg**	Tl	Pb	Bi	Po	As	Rn
0.8	0.9	**1.1**	**1.3**	**1.5**	**2.4**	**1.9**	**2.2**	**2.2**	**2.3**	**2.5**	**2.0**	2.0	2.3	2.0			
Fr	Ra	**Ac**															
		1.1															

Ce – Nd		**Sm – Er**		**Tm – Lu**	
1.1		**1.1**		**1.1**	
Th	**Pa**	**U**	**Np**	**Pu**	**Am - No**
1.3	**1.5**	**1.4**	**1.4**	**1.3**	**1.3**

[a] Electronegativites, rounded to one decimal place, are taken from A.L. Allred, *J. Inorg. Nucl. Chem.*, 1961, **17**, 215 or from L. Pauling, *The Nature of the Chemical Bond*, 3rd edn. Cornell University, Ithaca, New York, 1960, 93. The values are based on thermochemical data analysed using Pauling's approach for the elements in their 'normal' oxidation state *viz* M^{3+} for Sc, Y and La; M^{+} for Cu, Ag and Au, M^{2+} for the other d-block metals, Ln^{3+} for the lanthanides and An^{3+} for the actinides.

The f-block elements comprise two series with differing properties. The lanthanides from La to Lu show rather similar behaviour to one another with an aqueous chemistry dominated by Ln^{3+} ions which vary in size but otherwise essentially behave as if they were closed electron shell systems. Some of these elements are important medically because of their magnetic properties or long-lived luminescence. The second series of f-block elements is the actinides. The early members of this series out to americium show chemical properties intermediate between those of the d-block elements and the lanthanides but the later actinides are much like the lanthanides in chemistry. Actinides do not have significant rôles in biology, or yet in medicine, although fission of uranium is an important source of radionuclides.[7]

2.2 The Structures and Properties of Atoms

Under normal conditions a chemical reaction involves the electrons occupying the outermost shells, or valence shells, of the atoms involved. The inner or 'core' electrons do not take part in chemical reactions. Hence the chemical properties of an atom arise from its tendency to lose electrons from, or to attract electrons to, its valence shell. This tendency will depend upon the electronic structure of the atom and the nuclear charge experienced by the valence shell electrons. Thus, in order to explain the chemistry of an element, it is first necessary to consider the electronic structure of its atoms and how this influences the binding of its valence shell electrons and the directional properties of interactions between atoms.

2.2.1 Electron Configurations

In an atom of the simplest element, hydrogen, the nucleus consists of a single proton, and this is associated with a single electron. The proton is almost 2000 times heavier than the electron, which might be thought of as occupying an orbit around the proton rather like the moon around the earth. However the quantum theory shows us that an electron has both wave like and particle like qualities so that an electron in an atom is not associated with a fixed point in space but rather it is distributed around the nucleus (Figure 3).

An unconfined electron can be represented by a wave, the wavelength of which is determined by the momentum of the electron; the faster it is moving the more momentum it has and the shorter its wavelength. The amplitude of the wave at any point in space can be related to the probability that the electron might be found at that point. When an electron becomes confined through the attractive positive charge of a nearby nucleus the waveform which describes the

[7] A nuclide is a general term for an atom characterised by the number of protons and neutrons in its nucleus. The number of protons is its *atomic number* and determines which element the atom represents. The total number of protons and neutrons is the *mass number* of the nuclide. Isotopes are atoms with the same atomic number but different mass numbers, that is the same number of protons but different numbers of neutrons. The prefix 'radio' indicates an unstable nuclide or isotope which undergoes radioactive decay.

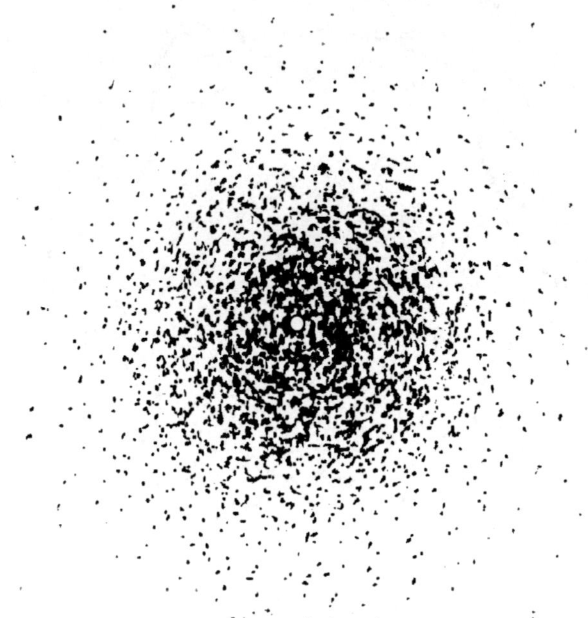

Figure 3 *A schematic representaion of the diffuse nature of the electrons in an atom, the density of dots representing the probability of finding the electron in that region. The white spot at the centre represents the nucleus, but not to scale; if an atom were the size of a Cathedral the nucleus would be about the size of a pea*

electron also becomes confined. Thus the 'orbits' which an electron may occupy around the nucleus have to meet certain criteria. The mathematical expression which represents the waveform of the electron is known as a 'wavefunction', and this describes the 'orbit' of the electron in the field of the positively charged nucleus. The wavefunction must only have one value at any point, this value must be finite and the wavefunction must be continuous. As an analogy consider a wave formed in a circle round a point representing the nucleus (Figure 4). Only a particular wavelength will give rise to a continuous wave at a given distance from the nucleus. A different wavelength would not fit the circle and would result in a discontinuity and is an unacceptable model for the behaviour of the electron. In 3-dimensional space, the waveforms are more complicated but the wavefunction must still meet exact criteria to be a valid representation of the electron in the positive charge field of the nucleus.

In this wave model the distribution of each electron within an atom may be uniquely defined by four *quantum numbers* and it is a requirement of the theory that no two electrons in an atom may share the same four quantum numbers. The *principal quantum number*, n, defines which shell around the nucleus is occupied by the electron. The values allowed for n are positive integers other than 0. The *azimuthal quantum number*, l, defines which subshell the electron occupies within the shell defined by the principal quantum number. The values allowed for l are the positive integers including 0 up to n–1. The *orbital magnetic quantum*

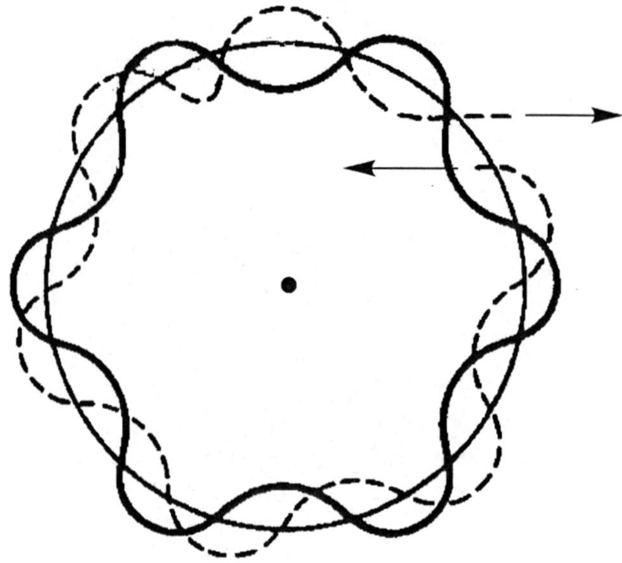

• Represents the nucleus

Figure 4 *An illustration of a continuous wave (solid line) and a discontinuous wave (broken line) in 2-dimensions around a nucleus*

number, m_l, defines which individual *orbital* an electron occupies within a subshell and m_l may take integer values from $-l$ to $+l$. Finally each electron has a *spin magnetic quantum number*, m_s, which may have one of only two values, $+1/2$ or $-1/2$. Provided they have different m_s values, two electrons may occupy the same orbital defined by n, l and m_l (Figure 5). Based on these quantum numbers atoms can be built up as electrons are added to the atom to cancel the positive charge on the nucleus (Figure 6). Following this methodology, known as the '*Aufbau principle*', atoms of all the elements in the Periodic Table can be constructed in such a way that none of the electrons share the same four quantum numbers within an atom. As an example carbon has the electron configuration $1s^22s^22p^2$. The structure of the atom comprises of a 1s shell filled by two electrons (only s with $l = 0$ is possible for $n = 1$), a 2s subshell ($l = 0$) filled by two electrons and a 2p subshell ($l = 1$; $m_l = -1, 0, +1$) with a capacity for six electrons, but only partly occupied by two electrons in the case of carbon, to bring the total to six electrons balancing the nuclear charge of 6+ (Figure 7).

2.2.2 The Spacial Distribution of Electrons in Atoms

So far we have considered the added shells or subshells of electrons rather as if they were just layers as in an onion. It is true that an electron in the 1s shell will, on average, be closer to the nucleus than one in a 2s subshell and that in turn closer to the nucleus than one in a 3s subshell (Figure 8). However, the electron density in the 2s and 3s subshells does not fall off uniformly between the outer

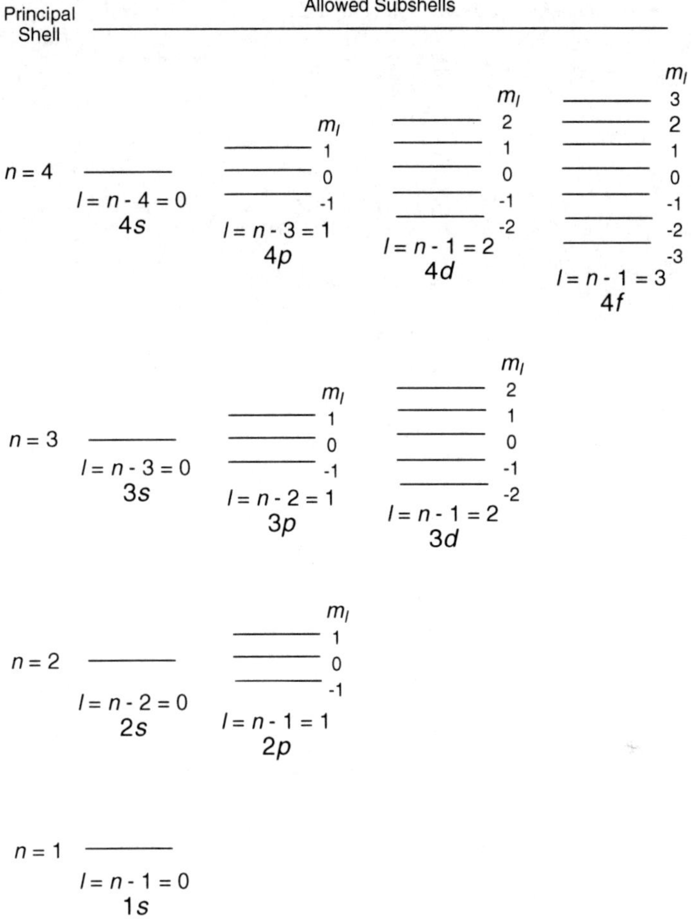

Figure 5 *A diagram representing the hierarchical structure of the atom and the quantum states up to $n = 4$ available to an electron in an atom. For $n = 1$ only $l = 0$ is allowed; for $n = 2$, $l = 0$ and $l = 1$ are allowed; for $n = 3$, $l = 0$, $l = 1$ and $l = 2$ are allowed; for $n = 4$, $l = 0$, $l = 1$, $l = 2$ and $l = 3$ are allowed. Each horizontal line represents a single orbital (defined by n, l, m_l) with a capacity to hold up to two electrons provided they have different m_s values*

part of the subshell and the nucleus, a significant proportion of the electron density is present in the interior of the atom. Similar behaviour is found for the p, d and f subshells, although the d subshell does not appear until the 3rd principal shell ($n = 3$) is reached and the f subshell until $n = 4$ is reached. All this has an important bearing on the ways in which the negatively charged electrons in one subshell shield the electrons in another subshell from the positive nuclear charge. Although an electron in a 3s subshell is predominantly near the outer surface of the atom, it penetrates the 3d and 3p subshells to some degree. The 3d subshell is somewhat buried in the filled 3s and 3p subshells,

Element	Nuclear charge Z	Electron Quantum numbers	Subshell	Electron Configuration
H	1	$n = 1, l = 0, m_l = 0, m_s = +{}^1/_2$	$1s$	$1s^1$
He	2	$n = 1, l = 0, m_l = 0, m_s = +{}^1/_2$	$1s$	
		$n = 1, l = 0, m_l = 0, m_s = -{}^1/_2$		$1s^2$
Li	3	$n = 1, l = 0, m_l = 0, m_s = +{}^1/_2$	$1s$	
		$n = 1, l = 0, m_l = 0, m_s = -{}^1/_2$		
		$n = 2, l = 0, m_l = 0, m_s = +{}^1/_2$	$2s$	$1s^22s^1$
Be	4	$n = 1, l = 0, m_l = 0, m_s = +{}^1/_2$	$1s$	
		$n = 1, l = 0, m_l = 0, m_s = -{}^1/_2$		
		$n = 2, l = 0, m_l = 0, m_s = +{}^1/_2$	$2s$	
		$n = 2, l = 0, m_l = 0, m_s = -{}^1/_2$		$1s^22s^2$
B	5	$n = 1, l = 0, m_l = 0, m_s = +{}^1/_2$	$1s$	
		$n = 1, l = 0, m_l = 0, m_s = -{}^1/_2$		
		$n = 2, l = 0, m_l = 0, m_s = +{}^1/_2$	$2s$	
		$n = 2, l = 0, m_l = 0, m_s = -{}^1/_2$		
		$n = 2, l = 1, m_l = 1, m_s = +{}^1/_2$	$2p$	$1s^22s^22p^1$
C	6	$n = 1, l = 0, m_l = 0, m_s = +{}^1/_2$	$1s$	
		$n = 1, l = 0, m_l = 0, m_s = -{}^1/_2$		
		$n = 2, l = 0, m_l = 0, m_s = +{}^1/_2$	$2s$	
		$n = 2, l = 0, m_l = 0, m_s = -{}^1/_2$		
		$n = 2, l = 1, m_l = 1, m_s = +{}^1/_2$	$2p$	
		$n = 2, l = 1, m_l = 1, m_s = -{}^1/_2$		$1s^22s^22p^2$
N	7	$n = 1, l = 0, m_l = 0, m_s = +{}^1/_2$	$1s$	
		$n = 1, l = 0, m_l = 0, m_s = -{}^1/_2$		
		$n = 2, l = 0, m_l = 0, m_s = +{}^1/_2$	$2s$	
		$n = 2, l = 0, m_l = 0, m_s = -{}^1/_2$		
		$n = 2, l = 1, m_l = 1, m_s = +{}^1/_2$	$2p$	
		$n = 2, l = 1, m_l = 1, m_s = -{}^1/_2$		
		$n = 2, l = 1, m_l = 0, m_s = +{}^1/_2$		$1s^22s^22p^3$

Figure 6 *The construction of the first two periods of the Periodic Table showing the quantum numbers of the individual electrons and their relationship to the subshells of electrons within the atom for elements from H to Ne*

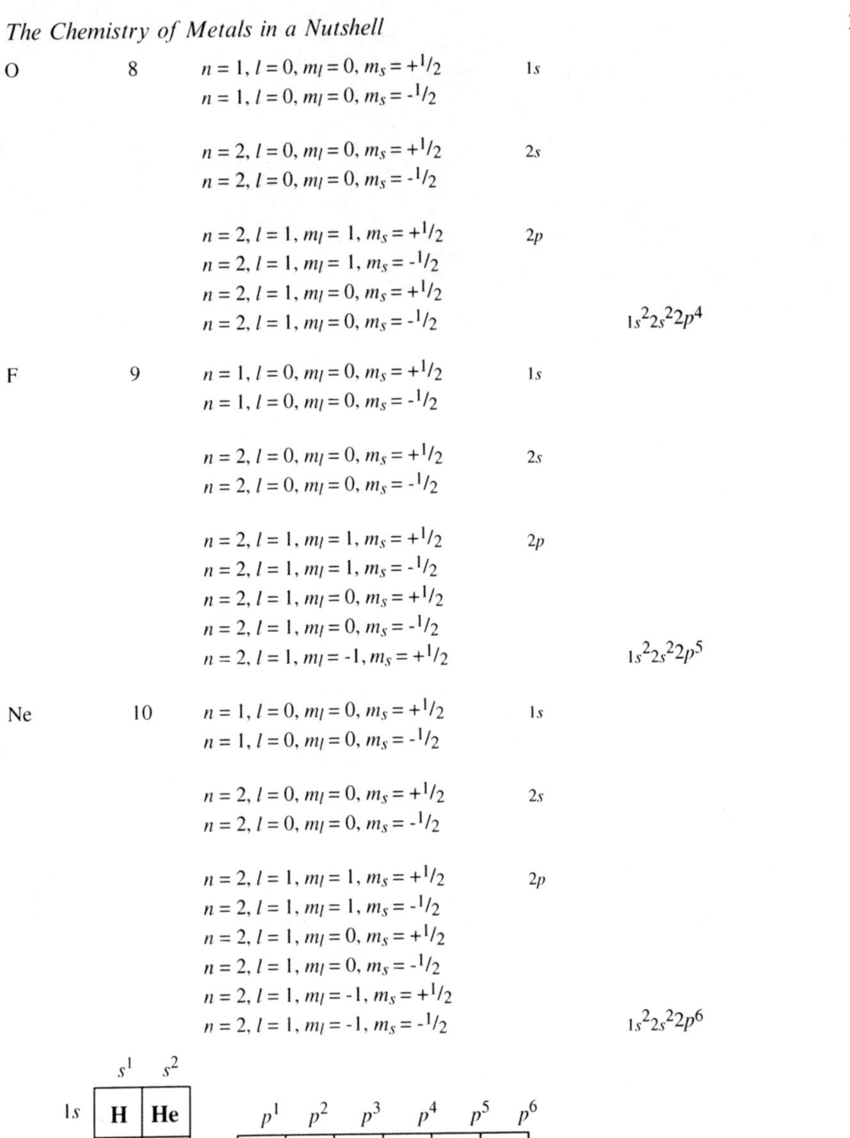

Figure 6 (*Continued*)

so electrons in this subshell will shield electrons in 3s and 3p subshells from the nuclear charge to some extent (Figure 9). This complicated interplay between shielding and penetration for electrons in different subshells has a pronounced effect on the nuclear charge experienced by electrons in the different subshells, and so on their energies. In a hydrogen atom the subshells of equal l within a principal shell all have the same energy because only one electron is present. In multi-electron atoms the effects of shielding, penetration and repulsion between electrons lead to a more complicated situation. It is convenient to represent

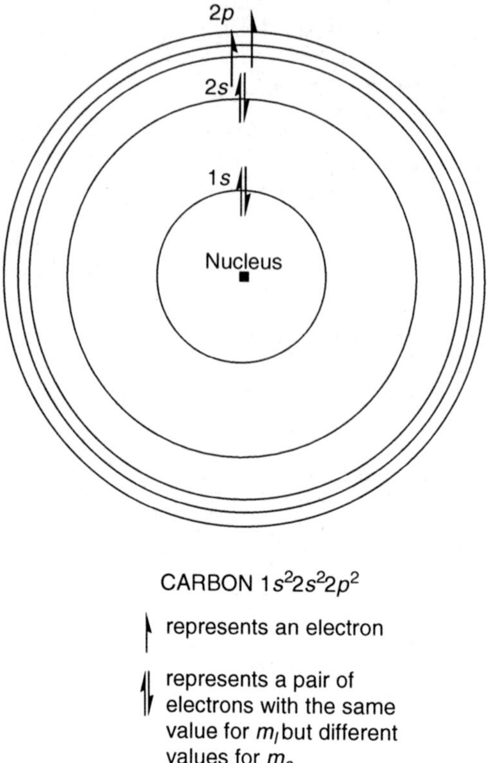

Figure 7 *A schematic representation of the structure of a carbon atom in cross section showing subshells of electrons as if layers in an onion. (Note that the 1s subshell is in effect the 1st electron shell as no other subshells arise for n = 1. The 2nd electron shell contains the 2s and 2p subshells)*

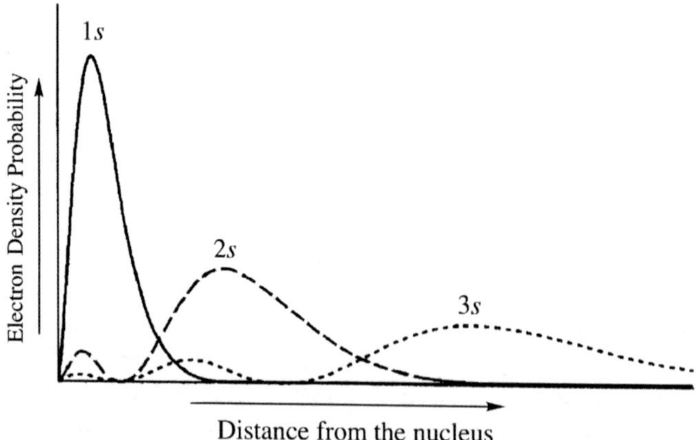

Figure 8 *Plots showing how the probability of finding an s subshell electron varies with distance from the nucleus for the 1s, 2s and 3s subshells*

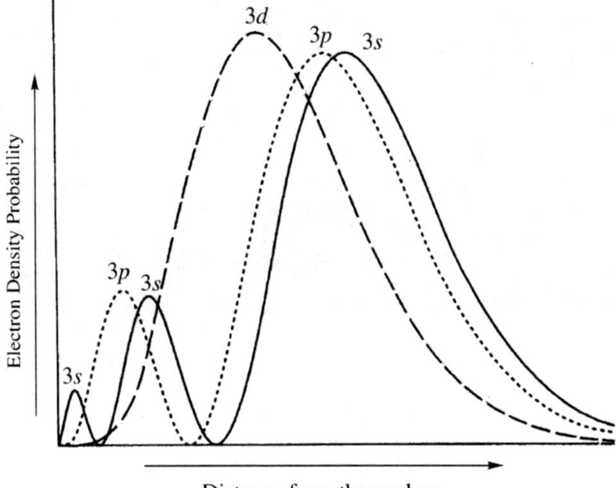

Figure 9 *Plots showing how the probability of finding an electron varies with distance from the nucleus for the 3s, 3p and 3d subshells*

the energies of electrons in different subshells by an energy level diagram (Figure 10). The electrons at highest energy are those most easily removed from the atom to form cations. Where the atom contains partly filled subshells, which are not strongly shielded from the nuclear charge, the atom will have a strong tendency to attract electrons into the lowest energy subshell with a vacancy, and so form an anion. Thus in the case of the sodium atom there is one electron in the 3s subshell outside a core of electrons corresponding with the neon electron configuration, [Ne]. This 3s electron is easily lost to form Na^+ which has the closed shell electron configuration of neon. (The electron configuration of Na could be represented by $[Ne]3s^1$ for electron book-keeping purposes).

In the chlorine atom there are five electrons in the 3p subshell and an electron entering the remaining vacant space will experience the stabilising effect of the nuclear charge of the atom which is incompletely shielded by the other electrons. Thus chlorine readily gains an electron to form Cl^-. The addition of further electrons is unfavourable, not only because of the negative charge on Cl^-, but also because the closed shell electron configuration of argon, [Ar], has been attained. Thus there are no spaces left for electrons to be added at favourable energies. Since sodium has a half filled 3s subshell it is reasonable to ask whether Na^- might be formed with an $[Ne]4s^2$ electron configuration. In fact the Na^- anion can be obtained but it reacts violently with traces of oxygen or moisture.

The 3-dimensional probability distribution describing an electron in an atom varies, not only with distance from the nucleus but also, other than in s subshells, with direction in space. The fact that electrons in the different orbitals within a subshell of an atom occupy different regions of space can have important consequences for the structures adopted by assemblies of atoms. In an approximate sense the space occupied by electrons in orbitals

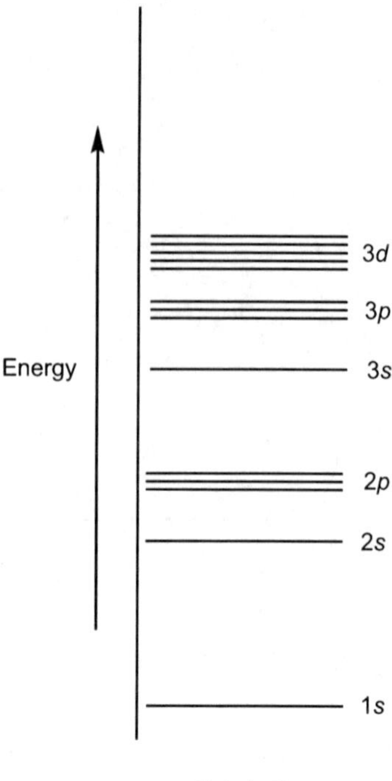

Figure 10 *An energy level diagram in which horizontal lines represent the energies of electrons in the 1s to 3d subshells. Each horizontal line represents a single orbital with given n, l and m_l values having a capacity to hold two electrons with different m_s values. The three p orbitals have the same energy but are represented by three closely stacked lines to make the diagram more compact. Similarly the five d orbitals are represented by five closely stacked lines*

defined by different l and m_l values can be represented in cartoon form. Geometrically s is the simplest orbital being represented by a spherical distribution [Figure 11(a)], although, as seen earlier, a radial cross section shows the electron density is not uniform throughout the interior of the sphere (Figure 8).

Similar cartoons might be drawn for electrons in p, d or f orbitals but in these cases the probability of finding the electron is not only dependent on distance from the nucleus but also on direction in space. Furthermore these orbitals contain a node at the nucleus, that is the probability of finding an electron in one of these orbitals falls to zero at the nucleus. This arises from the wave like character of the electron which changes phase from positive to negative amplitude at the nucleus. Thus the three p orbitals of a p subshell ($n \geq 2$, $l = 1$, $m_l = +1, 0, -1$) are each oriented along a different Cartesian axis and each contains two 'lobes', of different phase [Figure 11(b)], there being a nodal plane, containing the nucleus, in the electron density [Figure 11(c)]. These three p orbitals are given the labels p_x, p_y and p_z to show their differing orientations in space. In the case of the d subshell ($n \geq 3$, $l = 2$, $m_l = +2, +1, 0, -1, -2$) five orbitals arise, two of which have electron density oriented along Cartesian axes and are given the

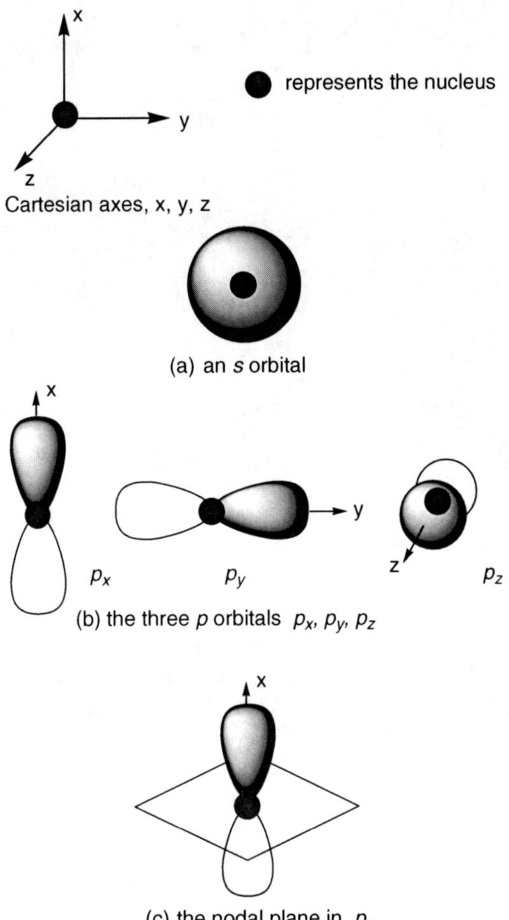

Figure 11 *'Cartoon' representations showing approximately how the probability of finding an electron varies with direction in space for (a) an s orbital (b) three p orbitals in a subshell and (c) showing the nodal plane in the electron density plot for a p orbital*

labels, d_{z^2} and $d_{x^2-y^2}$ (Figure 12). In the remaining three d orbitals the electron density is located between the Cartesian axes and oriented in each of the three planes defined by the pairs of axes *xy*, *yz* and *xz*; so these orbitals are given the labels d_{xy}, d_{yz} and d_{xz}. There are now two nodal planes present.

The situation becomes more complicated still with the f subshell ($n \geq 4, l = 3$, $m_l = +3, +2, +1, 0, -1, -2, -3$) where seven orbitals arise. Fortunately it is rarely necessary to consider the spacial orientation of electron density in the f orbitals of the lanthanides because the f subshell is almost completely buried within the filled 5^2s, $5p^6$ subshells of the [Xe] core of these atoms (Figure 13). As a result the 4f-electrons behave largely like core electrons and have little directional influence on interactions involving lanthanide metal ions. The

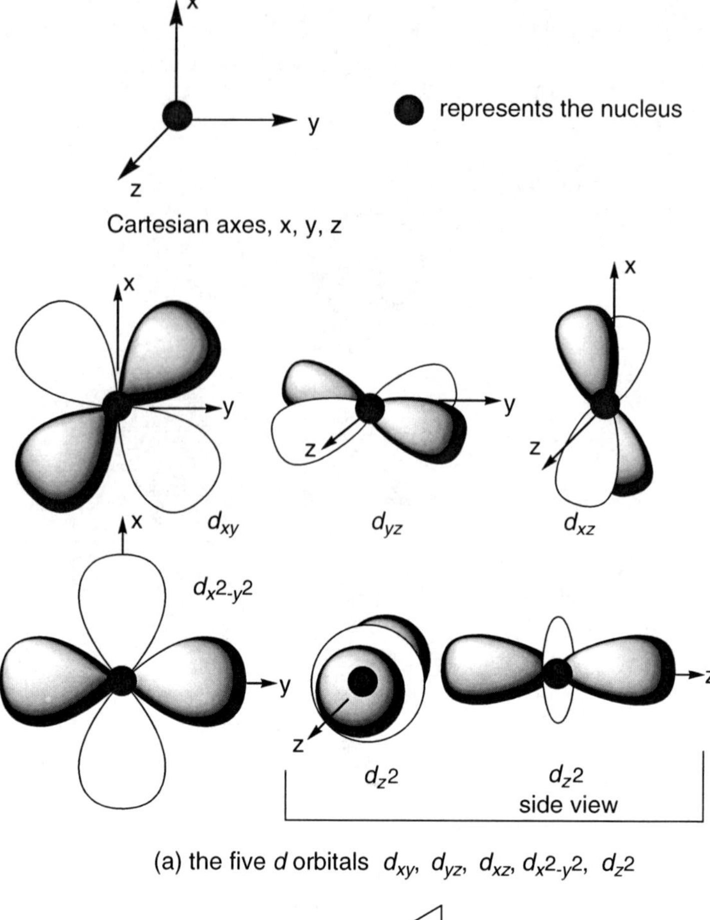

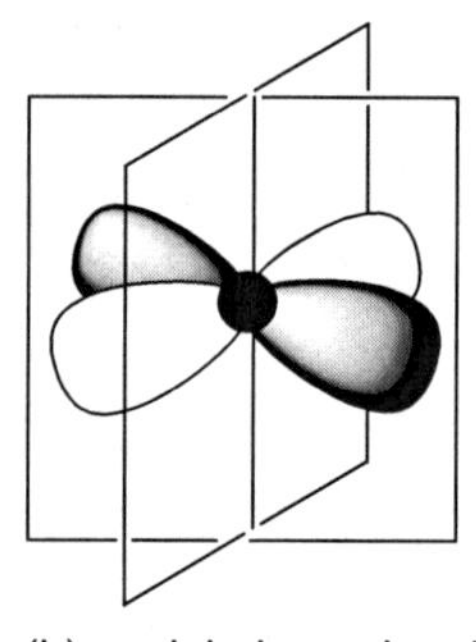

Figure 12 *'Cartoon' representations showing (a) how the probability of finding an electron varies with direction in space for the five d orbitals in a subshell and (b) showing the nodal planes in the electron density plot for a d orbital*

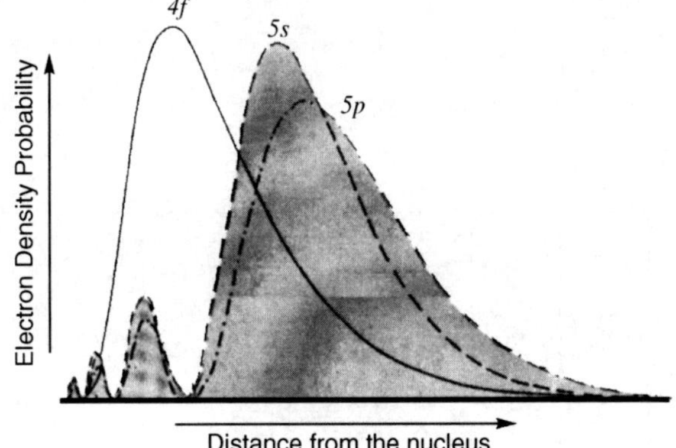

Figure 13 *Plots showing how the probability of finding an electron varies with distance from the nucleus for the 4f, 5s and 5p subshells. The electron density in the 5s and 5p subshells effectively forms the atom exterior of a lanthanide metal ion, Ln^{3+}, so that the 4f orbitals have virtually no interaction with adjacent atoms*

5f-electrons of the early actinides are less 'core-like' but again it is not usually necessary to consider the directional properties of the 5f orbitals.

2.2.3 The Electronic Properties of Atoms and Ions

The positive nuclear charge of an atomic nucleus attracts negative electrons into shells around it to achieve neutrality. The confinement of the wave like electrons to a region of space close to the nucleus imposes a particular internal structure on the distribution of electron density within the atom. This influences the way in which the electrons at the surface of the atom experience the attraction of the nuclear charge and the repulsions between one another. This in turn determines the ease with which an atom will accept or lose electrons. The energy input required to remove an electron from an atom is known as its *ionisation energy*. Similarly the energy change when an electron is gained by an atom or ion is known as its *electron gain energy* or electron affinity.[8]

Normally an energy input is required to remove an electron from a neutral atom. The ionisation energy is thus given a positive sign to show that energy is absorbed in the process of removing an electron to leave a positively charged ion. The Group 1 and 2 metals show relatively low ionisation energies and generally ionisation energies tend to increase with increasing atomic number across a block of the Periodic Table (Table 4). In many cases the addition of an electron to a neutral atom leads to a release of energy and, in such cases, the electron gain energy is given a negative sign. Only in cases where an energy

[8] The term energy has a rather specific meaning but for the sake of simplicity the terminology is used more loosely here. Systems spontaneously move from a state of higher energy to one of lower energy releasing energy in the process, in the form of heat for example. Work must be done to reverse a spontaneous process. Thus energy may be released on adding an electron to a positive ion to form a neutral atom, but an energy input would have been needed to form the positive ion from the atom in the first place.

Table 4 *The list ionisation (eV)*[a] and 1st electron gain energies of the elements[a]

1	2	3	4	5	6	7	8	9	10	11	12	13	14	15	16	17	18
H **13.6** *-0.75*	He **24.6** *0.5*																
Li **5.32** *-0.62*	Be **9.32** *0.5*											B **8.30.** *-0.28*	C **11.3.** *-1.26*	N **14.50** *.07*	O **13.6.** *-1.46*	F **17.4** *-3.40*	Ne **21.6** *1.2*
Na **5.14** *-0.55*	Mg **7.64** *0.4*											Al **5.98** *-0.44*	Si **8.15** *-1.39*	P **10.5** *-0.75*	S **10.4** *-2.08*	Cl **13.0** *-3.62*	Ar **15.8** *1.0*
K **4.34** *-0.50*	Ca **6.11** *0.03*	Sc **6.54** - *0.19*	Ti **6.82** - *0.08*	V **6.74** - *0.53*	Cr **6.77** - *0.67*	Mn **7.43** *0.52*	Fe **7.87** - *0.16*	Co **7.86** - *0.66*	Ni **7.64** - *1.16*	Cu **7.73** - *1.23*	Zn **9.39** *0.6*	Ga **6.00** *-0.3*	Ge **7.90** *-1.2*	As **9.812** *.-0.81*	Se **9.75** *-2.02*	Br **11.8** *-3.37*	Kr **14.0** *1.0*
Rb **4.18** *-0.49*	Sr **5.69** *0.3*	Y **6.38** - *0.31*	Zr **6.84** - *0.43*	Nb **6.88** - *0.89*	Mo **7.10** - *0.75*	Tc **7.28** - *0.6*	Ru **7.37** - *1.05*	Rh **7.46** - *1.14*	Pd **8.34** - *0.56*	Ag **7.58** - *1.30*	Cd **9.00** *0.7*	In **5.79** *-0.3*	Sn **7.34** *-1.2*	Sb **8.642** *.-1.07*	Te **9.01** *-1.97*	I **10.5** *-3.06*	Xe **12.1** *0.8*
Cs **3.89** *-0.47*	Ba **5.21** *0.3*	La **5.58** - *0.5*	Hf **6.78** - *0.1*	Ta **7.89** - *0.32*	W **7.98** - *0.82*	Re **7.88** - *0.1*	Os **8.71** - *1.1*	Ir **9.12** - *1.57*	Pt **9.02** - *2.13*	Au **9.22** - *2.31*	Hg **10.4** *0.5*	Tl **6.11** *-0.2*	Pb **7.42** *-0.36*	Bi **7.292** *-0.95*	Po **8.42** *-1.8*	At **9.64** *-2.8*	Rn **10.7** *0.7*
Fr -	Ra **5.28**	Ac **5.17** -															

Ce **5.47** *-0.5*	Pr **5.42** *-0.5*	Nd **5.49** *-0.5*	Pm **5.55** *-0.5*	Sm **5.63** *-0.5*	Eu **5.67** *-0.5*	Gd **6.14** *-0.5*	Tb **5.85** *-0.5*	Dy **5.93** *-0.5*	Ho **6.02** *-0.5*	Er **6.10** *-0.5*	Tm **6.18** *-0.5*	Yb **6.25** *-0.5*	Lu **5.43** *-0.3*
Th **6,08** -	Pa **5.89** -	U **6.19** -	Np **6,27** -	Pu **6.06** -	Am **5.99** -	Cm **6.02** -	Bk **6.23** -	Cf **6.30** -	Es **6.42** -	Fm **6.50** -	Md **6.58** -	No **6.65** -	Lr **4.6** -

[a] The 1st ionisation energy, in **bold** type, and the 1st electron gain energy, in *italic* type, are shown and the units for both are eV (1 eV – 96.485 kJ mol^{-1}). Ionisation energy values are taken from D.F, Shiver and P.W. Atkins, *Inorganic Chemistry*, 3rd Edn, OUP, Oxford, 1999, and electron gain energy values from H. Hotop and W.C. Lineburga, *J. Phys. Chem. Ref. Data*, 1985, **14**, 731 or S.G. Bratsch and J.J. Lagowski, *Polyhedron*, 1986, **5**, 1763.

input is required to attach an electron to an atom or ion will a positive number arise for the electron gain energy. The magnitudes of ionisation energies and electron gain energies vary considerably across the Periodic Table (Table 4) and some metals have a negative electron gain energy.

The electronegativity of an element, given the symbol χ, (Table 2) expresses the combined effects of the ionisation energy and electron gain energy of an element. Values of the electronegativity χ of an element can provide a useful insight into its chemical properties. The lighter early d-block metals, the lanthanides and the actinides show the lower electronegativity values. These fall in the range 1.0–1.5, which may be compared with values of 0.8–1.0 for the alkali metals and 1.3 for magnesium. The so-called 'platinum metals' (Ru, Rh, Pd, Os, Ir, Pt,) together with molybdenum and tungsten, show larger values, in the range 2.2–2.4, which may be compared to values of 2.0 for phosphorus, 2.2 for hydrogen and 2.6 for carbon. This suggests that the polarity of bonds between these metals and elements such as carbon, hydrogen and phosphorus will be lower than for the early d-block metals, the lanthanides or the actinides. In other words the contribution of covalency to bonding in platinum metal, molybdenum or tungsten compounds will be greater.

2.2.4 The Radii of Atoms and Ions

Another property of atoms and ions which has an influence on their chemical behaviour is their radius. As might be expected the radii of atoms increase as further shells of electrons are added. However, this increase is not uniform throughout the Periodic Table. Only the s subshell orbitals are spherically symmetric. Electrons in the other subshell orbitals, particularly the d and f subshells, occupy different regions of space if they have different m_l values. Consequently they shield one another from the nuclear charge incompletely. Thus the effective nuclear charge felt by the added electrons increases and the atoms shrink as the outermost electrons experience this increasing nuclear attraction. The effect is most pronounced in the lanthanide series of the f-block elements and this has an effect on the chemistry of the d-block elements. The increase in radius after silver and cadmium on moving from the principal shell with $n = 5$ (In to Xe) to that with $n = 6$ is compensated by the contraction experienced while filling the 4f subshell from cerium to lutetium. Therefore, the radii of the ions from hafnium to platinum are very similar to those of the corresponding ions in the series zirconium to palladium. This results in greater similarities in chemical behaviour between the second (Zr to Ag) and the third (Hf to Au) rows of the d-block elements than between the first (Ti to Cu) and second rows.

The size of an atom in a compound may be reported in one of three different ways. Where two atoms touch without any chemical interaction the radius is best represented by a value which is known as the Van der Waals radius.[9] In compounds where there is a chemical bonding interaction between the atoms

[9] When two non-bonded atoms are in close contact a weak force of attraction arises between them, the distance between the atoms at which this attractive force between them is balanced by the repulsive force between their electron shells is the sum of their Van der Waals radii.

covalent or ionic radii are used. In compounds where the bonding is predominantly covalent in nature a self consistent set of covalent radii can be devised by examining a large number of distances between bonded atoms. The best model for apportioning the internuclear distance between two atoms to each of the atoms gives a value for the covalent radius of each atom. Where the bonding is predominantly ionic in nature, or when dissolved solvated ions are being considered, it is more appropriate to refer to an ionic radius. The ionic radius of a monatomic ion in a compound is the distance from its nucleus at which the attraction due to electrostatic forces between it and an adjacent atom or ion is balanced by the repulsive forces between the electron shells surrounding the two ions. The sum of the ionic radii of an adjacent cation and anion is the internuclear distance between them. If ionic radii are to be compared the structural environments of the two ions involved must be comparable. That is they should both be in contact with the same number of nearest neighbour atoms. The number of atoms, or monatomic ions, in direct contact with an ion is known as its coordination number (CN) and ionic radii vary with CN. Larger CNs being associated with larger radii. As might be expected the ionic radii of anions are larger than those of cations having similar atomic number and increasing the positive charge on a cation reduces its ionic radius.

2.3 The Formation of Compounds

2.3.1 Chemical Bonds

In the previous section an outline account was given of the way in which atoms are constructed and how this affects their tendency to gain or lose electrons. The next stage in explaining the chemistry of the elements requires an examination of how atoms interact with one another to form compounds. In order for one atom to become attached to another through the formation of a chemical bond, some change in the distribution of the electron density in the valence shells of the two atoms must occur which lowers the energy of the system. This could result from one atom transferring one or more electrons to the other. In this way a pair of oppositely charged ions is formed and these can become bound together by electrostatic attraction in what is known as ionic bonding. However, the atoms could simply share electrons with one another without a complete transfer occurring. If this sharing results in the total energy of the system being reduced the atoms will be held together through what is known as covalent bonding. If the elements at each end of the bond are different this electron sharing may be quite unequal. This particularly applies when positively charged metal ions interact with an electron donor atom such as oxygen in water or nitrogen in ammonia. This type of interaction is an example of a coordinate bond. Other weaker bonding interactions between atoms are also possible through interactions involving hydrogen atoms in what are called 'hydrogen bonds'.

2.3.2 The Ionic Bond

One way to form a chemical bond between two atoms is simply to remove an electron from one atom (symbol M) and give it to the other (symbol E). The two ions so formed, M^+ and E^-, are then attracted to one another by the electrostatic force between their opposite charges to give an ionic bond. This bond is most apparent in the solid state where the two ions can pack in an ordered array of alternating cations and anions, as in Na^+Cl^- seen earlier [Figure 2(b)]. The strength of the ionic bond in the solid will depend upon the separation between the ions M^+ and E^-, therefore on their sizes, on the geometric arrangement of the ions in the solid state lattice and on their charges. Thus in a compound ME_2 containing M^{2+} ions and twice as many E^- ions the strength of the bonding would be different from that in ME. In order to convert the elements M and E to the ionically bonded compound M^+E^- it is first necessary to obtain isolated atoms in the gas phase, $M_{(g)}$ and $E_{(g)}$, from M and E from their usual form. This might be as a solid, liquid or gas in which two or more atoms of the element are attached together. In solid metallic sodium for example, [$Na_{(s)}$], the atoms are packed together in an ordered array like balls in a box. Energy is required to separate these atoms and evaporate them into the gas phase. In contrast chlorine exists as a gas made up of molecules in which two chlorine atoms are bound together [$Cl_{2(g)}$]. Energy must again be supplied to split each of these $Cl_{2(g)}$ molecules into two isolated gaseous $Cl_{(g)}$ atoms. Thus in general an initial energy input is required to convert M_x and E_y from their normal forms into $xM_{(g)}$ and $yE_{(g)}$. Energy is also needed to form M^+ by ionising atoms of M. Some of this energy may come from attaching an electron to E forming E^- but much of the energy needed will be obtained from the condensation of the separate gaseous ions, $M^+_{(g)}$ and $E^-_{(g)}$, into a solid lattice $M^+E^-_{(s)}$. The process may be represented by Equations (1) and (2).

$$1/xM_x + 1/yE_y = M^+_{(g)} + E^-_{(g)} \text{ energy absorbed} \quad (1)$$

$$M^+_{(g)} + E^-_{(g)} = M^+E^-_{(s)} \text{ energy released} \quad (2)$$

If the energy released by the second process is more than that absorbed in the first, the reaction to form ME from $1/xM_x$ and $1/yE_y$ will be spontaneous and can proceed to form ME with the evolution of heat. Otherwise some external energy input would be needed to drive the reaction, or else the reaction will not occur.

When solid ME is dissolved in a solvent such as water, the two ions M^+ and E^- become separated and the attractive force between them is reduced through the increase in distance between them [Figure 2(c)]. This process of disrupting the solid state lattice of ME requires a significant amount of energy which must come from the interactions between the ions, M^+ and E^- and the molecules of solvent. Thus when NaCl dissolves in water the Na^+ ions become surrounded by the negatively polarised oxygen atoms of the H_2O molecules and the negatively charged Cl^- ions become associated with the positively polarised hydrogen atoms in H_2O. If surrounding M^+ and E^- with solvent molecules (solvation) releases more energy than is needed to disrupt the solid lattice of M^+E^- ions dissolution will be spontaneous. If not an external energy input will be needed so, for example, the solution may become cooler as heat energy is absorbed. In some cases the compound may simply be insoluble.

2.3.3 The Covalent Bond

At the other extreme of chemical bonding lies the situation where the bond is not polar and no net transfer of electrons occurs; the pair of valence electrons associated with forming a bond is shared equally between both bonded atoms. This situation arises with the H–H bond in a molecule of gaseous dihydrogen, H_2, for example. In this case, transferring an electron from one atom to another to produce H^+H^- does not seem sensible as there is no electronegativity difference between the atoms. A different model is required to describe the bonding and it is necessary to consider how the atomic orbitals (AOs) in the valence shells of two separated hydrogen atoms may interact to allow a lower energy H_2 molecule to form. The wave theory of the electron suggests that when two electrons are brought together their individual wavefunctions may interact. Where this interaction is in phase the amplitude increases and with it the probability of the electron being found in that region [Figure 14(a)]. Where the interaction is out of phase, the reverse situation arises; the amplitude decreases and the electron is unlikely to appear in that region [Figure 14(b)]. Since orbitals cannot be created or destroyed in such interactions the two 1*s* orbitals of the two hydrogen atoms can interact to form two new orbitals associated with the dihydrogen molecule each containing contributions from both AOs. These are known as molecular orbitals (MOs). One orbital at lower energy results from the *in phase* combination of the two AOs in the region between the nuclei [Figure 14(c)]. The other, higher energy MO results from the *out of phase* combination of the two AOs in the region of contact [Figure 14(c)]. In the *in phase* combination orbital the electron density is increased between the atomic nuclei and attracts the nuclei together. Also, being under the influence of two positive nuclei, rather than one in the atom, electrons in the *in phase* combination MO are at a lower energy than in the free H atom. Thus combining the two AOs to form the MO and moving the two hydrogen valence electrons into this lower energy orbital leads to a more energetically favourable situation giving a bonding interaction.

A single two electron bond of this type is known as a sigma (σ) bond. In the higher energy *out of phase* combination MO, there is a node between the nuclei as the wavefunctions cancel one another in this region If this orbital were occupied the electron density between the nuclei would be depleted leading to increased internuclear repulsion and a higher energy situation. This orbital is described as an anti-bonding (σ^*) orbital. Provided this orbital remains unoccupied, a bonding situation arises. If instead of two hydrogen atoms, two helium atoms were brought together to form He_2, the four valence electrons would fill both the σ and σ^* MOs, the net energy of the system would remain that of the two helium atoms and no bond would form. Hence, while hydrogen

(a) *In phase* constructive interference of waves leading to increased amplitude

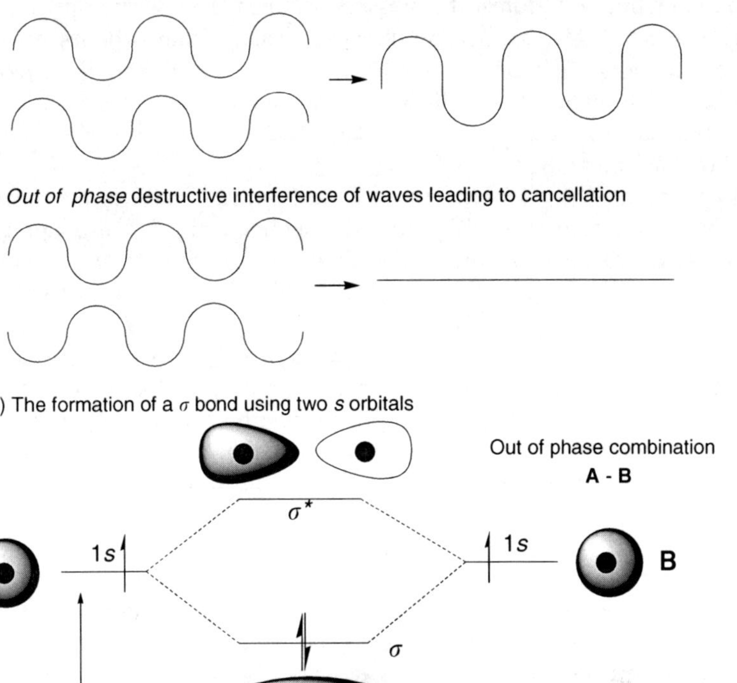

Figure 14 *The formation of a covalent σ-bond between two H atoms through the mixing of the two H(1s) orbitals to make two molecular orbitals (σ-bonding and σ*-anti-bonding) for H_2*
The illustration shows in phase and out of phase interactions between two wave forms with the resulting combination wave form showing (a) constructive interference leading to amplitude reinforcement or (b) destructive interference leading to amplitude cancellation. In (c) the in phase and out of phase interactions between two s orbital wave functions to form a σ-bonding MO and a σ-anti-bonding MO are shown. The figures provide an approximate representation of the regions of space occupied by electrons when present in the orbital depicted. Shading represents one phase of the wave function and the absence of shading the opposite phase. The horizontal lines represent the relative energies of the two AOs and MOs involved. Higher energy is associated with the uppermost lines, each orbital can accommodate two electrons with different m_s values and valence electrons are added to fill the lowest energy orbitals first. If there are more electrons in bonding MOs compared to their anti-bonding counterparts, so that the total energy of electrons in the MOs is less than that in the original AOs, a bonding interaction can arise*

forms a diatomic gas containing H_2 molecules, helium is monatomic gas containing only He atoms. It is possible to form He_2^+ where only one electron occupies the σ^* MO as there is an energy benefit from two electrons being at lower energy but only one at higher energy, although the bond is much weaker than in H_2. Similarly H_2^- will show some bonding having the same electronic structure as He_2^+, though with smaller nuclear charges, but H_2^{2-} would show no bonding interaction.

In elements where the second or higher principal shell is the valence shell, p orbitals also become available for bonding interactions. A σ-bonding arrangement can arise from two p orbitals oriented along the internuclear axis in a similar manner to that arising from two s orbitals [Figure 15(a)]. Again there is a lower energy σMO and a higher energy anti-bonding σ^*MO. Thus in gaseous chlorine Cl_2 molecules are present and the interaction of 2p orbitals contributes to the bonding of the two chlorine atoms. However, a second possibility arises if the p orbitals are oriented perpendicular to the internuclear axis. Again in phase and out of phase combinations can arise but in both cases there is a node, which contains the two nuclei in the wavefunctions of the MOs [Figure 15(b)]. A bonding interaction of this type is known as a pi (π) bond and involves the formation of a lower energy π MO together with the corresponding out of phase anti-bonding (π^*) combination orbital at higher energy. This π^* orbital has a further nodal plane perpendicular to the internuclear axis. In combination with a σ-bond a π-bond leads to two electron pairs being involved in binding two atoms together by means of a double bond. Thus ethane, $H_3C–CH_3$, contains a single C–C σ-bond while ethene, $H_2C=CH_2$, contains a double C=C bond consisting of one σ and one π component. The third of the three p orbitals may also be used to form a π-bond in a plane perpendicular to that of the π interaction described above and this can lead to a triple bond involving three pairs of electrons as seen between the carbon atoms of ethyne, $HC\equiv CH$. In each of these bonds the two electrons in the bonding σ *or* π MO are equally shared between the two carbon atoms.

2.3.4 The Coordinate Bond

The two types of bonding described above represent extreme views. If bonding is entirely ionic one or more electrons are considered to be completely transferred from one atom, M, to another, E, to form a compound ME_x. If bonding is entirely covalent one valence electron from each of the two bonded atoms is considered to be equally shared between two atoms in a non-polar bond, as in the C–C bond in ethane $H_3C–CH_3$. These extreme descriptions of bonding are inappropriate in many situations where the bonding is neither purely ionic nor purely covalent. This is particularly so in the case of many metal compounds where a metal ion binds to other atoms such as the oxygen in water or the nitrogen in ammonia to form what are known as complex compounds or 'metal complexes'. As an example, when the salt cobalt dichloride, which can be thought of as ionically bonded $Co^{2+}(Cl^-)_2$, is crystallised from water the product obtained is $CoCl_2.6H_2O$ in which the water molecules are bound to the Co^{2+} ions to give the 'complex' $[Co(H_2O)_6]^{2+}$ which now forms the cationic

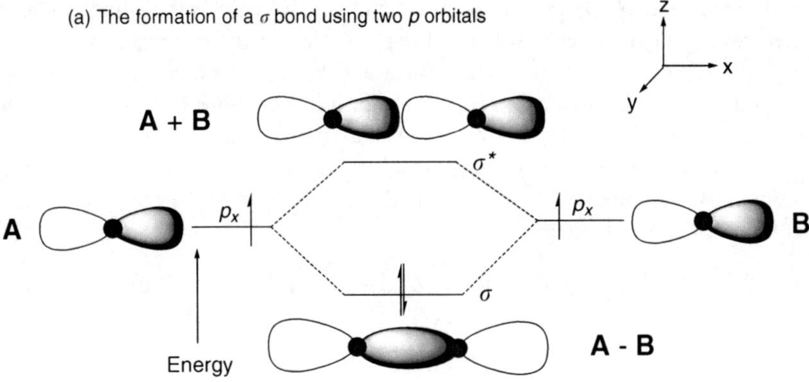

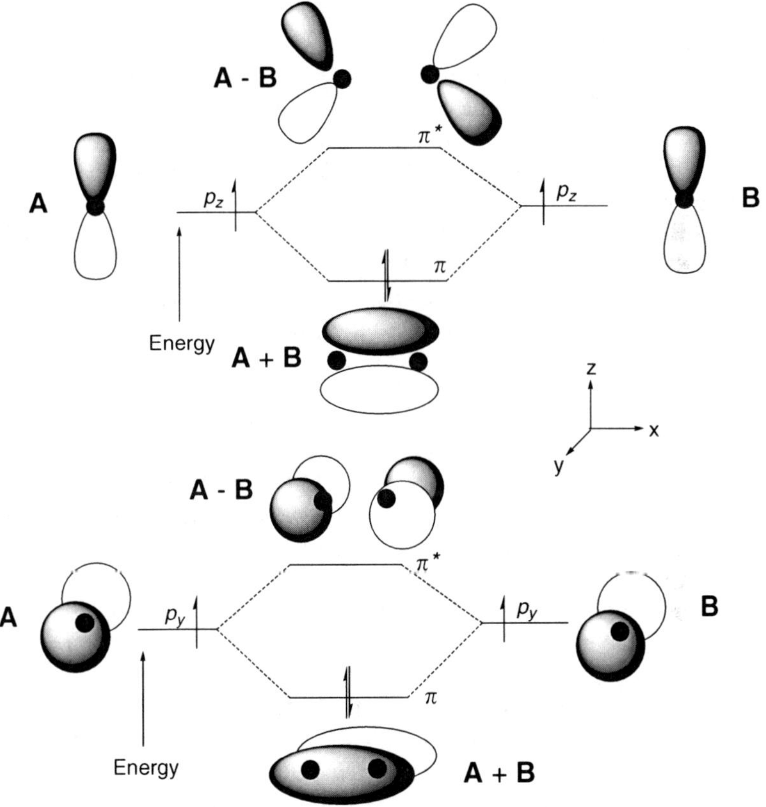

Figure 15 *The formation of a covalent σ-bond and two π-bonds between two atoms using three p orbitals on each. The figure shows interactions between pairs of p subshell AOs to form two MOs which are (a) σ-bonding and σ*-anti-bonding in character, or (b) π-bonding and π*-antibonding in character. If the bonding MOs are fully occupied by electrons and the anti-bonding MOs unoccupied, the bonding interaction is optimised*

part of the salt $[Co(H_2O)_6]^{2+}(Cl^-)_2$. In forming $CoCl_2$ the cobalt has given up two electrons, one to each chlorine, forming ionic bonds. Then, in forming $[Co(H_2O)_6]^{2+}$, each of the six water molecules donates an electron pair to the Co^{2+} ion in what is known as a coordinate bond. Although both the electrons in the cobalt to water bond are considered to arise from the oxygen of the water molecule, no formal electron transfer to Co^{2+} is involved. The three bonding situations may be represented in simplified form as follows where · represents an electron and : an electron pair.

$$M^{\bullet} + E^{\bullet} = M^{+}E:^{-} \quad M^{z+} + :L = M^{z+} \leftarrow :L \quad H_3C^{\bullet} + {}^{\bullet}CH_3 = H_3C\text{-}CH_3$$

IONIC COORDINATE COVALENT

Species such as Co^{2+}, or M^{z+} above, which can accept a pair of electrons to form a coordinate bond are known as Lewis acids and species which can donate an electron pair, such as H_2O: above, as Lewis bases. A special case arises when a hydrogen atom forms a coordinate bond in what is known as a hydrogen bond. In water, for example, the hydrogen atoms have a positive polarisation and the oxygens a negative polarisation. The resulting $H_2O{:}^{\delta-}\text{- -}\,^{\delta+}H\text{–}OH$ interaction is quite strong and holds the water molecules together, keeping water liquid at room temperature. The related compounds H_2S or H_2Se have much weaker H– –S or H—Se interactions and are gaseous at room temperature.

The coordinate bond is, in effect, intermediate in character between the ionic and covalent bond, the pair of electrons producing the bond are not shared equally between the two bonded atoms but no formal transfer of electrons from one atom to the other is assumed. The MO bonding model offers a more general description of bonding which naturally allows for the range of bonding situations from ionic through coordinate to fully covalent.

2.3.5 Polarity in Bonding

In practice most chemical bonds will be intermediate in character between the ionic and covalent extremes described above. In a bond between atoms of two different elements there will usually be some degree of sharing electrons, and so a covalent contribution, as well as some polarity in the bond, and so some ionic contribution. Depending on the electronegativities of the two elements the bond may be more or less polar. This can be taken into account by considering the way in which the electrons behave in the presence of more than one atomic nucleus. That is instead of electrons occupying AOs they are placed in MOs made up by mixing AOs. The MO bonding model can allow for ionic contributions to bonding by allowing one AO wavefunction to contribute more than another to the MO wavefunction constructed by combining the AO wavefunctions. This then allows the electron density to be distributed unequally between the two atoms so that the MO model can accommodate bonds ranging from purely covalent in character to almost purely ionic.

In order that two AOs may interact significantly to form two MOs it is necessary for the energies of the AOs to be similar and for there to be some overlap between them, that is there must be a region of space to which both wavefunctions can contribute significant electron density. A further requirement is that the AOs are of the same symmetry. This has a rather specific meaning, which cannot be fully explained here, but a simple example is provided by the overlap of an s orbital on one atom with a p orbital on another. If the p orbital is oriented along the internuclear axis, the s and p orbitals have the same symmetry allowing a σ-bonding situation to arise [Figure 16(a)]. However, if the p orbital is oriented perpendicular to the internuclear axis the s and p orbitals have different symmetries [Figure 16(b)]. In this case one part of the interaction between the s and p orbital wavefunctions is in phase but cancelled by an equal part which is out of phase leading to no net interaction and a non-bonding situation. A similar situation arises for two p orbitals, which are not of the same orientation [Figure 16(c)].

The H–F bond in hydrogen fluoride provides a simple example of how the MO description of bonding may be applied. The valence shell of the H atom consists of the 1s orbital and that of the F atom the 2s and the three 2p orbitals. The $1s^2$ [He] core electrons of the F atom are at lower energy and beneath the surface of the atom so may be neglected in this bonding scheme. The higher nuclear charge of the F atom means that its valence shell electrons are at lower energy than that of the H atom. Two symmetry allowed interactions are possible, one between the H 1s and the F 2s orbitals and the other between the H 1s orbital and one 2p orbital. Because the F 2s orbital is much lower in energy than the H 1s, their interaction is small and the F $1s^2$ electrons can be regarded as occupying an essentially non-bonding (*nb*) MO which has a wavefunction similar to that of H 1s. A larger interaction occurs between the H 1s and the F 2p orbital of appropriate symmetry ($2p_x$, if x is defined as the internuclear axis) leading to bonding σ and anti-bonding σ^* MOs (Figure 17).

The remaining F 2p orbitals are not of suitable symmetry to interact with the H 1s orbital and are included in the MO scheme as two non-bonding orbitals, which have the wavefunctions of the F $2p_y$ and F $2p_z$ orbitals. The bonding scheme is represented in an energy level diagram in which each orbital is represented by a horizontal line signifying its relative energy. The available valence electrons are added to the MOs starting from the lowest energy level. Provided there is an excess of electrons in bonding compared to anti-bonding orbitals, a bonding situation will arise. Electrons in non-bonding orbitals make no contribution to bonding. In HF the σ orbital is closer in energy to the F 2p orbital and has more fluorine character while the σ^* orbital is closer in energy to the H 1s orbital and so has more hydrogen character. Thus the bond between H and F is polarised, $H^{\delta+}–F^{\delta-}$, in effect containing both covalent and ionic contributions. The presence of two electrons in the σMO and none in the σ^* MO is in accord with the acidic character of HF which can dissociate on dissolution in water to give solvated H^+ and F^- ions.

The MO model can be extended to polynuclear molecules by considering the atoms present in groups which are symmetry related. In water, for example, the two hydrogen atoms are equivalent to one another but the oxygen atom is unique. Just as the two H 1s orbitals can mix to form two new in phase and out of phase combination orbitals in the H_2 molecule, so in H_2O they can be

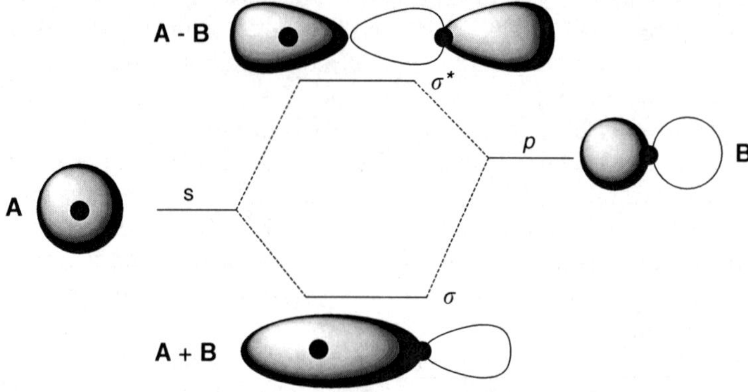

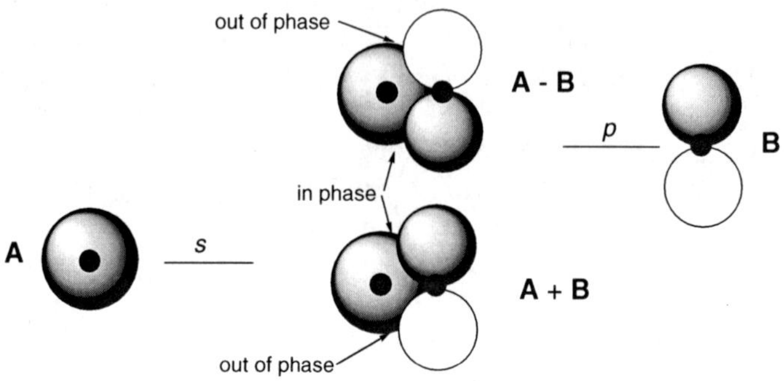

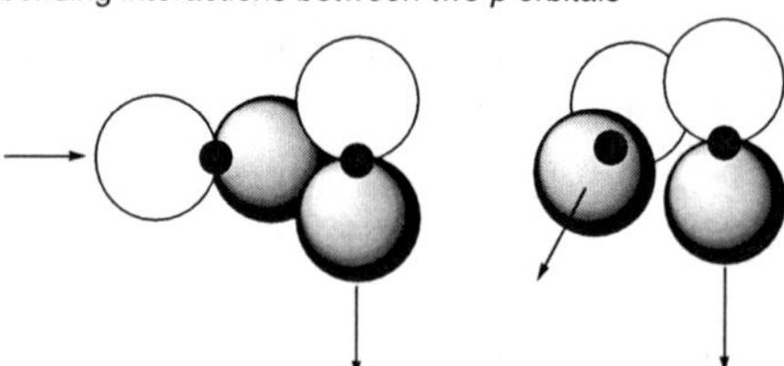

Figure 16 *The importance of symmetry in forming MOs from AOs. (a) An illustration of the interactions between an s subshell AO and a p subshell AO to form MOs which are σ-bonding and σ*-antibonding in character. (b) An illustration of how an s subshell AO and a p subshell AO of differing symmetry fail to produce a net interaction as constructive interference in one region is cancelled by destructive interference in another. (c) An illustration of how two p subshell AOs of differing symmetry fail to produce an overall interaction*

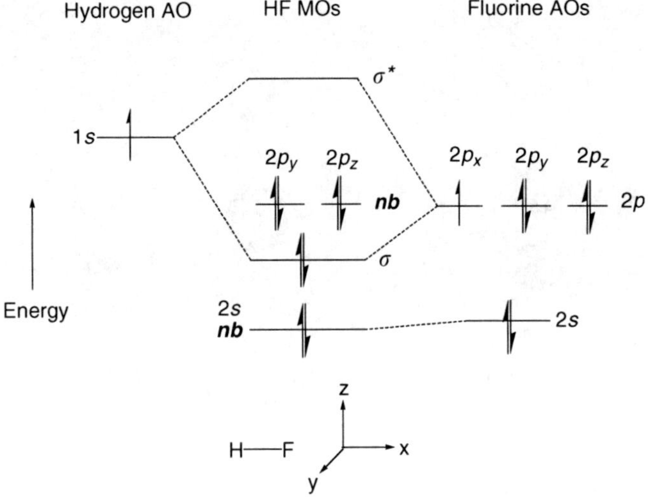

Figure 17 *An energy level diagram representing the formation of MOs from AOs in a molecule of HF. The different AOs of H and F combine to form MOs which, when occupied by electrons, contribute to bonding (σ), oppose bonding (σ*) or make no contribution to bonding (nb). Each horizontal line represents an orbital capable of accommodating two electrons with differing m_s quantum numbers*

described in a similar way, although there is now little direct overlap as they are too far apart (Figure 18). In the symmetry defined by the shape of the water molecule the in phase combination of hydrogen 1s orbitals is of a symmetry designated by the label A_1, so the combination orbital is given the lower case label a_1. The out of phase combination is of a symmetry designated by the label B_1 so the combination orbital is given the label b_1. The symmetries of the oxygen valence shell orbitals in water are such that they are given labels as follows:[10]

$$2s\ a_1;\ 2p_z\ a_1;\ 2p_x\ b_1,\ 2p_y\ b_2$$

[10]These symmetry labels arise from Group Theory, a mathematical treatment of the symmetries of groups of objects in space. In Group Theory the fundamental symmetry properties of an object or group of objects are given symbols such as A_1, B_1, T_{2g}. These define the properties of the particular object or group of objects with respect to the symmetry of the total collection of objects under consideration. In a water molecule, for example, the group of three atoms contains two planes of symmetry and a 2-fold axis of rotation, these are known as symmetry elements, each of which is associated with one or more symmetry operations. Here the operations are reflections through a plane and a rotation about an axis. The two equivalent hydrogen atoms form a group of symmetry related objects and the oxygen atom constitutes a different unique object. The orbitals contained within the atoms behave in particular ways towards the symmetry elements associated with the water molecule. The p_z orbital on oxygen is unchanged by any of the symmetry operations associated with the symmetry elements of the water molecule, behaviour corresponding with the symmetry symbol A_1. However, the lower case symbol a_1 is used to denote an AO which has the symmetry properties associated with A_1. The p_x orbital behaves differently. Reflection through one of the mirror planes produces no change but the other mirror plane reverses the phases of the orbital essentially making it into minus itself. A 180° rotation about the 2-fold rotation axis has the same effect. Thus, although the symmetry properties of p_x and p_z are identical in an isolated oxygen atom, they are quite different in a water molecule. AOs may only mix to form MOs if they have the same symmetry as defined by labels such as a_1.

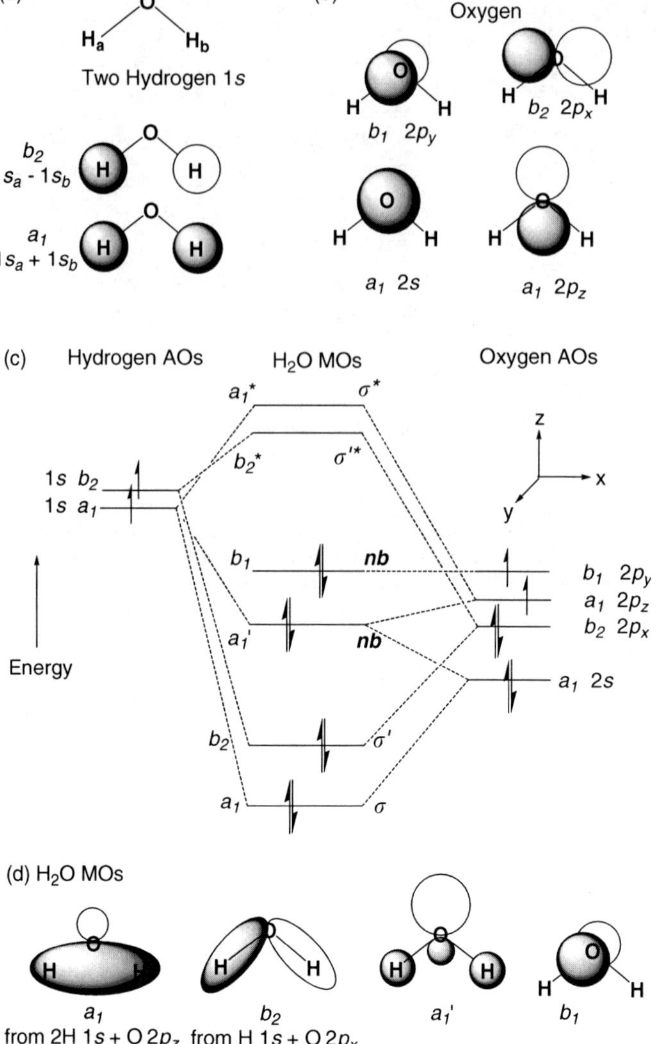

Figure 18 *An illustration of the bonding arrangement in a molecule of water showing (a) how the group of two hydrogen 1s AOs can give rise to new combination orbitals of differing symmetry, (b) the symmetry labels of the oxygen atom valence shell orbitals, (c) the energy level diagram representing the formation of MOs in a molecule of H_2O from interactions between AOs or group orbitals of the same symmetry and (d) a cartoon representation of the distribution of electron density associated with the occupied bonding and non-bonding MOs of different symmetry*

Only AOs of the same symmetry, and so having the same symmetry label, are allowed to mix to form new combinations leading to MOs. As there are three AOs with a_1 symmetry labels in the bonding scheme for H_2O these can mix to form three a_1 MOs. The contributions of each AO to each MO will vary depending on the overlap and relative energies of the various AOs. In the case of water this results in a low energy σ-bonding MO, an intermediate energy essentially non-bonding MO and a high-energy anti-bonding σ^* MO, all of a_1 symmetry. These are given the labels a_1, a_1' and a_1^* to indicate that they are of A_1 symmetry but differ in energy. The b_1 symmetry combination orbital derived from the hydrogen 1s orbitals can mix with the p_x orbital on oxygen to form two MOs, one σ-bonding at lower energy (b_1) and one high-energy anti-bonding σ^*MO (b_1^*). The remaining p_y orbital on oxygen is of B_2 symmetry and so cannot interact with the hydrogen 1s orbitals and remains as a non-bonding MO (b_2). Placing the eight valence electrons from the oxygen and the two hydrogen atoms into the MO energy level diagram leads to the formation of two σ-bonds through the occupancy of the low energy a_1 and b_1 orbitals. Above these in energy are the two filled non-bonding orbitals a_1' and b_2. The bonding situation corresponds with the formation of two O–H bonds but in this MO model neither is localised between O and one H atom rather they are both distributed over the H–O–H framework.

The same basic approach can be applied to more complicated molecules. If a water molecule approaches a metal ion with an empty valence shell orbital, provided symmetry allows it, a non-bonding orbital on the oxygen of water may mix with the empty orbital on the metal. This should lead to new combination orbitals at lower energy and at higher energy compared to the MO in free water. The pair of electrons in the water MO can move to lower energy in the new combination orbital leading to chemical bonding between the water and the metal ion. If the metal orbital is filled there will be no bonding because the higher energy orbital will also become filled and cancel out the energy benefit of filling the lower energy orbital. This is in effect the same situation as that described for H_2 and He_2, except that in this description, because oxygen is more electronegative than the metal, two electrons are assumed to arise from the water oxygen rather than one each from water and the metal ion.

2.4 The Chemistry of Metals – An Overview

2.4.1 The Oxidation Numbers and Electron Configurations of Metal Ions

The number of electrons in the valence shell of an element has an important influence on the type of compound it can form. Some elements typically involve all of their valence shell electrons in bond formation but others, particularly the transition elements, may not. It is useful, therefore, to have a measure of the number of valence electrons an element uses in bond formation. One way to determine this number for an atom in a compound containing two elements is to assume a purely ionic bonding model, regardless of the true bonding situation, and transfer electrons from the least electronegative atom to the most electronegative atom to fill its valence shell. The charges on the two ions so formed are known as their *oxidation numbers* or *oxidation states*.[11] As an example, consider the compounds $FeCl_2$ and CrO_3. As the most electronegative of the two elements, the chlorine in $FeCl_2$ will be in the form of Cl^- with the [Ar] electron configuration and an oxidation number of (−1). In order to achieve the neutral charge of the compound the iron would need to be present as Fe^{2+} with an oxidation number of (+2). In an electron counting sense the iron would only be using two of its eight valence electrons in bond formation. In the case of CrO_3, oxygen will be present as O^{2-} with the [Ne] electron configuration and an oxidation number of (−2), the chromium would need to be present as Cr^{6+} with an oxidation number of (+6) requiring use of all its valence electrons. In text oxidation numbers are written after the element name or symbol *e.g.* platinum(+2) for Pt^{2+}, sometimes Roman numerals are used as in platinum(II). In formulae superscript roman numerals may be used to denote oxidation number in place of a charge *e.g.* $Cr^{VI}O_3$.

The term oxidation number (or oxidation state) only has a clear chemical meaning in fully ionic compounds containing elements of very different electronegativity. In NaCl, for example, oxidation states of +1 for Na^+ and −1 for Cl^- may be assigned without difficulty. However, in a compound where the two elements have very similar electronegativity values the situation is not so clear-cut. For example, the bonding in CrO_3 will involve significant covalency and it is not safe to assume that the chromium is actually present as Cr^{6+} even though it may be assigned an oxidation state of +6. A Cr^{6+} ion would be very polarising and able to distort the electron density around an anion to

[11] In assigning an oxidation number to an atom of an element in a compound it is necessary to assume an extreme ionic bonding model, even though this may not be chemically reasonable. In the case of the metallic elements, it is usual to consider the metal as the least electronegative element in the compound and so assign the other atoms or groups bonded to the metal a negative or, for uncharged groups such as H_2O, NH_3 or CO, zero oxidation number. The total of these numbers when subtracted from the charge on the compound gives the oxidation number of the metallic element. In salts such as $Na_2[PtCl_4]$ or $[Co(NH_3)_6]Cl_3$ it is necessary to separate the metal complex from the counterions; in these examples it gives $[PtCl_4]^{2-}$ and $[Co(NH_3)_6]^{3+}$, respectively (see footnote 12, p. 55).

Some examples

$[Co(NH_3)_6]Cl_3$ — the cobalt complex is $[Co(NH_3)_6]^{3+}$ so, assigning a normal charge of zero to the ammonia group:
$+3-(6 \times 0) = +3$ *i.e.* $Co^{3+}.6NH_3$ and so Co(+3).

$K_2[PtCl_6]$ — the platinum complex is $[PtCl_6]^{2-}$ so assigning a normal oxidation number of −1 to chloride gives, for a −2 charge on the complex:
$-2-\{6 \times (-1)\} = 4+$ for platinum *i.e.* $Pt^{4+}.6Cl^-$ and so Pt(+4).

$[PtCl_2(NH_3)_2]$ — assigning a normal oxidation number of −1 to chloride and a zero charge to ammonia gives, for a zero charge on the complex:
$0-\{2 \times (-1)\}-(2 \times 0) = 2+$ for platinum so Pt(+2).

Tc_2S_7 — assigning sulfide its normal oxidation number of −2 gives, for a zero charge on the compound:
$0-\{7 \times (-2)\} = 14+$ for two technetium ions *i.e.* $2\ Tc^{7+}.7S^{2-}$ and so Tc(+7).

produce a covalent contribution to the bonding. Furthermore there may be an ambiguity in the way the compound is formulated, should it be considered an oxide containing $3O^{2-}$ and Cr^{6+} or a peroxide containing one $O_2{}^{2-}$, one O^{2-} and Cr^{4+}? Since a peroxide formulation would require two of the three oxygen atoms to be close together, knowledge of the exact structure of CrO_3 should distinguish between these possibilities. Despite such difficulties, oxidation state remains a useful formalism in chemical discussions and provides an aid in electron 'book-keeping'. The oxidation number allows the remaining valence shell electrons of a metal ion to be calculated. Thus K^+ or Ca^{2+} have no electrons remaining in their valence shells whereas Fe^{2+} has six electrons left which are all presumed to be associated with the d subshell. As a general rule all the valence shell electrons of a d-block element are assumed to be in the d subshell when it is present as an ion or in a compound. There are two ways of counting the electrons in a d-block element. One assumes a purely ionic bonding model, assigns an oxidation number and allocates the remaining valence shell electrons to the d subshell. The other assumes a purely covalent bonding model, treats the metal as a neutral atom and allocates all of the valence shell electrons to the d subshell. These approaches are revisited later in the discussion of bonding in d-block metal compounds (Section 2.5). The addition of one or more electrons to a metal ion to reduce its oxidation number is a process known as *reduction*. The reverse process of removing electrons to increase the oxidation number is known as *oxidation*.

Oxidation states provide a convenient basis for summarising the chemistry of the metallic elements. Starting with the s-block elements, the alkali metals Na to Cs with one valence electron all readily form monocations with the filled shell core of the preceding noble gas, *e.g.* [Ne] for Na^+, [Ar] for K^+ and [Kr] for Rb^+. Similarly the alkaline earth metals, Mg to Ba, with two valence electrons all readily form dications with the filled shell core of the preceding noble gas, *e.g.* [Ne] for Mg^{2+}, [Ar] for Ca^{2+} and [Kr] for Sr^{2+}. These elements all have low electronegativity values, readily lose their valence shell electrons and show ionic bonding. Lithium and beryllium similarly form Li^+ and Be^{2+}, respectively but these are small ions with a large charge to radius ratio and are highly polarising leading to more covalent behaviour. Their chemistry is thus rather different from that of the other elements in their groups.

The p-block elements vary in properties between metals and non-metals. The metal non-metal transition extends diagonally across the block so that Al, Sn and Bi are metallic while B, P, Se are non-metallic, the intervening elements showing intermediate chemical behaviour. Under mild conditions the halogens (X) act as electron acceptors forming the monoanionic X^-. However, under more forcing chemical conditions positive oxidation state compounds form with oxygen or fluorine. Typical examples being $XO_3{}^-$, and $XO_4{}^-$ containing X with oxidation numbers +5 and +7, respectively. These are reactive compounds able to act as oxidants (electron acceptors) or sources of oxygen in chemical reactions. Having filled valence shells the noble gases He to Xe do not form compounds under mild conditions. However, the outermost electrons of the larger of these elements can be removed at chemically accessible energies so some compounds of Xe and a few of Kr are known, but these are very reactive.

Among the p-block elements there is an increasing tendency for two valence electrons to be retained in the outer s subshell of the ion on descending a group. Thus whist the chemistry of aluminium is dominated by the formation of Al^{3+}, the chemistry of thallium has a much greater contribution from oxidation number +1. Thus Tl^{3+} is quite a strong oxidant and the stable form for thallium in aqueous media is Tl^{+}. Similarly in group 14, the chemistry of lead is dominated by Pb^{2+} although Pb^{4+} compounds are known. In the case of tin, Sn^{2+} is a good reducing agent, capable of delivering two electrons to a substrate, but oxidation numbers +2 and +4 are well represented in the chemistry of tin. Bismuth is most commonly found in the form of Bi^{3+} compounds. The reason for these observations is the reduced energy available from bond formation with the heavier elements. In an ionic bonding model, the energy required to completely ionise the valence shell of an atom must come from the formation of chemical bonds. Larger atoms tend to form weaker bonds so there is less energy available to reach the higher oxidation numbers. Only with the most electronegative elements, oxygen and fluorine, is the highest, or Group, oxidation number attainable. Where all the valence shell electrons are ionised the ion is left with the electron configuration of the preceding noble gas. If all but the two valence subshell s electrons ionise the metal ion retains the spherical ns^2 electron configuration with the np subshell remaining unoccupied and available for involvement in bonding interactions.

The oxidation numbers of the transitional elements show even more variation and are summarised in Table 5. Among the individual elements, the least tendency to variable oxidation number is found at the beginning and end of the d-block and among the lanthanides. Zinc, cadmium and mercury form oxidation number +2 ions having filled d^{10} subshells which cannot easily be ionised further, although mercury also forms mercury(+1) in the form of the $Hg_2{}^{2+}$ ion found in Hg_2Cl_2, for example. Scandium, yttrium, lanthanum and the lanthanides are electropositive elements for which the most stable oxidation number is +3. In Sc, Y and La this arises from the loss of all the valence electrons while for Ce to Lu the energy available from bond formation is usually only sufficient to drive three ionisations giving compounds of Ln^{3+} (Ln being a generic symbol for a lanthanide element or La). In some cases higher or lower oxidation states may be found, important examples being Ce^{4+} ($4f^0$) and Eu^{2+} ($4f^7$), which can be obtained in aqueous solution but are reactive and readily return to their respective Ln^{3+} ions. Despite the relatively high charge on these ions their larger size prevents substantial polarisation of bonded anions so that bonding with these metals is essentially ionic in character. Furthermore the 4f orbitals of the lanthanide ions are concealed within the [Xe] core electron density so that the partly filled 4f orbitals are not able to overlap to any significant extent with the orbitals of adjacent atoms. This contributes further to the ionic nature of the bonding for these elements. The outer subshells are in fact the filled $5s^2$ and $5p^6$ subshells of the unreactive [Xe] core.

Table 5 *Oxidation number of the transitional elements*[a]

Sc	Ti	V	Cr	Mn	Fe	Co	Ni	Cu	Zn
3	2, 3, 4	1, 2, **3**, **4**, **5**	1, **2**, **3**, 4, 5, **6**	1, **2**, 3, 4, 5, 6, **7**	1, **2**, **3**, 4, 6	1, **2**, **3**, 4	1, **2**, 3, 4	**1**, **2**, 3	**2**

Y	Zr	Nb	Mo	Tc	Ru	Rh	Pd	Ag	Cd
3	2, 3, **4**	2, 3, 4, **5**	2, 3, 4, 5, **6**	1, 2, 3, **4**, **5**, 6, **7**	2, **3**, 4, 5, 6, 7, 8	**1**, 2, **3**, 4, 5, 6, 7, 8	**2**, 4	**1**, 2, 3	**2**

La	Hf	Ta	W	Re	Os	Ir	Pt	Au	Hg
3	2, 3, **4**	2, 3, 4, **5**	2, 3, **4**, **5**, **6**	1, 2, 3, **4**, **5**, 6, **7**	2, 3, **4**, 5, 6, 7, 8	1, 2, **3**, **4**, 5, 6	**2**, **4**, 5, 6	**1**, **3**, 5	**1**, **2**

Ac
3

Ce	Pr	Nd	Pm	Sm	Eu	Gd	Tb	Dy	Ho	Er	Tm	Yb	Lu
3, 4	**3**	**3**	**3**	2, **3**	2, **3**	**3**	**3**	**3**	**3**	**3**	**3**	2, **3**	**3**

Th	Pa	U	Np	Pu	Am	Cm	Bk	Cf	Es	Fm	Md	No	Lr
4	3, 4, **5**	3, 4, 5, **6**	3, 4, **5**, 6, 7	3, **4**, 5, 6, 7	3, 4, 5, 6	**3**, 4	**3**	**3**	**3**	**3**	**3**	2, **3**	**3**

[a]The more common oxidation numbers likely to be stable in aqueous media are shown in **bold** type, other values refer to other known oxidation numbers in compounds which may not be stable in aqueous media. The list does not include every known oxidation number and zero or negative oxidation number have been excluded, although these are sometimes encountered in compounds involving metal to carbon bonds.

Within the d-block the *n*d subshell of the metal ions may remain partly occupied and, depending upon the metal and its oxidation number, the d orbitals may be sufficiently exposed at the surface of the ion to become involved to some extent in more covalent bonding interactions with adjacent atoms. This leads to important differences in properties when compared to s-block metal ions and to the lanthanides (Table 5). The earlier d-block elements, on the left of the Periodic Table can form compounds with oxygen or fluorine in which they attain their Group oxidation number, in an ionic bonding model they would be supposed to lose all their valence shell electrons. However, this implies that these higher oxidation state compounds would contain highly charged cations which would be strongly polarising leading to more covalent behaviour. Thus $TiCl_4$ is actually a liquid at room temperature rather than an ionic solid like NaCl. Similarly $KMnO_4$ and $K_2Cr_2O_7$ dissolve in aqueous solution as K^+ ions and discrete MnO_4^- or $Cr_2O_7^{2-}$ ions rather than dissociating completely as might be expected for a purely ionic compound. In fact these high oxidation number oxyanions (containing both metal and oxygen) can oxidise water to dioxygen but the rate of reaction is sufficiently slow that aqueous solutions can be prepared and stored for some time. In general, in aqueous solution metal ions with oxidation numbers beyond +4 are not found in the form of simple metal ions but only in oxo-species, exemplified by MoO_2^{2+}, VO^{2+} and MnO_4^-, in which π-bonding between oxygen and the metal ion compensates for the electron deficiency which would otherwise be associated with the high charge on the metal ion. As the ionisation energies increase towards the right of the d-block the energy available from bond formation becomes insufficient to generate the higher oxidation states, even with oxygen or fluorine. Thus towards the right side of the d-block the +2 oxidation state becomes dominant. The oxidation numbers observed in compounds of the metals in the centre of the d-block vary over a wide range but in aqueous solutions oxidation numbers +2 and +3 predominate.

The actinide elements show a chemistry intermediate between that associated with the d-block metals and the lanthanides. The aqueous chemistry of thorium is limited to Th^{4+} but the early actinides, Pa to Pu, show some variation in oxidation number. All can attain the maximum oxidation number for their group but oxidation number +3 dominates the chemistry of the later actinides, Am to Lr.

2.4.2 The Structures of Metal Compounds

Other than in a gaseous plasma state, metal ions do not exist in isolated form but are surrounded by other atoms or ions. In solid NaCl the Na^+ ions are surrounded by Cl^- ions which are themselves surrounded by Na^+ ions (Figure 2). In solution the metal ions are surrounded by solvent molecules or may be attached to other ions or molecules in what are known as complex compounds, *i.e.* metal complexes. In these complexes the groups attached to the metal ion are known as *ligands*. When not attached to the metal such species are also often referred to as ligands but are more correctly called *proligands* in the absence of the metal. Since the applications of metals in medicine are often concerned with complex compounds rather than solid salts such as NaCl the remainder of this discussion will focus on metal complexes.

2.4.2.1 Geometric Structure

In water cobalt chloride, $CoCl_2$, may dissolve to form solvated Co^{2+} ions in the form of the aqua complex $[Co(H_2O)_6]^{2+}$.[12] However, one or more of the chloride ions may become associated with the cobalt (+2) ions in the solution to form chloride complexes such as $[CoCl(H_2O)_5]^+$ or $[CoCl_2(H_2O)_6]$. In the presence of high concentrations of chloride ion $[CoCl_4]^{2-}$ may form. These complexes are characterised by several features. One is the oxidation number of the metal, +2 in each of these examples. Another is the number of atoms to which the metal ion is attached, known as the *coordination number* (CN) of the metal ion, this is six in $[CoCl(H_2O)_5]^+$ and $[CoCl_2(H_2O)_6]$ but only four in $[CoCl_4]^{2-}$. The arrangement in space of the atoms attached to a metal in a complex is a further important feature and is known as the *coordination geometry* of the complex. The common coordination geometries associated with particular CNs are shown in Figure 19. These diagrams include two different representations of the coordination environment of a metal ion. One shows the links between the metal ion and the adjacent ligand atom to which it is attached, sometimes called the *donor atom*. In these representations a solid line represents a bond which lies in the plane of the paper, a wedge indicates a bond projecting out the plane of the paper and a dashed line a bond directed into the plane of the paper. The other representation shows only lines linking the donor atoms attached to the metal ion forming what is known as the coordination polyhedron *i.e.* the three dimensional geometric figure with vertices corresponding with donor atom positions. These polyhedra are usually shown in their most regular form as if all the edges were of equal length, although in reality this is an idealised representation of a structure which is usually not fully regular.

Coordination compounds, or complexes, of metals typically contain inorganic donor atoms such as O, N, S, P, F or Cl. In the particular case where one or more donor atoms in the complex is carbon, and there is a direct metal to carbon bond, the metal complex is said to be an *organometallic* compound. In order for such compounds to be unreactive towards air and moisture it is necessary for their metal-ligand bonding to have significant covalent character. Organometallic compounds in which the bonding is ionic in character are usually rather reactive. As an example diethyl zinc, $Zn(C_2H_5)_2$, is spontaneously flammable in air.

2.4.2.2 Isomerism

In some cases different structural arrangements of atoms are possible for a particular metal complex. As an example the ligands in a complex $[ML_4]$ (L is a monodentate ligand, *i.e.* bound to the metal through only one donor atom) containing a metal ion with CN four might be arranged at the vertices of a

[12] When the formula of a complex is written in square brackets, these enclose the full list of ligands directly attached to the metal centre. The symbol for the metal appears first in the formula followed by those of the, ligands usually in alphabetical order. If the complete set of ligands bound to the metal is not listed, square brackets should not be used.

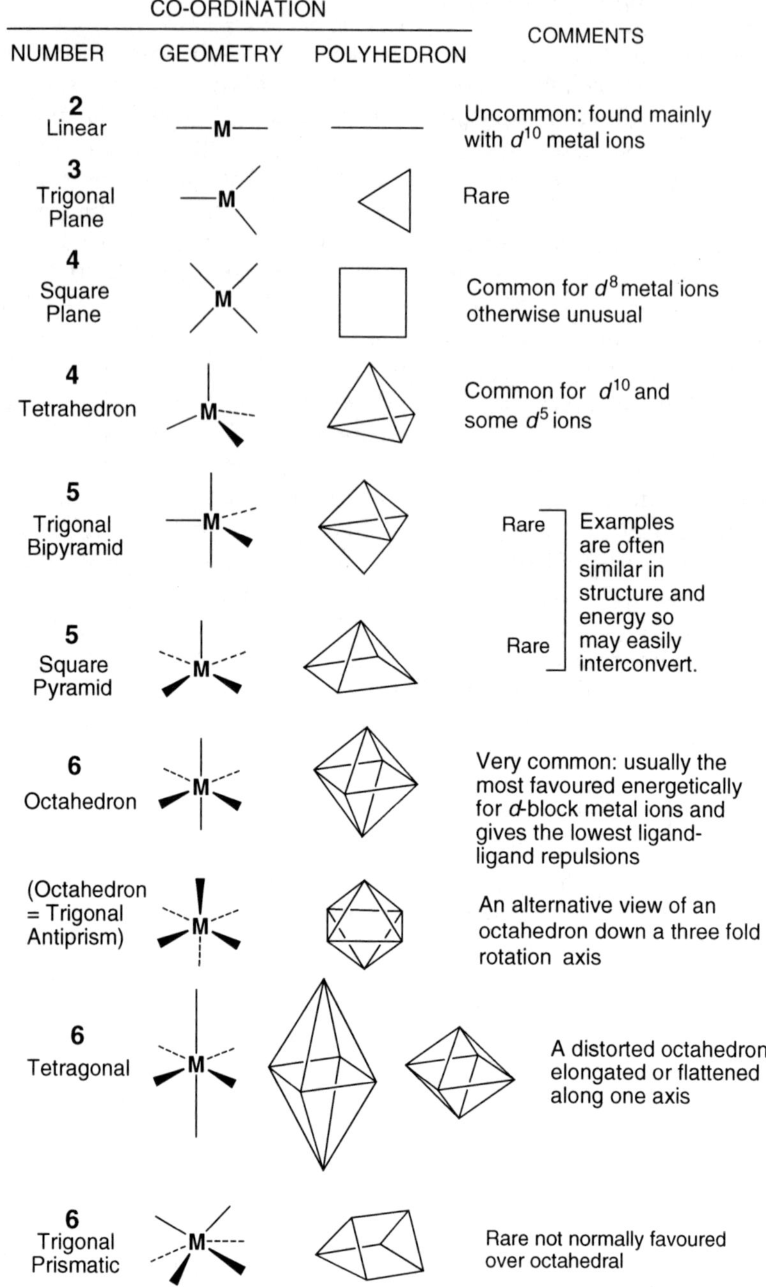

Figure 19 *Representations of the idealised coordination geometries and polyhedra associated with the different coordination numbers of metal ions in complexes*

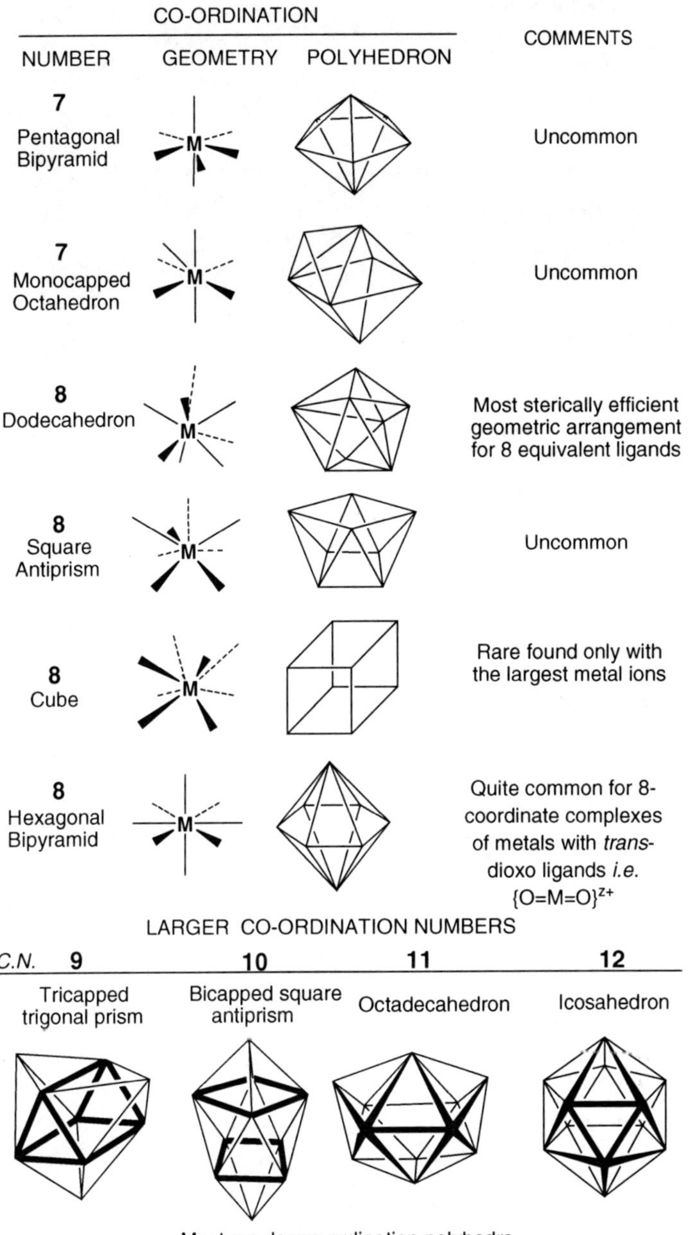

Figure 19 (*Continued*)

tetrahedron or of a square. Such complexes in which different structures arise for the same formula are known as *isomers* (Figure 20). Many types of isomerism are possible and only a few examples are discussed here. In a CN four complex of formula $[ML_2X_2]$, where L and X are different monodentate

(a) Square Planar

$[ML_2X_2]$ $[ML_2XY]$

cis *trans* *cis* *trans*

(b) Octahedral

$[ML_4X_2]$ $[ML_3X_3]$

cis *trans* *fac* *mer*

Isomers Isomers

(c) Chiral complexes

$[M(L\text{-}L)_3]$ {chelating ligand L-L} $[M(L\text{-}L)_2X_2]$

Λ *enantiomers* Δ Λ *enantiomers* Δ

Isomers Isomers

for L-L = en or represent

(d) Chiral carbon centres

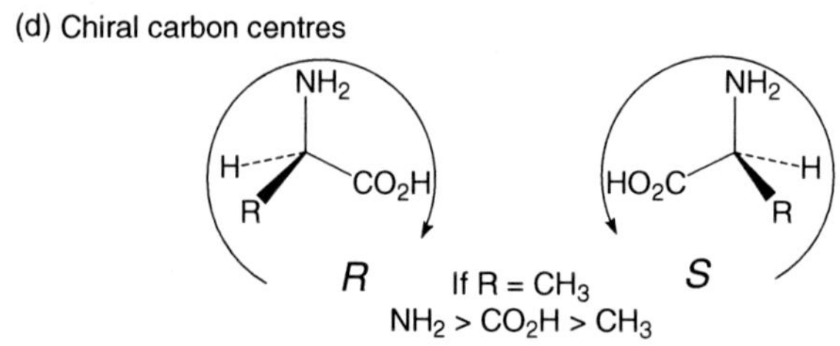

Figure 20 *Important isomeric structures of (a) square planar complexes, (b) octahedral complexes, (c) chiral octahedral complexes and (d) chiral carbon centres, the labels R and S being assigned according to the Cahn/Ingold/Prelog convention. This assigns priority among the groups attached to the chiral carbon to atom according to the atomic number of the bonded atom followed by the next neighbour atoms where a common bonded atom is present. e.g. O > N > C and CO_2 > CH_2O > CH_2N. When arranged with the lowest priority group at the back a priority order decreasing clockwise is given the symbol R and the reverse arrangement the symbol S*

ligands, the tetrahedral geometry has only one isomer. However the square planar geometry can have two isomers, the *cis*-isomer with the two X ligands adjacent to each other and the *trans*-isomer with the two X ligands opposite to each other [Figure 20(a)]. This situation, known as *geometric isomerism*, also arises in octahedral complexes of general formula $[ML_4X_2]$. Similarly an octahedral complex of formula $[ML_3X_3]$ can have *fac* (facial) or *mer* (meridional) isomers depending upon whether or not the three X ligands occupy a triangular face of the octahedron [Figure 20(b)].

So far only ligands which bind to the metal ion through one donor atom have been considered. However, many ligands are known which bind through more than one donor atom. These are called *polydentate* ligands, the 'denticity' indicating the number of donor atoms in a single ligand which bind to the metal ion. The prefixes for denoting ligand denticity are mono-, di-, tri, tetra-, penta- and hexa-, respectively for one, two, three, four, five and six donor atoms. Less commonly the respective prefixes uni-, bi-, ter-, quadri-, quinqui- and sexi- may be encountered. A simple example of a didentate ligand is provided by 1,2-diaminoethane, $H_2NCH_2CH_2NH_2$ (often abbreviated to en in formulae). This can bind to a single metal ion *via* the two amine nitrogen atoms to give what is called a *chelate* complex in which the ligand grips the metal through two donor atoms behaving as if it were a 'claw'. In general didentate ligands may be represented by the symbol L–L in formulae. Continuing this series, bis(2-aminoethyl)amine, $HN(CH_2CH_2NH_2)_2$, also known as diethylene triamine or dien, is a tridentate ligand able to bind to a metal ion through three nitrogen donor atoms.

Chelating ligands of this type can introduce further types of isomerism. One important example, known as *optical isomerism*, arises with octahedral complexes of general formula $[M(L–L)_3]$ [Figure 20(c)]. If three didentate ligands are bound to a metal ion in an octahedral geometry the complex takes on the form of a propeller and two isomers are possible which are related as mirror images and cannot be superimposed on one another. Such isomers are called *enantiomers* and the complex is said to be *chiral* having right- and left-handed forms. A similar situation can arise if two didentate and two monodentate ligands are present in a complex of general formula $[MX_2(L–L)_2]$. Chirality is more commonly associated with carbon centres carrying four different substituents exemplified by α-amino acids, $CH(R)(NH_2)(CO_2H)$, which can exist as enantiomers. The carbon atoms in the two different chiral arrangements are distinguished by the labels R and S [Figure 20(d)] according to the Cahn/Ingold/Prelog convention. The labels d and l are also used to identify chiral compounds on the basis of the direction in which they rotate the orientation of plane polarised light but there is no direct correlation between d or l. and absolute configuration defined by R *or* S. *Diastereomers* are chiral compounds which are not enantiomers. Where they contain two chiral centres d, l and *meso* forms are possible depending on whether both centres are d, or l in character or, in the *meso* form, one d and one l centre are present.

Another type of polydentate ligand which may be encountered in metal complexes is the *macrocyclic* ligand. Such ligands are cyclic molecules capable of encircling a metal ion. To do this the cyclic molecule must have a ring size of at least nine atoms of which at least three must be donor atoms positioned so that they may bind to a metal ion. Macrocyclic ligands are very important in biology, examples being provided by porphyrin or corrin derivatives [Figure 21]. Representing the structures of such complicated molecules can present problems. If all the atoms are labelled the diagram can become confusing, so a shorthand notation is used in which lines define the carbon skeleton and only those groups which are not carbon and its associated hydrogen are labelled [Figure 21(a)]. The problem becomes even greater when representing biomolecules such as proteins (Figure 22) or nucleic acids (Figure 23). In these cases 'cartoon' representations may be used in which only the gross structural features are presented and the underlying chemical structure is largely omitted, though implied as shown in Figures 22, 23 and 24.

In complexes of the s-block metal ions or the lanthanide ions Ln^{3+} there is little or no directionality in the way in which the metal ion interacts with adjacent atoms. A ligand attached to the metal ion experiences a closed shell electron configuration, [Ne] for Na^+ or Mg^{2+} down to [Xe] for Cs^+ or Ba^{2+}. In the case of the Ln^{3+} ions, although the 4f valence subshell is partially occupied, the electron density associated with this subshell is buried beneath the $5s^2$ and $5p^6$ subshells of the [Xe] core (Figure 13). The unoccupied 5d 6s and 6p orbitals are at higher energy so that the Ln^{3+} ions, for the most part, behave as if they had a closed shell [Xe] electron configuration. Consequently the geometrical arrangement of the atoms in contact with these metal ions is largely determined by the need to minimise the energy of interactions between the ligands attached to the metal. Depending on the variety and size of the ligands attached to the metal ion, its coordination geometry may be more or less regular. In complexes of the s-block and lanthanide metal ions the CN depends primarily on the ionic radius of the metal ion and the sizes of the ligands attached to it. In the case of the d-block metal ions the electron configuration of the metal ion can have a significant influence on the CN and geometry found in its complexes. Certain geometrical arrangements may result in a lower energy situation for the metal atom and so favour a particular structure for its complexes. In order to understand how these effects arise it is necessary to consider the bonding arrangements between metal ions and their associated ligands.

2.5 Bonding in Metal Complexes

2.5.1 The Crystal Field Theory and Octahedral Complexes

Evidence for bonding interactions which depend on the electron configuration of a d-block metal ion can be seen in a comparison between the *hydration energies* of the 1st row d-block metal dications and the lanthanide Ln^{3+} ions (Figure 25).[13] Provided the charge on the metal ion remains the same, the

[13] The hydration energy of a metal ion represents the energy change associated with moving the metal ion from isolation in the gas phase to an aqueous solution in which the metal ion is surrounded by water molecules, some directly bonded to the metal ion.

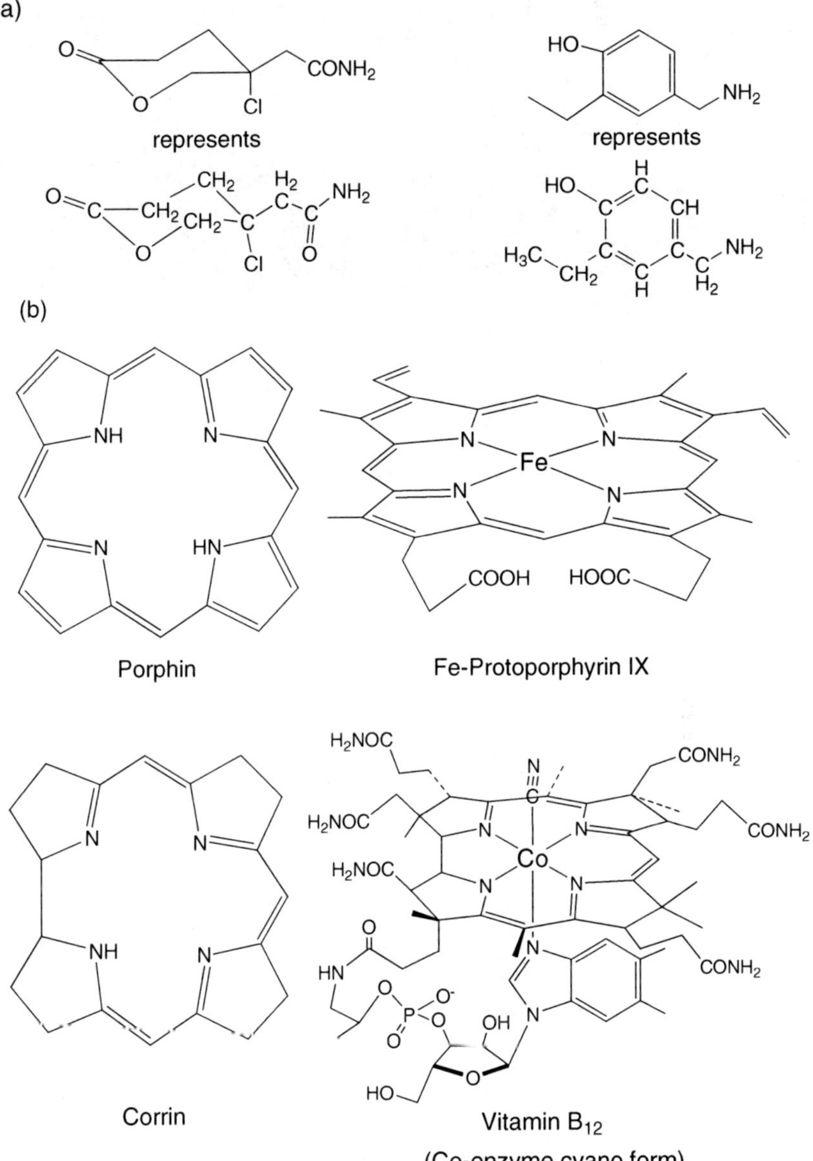

Figure 21 *Some ways of representing chemical structure. (a) The chemical structures shown employ a shorthand notation for the carbon skeleton in which carbon atoms and their associated hydrogens are represented by points at the intersection of lines representing bonds as shown. (b) Examples of two types of macrocyclic ligands found in biology and their metal complexes*

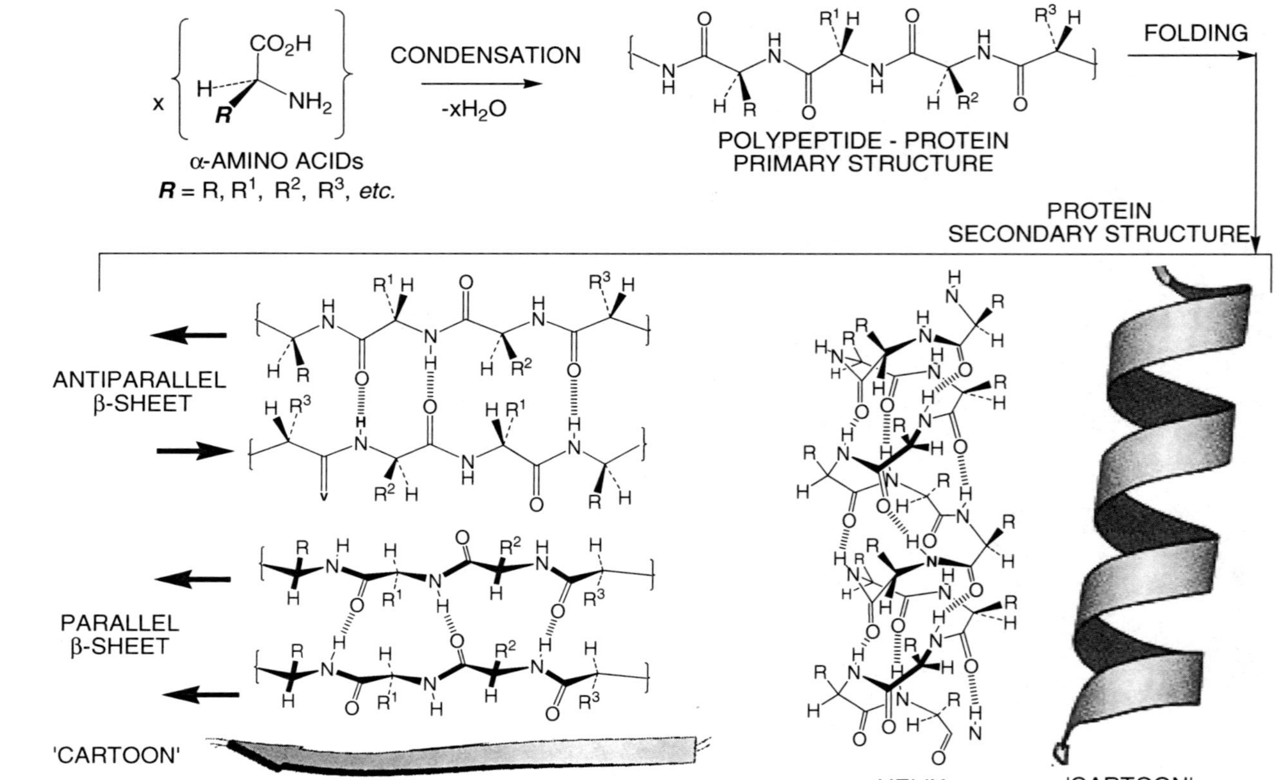

Figure 22 *(a) The formation of a polypeptide from α-aminoacids and (b) ways of representing protein structures using "cartoon" representations of β-sheets and α-helices within proteins*

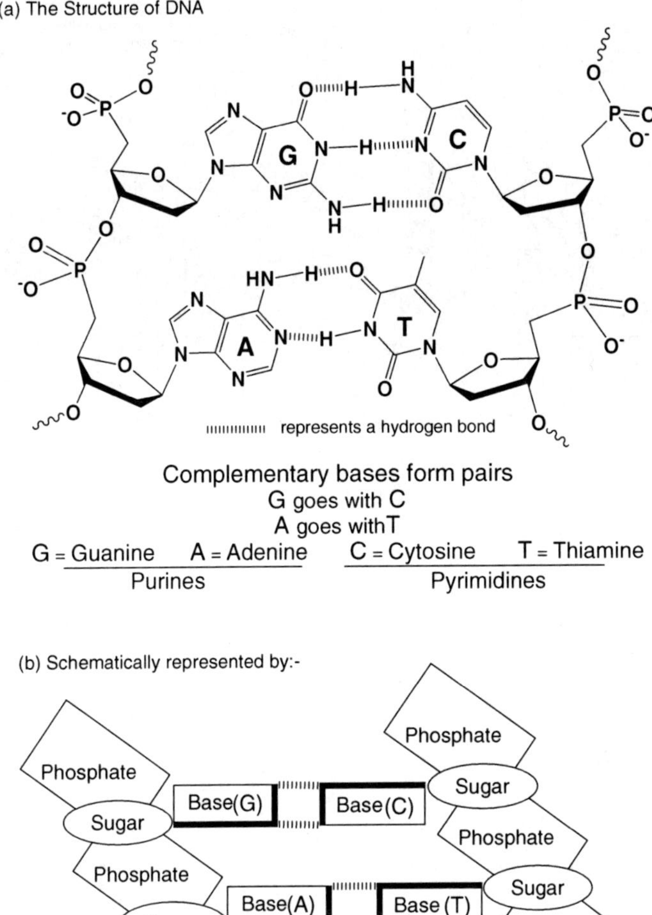

Figure 23 *The structure of DNA showing (a) the complementary base pairing of the bases, Adenine with Thymine and Guanine with Cytosine and (b) a schematic representation of the 'double strand' structure*

magnitude of the hydration energy will increase as the metal ion becomes smaller and so more polarising through having a higher charge/radius ratio. This leads to a stronger interaction between the metal ion and the water molecules and a more negative hydration energy. If the size of the metal ion is the primary factor determining hydration energy there should be a steady change with increasing atomic number across a period. This effect can be seen for the lanthanide series Ln^{3+} ions as shown in Figure 25(a). In the case of the 1st row d-block M^{2+} ions there is a similar overall trend with Ca^{2+}, d^5 Mn^{2+} and d^{10} Zn^{2+} defining a uniform change in hydration energy with ionic radius [Figure 25(b)]. However, unlike the situation with the lanthanide series, two

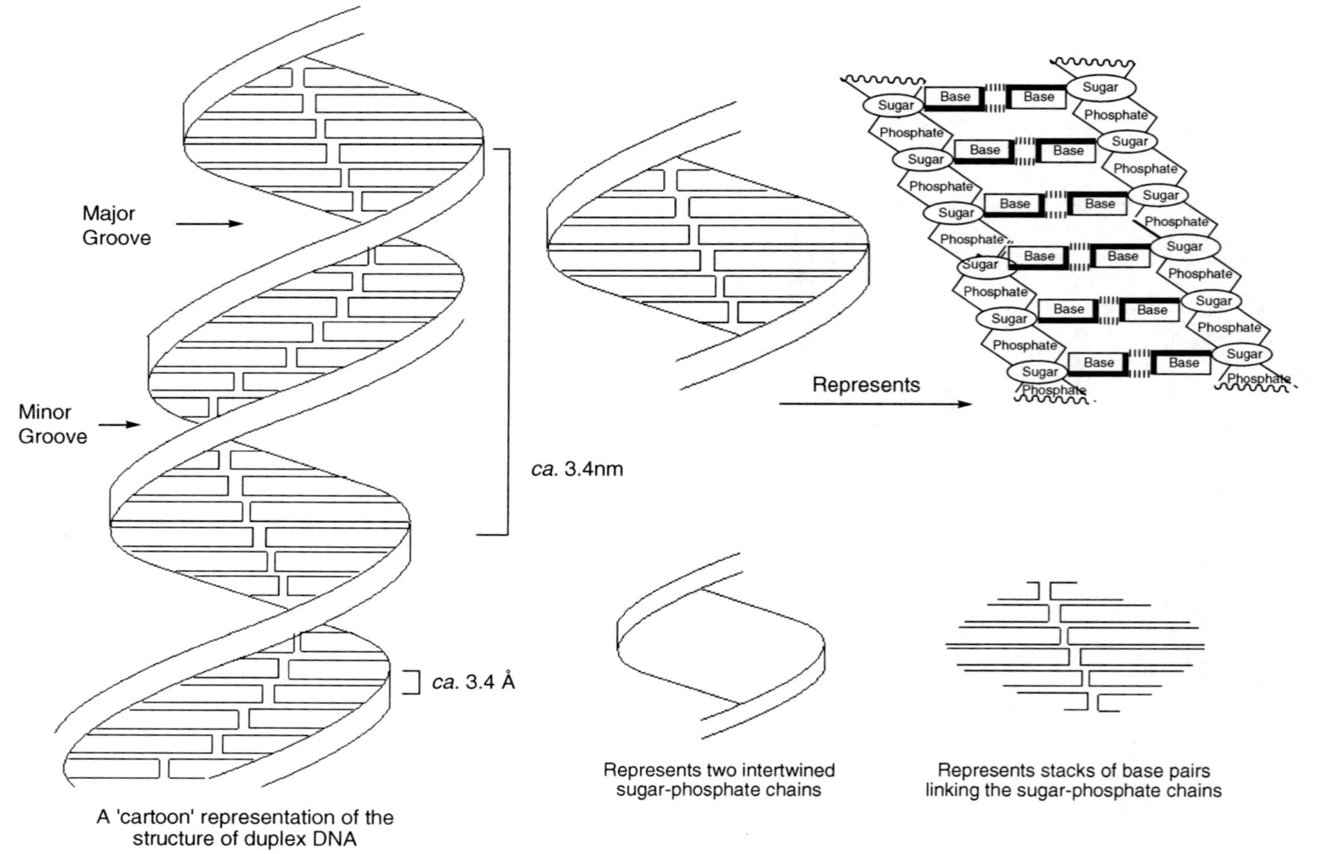

Figure 24 *A 'cartoon' representation of a section of DNA double helix showing the presence of major and minor grooves in the double helix structure. (1 Å = 0.1 nm = 100 pm)*

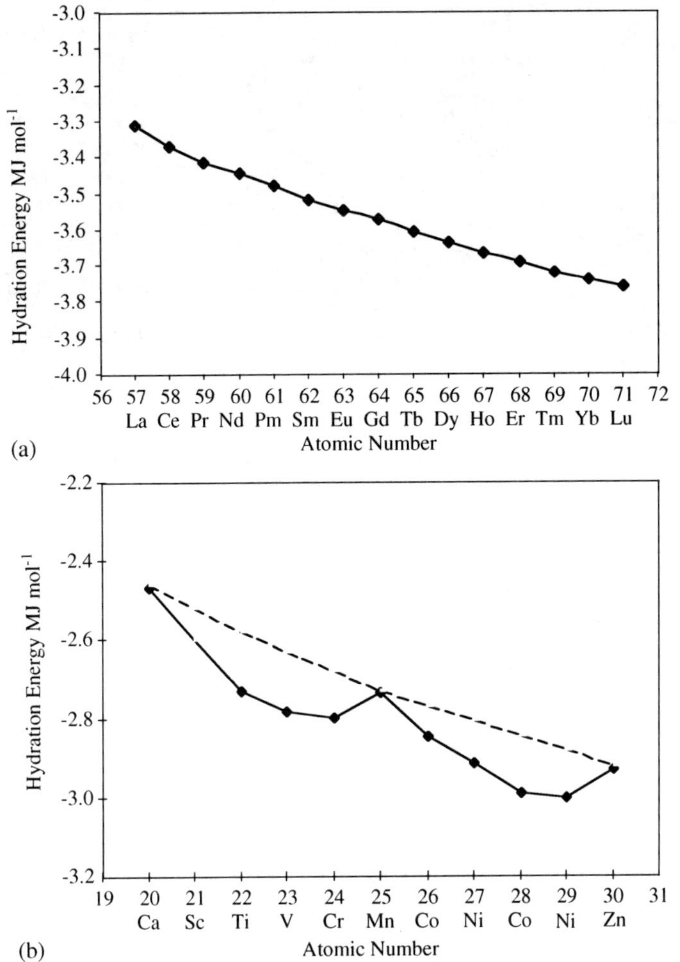

Figure 25 *Plots of hydration energy against atomic number for (a) the lanthanide(+3) ions Ln^{3+} and (b) the first row d-block metal ions M^{2+}*

minima appear in the plot. In order to understand how this comes about it is necessary to consider how the electron density in the various d orbitals is arranged in space and how this may interact with the electron density of the atoms adjacent to the metal ion.

In aqueous solution the dications of the 1st row d-block metal ions are present as the complexes $[M(H_2O)_6]^{2+}$ in which the water ligands of the six coordinate metal ion adopt an octahedral geometry. This means that the oxygen atoms of the bound water molecules represent regions of negative electron density at the vertices of an octahedron surrounding the metal ion.

If an isolated metal ion were removed from a vacuum into the electrostatic field created by a spherical shell of electron density equivalent to that of the six ligands in an octahedral complex, any electrons in the d orbitals will move to

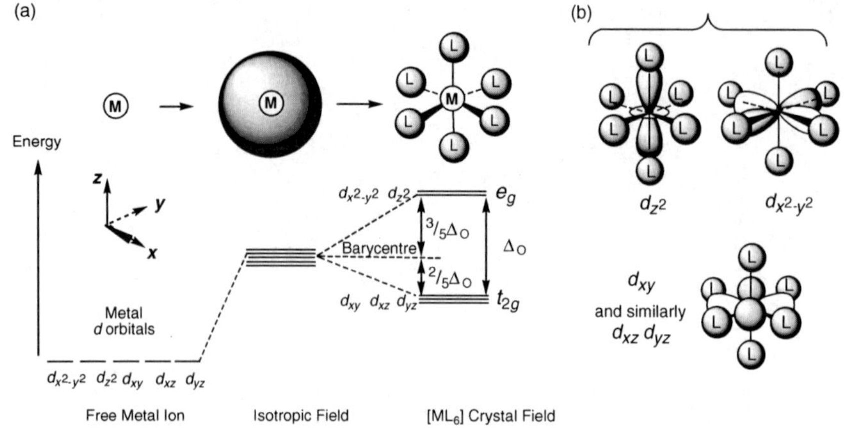

Figure 26 *(a) Energy level diagrams showing the effect, on the d orbitals of a transition metal, of a spherical shell of negative charge and its conversion to six point charges representing six ligands in an octahedral structure. The five co-linear lines for the free metal ion represent five d orbitals with the same energy. The five closely stacked lines for the metal ion in an isotropic field also represent five d orbitals with the same, higher, energy but in a more compact representation. In the octahedral complex the orbitals split into three (t_{2g}) with the same lower energy and two (e_g) with the same higher energy. (b) An illustration of how electrons in the d_{z^2} and $d_{x^2-z^2}$ lie closer to electrons in the ligand donor atoms than those in the d_{xy}, d_{xz} or d_{yz} orbitals*

higher energy because of the repulsive force produced by the surrounding shell of negative charge. In an energy level diagram the energies of the five degenerate d orbitals may be represented by five lines indicating that they have the same energy [Figure 26(a)].[14] If the spherical charge is then redistributed to six equivalent points at the vertices of an octahedron the d orbitals no longer experience the negative charge of the ligands to the same extent. This happens because electrons in the d_{z^2} and $d_{x^2-y^2}$ orbitals will, on average, spend more time closer to the six regions of ligand electron density than those in the d_{xy}, d_{xz} and d_{yz} orbitals which are oriented between the ligands [Figure 26(b)]. Thus, in an energy level diagram, the d_{z^2} and $d_{x^2-y^2}$ orbitals would be placed higher in energy than the d_{xy}, d_{xz} and d_{yz} orbitals. The difference in energies between the two sets of orbitals is known as the *crystal field splitting* and is represented by the crystal field splitting parameter Δ. The more specific symbol Δ_O may be used in the particular case of an octahedral complex. In the symmetry of an octahedron the three Cartesian axes are equivalent and cannot be distinguished. As a result the d_{z^2} and $d_{x^2-y^2}$ orbitals cannot be separated and must be treated as a pair sharing the same energy. Such orbitals are said to be degenerate. Similarly the three

[14] In some diagrams such lines are drawn side by side at the same level to show they have the same energy but, in order to give a more compact presentation, in others they are drawn in a stack of closely spaced lines intended to indicate that they are of the same energy. Both versions are illustrated in Figure 26(a). Each line represents a single orbital capable of accommodating up to two electrons provided they have different m_s quantum numbers.

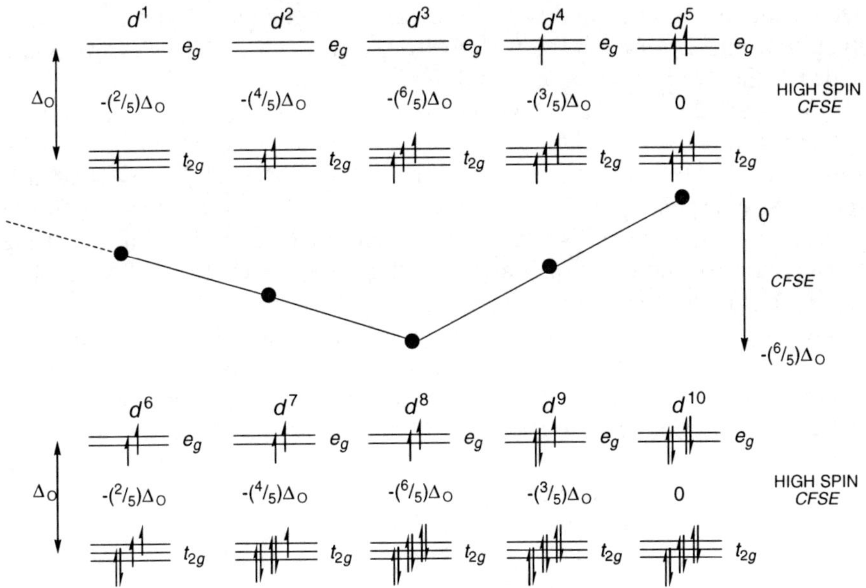

Figure 27 *Crystal field splitting diagrams showing how the energy of the metal ion varies with number of d electrons in an octahedral crystal field*

orbitals d_{xy}, d_{xz} and d_{yz} cannot be separated in octahedral symmetry and form a degenerate set of three orbitals sharing the same energy level [Figure 26(a)]. The redistribution of the charge within the initial spherical shell of electron density to the vertices of an octahedron does not produce any change in the total energy of the system. As a consequence the total energy of the set of d orbitals does not change and the 'energy centre of gravity' known as the *barycentre*, must remain fixed. To achieve this the energy of the two orbitals of the e_g set will increase by $3/5\Delta_O$ and that of the three orbitals of the t_{2g} set orbitals will decrease by $(2/5)\Delta_O$. This means that, in a filled subshell d^{10} ion, four electrons will move to higher energy by $(3/5)\Delta_O$ each and six electrons will move to lower energy by $(2/5)\Delta_O$ each. The result for the d^{10} closed subshell system is no net change in energy $\{[(4 \times 3/5)-(6 \times 2/5)]\Delta_O = 0\}$. A similar result would be obtained for a d^0 or a d^5 electron configuration with all electrons unpaired, so that the energies of Sc^{3+}, Mn^{2+}, Fe^{3+} and Zn^{2+} ions would not be affected by the field of the ligands. However, in a d^3 ion, such as Cr^{3+}, three electrons would occupy the lower energy t_{2g} orbital, so that the energy of the ion in an octahedral crystal field would be less than in an equivalent spherical field by $3 \times (2/5)\Delta_O$; *i.e.* an energy change of $-(6/5)\Delta_O$. This additional stabilisation of the complex, compared to an equivalent closed subshell metal ion, is known as the *Crystal Field Stabilisation Energy* (*CFSE*) (Figure 27). In the case of an octahedral metal complex with the electron configuration $t_{2g}{}^x e_g{}^y$ and a crystal field splitting Δ_O the *CFSE* is given by Equation (3)

$$CFSE = \{y(3/5) - x(2/5)\}\Delta o \qquad (3)$$

This model provides a good basis for explaining anomalies in the variation of hydration energies across the first row d-block metal ions. As electrons are added to the t_{2g} orbitals going from d^0 ions to d^3 ions, the *CFSE* increases from zero to $(-6/5)\Delta_O$, leading to a larger hydration energy. However, as electrons are added to the e_g orbital this effect is reduced then cancelled out, with *CFSE* values of $(-3/5)\Delta_O$ for d^4 and zero for d^5. Between d^5 and d^{10} this pattern is repeated. When this pattern of changing *CFSE* values is superimposed on a hydration energy, which is steadily increasing in magnitude with increasing atomic number and decreasing atomic radius, the general form of the plot in Figure 25(b) emerges.

Creating a pair of electrons in a single orbital, rather than adding them to two different degenerate orbitals, requires an energy input so that the three electrons of Cr^{3+} in an octahedral complex will enter t_{2g}, one in each of the three orbitals. Once three electrons have been added a further electron must either enter the higher energy e_g orbitals or form a pair with an electron in the t_{2g} set of orbitals. Forming this electron pair requires an energy input since two electrons will occupy the same region of space differing only in their m_s values. If this energy, known as the *pairing energy* (*PE*) is larger than Δ_O the electron will enter the e_g orbital set, as shown in Figure 28(a). If the interaction between the metal ion and the ligands is sufficiently large that $\Delta_O > PE$, after the d^3 electron configuration, it becomes more energetically favourable to add electrons to the t_{2g} orbital set before filling the e_g orbital set. Ligands which produce this effect are known as *strong-field ligands* ($\Delta_O > PE$) and the complexes they form are called *low spin complexes*. Ligands for which $\Delta_O < PE$ are known as *weak-field ligands* and form *high spin complexes*. When calculating the *CFSE* of a strong-field low spin complex it is necessary to include the pairing energies associated with the additional electron pairs which must form. In such cases Equation (3) must be modified to Equation (4) in which Π represents the number of additional electron pairs and $(CFSE)_L$ signifies a value of *CFSE* for a low spin case [Figure 28(b)].[15]

$$(CFSE)_L = \{y(3/5) - x(2/5)\}\Delta_O + \Pi PE \qquad (4)$$

High and low spin behaviour is typical of the 1st row d-block metal ions, but complexes of 2nd and 3rd row d-block metal ions tend to be low spin for the most part. A metal ion in a high spin complex may show a different radius from that found in a low spin complex. Electrons in the e_g orbitals of an octahedral complex are in closer proximity to the ligand electrons than those in the t_{2g} orbitals. As a consequence, ligands approaching along the Cartesian axis directions will encounter metal electron density in the e_g orbitals earlier than electron density in the t_{2g} orbitals [Figure 29(a)]. This leads to greater electron–electron repulsion and increased shielding from the positive nuclear charge for ligand electrons approaching along the Cartesian axes. Thus a metal ion with

[15] In the d^6 case there is already one electron pair present even in the high spin situation so this does not need to be included as the energy price was paid before the complex was formed. Similar arguments apply to other electron configurations above d^5.

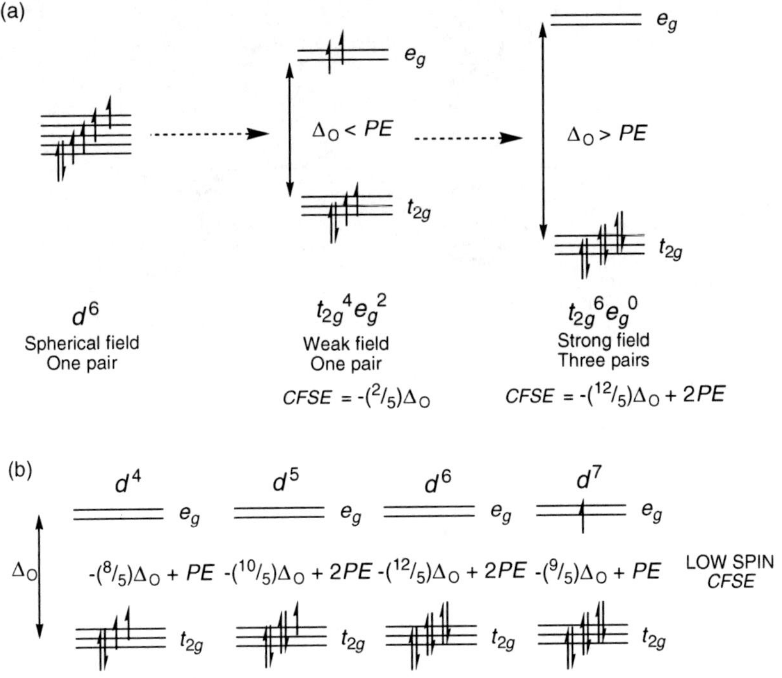

Figure 28 *Energy level diagrams showing (a) the effect of the relative magnitudes of the pairing energy (PE) and crystal field splitting (Δ_o) on the spin state of a d^6 metal ion, (b) the effect of the PE on the Crystal Field Stabilisation Energies of some d-block metal ions*

occupied e_g orbitals will appear to have a larger radius than one in which only the t_{2g} orbitals are occupied. The first three electrons in a high or a low spin complex enter t_{2g}, so between d^1 and d^3 the normal trend of decreasing ionic radius with increasing atomic number is found. However, the fourth and fifth d electrons of a high spin complex enter e_g, counteracting the normal decrease in radius. Subsequently the sixth, seventh and eighth electrons go into t_{2g} so the radii decrease again slightly but, as the last two electrons are added to e_g, for the d^9 and d^{10} configurations the radii increase again. In a low spin complex electrons continue to enter t_{2g} until a d^6 configuration is reached, so the radii continue to decline beyond d^3 and only begin to increase from d^7 onwards as the e_g orbitals become occupied [Figure 29(b)].

Although the effect of spin state on ionic radius might seem a rather esoteric topic, it is of some importance in human biochemistry. The Fe^{2+} ions in the oxygen transport protein hemoglobin are bound within a macrocyclic ligand derived from porphyrin. In the deoxygenated state the Fe^{2+} ions are high spin and too large to fit easily into the cavity within the porphyrin macrocycle. When dioxygen binds to the Fe^{2+} ion it switches to a low spin state which, being smaller, fits into the cavity in the porphyrin ligand leading to a structural change. This is thought to be the trigger responsible for the cooperative effect in

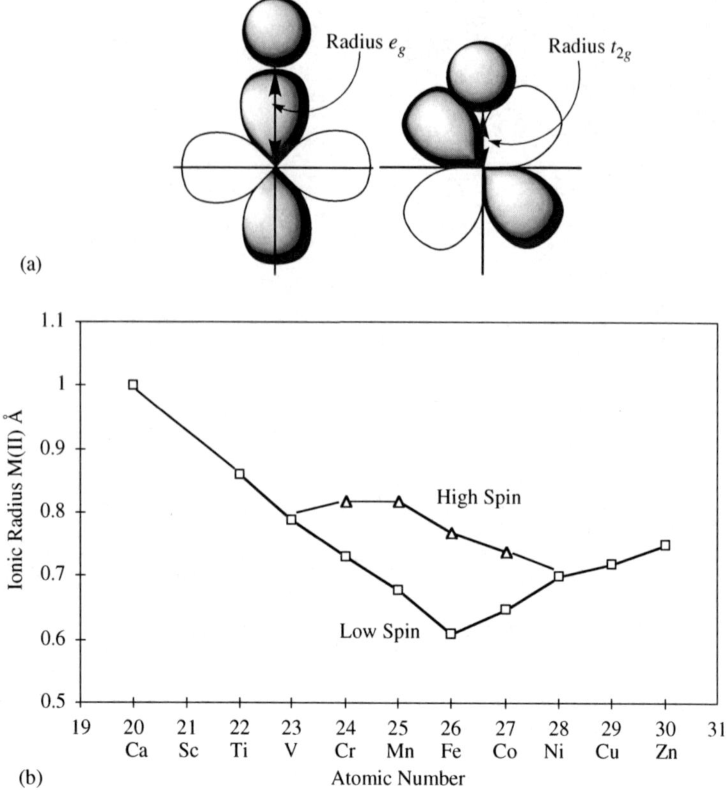

Figure 29 *(a) The difference between e_g and t_{2g} set orbitals in their interaction with a ligand in an octahedral complex. (b) The variation in ionic radius of first row d-block metal ions, M^{2+}, with spin state and atomic number*

hemoglobin whereby, after the binding of the first O_2 molecule, subsequent O_2 molecules are bound more strongly. This improves the efficiency of the protein in collecting and releasing the four O_2 molecules it can carry.

2.5.2 The MO Theory and Octahedral Complexes

The crystal field model of bonding in d-block metal complexes takes account of the repulsion between the ligand electrons and the metal electrons and is, in essence, an electrostatic or ionic type of bonding model. This provides a useful basis for describing many of the properties of transition metal ions, particularly predicting the numbers of unpaired electrons. However, it does not naturally take account of covalency in the metal-ligand bonding. It is possible to modify the theory to do this but a *molecular orbital* (MO) theory bonding model more naturally takes account of covalency.

In the MO model for bonding in an octahedral metal complex, the six ligand orbitals available for forming σ-bonds to the metal are treated as a group. In octahedral symmetry the six orbitals in this group may be mixed to produce six combination orbitals known as *group orbitals*. Each group orbital has a capacity of two electrons but contains contributions from two or more of the original ligand orbitals. The six group orbitals form three sets, a unique high symmetry a_{1g} orbital with A_{1g}, symmetry, two degenerate orbitals in an e_g set with symmetry E_g and three degenerate orbitals in a t_{1u} set with T_{1u} symmetry (See Footnote 10 p. 47). These orbitals are of appropriate energies to interact with metal valence shell orbitals of the same symmetry. In an octahedral structure the 4s orbital of the metal ion is of A_{1g} symmetry and is given the group orbital label a_{1g}, the three 4p orbitals form a set of T_{1u} symmetry and are labelled t_{1u}. The five 3d orbitals form two sets, one of E_g symmetry derived from the d_{z^2} and $d_{x^2-y^2}$ orbitals and given the label e_g, the other of T_{2g} symmetry derived from the d_{xy}, d_{xz} and d_{yz} orbitals and labelled t_{2g}.

A qualitative energy level diagram can be constructed for $[ML_6]$ in a similar manner to that described earlier for HF or H_2O, although the situation is more complicated for $[ML_6]$ (Figure 30). Six σ-bonding MOs, together with their anti-bonding counterparts, can form from the interaction between the a_{1g}, e_g and t_{1u} metal valence shell orbitals and ligand group orbitals of appropriate symmetry. There are no σ-bonding ligand group orbitals of suitable symmetry to interact with the t_{2g} set of d orbitals so, in this σ bonding model, these remain as a set of non-bonding orbitals associated only with the metal ion. The MO bonding model naturally allows for varying degrees of covalency in the metal ligand σ-bonds but typically the bonding will have a large ionic component.

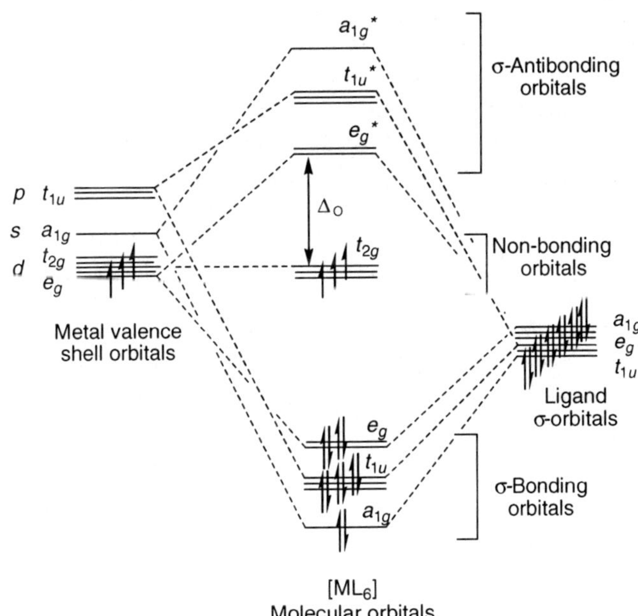

Figure 30 *An energy level diagram showing the MOs arising from ligand σ orbitals and the valence shell orbitals of a d-block metal*

Thus the σ bonding MOs will be largely ligand in character and the e_g* anti-bonding orbital will be mostly metal in character. Each ligand can be thought of as donating two electrons to the metal-ligand bonding orbitals leaving the metal valence shell electrons in the t_{2g} and e_g* orbitals. In fact the t_{2g} and e_g* orbitals comprise the MO equivalent of the crystal field theory bonding model. Thus the crystal field model has, in essence, extracted from the MO scheme those orbitals most closely associated with the metal ion and, where the bonding is predominantly ionic provides a satisfactory but simpler bonding description of the bonding.

In the σ-bonding model for octahedral metal complexes the ligands are seen as donating charge to a metal cation. According to the electroneutrality principle, proposed by Linus Pauling, the extent of charge transfer is expected to be that required to attain near neutrality at the metal atom. In effect the actual charge on the metal centre needs to be less than *ca.* +1 even though the oxidation number may be much higher. Thus with a complex of M^{3+} the six ligands between them might be expected to transfer to the metal, electron density equivalent to between two and three electrons. Problems arise with this concept if a low oxidation number is associated with the metal ion. In particular it is possible to make stable complexes with the neutral ligand carbon monoxide in which the transition metal has an oxidation number of zero. An ionic bonding model is not appropriate here since neither the metal nor the ligand are charged. Any transfer of charge from ligand to metal would lead to the appearance of a negative charge on the metal centre but detailed studies of metal complexes show that this does not happen. Instead electronic charge is transferred back from the metal ion to the ligands. This cannot happen through the σ-bonding MOs but the non-bonding metal centred t_{2g} orbitals of the σ-bond model can enter into a π-bonding interaction if the ligands have unoccupied orbitals of appropriate symmetry.

In the case of the six CO ligands there are anti-bonding π* orbitals available which include a set of ligand group orbitals having t_{2g} symmetry. The metal ion can back-donate electronic charge from the metal valence shell t_{2g} orbitals to the ligand group orbitals of t_{2g} symmetry reducing the electron density at the metal centre [Figure 31(a)]. This strengthens the metal-ligand bonding interaction but weakens the C–O bond in the ligand itself because π* orbitals, which are anti-bonding with respect to the C–O interaction, are becoming partly occupied. This type of bonding, in which donation of electrons from ligand to metal through a σ-bond is reinforced by donation of electrons from metal to ligand through a π-bond, is known as *synergic bonding*. Ligands which can accept charge from the metal through π donation from metal to ligand are known as *π-acceptors*. Examples of such ligands are provided by CO, NO, N_2, $H_2C{=}CH_2$, CNR [R=hydrocarbyl *i.e.* a hydrocarbon substituent such as CH_3, $C(CH_3)_3$ or C_6H_5]. In an octahedral complex $[ML^a{}_6]$ (where L^a represents a monodentate π-acceptor ligand) the effect of this type of π-acceptor behaviour is to lower the energy of the t_{2g} orbitals. This makes the electrons in these orbitals more difficult to ionise and so favours lower oxidation numbers [Figure 31(a)].

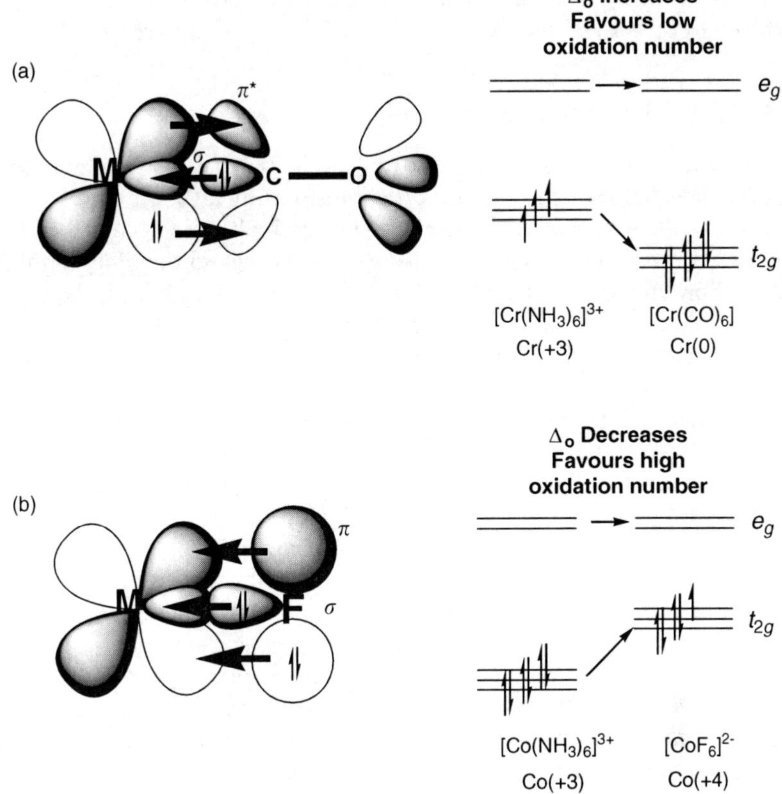

Figure 31 *(a) Synergic Bonding – metal back-donation to a π-acceptor ligand. σ Donation from a filled σ orbital on CO to an empty d orbital on M is reinforced by π back-donation from a filled d orbital on M to an empty π* orbital on CO (for clarity the 2nd out of plane π interaction is not shown) (b) Ligand to metal π-bonding in which σ donation from a filled σ orbital on F to an empty d orbital on M is reinforced by π donation from a filled p orbital on F to an empty d orbital on M (for clarity the 2nd out of plane π interaction is not shown)*

Interactions between the metal valence shell t_{2g} orbitals and filled π-type orbitals on the ligand donor atom are also important and can stabilise metal ions with high oxidation numbers. In such cases the metal t_{2g} orbitals may be unoccupied and available to accept charge from filled p orbitals on ligands such as oxide or fluoride [Figure 31(b)]. This alleviates the high positive charge associated with the metal ion so that *π-donor* ligands such as O^{2-} and F^- stabilise high oxidation number complexes. In an octahedral complex such as $[CoF_6]^{2-}$ the effect of this type of π-donor behaviour is to raise the energy of the t_{2g} orbitals. This makes the electrons in these orbitals more easily ionised and so favours higher oxidation numbers [Figure 31(b)]. As an example the typical maximum oxidation number of cobalt in complexes is Co(+3) (d^6) but in $[CoF_6]^{2-}$, containing the π-donor F^- ligand, it is Co(+4) (d^5). This effect is also seen in the formation of oxometal ions such as $Cr_2O_7^{2-}$ [Cr(+6)], MO_4^{2-} [M=Mo(+6) or W(+6)] and MO_4^- [M=Mn(+7), Tc(+7), Re(+7)].

2.5.3 The Eighteen-Electron Rule

In cases where there is significant covalency in the metal-ligand bonding it might be expected that the full use of a transition metal's s, p and d valence subshell orbitals would favour a situation in which all were occupied so that a total of 18 electrons would be associated with metal-ligand bonding. In the MO model of bonding in an octahedral complex this corresponds with the electron configuration a_{1g}^2, t_{1u}^6, e_g^4, t_{2g}^6 (Figure 30 shows a 15-electron complex formed from a d^3 metal ion). This expectation gives rise to the *eighteen-electron rule* which requires that, to form a stable complex, a metal atom will need to have eighteen valence electrons in its valence shell. This requirement is met in many organometallic complexes, and those involving π-acceptor ligands, but exceptions are not unusual in compounds which involve more ionic bonding.

In calculating the number of valence shell electrons in a metal complex, it is usual to assume a fully covalent model with a neutral metal atom contributing all its valence shell electrons. In this model the ligands are also taken as being neutral species so that each donor atom in a ligand which is neutral as a proligand contributes two electrons; *e.g.* each nitrogen in NH_3 or $NH_2CH_2CH_2NH_2$ can donate two electrons. The neutral ligands CO or CNR similarly act as two electron donors. However, a chlorine atom ligand, normally anionic Cl^- in the free proligand form, is treated as a neutral Cl atom capable of forming a single covalent bond as in HCl or CCl_4 and is one electron donor ligand. Similarly groups such as OH, OR (R=hydrocarbyl, *i.e.* a hydrocarbon substituent such as CH_3 or C_6H_5), SR or NHR would act as one electron donor ligands. Some examples of this method of electron counting are as follows:

[Cr(CO)$_6$]
Cr d^6
6(CO) $6 \times 2 = 12$ electrons
Total 18 electrons (6 + 12)

[Tc(CNCH$_3$)$_6$]$^+$
Tc d^7
6($CNCH_3$) $6 \times 2 = 12$ electrons
+1 charge −1 electron
Total 18 electrons (7 + 12 − 1)

[FeCl$_2$(NH$_3$)$_4$]
Fe d^8
4(NH_3) $4 \times 2 = 8$ electrons
2Cl $2 \times 1 = 2$ electrons
Total 18 electrons (8 + 8 + 2)

[$FeCl_3(NH_3)_3$]
Fe d^8
3(NH_3) 3 × 2 = 6 electrons
3Cl 3 × 1 = 3 electrons
Total 17 electrons (8 + 6 + 3)

[$CoCl_2(NH_3)_4$]
Co d^9
4(NH_3) 4 × 2 = 8 electrons
2Cl 2 × 1 = 2 electrons
Total 19 electrons (9 + 8 + 2)

[$CoCl_3(NH_3)_3$]
Co d^9
3(NH_3) 3 × 2 = 6 electrons
3Cl 3 × 1 = 3 electrons
Total 18 electrons (9 + 6 + 3)

From these calculations it can be seen that [$Cr(CO)_6$], [$Tc(CNCH_3)_6$]$^+$, [$FeCl_2(NH_3)_4$] and [$CoCl_3(NH_3)_3$] might be expected to be relatively stable complexes, [$CoCl_2(NH_3)_4$] should easily lose an electron to form [$CoCl_2(NH_3)_4$]$^+$ and [$FeCl_3(NH_3)_3$] should easily gain an electron to form [$FeCl_3(NH_3)_3$]$^-$. Complexes of d^8 metals with strong field ligands have one d orbital at high energy which tends to remain unoccupied and so typically have sixteen electrons associated with the metal valence shell.

2.5.4 The Crystal Field Theory and Complexes of CN 4

Although the majority of 1st row d-block transition metal complexes contain six coordinate metal ions with octahedral coordination geometries, examples of complexes with lower CNs are known. In particular CN four complexes are well known and may adopt tetrahedral or square planar geometries. The crystal field theory model for a tetrahedral complex shows the d orbitals splitting into two groups as was found for the octahedral case. However, in the tetrahedral case the d_{z^2} and $d_{x^2-y^2}$ orbitals (symmetry label e) form the lower energy set and the d_{xy}, d_{xz} and d_{yz} orbitals (symmetry label t_2) form the higher energy set (Figure 32). The crystal field splitting parameter, Δ_T, for tetrahedral [ML_4] is approximately half that for the corresponding octahedral complex, [ML_6]. As a result of this essentially all tetrahedral complexes of 1st row transition metals are high spin since Δ_T will be smaller than *PE*.

CFSE values can be calculated for tetrahedral complexes in a similar manner to octahedral complexes but, because $\Delta_T = (4/9)\Delta_O$, the values obtained are smaller for the tetrahedral case. Consequently tetrahedral coordination is not normally preferred and is only observed if certain factors apply. Firstly, electron configurations for which there is little or no *CFSE* difference between

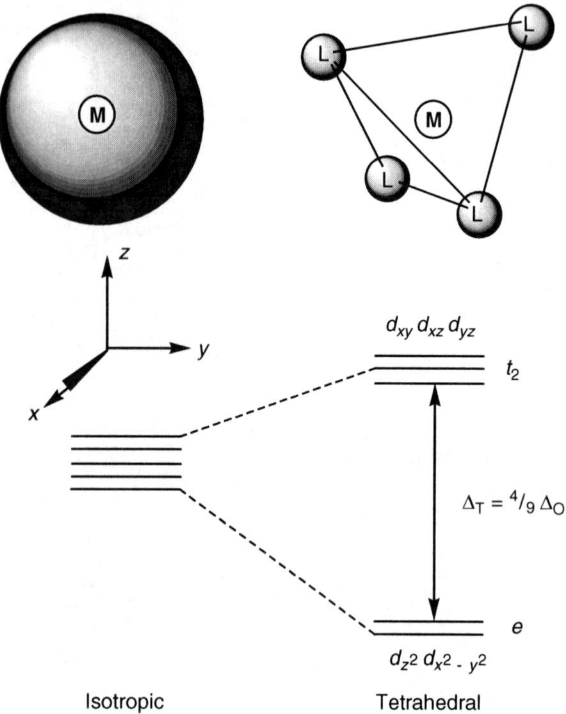

Figure 32 *An energy level diagram showing the crystal field splitting for a d-block metal ion in a tetrahedral crystal field. The MO energy level diagram for such a complex also contains this feature*

octahedral and tetrahedral structures will be more likely to show tetrahedral coordination. Secondly, larger anionic ligands will also favour the lower CN as this will reduce ligand–ligand repulsions. Typical examples of terahedral complexes arise with weak field ligands and d^5 or d^{10} electron configurations for which the *CFSE* will be zero *e.g.* $[MnCl_4]^{2-}$, $[FeCl_4]^-$ and $[ZnCl_4]^{2-}$. Another example is provided by $[CoCl_4]^{2-}$ where the *CFSE* advantage expected for $[CoCl_6]^{4-}$ is overcome by the ligand–ligand repulsions between the large negatively charged Cl^- so that the octahedral complex is unstable.

The other geometry associated with CN four is square planar, found primarily with d^8 metal ions bonded to strong field ligands. To see why this structure is associated with d^8 metal ions, it is simplest to begin with an octahedral structure and consider the effect on the energies of the d orbitals of removing two axial ligands to infinity (Figure 33). If z is chosen as the axis along which the ligands are removed, as the ligand electron density is removed the d_{z^2} orbital is most strongly affected and its energy falls. The $d_{x^2-y^2}$ orbital is more exposed to the electronic charge in the remaining four ligands and becomes relatively higher in energy. Similar effects occur within the d_{xy}, d_{xz} and d_{yz} orbital set but are of little consequence because these orbitals are fully occupied in a d^8 metal ion so that the net energy change within the set is zero.

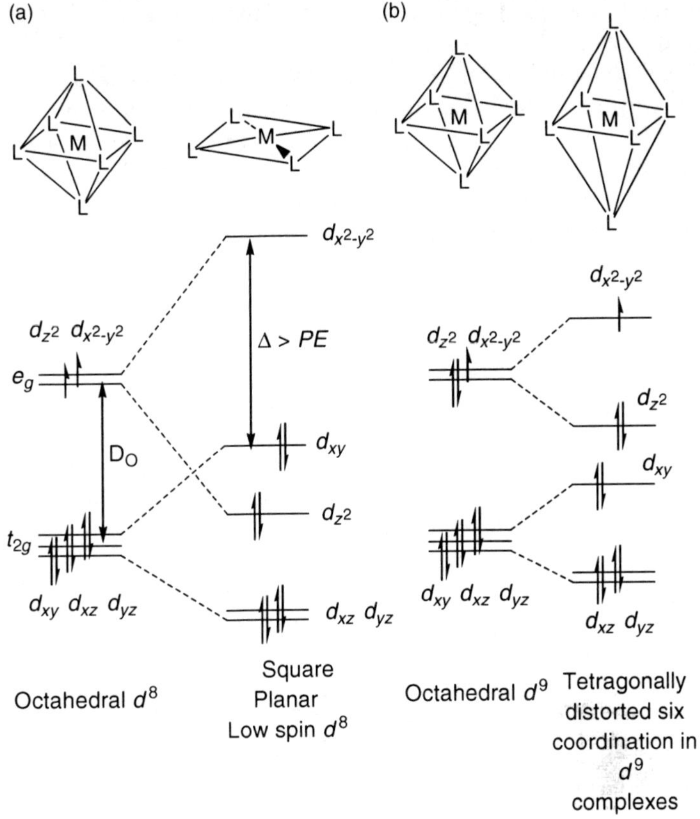

Figure 33 *Qualitative energy level diagrams describing the effect on relative d orbital energies of a metal ion bonded to (a) a square planar set of ligands in a four coordinate complex and (b) a tetragonally distorted set of ligands in a six coordinate complex*

If the ligands are sufficiently strong field to induce a splitting, Δ which is greater than *PE*, it will be energetically more favourable for the complex to adopt a square planar rather than an octahedral coordination geometry.

In a six coordinate complex containing a low spin d^9 metal ion, similar arguments lead to an expectation that the structure will distort from an ideal octahedron to a tetragonally distorted octahedral structure in which the two axial ligands are further from the metal than the equatorial ligands. By stretching the complex along the *z* axis the energy of one orbital (d_{z^2}), originating from the e_g set and containing two electrons, is reduced while the remaining singly occupied orbital from the e_g set moves to higher energy (Figure 33). This results in the tetragonal structure being lower in energy than the regular octahedral structure and is known as a *Jahn–Teller distortion*. This effect is often observed in complexes of the d^9 Cu^{2+} ion. A similar situation would arise with a low spin d^7 metal ion for which the single electron in the

e_g set would move to lower energy, the higher energy orbital remaining unoccupied.

MO bonding descriptions for tetrahedral, square planar and tetragonal complexes can be developed as for the octahedral case. However, the crystal field model offers a simpler description, which explains the main features of these structures. Nonetheless covalency in bonding is often important, particularly in compounds such as $[MoO_4]^{2-}$, $[TcO_4]^-$ or $[ReO_4]^-$ which are surprisingly unreactive considering the high oxidation numbers of the metal centres present.

2.6 The Magnetism and Spectroscopy of Metal Complexes

The formation of compounds which may be coloured or which may have magnetic properties is a distinctive feature of transition element chemistry. If a complex absorbs light in the visible region of the spectrum a colour is observed which arises from the light which is not absorbed. A blue colour corresponds to the absorption of light towards the red end of the visible spectrum and, conversely, a red colour corresponds with the absorption of light at the blue end of the spectrum. A plot of the extent of light absorption against either wavelength or frequency produces the *electronic spectrum* of the complex. This will typically contain a series of *absorption bands* whose energies correspond with the energy differences between particular MOs within the molecule. The magnetic properties of metal complexes arise from the presence or absence of unpaired electrons. The number of unpaired electrons also relates to the electronic structure of the metal ion as described by the crystal field or MO bonding models. Thus the electronic spectra and magnetic properties of complexes provide an insight into their electronic structures. These special properties of metals can also find applications in medicine.

2.6.1 The Electronic Spectra of Metal Complexes

The absorption of light energy by a metal complex is associated with the excitation of the complex from its lowest energy form, or *ground state*, to a higher energy *excited state*. This involves the movement of an electron from one MO to another higher energy MO. The energy at which light is absorbed by a metal complex provides a measure of the energy differences between the MOs, which constitute the ground state and the excited states associated with the electronic transition. Since MOs have varying degrees of metal or ligand character, some electronic transitions may correspond with charge redistribution between the metal and the ligands whereas others may be primarily confined to ligand-based or to metal-based MOs. As a consequence, several different types of electronic transition are possible within a metal complex. *Intra-ligand transitions* are associated with polyatomic ligands which have electronic spectra in their own right and so contribute to the spectrum of the complex. *Metal to ligand charge transfer* (MLCT) transitions arise where light absorption causes an electron to be excited from an orbital based largely on the metal to an orbital based largely on the ligand. Transitions of this type can be exploited in the conversion of light energy into chemical energy through subsequent reactions of the reduced ligand. *Ligand to metal charge transfer* (LMCT) occurs when light absorption causes an electron to be excited from an orbital based largely on the ligand to an orbital based largely on the metal. Finally, there are metal-based electronic transitions in which light absorption leads to an electronic transition between orbitals which are largely metal in character. In the visible spectrum a d–d absorption band may be observed for a d-block metal ion and an f–f transition for an f-block metal ion. In some cases transitions between the nf and $(n + 1)$d orbitals are also observed in the spectra of f-block metal complexes.

The electronic transitions in a metal complex are governed by two important *selection rules* which require certain criteria to be met for the transition to occur. The first is the spin selection rule which requires that there must be no change in the total spin quantum number S during the transition.[16] In practice spin-forbidden transitions may be observed, but will be very weak. The second selection rule, known as the *Laporte rule*, requires that, for ions in an environment with a centre of symmetry such as at the centre of an octahedron, the parity of the orbitals between which the electron moves must differ. The symmetry labels $_g$ and $_u$ in e_g, t_{2g} and t_{1u} refer to this parity with respect to a centre of symmetry, the subscript $_g$ means the object is symmetric with respect to inversion through a centre of symmetry and u means that the object is antisymmetric. As an example consider the d orbitals and p orbitals of a metal ion in an octahedral environment. On inversion through the centre of symmetry, the d orbitals, e_g, and t_{2g}, retain their phases but the p orbitals, t_{1u}, reverse their phase as shown in Figure 34. This means that in an octahedral complex d–d transitions are Laporte forbidden. In practice, vibrations occurring within the complex can reduce its symmetry on the timescale of the electronic transition so that weak absorptions are observed. In tetrahedral complexes there is no centre of symmetry, so the Laporte rule is relaxed and d–d absorption bands are more intense. In the case of complexes of the lanthanide ions, there is little interaction between the 4f subshell and the ligands so that the metal ion in effect experiences a near spherically symmetrical environment. This means that the f–f transitions are Laporte forbidden through the apparent presence of a centre of symmetry and, where observed, are very weak. In MLCT and LMCT transitions only one of the two orbitals involved is localised on the metal ion, so a change in parity becomes possible. As a result, when present, these charge transfer transitions give far more intense absorptions than d–d or f–f transitions.

[16] S is the sum of all the m_s values of the electrons in the compound. Each electron pair counts zero as the two electrons must have $m_s = +1/2$ for one and $m_s = -1/2$ for the other.

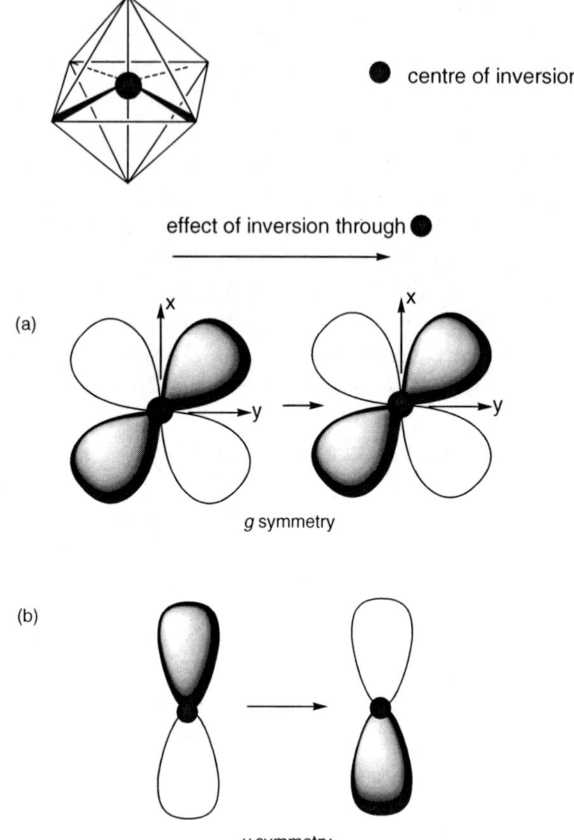

Figure 34 *Diagrams illustrating the difference in symmetry properties denoted by the symmetry symbols* $_g$ *(gerade) and* $_u$, *(ungerade) the former (a) being exemplified by a d orbital in an octahedral complex and the latter (b) by a p orbital in an octahedral complex*

2.6.2 Lanthanide Luminescence

An important property of lanthanide ions is their *luminescence*, that is their ability to re-emit light energy from their excited states. Normally the light energy absorbed by a metal complex is very rapidly lost through radiationless relaxation processes, such as producing atomic motions and bond vibrations leading to the conversion of the light energy into heat. However, because excited state lanthanide ions behave more like isolated metal ions there is more opportunity for direct reversion to the ground state through the emission of light in an f–f transition converting an excited state ion to its ground state. Because this is not an allowed process, the timescale of the emission can be relatively long, up to *ca.* 2 ms or more in some cases. This means that lanthanide compounds can be important in time resolved fluorescence assays. Normally the excitation of a luminescent organic compound by a powerful light source such as a laser results in a very short-lived fluorescence. This is subject to interference from light emitted from other organic compounds in the sample. If a lanthanide complex is used as a luminescent reporter in the assay the measurement of emitted light can be delayed until the background fluorescence due to the organic matrix has decayed away. The longer lived lanthanide luminescence can then be measured more accurately with higher sensitivity. All Ln^{3+} ions except $4f^0$ La^{3+} and $4f^{14}$ Lu^{3+} show some luminescent emission; Eu^{3+} and Tb^{3+} in particular show strong emissions.

Direct excitation of Ln^{3+} ions is possible but the low intensities and narrowness of the f–f absorption bands mean that intense radiation sources, *i.e.* lasers, are required for effective excitation. An alternative excitation mechanism is provided by the indirect excitation of Ln^{3+} through energy transfer from an excited state of a ligand which has a broad and intense intraligand absorption band. Quenching of luminescence occurs when the radiationless relaxation processes are fast enough to compete with the light emission process. This is a particular problem in aqueous solutions where energy transfer from excited Ln^{3+} to overtones of O–H vibrations provide an efficient non-radiative relaxation pathway for the Ln^{3+} excited state. However, use of non-aqueous solvents and ligands which exclude water from the lanthanide co-ordination sphere provide conditions which allow luminescence to occur.

2.6.3 The Magnetic Properties of Transitional Element Complexes

The magnetic properties of certain metal ions have found important applications in Magnetic Resonance Imaging. There are two ways in which a material may respond to the presence of an applied magnetic field. Where only paired electrons are present in the atoms, ions or molecules comprising the material it will be weakly repelled by the magnetic field and exhibit what is known as *diamagnetism*. All materials have some diamagnetic properties but in some cases these may be masked by a larger *paramagnetism* through which the material is attracted into the applied magnetic field. Paramagnetic effects are typically some 10–10^4 times larger than diamagnetic effects and arise from the presence of unpaired electrons. The paramagnetism of a metal ion can be quantified by its *magnetic moment*, which depends upon the total number of unpaired electrons and the orbitals they occupy. Assuming that there is no coupling between the spin and orbital angular momenta of a metal ion denoted by m_s and m_l respectively, the magnetic moment, μ_{SL}, of an atom or ion is related to the total spin quantum number S ($S = \Sigma m_s$) and the total orbital angular momentum quantum number L ($L = \Sigma m_l$)[17] by Equation (5). In this equation, g represents the gyromagnetic ratio or Lande splitting parameter which, for a free electron, is 2.00023. The units of μSL, are Bohr Magnetons (BM).

$$\mu_{SL} = \sqrt{g^2 S(S+1) + L(L+1)} \qquad (5)$$

[17] L is the total m_l value for the ion, half filled or filled subshells have $L = 0$.

In practice, spin orbit coupling is small for 1st row d-block metal ions, so that Equation 5 may be applied. Also, because, the d orbitals of a transition metal ion show significant interactions with the ligand orbitals the orbital angular momentum contribution is usually quenched so that $g^2S(S+1) >> L(L+1)$. In such cases, Equation (5) can be simplified to the *spin-only formula* shown in Equation (6), where μ_S is known as the spin-only magnetic moment.

$$\mu_S = g\sqrt{S(S+1)} \tag{6}$$

The observed magnetic moment, μ_{obs}, of the 1st row transition metal complex is generally found to lie between μ_S and μ_{SL}. In the case of high spin d^5 ions such as Mn^{2+} or Fe^{3+}, $L = 0$ so that $\mu_S = \mu_{SL}$. The crystal field model, and a knowledge of whether a complex is high or low spin, allows the number of unpaired electrons in a complex, and so its magnetic moment, to be calculated. The magnetic behaviour of paramagnetic 2nd and 3rd row d-block metal ions is less simple, as spin-orbit coupling and the effects of covalency in bonding are more pronounced. Magnetic measurements can be used to obtain information about the electronic structures of metal complexes. An example is provided by the binding of dioxygen to haemoglobin, which contains d^6 Fe^{2+}. On binding O_2, the iron centre switches from a paramagnetic high spin state with four unpaired electrons to a low spin diamagnetic state. This information has been used to evaluate different bonding models for the $Fe^{2+} - O_2$ interaction, which is chemically unusual because of its reversible nature. Normally, Fe^{2+} is irreversibly oxidised to Fe^{3+} by dioxygen.

The magnetism of f-block metal ions is complicated by coupling between the spin and orbital angular momenta of the electrons. In effect this means that the quantum numbers m_s and m_l can no longer be used as discrete individual values but must be replaced by a new quantum number j, where $j = m_s + m_l$. The magnetic moment of a paramagnetic lanthanide ion, μ_J, depends upon the total spin and orbital angular momentum of the electrons, J where $J = \Sigma j$ and is given by Equation (7), in which the value of g_J is given by Equation (8)

$$\mu_J = g_J\sqrt{J(J+1)} \tag{7}$$

$$g_J = \frac{3}{2} + \frac{S(S+1) - L(L+1)}{2J(J+1)} \tag{8}$$

The absence of significant interactions between 4f subshell electrons and the ligands in complexes of the lanthanide ions means the high or low spin situation does not arise. Thus the number of unpaired electrons can be simply predicted from the electron configuration of the ion. In practice, Equation (7) predicts the observed magnetic moments of tripositive lanthanide ion complexes quite well (Table 6). Two discrepancies arise at Sm^{3+} and Eu^{3+} because the first excited state of these ions is sufficiently close to the ground state that, at room temperature, thermal energy is sufficient to partly populate the excited state.

Table 6 *The magnetic moments of the lanthanide ions*

Element	*n in f^n for Ln^{3+}*	*Number of unpaired electrons*	*Calculated μ_J (BM)*	*Measured μ_{obs} (BM)*
La	0	0	0	0
Ce	1	1	2.54	2.5
Pr	2	2	3.58	3.5
Nd	3	3	3.62	3.6
Pm	4	4	2.68	--
Sm	5	5	0.84	1.5
Eu	6	6	0	3.4
Gd	7	7	7.94	8
Tb	8	6	9.72	9.3
Dy	9	5	10.6	10.6
Ho	10	4	10.6	10.4
Er	11	3	9.57	9.5
Tm	12	2	7.57	7.4
Yb	13	1	4.54	4.5
Lu	14	0	0	0

Since the J value of the 1st excited state differs from that in the ground state, the observed magnetic moment is no longer purely that of the ground state and the excited state magnetic moment becomes mixed into the value of μ_{obs}. If allowance is made for this effect the magnetic moments of the Eu^{3+} and Sm^{3+} ions can also be successfully predicted.

2.7 The Energetics of Metal Complex Formation

2.7.1 Equilibrium or Stability Constants

The use of a metal ion in a particular medical application will require that its form is controlled to suit the application. It may be necessary, in diagnostic imaging for example, to administer the metal in the form of a stable complex which does not release the metal to other potential ligands *in vivo*. Alternatively it may be necessary for the metal to lose some of its ligands and interact with a biological substrate, as in the interaction of a platinum anti-cancer drug with DNA. In order to understand in detail the *in vivo* behaviour of a metal complex it will often be necessary to quantify the strength of metal ligand interactions in terms of thermodynamics. The formation of a metal complex through the interaction between a metal ion and a proligand represents a form of chemical reaction. The extent to which this reaction proceeds may be defined by an *equilibrium constant* K. The value of K defines the relative concentrations of reactants and products which coexist at equilibrium.[18] As an example, in a reaction where two reactants A and B combine to form two products C and D,

[18] Strictly speaking concentrations can only define the equilibrium constant accurately under rather special circumstances but for most practical purposes K is expressed in terms of concentrations.

the value of K is given by Equation (9) where square brackets denote the concentrations of A, B, C and D.[19]

$$K = \frac{[C][D]}{[A][B]} \tag{9}$$

Complex formation is normally studied in solution, often in water, where the solvent can also act as a ligand. Furthermore, it cannot necessarily be assumed that the coordinated solvent will be inert to reaction with the proligand. This is particularly true in water where the reaction of an aquated metal ion (M^{z+}) with a proligand (B:) could result in the removal of H^+ from a coordinated water molecule by the proligand [Figure 35(a)], as shown in Equation (10):

$$[M(H_2O)_x]^{z+}{}_{(aq)} + B: \rightleftharpoons [M(H_2O)_{(x-1)}(OH)]^{(z-1)+}{}_{(aq)} + B{:}H^+{}_{(aq)} \tag{10}$$

In aqueous solutions water itself may act as a base to accept H^+ by forming aquated H^+ ions. An equilibrium constant (K_a), known as the acid dissociation constant, may be defined for this process as shown in Equation (11).

$$K_a = [H^+][[M(H_2O)_{(x-1)}(OH)]^{z+}]/[[M(H_2O)_x]^{(z-1)+}] \tag{11}$$

The larger the value of K_a, the larger the concentration of $[M(H_2O)_{(x-1)}(OH)]^{z+}$, and so of H^+, in equilibrium with the aquated complex. Since the value of K_w for the self ionisation of water is 10^{-14} mol^2 dm^{-6} the value of 6.3×10^{-3} for the K_a of aquated Fe^{3+} ions corresponds with the formation of an acidic solution. Just as hydrogen ion concentrations are often expressed in terms of pH so K_a values are often expressed as pK_a, which is defined in Equation (12) and exemplified in Equations (13) and (14).

$$pK_a = -\log_{10}K_a \tag{12}$$

$$e.g.\ [Fe(H_2O)_6]^{3+}{}_{(aq)} \rightleftharpoons [Fe(H_2O)_5(OH)]^{2+}{}_{(aq)} + H^+{}_{(aq)} \tag{13}$$

$$K_a = 6.3 \times 10^{-3}$$

$$pK_a = 2.20$$

$$[Pt(NH_3)_4(H_2O)_2]^{4+}{}_{(aq)} \rightleftharpoons [Pt(NH_3)_4(H_2O)(OH)]^{3+}{}_{(aq)} + H^+{}_{(aq)} \tag{14}$$

$$K_a \approx 10^{-2}$$

$$pK_a = 3.00$$

[19] Unfortunately square brackets have conventionally been used both to denote concentrations, [A] representing the concentration of **A** in mol dm^{-3}, and co-ordination complexes, *e.g.* $[Co(NH_3)_6]^{3+}$. This presents a potential source of confusion when defining equilibrium constants for co-ordination compounds since often only one set of brackets is used. Usually the function of the square bracket is clear from the context in which it is used but for clarity, in this text, two sets of brackets will be used when appropriate.

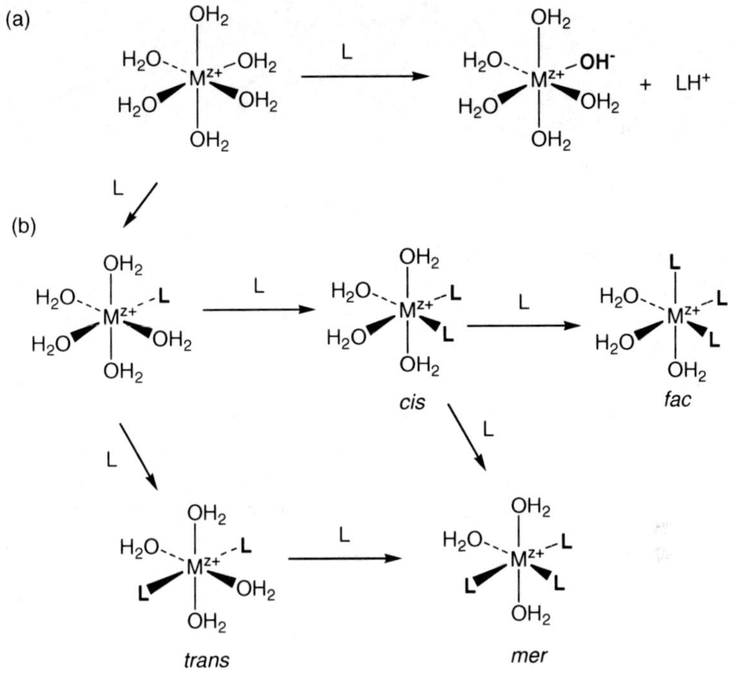

Figure 35 *A diagram illustrating some of the ways in which a hexa-aqua metal ion $[M(H_2O)_6]^{z+}$ may react with a proligand, L, to (a) deprotonate (acid dissociation) or (b) form complexes through the displacement of H_2O by L*

In the simplest case of a complex $[ML_n(H_2O)]^{z+}$ (where M represents a metal and L a ligand in a complex of overall charge z+) the first acid dissociation may be described by Equation (15).

$$[ML_n(H_2O)]^{z+}{}_{(aq)} + H_2O_{(aq)} \leftrightharpoons [ML_n(OH)]^{(z-1)+}{}_{(aq)} + H_3O^+{}_{(aq)} \quad (15)$$

The pH of a solution of a metal complex may be estimated as follows. A solution of $[ML_n(H_2O)]^{z+}$ with an initial concentration of c mol dm^{-3} will ionise to produce equal concentrations of $[ML_n(OH)]^{(z-1)+}$ and H^+ so that

$$[\{ML_n(H_2O)^{z+}\}] + [\{ML_n(OH)^{(z-1)+}\}] = \text{c mol dm}^{-3}$$

and

$$[H^+] = [\{ML_n(OH)^{(z-1)+}\}]$$

From Equation (11)

$$K_a = [H^+][\{ML_n(OH)\}^{z+}]/[\{ML_n(H_2O)^{z+}\}]$$

$$=[H^+]^2/(c - [H^+])$$

$$[H^+]^2 = K_a c - K_a[H^+]$$

$$[H^+]^2 + K_a[H^+] = K_a c$$

using the standard formula to solve a quadratic equation, or noting that:

$$([H^+] + K_a/2)^2 = [H^+]^2 + K_a[H^+] + K_a^2/4 = K_a c + K_a^2/4$$
$$\text{so } [H^+] = \sqrt{K_a c + K_a^2/4)} - K_a/2$$

and the value of the solution pH is given by $-\log_{10}[H^+]$.

Complex formation will involve the displacement of solvent from the metal coordination sphere by the incoming ligand [Figure 35(b)]. In complexation reactions occurring in aqueous solution, water is present in large excess so its concentration is essentially constant and can be omitted from the description of the equilibrium. In such cases, provided acid dissociation is not important, the formation of a metal complex from a metal ion and one or more proligands, L, according to Equation (16) can be described by the simplified Equation (17) in which the solvent and the charge on the complex are omitted from the formulae.

$$[M(H_2O)_x]^{z+}{}_{(aq)} + L_{(aq)} \rightleftharpoons [M(H_2O)_{(x-p)}(L)]^{z+}{}_{(aq)} + pH_2O_{(aq)} \quad (16)$$

$$M + L \rightleftharpoons ML \quad (17)$$

The equilibrium constant K_1, for the process shown in Equation (16) is called a *stability constant* and is given by Equation (18). (The terms *formation constant* and *binding constant* are also sometimes used).

$$K_1 = [ML]/[M][L] \quad (18)$$

Similarly, for reactions involving the addition of further L to M, stability constants K_2 to K_n may be defined as shown in Equations (19–22):-

$$\text{for } ML + L \rightleftharpoons ML_2 \quad (19)$$

$$K_2 \text{ is defined by: } K_2 = [ML_2]/[ML][L] \quad (20)$$

$$\text{and so on to } ML_{(n-1)} + L \rightleftharpoons ML_n \quad (21)$$

$$\text{with } K_n = [ML_n]/[ML_{(n-1)}][L] \quad (22)$$

These equilibrium constants, K_1 to K_n, are called *stepwise stability constants* and represent the equilibria involved in the stepwise addition of one ligand to the metal.[20] However, equilibrium constants could also be written for an overall

[20] Because of the competition between proteins and other potential metal binding agents *in vivo* metal complexes used in medical applications will usually need to have quite large stability constants to ensure that the metal centre reaches its site of action. Stability constants in excess of 10^{50} have been reported for some metal complexes but typical values for medical applications might lie in the range 10^{10}–10^{35}.

reaction involving more than one ligand. These are known as *overall stability constants* and are usually given by the symbol β with a suffix to indicate the number of ligands involved as shown in Equations (23–28).

$$\text{for } M + L \rightleftharpoons ML, \tag{23}$$

$$\beta_1 = [ML]/[M][L] \tag{24}$$

$$\text{for } M + 2L \rightleftharpoons ML_2, \tag{25}$$

$$\beta_2 = [ML_2]/[M][L]^2 \tag{26}$$

$$\text{and so on to } M + nL \rightleftharpoons L_n \tag{27}$$

$$\text{with } \beta_n = [ML_n]/[M][L]^n \tag{28}$$

the β_1 to β_n values are called overall stability constants and β_j is the product of the stepwise stability constants K_1 to K_j as shown by Equation (29).

$$\beta_j = K_1.K_2.\text{----}K_j \tag{29}$$

Normally the magnitudes of K_n decrease with increasing n. In part this is the result of statistical factors but the addition of ligands may also influence the electronic properties of the metal and so the binding of subsequent ligands. Decreasing positive charge on the complex, if the ligands are negatively charged, can also influence the stability constant. Another factor to consider is the presence of interactions between ligands. Six relatively small ligands such as the F^- donor atom can fit round an ion such as Co^{3+} to give in this case $[CoF_6]^{3-}$. However, six larger donor atoms such as Cl^- cannot fit so well around the Co^{3+} ion so that $[CoCl_6]^{3-}$ is not formed but instead the four coordinate Co(+2) complex $[CoCl_4]^{2-}$ is obtained. This is an example where what are known as *steric interactions* between the ligands limit the CN of the complex. In this case the size and shape of the ligands affects the number which can fit into the volume of space forming the first coordination sphere around the metal ion. Where the stepwise addition of ligands occurs, several metal complexes may coexist in solution depending upon the relative proportions of metal and ligand present.

Where a metal complex is formed from a ligand which loses H^+ during complex formation according to Equation (30), the acid dissociation constant of the ligand, K_L, contributes to the stability constant as shown in Equations (31–35).

$$M^{z+} + LH \rightleftharpoons \{ML\}^{(z-1)+} + H^+ \tag{30}$$

$$K^{LH} = [\{ML\}^{(z-1)+}][H^+]/[M^{z+}][LH] \tag{31}$$

$$\mathrm{LH} \rightleftharpoons \mathrm{L^-} + \mathrm{H^+} \tag{32}$$

$$K_\mathrm{L} = [\mathrm{L^-}][\mathrm{H^+}]/[\mathrm{LH}] \tag{33}$$

$$K_I = [\{\mathrm{ML}\}^{(z-1)+}]/[\mathrm{M}^{z+}][\mathrm{L^-}] \tag{34}$$

$$K^\mathrm{LH} = [\{\mathrm{ML}\}^{(z-1)+}][\mathrm{H^+}]/[\mathrm{M}^{z+}][\mathrm{LH}] = K_I \times K_L \tag{35}$$

Where Equation (34) is simply Equation (18) rewritten to contain more information on the charges of the species involved. In such cases the binding of the metal to the ligand becomes dependent upon the acidity, that is the pH, of the solution. A polydentate ligand LH_n with multiple binding sites, each of which can lose H^+, has a series of stepwise acid dissociation constants, K_{L1} to K_{Ln}, which contribute to the stability constant of the complex. In such cases a 'conditional stability constant', K^*, can be defined according to Equation (36) to take account of the effects of acidity on the metal-ligand binding.

$$K^* = K_1 /(1 + K_{L1}\,[\mathrm{H^+}] + K_{L1}\,K_{L2}[\mathrm{H^+}]^2 + \ldots\ldots K_{L1}\,K_{L2}\ldots\ldots K_{Ln}[\mathrm{H^+}]^n) \tag{36}$$

2.7.2 Hard and Soft Donors or Acceptors

Certain general trends emerge from the large number of stability constants which have been measured. Particular types of ligand donor atom form stronger complexes with certain metal ions and weaker complexes with others. That is, their complexes have higher stability constants with certain metal ions than with others. This allows a broad classification of metal ions according to the type of ligands with which they form the strongest complexes. Those metals which form their most stable complexes with oxygen or nitrogen donor atoms are called *class a*, or *hard*, metal ions and those which form their strongest complexes with sulfur or phosphorus donor atoms are called *class b*, or *soft*, metal ions. There is also a group of metal ions which show *borderline* behaviour. *Class a* metal ions are usually smaller, more highly charged, cations. They form their strongest complexes with *hard ligands* which are the smaller electronegative donor atoms O, N, F. At the other end of the scale are the *soft* metal ions, *i.e.* the larger more polarisable metal ions, often in their lower oxidation states. These form their strongest complexes with *soft ligands* which contain larger more polarisable and less electronegative donor atoms such as S, Se, P, As. This classification is summarised in Table 7.

The hard/soft classification of metal ions and ligand donor atoms is a useful qualitative concept in that it is a guide to predicting which ligands may be more suited to forming complexes with a particular metal ion. Typically thiolate, RS^- (R=hydrocarbyl), or phosphine, R_3P, ligands stabilise lower oxidation states such as Rh^+, Ir^+ or Cu^+, whereas fluoride or oxide ions are better suited

Table 7 *Examples of class a (hard) and class b (soft) metal ions and ligands*

Class a (hard)	*Borderline*	*Class b (soft)*
Metal Ions	Metal ions	Metal ions
Mn^{2+}, Sc^{3+}, Cr^{3+}, Fe^{3+}, Ti^{4+}, Sc^{3+}	Fe^{2+}, Co^{2+}, Ni^{2+}, Cu^{2+}	Hg^+, Hg^{2+}, Cu^+, Ag^+, Au^+, Pd^{2+}, Pt^{2+}, Rh^+, Ir^+
Ligands	Ligands	Ligands
F^-, R_2O, ROH, OH^-, RCO_2^-, SO_4^{2-}, NR_3, Cl^-, SC*N*$^{-a}$	Br^-, pyridine	R_2S, R_3P, R_3As, RNC, CO, CN^-, I^-, NC*S*$^{-a}$

[a] The donor atom is italicised in NCS^-, SCN^-.

to higher oxidation state metal ions. In some applications of metal complexing agents there is a need for ligands which show very large stability constants for particular metals. Examples are provided by the use of hydroxypyridinones as chelating agents to selectively remove iron from patients suffering iron overload because of repeated blood transfusions in the treatment of β-thalassaemia. In the treatment of Wilson disease either *d*-penicillamine, or the tetramine $NH_2CH_2CH_2NHCH_2CH_2NHCH_2CH_2NH_2$ (2,2,2-tet) can be used to selectively remove copper.

2.7.3 The Chelate and Macrocyclic Effects

The term 'chelate effect' refers to the finding that complexes containing chelating ligands usually show larger stability constants than their non-chelated counterparts. An example of this is provided by $[Cd(NH_2CH_2CH_2NH_2)_2]^{2+}$ containing the chelating $NH_2CH_2CH_2NH_2$ (en) ligand and for which $\log\beta_2 = 10.6$. The corresponding non-chelated complex containing the monodentate ligand $MeNH_2$ $[Cd(NH_2Me)_4]^{2+}$ has $\log\beta_4 = 6.52$. In both complexes Cd^{2+} is bound to a NH_2CH_2- group so that the strength of the metal ligand interaction is very similar. The main difference lies in the way in which the complexes are formed. The reaction of four NH_2Me proligands with $[Cd(OH_2)_4]^{2+}$ to form $[Cd(NH_2Me)_4]^{2+}$ releases four molecules of water so that the total number of particles stays the same, four plus one giving five. In the case of the chelate complex two molecules of en react with $[Cd(OH_2)_4]^{2+}$ to produce the complex and four molecules of water so that the total number of particles increases from three to five. This has an important effect on the energetics of the system as the release of the additional molecules in the form of water increases the degree of disorder in the system and so its entropy. An increase in the entropy of a system corresponds with a decrease in its energy, other things being equal, and so helps to drive the reaction. Usually the more chelate rings that are formed in a complex the greater will be the additional stability from the chelate effect. Thus the tetradentate ligand 2,2,2-tet will typically form a more stable complex than two en ligands.

The 'macrocyclic effect' refers to the finding that complexes of macrocyclic ligands have larger stability constants than those of their acyclic counterparts, examples being shown in Figure 36. It appears that the chelate effect is

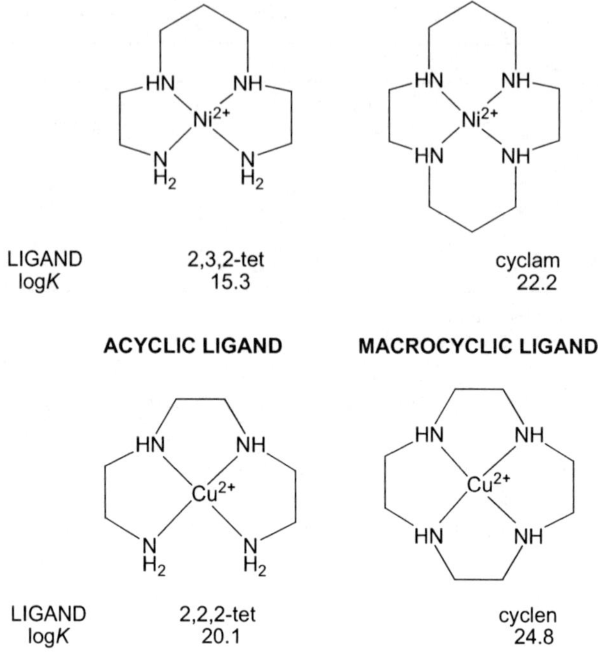

Figure 36 *Examples of the macrocyclic effect*

generally driven by entropy, the origin of the macrocyclic effect is not always so clear and may result from changes in entropy or in the strength of the metal ligand donor atom interaction. In a simple didentate chelating ligand the donor atoms may be free to move into the optimum position for bonding to the metal ion. In a macrocyclic ligand the cyclic ligand structure constrains the positions of the donor atoms so that they may not be optimal and, a stabilising entropy effect may be eliminated by an unfavourable change in the strength of the metal ion donor atom interaction. A further factor to consider is that when polydentate ligands are forming complexes, a significant amount of energy may be needed to reorganise the structure of the ligand from that found in the free state to that required by the complex. This restructuring on complexation can offer an obstacle to complex formation. If the ligand is already of a similar structure to that in the complex and little energy is required to adopt the structure found in the complex, the ligand is said to be *preorganised.* To some extent at least a macrocyclic ligand could be described as preorganised.

2.8 Reactions at Metal Centres

2.8.1 Thermodynamic and Kinetic Stability

The terms 'stable' and 'unstable' are often used in a general way to describe the reactivity of compounds. However, in the case of metal complexes these terms have a specific meaning. *Stable* implies a complex with a large overall stability constant while *unstable* implies a complex with stability constant less than one under the conditions concerned. The ligand will be unable to replace the solvent molecules bound to the metal ion to any significant extent. Thus stable complexes would not be expected to react with the solvent, or other proligands present, in the solution. In contrast unstable complexes could undergo ligand substitution reactions with solvent, or they may be unstable with respect to ligand substitution by other proligands present in the solution. The numerical value of the stability constant provides information on the relative proportions of solvated metal ion, free proligand and complex present in a solution at equilibrium but this may not always be a good guide to the chemical behaviour of the system. If some process removes the metal ion from the system, even a high stability constant may not serve to maintain the metal in the complex. If, for example, some other species binds to the metal and removes it, perhaps by forming a highly insoluble compound or transporting the metal to another region, possibly through a cell membrane, the original metal complex may continue to dissociate to try to maintain the equilibrium concentration of the metal ion required by the stability constant. As the metal is continuously being removed, it will ultimately be stripped from the complex despite the complex being stable in that it has a large stability constant.

Another factor to be considered is the rate at which a complex forms or dissociates. If the dissociation is rapid the equilibrium distribution between the free metal and the complex will be established rapidly. However, if the dissociation is very slow, even a small stability constant may be sufficient to retain the metal in the complex as metal will not be released from the complex quickly enough for a competing ligand to remove it efficiently. As the equilibrium constant does not provide any information about the rate at which the equilibrium is reached, two different terms are used to describe the rates at which metal complexes react to gain or lose ligands. Metal complexes which undergo rapid ligand exchange reactions are said to be *labile*, those which undergo slower reactions are said to be *inert*. The term labile is usually taken to mean reactions which are at least half complete within *ca.* 30 s or less at 298 K. Inert complexes react more slowly than this. A labile complex may be stable as judged by its stability constant and an inert complex may be unstable. This type of behaviour is exemplified by the reaction of aqueous $[Co(NH_3)_6]^{3+}$ with acid according to Equation (37), which has an equilibrium constant of 10^{25}.

$$[Co(NH_3)_6]^{3+} + 6H_3O^+ \leftrightharpoons [Co(H_2O)_6]^{3+} + 6NH_4^+ \qquad (37)$$

This means that the reaction of $[Co(NH_3)_6]^{3+}$ with acid should proceed very readily to give $[Co(H_2O)_6]^{3+}$ and NH_4^+. However $[Co(NH_3)_6]^{3+}$ is very inert and the reaction proceeds only very slowly so that for most practical purposes $[Co(NH_3)_6]^{3+}$ appears 'stable' to acid although in the strict use of the terms it is, in fact, unstable but inert. The observed chemistry of a metal ion and its complexes reflects this interplay between thermodynamic stability or instability and kinetically inert or labile behaviour.

2.8.2 Substitution Reactions at Metal Centres

In order to appreciate why the rates of reactions of metal complexes vary, it is necessary to consider the energy changes which occur during the reaction. When a metal complex $[ML_nX]$ (M represents a transition metal ion, L represents a ligand which does not undergo substitution and X a ligand which does undergo substitution) reacts with a proligand, Y, to form a more stable product $[ML_nY]$ (Equation (38)), the system must pass through an intermediate or transition state.

$$[ML_nX] + Y \leftrightharpoons [ML_nY] + X \quad (38)$$

This will be a modified structure which constitutes an intermediate stage in the progression from the reactant to the product, and an *activation energy* will be required to form the transition state structure. The relative concentrations of $[ML_nX]$ and $[ML_nY]$ which exist at equilibrium in the presence of M, X and Y are determined by the equilibrium constant K. The rate at which the reaction proceeds through the transition state species depends on the activation energy input needed to form the transition state as this constitutes an energy barrier to the progress of the reaction, and affects the rate at which the reaction proceeds to equilibrium. Factors which contribute to the magnitude of the activation barrier are the charge radius ratio of the metal ion and whether any loss in *CFSE* is associated with the formation of the intermediate. Such effects are apparent in the rates of exchange of water present as coordinated water in aquated metal complexes and as free water in the solvent. Water exchange reactions represent a special case of substitution at a metal centre in which the entering and leaving ligands are of the same chemical structure. Such a process might be studied using isotopically labelled water (Equation (39)):

$$[M(H_2O)_x]^{z+} + H_2{}^{17}O \leftrightharpoons [M(H_2O)_{(x-1)}(H_2{}^{17}O)]^{z+} + H_2O \quad (39)$$

Ions of s- or p-block metals do not show crystal field effects, and their rates of water exchange depend largely upon the polarising power of the metal ion, *i.e.* its charge to radius ratio. High charge to radius ratios tend to be associated with strong binding of the water molecules and a high energy input for removing water from the co-ordination sphere of the metal. Charge to radius ratio effects will also be apparent among d- and f-block metal ions but, in addition, any change in *CFSE* during the reaction process will be reflected in the activation energy barrier.

If it is assumed that the exchange of a water molecule in an octahedral d-block metal complex $[M(H_2O)_6]^{z+}$ proceeds by either of two extreme mechanisms, it is possible to make some predictions about the effect of electron configuration on reaction rate. If the reaction involves the addition of a seventh water molecule to form $[M(H_2O)_7]^{z+}$, a 7-coordinate intermediate will be present in the transition state before expelling one water molecule to form the product. Alternatively, the complex might expel a water molecule to form a 5-coordinate intermediate $[M(H_2O)_5]^{z+}$ before binding another water molecule to form the product. The change in the co-ordination geometry at the metal centre during the formation of the intermediate will result in a change in *CFSE* which will affect the activation energy barrier and so the rate of reaction. The structures of the intermediates involved in such reactions are not known with certainty but estimates have been made of the change in *CFSE* associated with forming a pentagonal bipyramidal 7-coordinate intermediate or a square pyramidal 5-coordinate intermediate. In a general sense these results correlate well with observation, in that ions which suffer a comparatively large loss in *CFSE* on forming the transition state structures, show slow exchange rates. Thus d^3 and low spin d^5 or d^6 ions, exemplified by Cr^{3+}, Ru^{3+}, Co^{3+} and Ru^{2+} show very slow exchange and are kinetically inert. In contrast, ions which have d^0, d^{10} or high spin d^5 electron configurations have no *CFSE* to lose, and show rapid exchange; *e.g.* Mn^{2+} and Zn^{2+}.

2.8.3 Mechanisms of Substitution at Metal Centres

Water exchange reactions represent a special case of substitution at a metal centre in which the entering and leaving ligands are of the same chemical structure. A more general case involves the substitution of a ligand X in an octahedral complex $[ML_5X]^{z+}$ by a proligand Y, where L_5 represents the remainder of the ligand set not involved in the substitution reaction. Two extreme mechanisms may be proposed for such substitution reactions (Figure 37). The first, known as an associative or ***A*** mechanism, implies that reaction with Y, involves the formation of a detectable 7-coordinate intermediate $[ML_5XY]^{z+}$ prior to the

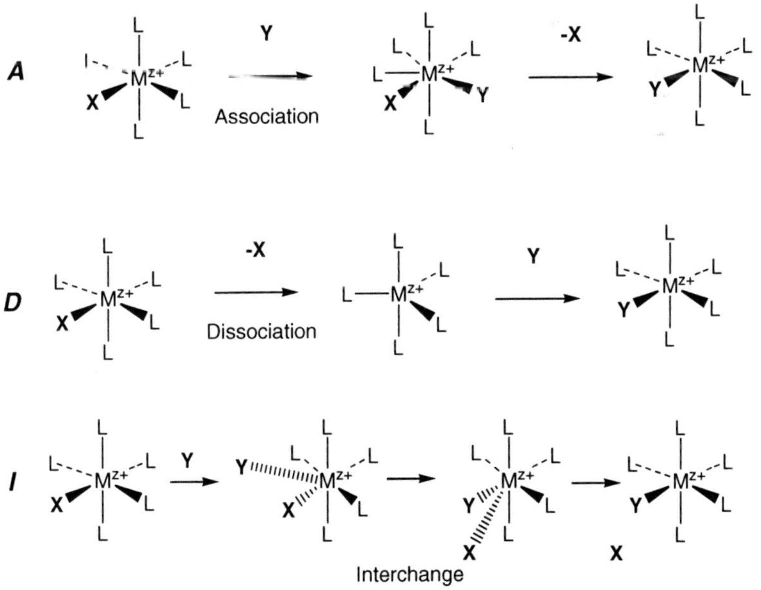

Figure 37 *Representations of the associative,* ***A****, dissociative,* ***D*** *and interchange,* ***I****, mechanisms for substitution at an octahedral metal centre*

expulsion of X. In contrast the second, known as a dissociative or ***D*** mechanism, involves the formation of a detectable 5-coordinate intermediate $[ML_5]^{z+}$ by dissociation of X prior to the addition of Y. A further possibility, known as the interchange or ***I*** mechanism exists. In this case an *encounter complex* is first formed in which Y resides in a loosely bound state at the surface of the reacting complex, $\{Y{:}[ML_5X]^{z+}\}$. The ligands Y and X then exchange places in a rate determining interchange step to form $\{X{:}[ML_5Y]^{z+}\}$ before X diffuses away. The ***A***, ***D*** and ***I*** mechanisms represent the detectable steps in a reaction sequence and are called *stoichiometric mechanisms.* The ***I*** mechanism can be subdivided into two types, depending upon whether bond making or bond breaking is more important in determining the reaction rate. The former involves an associative or ***a*** intimate mechanism, and the latter a dissociative or ***d*** intimate mechanism, giving rise to the I_a and I_d mechanisms.

The substitution of a ligand X in a metal complex $[ML_nX]$ by a proligand Y occurs in a solvent, often water, and there is a possibility that the solvent may participate in the reaction. An example of such a scheme is presented in Equations (40) and (41) where Sol represents a solvent molecule.

$$[ML_nX] + Sol \rightarrow [ML_n(Sol)] + X \quad (40)$$

$$[ML_n(Sol)] + Y \rightarrow [ML_nY] + Sol \quad (41)$$

Since the solvent is present in large molar excess it can become an effective competitor for the metal centre, even though a more thermodynamically stable complex is formed with Y. Thus the solvent can have an important effect on the substitution reaction. An example of this is provided by substitution at square planar Pt^{2+} centres. Square planar complexes typically arise with d^8 ions exemplified by Ni^{2+}, Pd^{2+} and Pt^{2+}. Since such complexes have an empty metal p orbital, and there is normally no steric problem with forming a 5-coordinate intermediate, they might be expected to undergo associative substitution. However, for Pt^{2+} complexes $[PtL_3X]$ reactions with a proligand Y are often found to proceed through two different reaction pathways. One involves an associative reaction between $[PtL_3X]$ and Y, according to Equation (42) overall:

$$[PtL_3X] + Y \rightarrow [PtL_3Y] + X \quad (42)$$

The second results from an associative reaction with solvent (Sol) followed by reaction with Y to give the product, according to Equations (43) and (44)

$$[PtL_3X] + Sol \rightarrow [PtL_3(Sol)] + X \quad \text{slow} \quad (43)$$

$$[PtL_3(Sol)] + Y \rightarrow [PtL_3Y] + Sol \quad \text{fast} \quad (44)$$

The rates of reactions of square planar complexes, *trans*-$[ML_2TX]$, containing a substitution-labile ligand X *trans* to a less labile ligand T, can show a marked

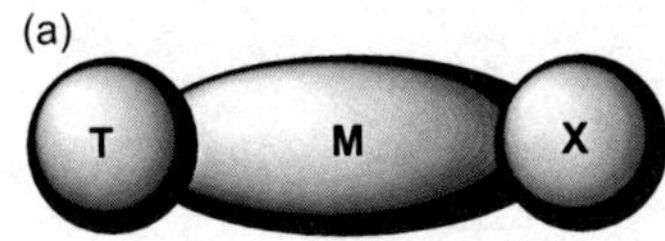

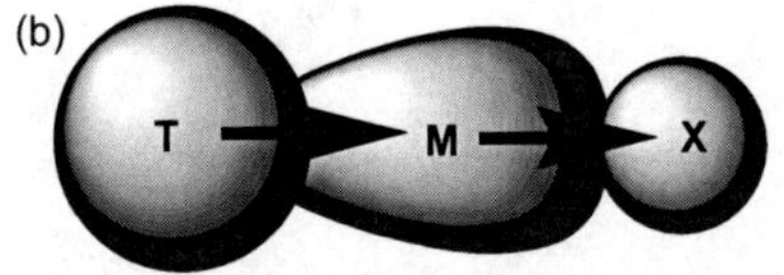

Figure 38 *The polarisation of a metal ion by a ligand T to increase the lability of a trans-ligand, X. (a) T does not have a strong trans-influence; (b) T does have a strong trans-influence increasing the electron density at the metal centre as perceived by X*

sensitivity to the nature of T. This ligand influences the bonding between M and X through what is known as the *trans-influence*. This is a ground state effect in which T influences the bonding between the metal and X and may be explained in terms of the polarisation of the M–T and M–X bonds. A more polarisable or stronger σ donor ligand, T, may more effectively satisfy the need for electro-neutrality at the metal, polarising the metal centre (Figure 38). Since the *trans*-oriented M–T and M–X bonds involve a common metal orbital, this polarisation exerts a repulsive effect on the electron density of the less polarisable weaker σ-donor ligand, X, weakening the M–X bond.

The variation in substitution rates for X with differing *trans*-ligands, T, is known as the *trans*-effect. This is the *kinetic* consequence of the *trans*-influence in determining the outcome of a substitution reaction. It is the effect of the coordinated ligand T upon the rate of substitution of the ligand, X, *trans* to it. If the *trans* influence of T increases the energy of the ground state relative to the transition state, as compared to some other similar complex, the activation energy barrier for the substitution reaction will be lower, and so the reaction rate higher. A ligand which reduces the energy of the transition state relative to the ground state will have a similar effect. Thus strong π-acceptor ligands can increase the stability of the 5-coordinate intermediate in the associative substitution of square planar complexes and accelerate the reaction. Ligands may be placed in order according to their labilising effect on a *trans* ligand to give a series in order of increasing *trans*-effect:

$$H_2O < OH^- < NH_3, \text{amines} < py < Cl^-, Br^- < SCN^-, I^-, NO_2^- < C_6H_5^-$$

Increasing *trans*-effect ⟶

$$C_6H_5^- < CH_3^- < H^-, PR_3 < C_2H_4 < NO, CO < CN^-$$

A good example of the exploitation of the *trans*-effect in synthesis is provided by the routes to *cis* and *trans*-isomers of the chloro-ammine complexes of Pt^{2+} shown in Figure 39. Some π-acceptor ligands exhibit a strong *trans*-effect, as shown by the series above, but only have a weak *trans*-influence. In these cases the *trans*-effect is very much attributable to the stabilisation of the transition state by the π-acceptor ligand.

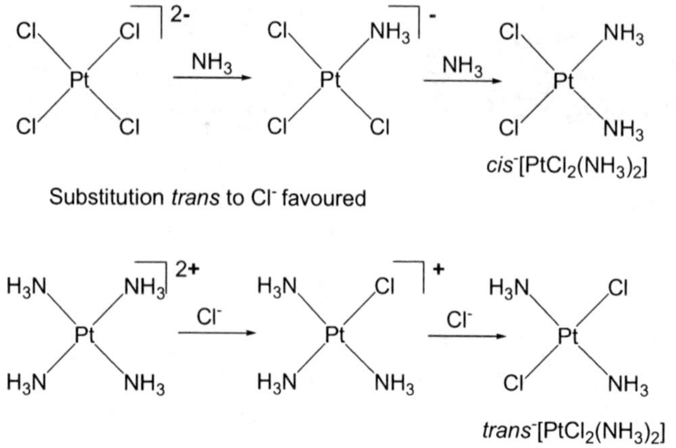

Cl Cl Pt Cl Cl $]^{2-}$ $\xrightarrow{NH_3}$ Cl NH_3 Pt Cl Cl $]^{-}$ $\xrightarrow{NH_3}$ Cl NH_3 Pt Cl NH_3

cis-$[PtCl_2(NH_3)_2]$

Substitution *trans* to Cl^- favoured

H_3N NH_3 Pt H_3N NH_3 $]^{2+}$ $\xrightarrow{Cl^-}$ H_3N Cl Pt H_3N NH_3 $]^{+}$ $\xrightarrow{Cl^-}$ H_3N Cl Pt Cl NH_3

trans-$[PtCl_2(NH_3)_2]$

Figure 39 *An example of the trans effect*

2.9 Redox Potentials

The ability to exist in more than one oxidation state under physiological conditions is an important feature of the chemistry of a number of d-block transition metals. The conversion from one oxidation state to another in solution involves the transfer of electrons and this process is, by convention, written as a reduction process in which an electron is added (Equation (45)). The standard electrode potential, $E^{\ominus}$, assigns a numerical value to the electrochemical potential at which the two oxidation states exist in equal concentrations at equilibrium in solution.[21]

$$M^{z+} + ne^- \leftrightharpoons M^{(z-n)+} \quad E^{\ominus} \qquad (45)$$

It is possible to construct electrode potential diagrams, sometimes called *Latimer diagrams*, relating the various oxidation states by their redox potentials, some examples are shown in Figure 40, which relates to aqueous solutions in 1 mol dm^{-3} acid. In cases where the reaction involves complexation to oxide or hydroxide ions, the reduction potential will be dependent upon pH. As an example, the reduction of $[MnO_4]^-$ to Mn^{2+} involves the addition of $5e^-$ and the consumption of $8H^+$ to produce $4H_2O$ from the oxide ions bound to the Mn(+7). Thus the equilibrium involves hydrogen ions, so that the $E^{\ominus}$ value for reduction will vary with hydrogen ion concentration. The effect of this is illustrated in Figure 40 for the case of manganese, where the Latimer diagrams for both acid and alkaline conditions are included. When the metal ions are complexed by a ligand the stability constants of the complexes formed from the

[21] The value $E^{\ominus}$ refers to a potential measured against a specific reference electrode (The Standard Hydrogen Electrode) under defined conditions. Often potentials are measured under less stringent conditions and against different reference electrodes and so will not be strictly comparable with $E^{\ominus}$. However, correction can easily be made for differences in the reference electrode used so that $E^{\ominus}$ values can be used with potential values obtained under less stringent conditions unless particularly accurate results are required.

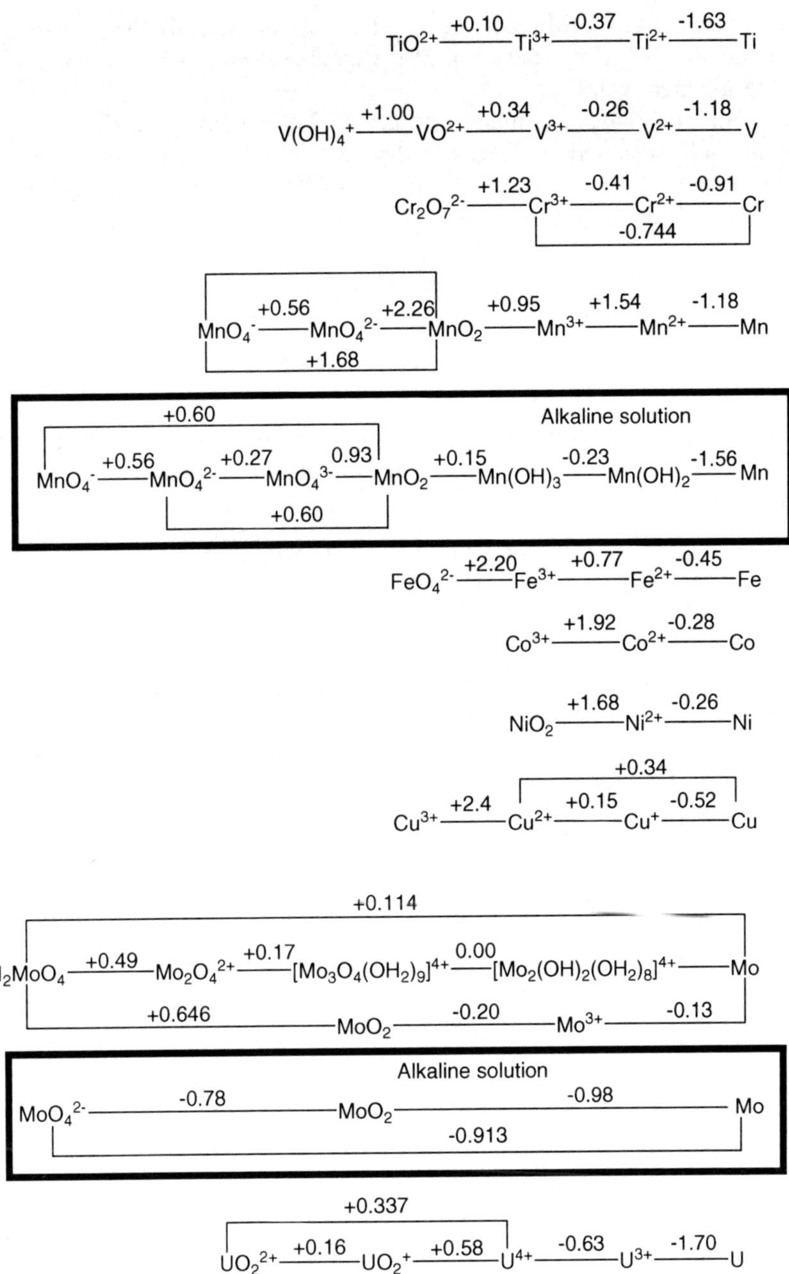

Figure 40 *Examples of Latimer diagrams showing the standard electrode potentials for the interconversion of different metal species in aqueous solutions*

oxidised and reduced forms of the metal ion are not usually the same. In such cases the $E^{\ominus}$ value for reduction of the complex will not be the same as that for the solvated metal ion.

The potential at which a metal complex, $[ML_x]^{z+}$, is reduced ($E^{\ominus}_{ML}$ Equations (46) and (47)) is related to the standard potential of the solvated metal ion ($E^{\ominus}$ Equation (45) above), and the stability constants of the oxidised (β_{ox} Equation (48)) and reduced (β_{rd} Equation (49)) forms of the complex by Equation (47).

$$[ML_x]^{z+} + ne^- \ [ML_x]^{(z-n)+} \ E^{\ominus}_{ML} \tag{46}$$

$$E^{\ominus}_{ML} = E^{\ominus} - (RT/nF) ln(\beta_{ox}/\beta_{rd}) \tag{47}$$

(R is the gas constant, T the absolute temperature in degrees Kelvin, 20 °C = 293 K; F the Faraday constant and n the number of electrons involved.)

$$\beta_{ox} = [ML_x]^{z+}/[M^{z+}][L]^x \tag{48}$$

$$\beta_{rd} = [ML_x]^{(z-n)+}/[M^{(z-n)+}][L]^x \tag{49}$$

An example is provided by $[Fe(CN)_6]^{3-}$ which may be reduced to $[Fe(CN)_6]^{3-}$ (Equation (50)). The difference between the reduction potential of this complex and that of aquated Fe^{3+} (Equation (51)) is 410 mV which corresponds with the stability constant for the Fe^{3+} complex being about 10^7 times larger than for the Fe^{2+} complex (Equation (52)). In fact the reported stability constants for $[Fe(CN)_6]^{3-}$ and $[Fe(CN)_6]^{4-}$ are respectively $\log\beta_{ox} = 24$ and $\log\beta_{rd} = 31$, a difference of 10^7.

$$[Fe(CN)_6]^{3-} + e^- \ [Fe(CN)_6]^{3-} \ E^{\ominus}_{ML} = +0.36 \text{ V} \tag{50}$$

$$Fe^{3+} + e^- \leftrightharpoons Fe^{2+} \ E^{\ominus} = +0.77 \text{ V} \tag{51}$$

$$log(\beta_{ox}/\beta_{rd}) = (+0.77 - 0.36)/2.303\ (RT/nF) = 6.95 \tag{52}$$

(A factor of 2.303 is needed to convert logarithms to the base 10 to natural logarithms and the expression 2.303 RT/F has a value of 0.059 V at 20 °C).

At an intuitive level it might be expected that the redox potentials of metal complexes with polyatomic ligands might be affected by changes in ligand structure. Since the nature of the donor atoms affects the stability constant for a particular metal this should also affect the redox potential. Thus, the presence of electron-withdrawing substituents might be expected to make reduction more difficult and oxidation easier. Other factors, such as the extent of unsaturation within the ligand, may also have an effect on redox potential. The greater the degree of unsaturation and conjugation within the ligand system, the better it is at tolerating added electrons, and the easier it is to reduce the complex.

2.10 Conclusions

Metallic elements have a wide range of properties making them useful in many different rôles. The simplest behaviour is found with the alkali or alkaline earth metals of Groups 1 and 2. These form cations with closed shell electron configurations and, with the exception of Li^{+}, Be^{2+} and to some extent Mg^{2+}, their chemistry is characterised by ionic behaviour and highly labile metal-ligand interactions. There is no directionality in the metal-ligand bonding but the metal ions vary in their ability to polarise ligands depending on their charge radius ratios. They show only their group oxidation numbers, +1 for Group 1 and +2 for Group 2. Perhaps the next simplest group of elements chemically is the lanthanide metals, the chemistry of which is dominated by Ln^{3+} ions. These also behave largely as closed electron shell systems but their higher charge aids complex formation, although ligand binding is still rather labile. Again there is no directionality in the metal-ligand bonding. Distinctive physical properties of the lanthanide ions are their magnetism and luminescence, which can have important medical applications. Variable oxidation state behaviour among the lanthanides is very limited. Other than the Ln^{3+} ions only Ce^{4+} and Eu^{2+} are likely to have any significant chemical presence in aqueous media.

The metallic p-block elements can show variable oxidation states in particular the Group oxidation state becomes less favoured on descending a Group as the retention of the valence shell ns^2 electrons becomes increasingly important because bond energies decrease for the larger atoms. When in their Group oxidation state, some of these metal ions can be powerful Lewis acids accepting electron pairs and strongly polarising attached ligand atoms. This property is particularly pronounced in the post d-block metal ions such as Ga^{3+} where there is incomplete screening of the nuclear charge by the filled d subshell leading to a higher effective nuclear charge for the ion than might otherwise be expected. This in turn makes the metal ion a more powerful electron pair acceptor, strengthening the metal-ligand interaction and more strongly polarising the donor atom attached to the metal ion. In water this effect promotes hydrolysis to form metal hydroxides.

The most versatile chemical behaviour is found for the d-block metals. These metals typically show variable oxidation numbers and CNs. The metal cations show Lewis acidity and can form complexes having high stability constants. Labile through to inert behaviour is found in some cases and the kinetics of ligand substitution are highly sensitive to oxidation number. The ability of metal cations to bring together several donor atoms in a defined geometry allows some metal ions to guide the construction of more complex structures in what may be described as metal directed self-assembly reactions. In a biological example a metalloprotein will typically contain several amino acid side chains which can bind to a metal, in bringing these together to form the metal complex the metal will have major effect on the structure of the protein. Metal ions such as Ca^{2+} are important structure directing agents in proteins but do not show strong directional control in their bonding or any particular reactivity. Where a d-block metal ion is involved the metal centre will have directional influences on the ligand binding arrangement and may also have an intrinsic reactivity which goes beyond any structural rôle. One example is provided by zinc in carbonic anhydrase (*CA*) which catalyses the hydration of carbon dioxide at physiological pH according to Equation (53). This is of vital

$$CO_2 + H_2O \overset{CA}{\rightleftharpoons} HCO_3^- + H^+ \tag{53}$$

importance for dissolving CO_2 produced during respiration as this reaction is normally very slow at physiological pH. The Lewis acid character of the zinc ion, combined with the nature of the protein environment of the zinc, allow a million fold acceleration of the hydration process. Another example is provided by iron in cytochrome P_{450} in which the iron atom functions as oxygen transfer reagent towards a substrate (SubH), which is to be oxidised. In effect the iron centre catalyses the reaction shown in Equation (54). The ability of the iron to adopt

$$SubH + O_2 + 2H^+ + 2e^- \overset{Cyt\,P_{450}}{\rightleftharpoons} SubOH + H_2O \tag{54}$$

different oxidation numbers and to bind in different ways to oxygen is a crucial aspect of this process.

Distinctive metal centred reactivity is one aspect of the chemistry of metals which has applications in medicine. Different metals offer different reactivity so that platinum finds use in treating cancer, gold for rheumatoid arthritis, vanadium for diabetes and iron or ruthenium for the management of nitric oxide levels *in vivo*. Beyond chemical properties, physical properties of metal ions such as magnetism luminescence and radioactivity also have important applications in medicine. However, the potential utility of metals in pharmaceutical formulations can only be realised through the careful design of compounds so that they exhibit the appropriate properties under clinical conditions. The subsequent chapters provide examples of how coordination chemistry has been exploited to good effect in the creation of new pharmaceuticals for use in diagnostic and therapeutic medicine.

CHAPTER 3

Diagnostic Medicine

3.1 Introduction

Coordination compounds have a variety of applications in the diagnosis of illness or injury. The area attracting the greatest current research and development interest concerns the *in vivo* use of metal complexes for non-invasive diagnostic imaging procedures. Metal complexes are also used in some *in vitro* diagnostic tests, which may exploit the luminescence or radioactivity of the metal. However, to a large extent, the coordination chemistry underlying these *in vitro* applications is encompassed by that used in the development of *in vivo* applications.

The ability to create, from outside the body, images of structures hidden from sight within the body offers an extremely powerful clinical tool. Two main types of imaging procedure are possible. The first is anatomical in nature and reveals internal structures through differences in the nature of the material within the body. An obvious example is provided by the use of X-rays to create images of broken bones. The high concentration of calcium and phosphorus in bone compared to soft tissues, which contain mainly the lighter elements carbon, hydrogen, oxygen and nitrogen, makes bone the stronger absorber of X-rays. Thus the difference in composition between the bone and soft tissue allows a conventional X-ray image to be obtained in which bone appears as areas of contrasting high X-ray absorption compared to the soft tissue. In this way damage to the bone structure can be visualised. More subtle differences within soft tissue can also be detected by careful control of measurement conditions combined with expert image interpretation. One problem with a conventional X-ray image is that all the structures in the path of the X-ray beam are overlaid in the 2-dimensional image obtained. Computer Aided Tomography (CAT or CT) scans allow a 3-dimensional representation of the subject to be created. This is achieved by taking X-ray absorption measurements from a variety of different directions and using computational methods to process the results to form a 3-dimensional model of the internal structure of the subject.

In some applications contrast agents which strongly absorb X-rays can be used to enhance the information available from the X-ray images. A good example is provided by the use of barium sulfate to obtain images of the gastrointestinal (GI) tract. The relatively high atomic number of barium means that it is a strong absorber of X-rays. Thus the barium in a 'barium meal' ingested by a patient can reveal the structure of the GI tract through the strong absorption of X-rays in regions accessible to the barium preparation. In this way it is possible to visualise abnormalities in structure such as defects in the stomach wall caused by ulcers. Barium sulfate is currently approved for human use in GI tract imaging. The other type of X-ray contrast agent currently approved for human use involves the tri-iodobenzene moiety in which the three iodine atoms act as strong X-ray absorbers. A wide variety of derivatives have been developed based on this unit and around 20 million procedures involving these agents are performed annually in the USA alone.

In recent years another anatomical imaging method has become important. This exploits the phenomenon known as nuclear magnetic resonance (NMR) and is known as Magnetic Resonance Imaging (MRI). An MRI scan is carried out by placing the patient in a strong magnetic field. Pulses of radiofrequency energy are then applied to excite the hydrogen nuclei present in the water molecules within the subject. As these nuclei relax back to their ground state, they emit radiofrequency signals which are detected externally. These contain information about the distribution of water within the subject and can be computationally converted to an image based on the relative amounts of water in different locations. This reveals internal structures within the body with a resolution of about 1 mm, similar to X-ray methods. The contrast of an MRI image can be improved if a suitable paramagnetic compound is present to change the timescale in which the excited hydrogen nuclei emit radiation. Complexes of lanthanide elements or transition metals offer a means of introducing such paramagnetic 'contrast agents'.

In addition to anatomical imaging there is a second type of imaging procedure known as functional imaging. This approach involves administering, usually through an injection, a substance which can be detected from outside the body. Its distribution within the body is then tracked by external detectors. Since the distribution of the substance depends on the way in which it is processed by the body, such imaging agents can be said to reveal structure through function. In order for the external detection of the substance to be possible it is usual to incorporate in the substance a radioactive element, which emits penetrating γ–rays which can easily be detected from outside the body. Because radiation detectors are so sensitive and γ–rays so penetrating, this is possible while maintaining the radiation dose to the subject at acceptably low levels. Technetium is one element in particular which has come to dominate this area of application because of the near ideal nuclear properties of its radionuclide ^{99m}Tc, together with a rich coordination chemistry allowing the preparation of a wide variety of stable complexes offering a range applications. Another possible, but not yet well developed, approach to functional imaging is to use paramagnetic contrast agents with suitable *in vivo* behaviour to perform 'functional' MRI scans.

3.2 Anatomical Imaging

3.2.1 Magnetic Resonance Imaging Contrast Agents

3.2.1.1 Nuclear Magnetic Resonance

MRI is a form of computed tomography based on a measurement technique familiar to chemists and known as Nuclear Magnetic Resonance or NMR. This exploits the effect of a strong magnetic field on the nuclei of atoms, in the case of MRI on the nuclei of hydrogen atoms in the water molecules within the body. In order to create an MRI image the subject is placed within a large cylindrical chamber so that a strong magnetic field can be applied to the region of interest (Figure 1). Once the magnetic field is applied the subject is investigated using pulses of radiofrequency radiation. These induce radiofrequency emissions (echo signals) from the subject which can be detected externally. The frequencies of the echo signals can be analysed and the results converted into a 3-dimensional image of the distribution of water within subject. Image contrast arises from the differing water contents of fat, muscle and bone as well as from other more subtle effects. The technology involved in MRI imaging is both elegant and complex but some appreciation of the underlying principles is necessary in order to better understand the role of coordination chemistry in MRI imaging applications. To this end a very much simplified account of the MRI measurement follows to explain how introducing paramagnetic metal compounds into the subject can improve the contrast in MRI scans.

The nuclei of hydrogen atoms are protons and these possess a magnetic moment rather like a simple bar magnet. In a strong magnetic field these magnetic moments precess about the magnetic field direction and, for a proton, quantum mechanics allows only two possible energy states to exist. In simple terms the protons must align with the magnetic field or against it. To attain

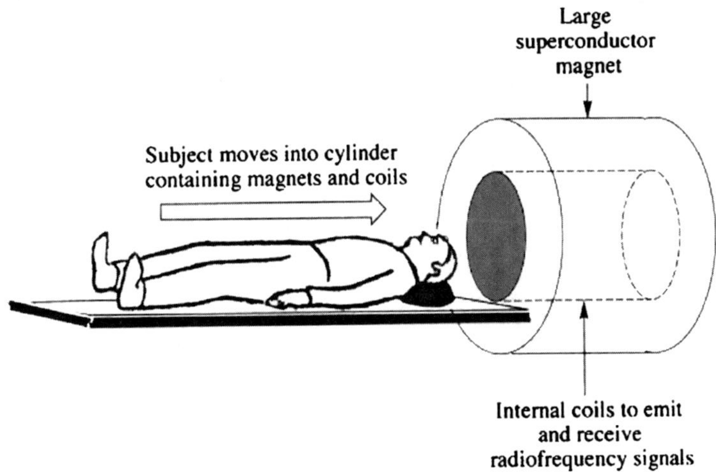

Figure 1 *The physical arrangements for performing an MRI scan*

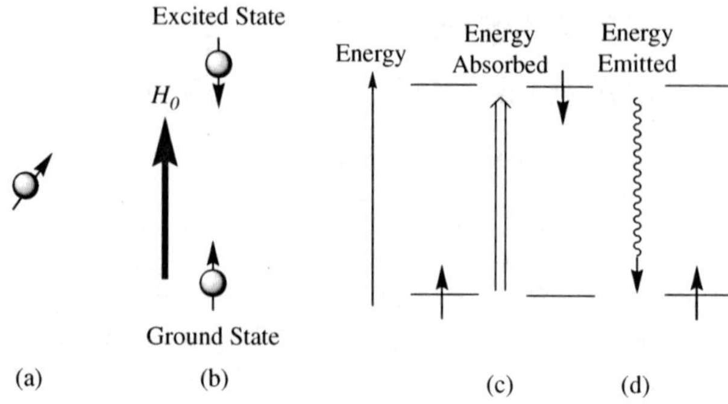

Figure 2 *(a) A representation of a proton with an arrow showing the orientation of the magnetic moment. (b) The ground and excited state orientations of a proton in an applied magnetic field H_o. (c) An energy level diagram showing the excitation of a proton through the absorption of radiofrequency radiation. (d) An energy level diagram showing the emission of radiofrequency radiation through the radiative transition of a proton from an excited state to its ground state*

their lowest energy, or ground state, configuration the protons align their magnetic moments with the field direction. The other possible arrangement has the magnetic moments oriented in opposition to the magnetic field, but this is a higher energy, or excited, state (Figure 2). If radiation of the energy needed to excite the proton from the ground state to the excited state is absorbed, the proton may be promoted into the higher energy state (Figure 2c). The frequency of the energy required to do this is usually referred to as the Larmor precession frequency. The proton can subsequently relax back into its ground state by emitting radiation of an energy equivalent to the energy difference between the ground and excited states (Figure 2d), or through losing energy to its surroundings. The energy difference between the ground and excited states depends upon the magnetic field, H_0, experienced by the proton. This field is made up from the local magnetic field within the water molecule and any externally applied magnetic field. The energy differences attainable using practicable magnetic fields are small and correspond with the radiofrequency region of the electromagnetic spectrum. At room temperature both the ground and excited states of the hydrogen nuclei are populated but there is a very small excess of protons in the ground as compared to the excited state. Thus if radiofrequency radiation of the correct energy is supplied some is absorbed, inverting the relative populations of the ground and excited states. If the radiofrequency energy source is removed the nuclei relax back to their ground states, in some cases emitting the absorbed radiofrequency radiation.

If an object containing water is placed in a magnetic field gradient, the value of H_0, experienced by the protons in the water will vary with position (Figure 3). Thus when the a proton relaxes back to the ground state the energy, and so the frequency, of any radiation emitted will correspond with the position of the proton in the field gradient. If a patient is placed in a strong magnetic field with gradients in three directions, and exposed to a broad spectrum pulse of

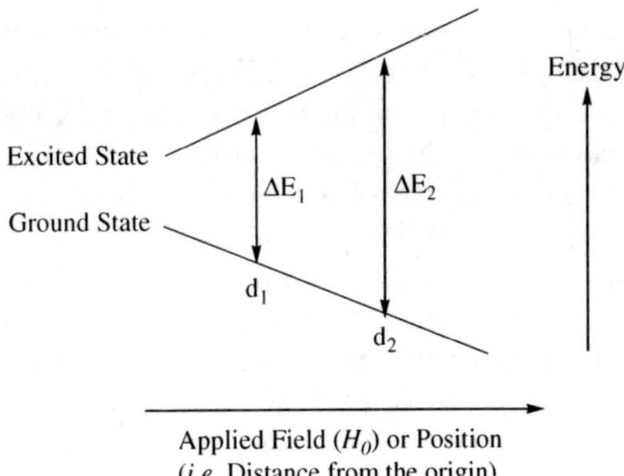

Figure 3 *In a field gradient different energies, ΔE_1 and ΔE_2, for emitted photons correspond with different respective distances d_1 or d_2 from the origin*

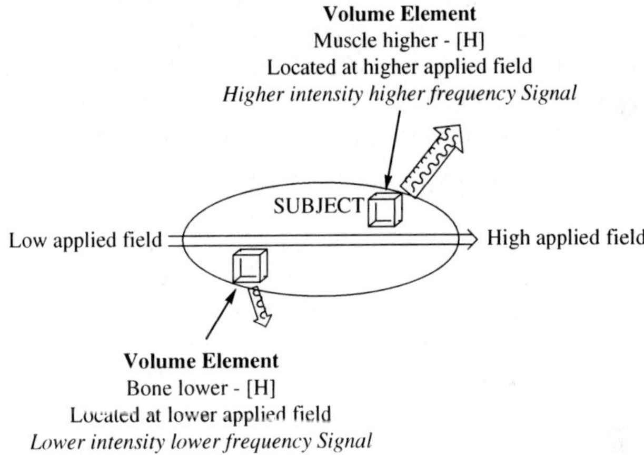

Figure 4 *A schematic representation of the basis for an MRI measurement in which [H] represents the local concentration of water H atoms in the tissue*

radiofrequency radiation, protons in the water within the body are promoted to their excited states. When these protons subsequently relax down to their ground states and emit radiation, this will be at frequencies characteristic of their location. Measurement of the emitted radiation coupled with computed tomography, then allows a 3-dimensional image of the distribution of water within the subject to be constructed. In practice the subject is divided into a number of volume elements and the signal intensity from each is determined. An image of the human head might, for example, be made up from over 65,000 volume elements each 1 mm by 1 mm by 5 mm in size. A sequence of operator defined excitation pulses is applied, each with a duration of about a microsecond, and the emitted radiation is measured during an operator specified time period between consecutive pulses. Resolutions of down to about 1 mm can be achieved in this way (Figure 4).

The contrast in an MRI image arises from the differing signal intensities (*SI*) arising from the different volume elements. These SI are related to the concentrations of water protons, [H], in the different tissues in the body so that muscle, fat and bone, for example, might produce signals of differing intensities. However, signal intensity is not only dependent on the concentration of water, it also depends upon the rate at which the protons concerned relax from the excited state to the ground state as a result of interactions with their surroundings and with other protons. A 'longitudinal' relaxation time, T_1, is associated with the transfer of energy from the excited protons to their surroundings and a 'transverse' relaxation time, T_2, is associated with the exchange of energy between ground state and excited state protons. The reciprocals of these times are the relaxation rates for the two processes. One further complication in the measurement is the need to allow for the movement of water into or out of the volume element during the time the measurement is being made. This is expressed in a parameter H_v. The MRI machine operator can optimise the image produced by varying two operator defined parameters. These are *PR* and *SE,* which respectively represent the applied radiofrequency pulse repetition time and the time during which the emitted radiation is detected, *i.e.* the spin echo time. Taking these various parameters into account an approximate expression for the signal intensity from a given volume element is given by Equation (1).

$$SI = [\mathrm{H}]Hv[\exp(-SE/T_2)\{1-(-PR/T_1)\}] \tag{1}$$

Signal intensity tends to increase with increasing $1/T_1$ and decrease with increasing $1/T_2$ but, for most tissues, the average T_2 values are only a fraction of the corresponding T_1 values. As an example values of 747 ms for T_1 and 71 ms for T_2 were found in Putamen tissue in the brain. In a magnetic field of 1.4 T (corresponding with a frequency of 60 MHz) values for T_1 typically lie in the range 200–500 ms for different tissue types. The operator can select radiofrequency pulse sequences to emphasise the effect of changes in $1/T_1$ compared to $1/T_2$ (T_1 weighted) or, conversely, to emphasise $1/T_2$ compared to $1/T_1$ (T_2 weighted).

3.2.1.2 *Contrast Agents for Magnetic Resonance Imaging*

In clinical applications of MRI it is often useful if the contrast between regions in the image can be altered beyond that possible using adjustments of the scanning equipment. A simple way to do this would be to change the relative signal intensity by changing the concentration of water in one region compared to another. As an example, attempts to displace liquid containing water from the GI tract using $C_8F_{17}Br$ (Imagent®), an inert liquid free of water protons, showed the procedure to be possible. However, the approach was impracticable because of the cost and side effects of the treatment. Similarly replacement of water by deuterium oxide (heavy water) is precluded by the toxicity and quantity of heavy water required. Fortunately there is another way in which image contrast might be manipulated. The magnitude of *SI* depends, not only upon [H], but also upon T_1 and T_2 so that compounds capable of modifying relaxation times could be used to modify *SI* and hence image contrast. Dissolved or colloidal paramagnetic materials can have a profound effect on the values of T_1 and T_2 and so can change the signal intensity in an MRI scan. If a paramagnetic compound is introduced into the subject and enters one region or tissue type in preference to another, contrast enhancement should result. Materials which can produce this effect are known as MRI contrast agents.

Contrast agents can increase $1/T_1$ and $1/T_2$ to varying degrees depending on the applied magnetic field and the nature of the agent. As examples, contrast agents containing Gd^{3+} complexes tend to increase $1/T_1$ and $1/T_2$ by broadly similar amounts and T_1 weighting typically gives the best images with these because of the dominance of the $1/T_1$ term in determining the magnitude of the signal intensity. In contrast iron oxide particles tend to have a much larger effect on the$/T_2$ term and so are usually best used with T_2 weighted imaging.

Paramagnetic materials are those in which the atoms contain unpaired electrons. Some non-metal compounds of this type (known as radicals) can be obtained but they tend to be highly reactive and toxic. The nitroxide radical provides an example of one of the more stable groups which may appear in non-metal compounds of this type (*e.g.* **1**). Apart from the problems of reactivity associated with non-metal radicals, such compounds tend to have only one unpaired electron. In contrast compounds of transition metals or lanthanides often contain more than one unpaired electron and do not normally show the reactivity associated with radicals. As examples Cr^{3+} contains 3 unpaired electrons, Mn^{2+} and Fe^{3+} in their high spin states contain 5 unpaired electrons and Gd^{3+} contains 7 unpaired electrons (Figure 5). Complexes of such metal ions, which also show suitable biodistribution and pharmokinetic behaviour, have potential for use as MRI contrast agents through their effects on T_1 and T_2. However, the design of the ligands used to complex the metals is crucial if the necessary combination of stability and *in vivo* behaviour is to be obtained. The choice of ligand also has an important effect on the ways in which the complex can influence T_1 and T_2.

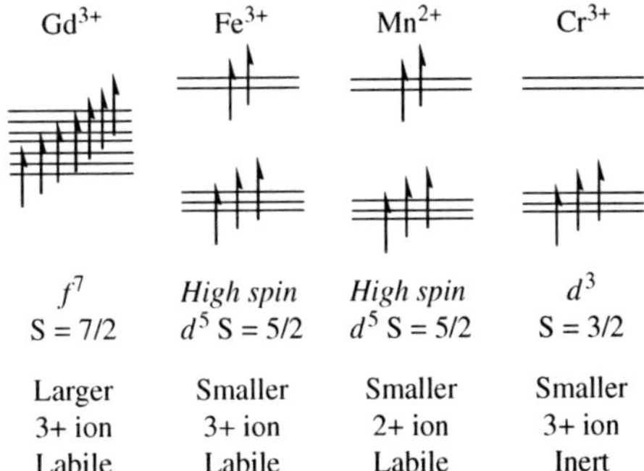

Figure 5 *Crystal field splitting diagrams to show the number of unpaired electrons in Gd^{3+}, high spin Fe^{3+}, high spin Mn^{3+} and Cr^{3+}*

1

The mechanisms through which a paramagnetic metal ion can influence the relaxation times of a proton in a water molecule are quite complicated. At the simplest level it might be expected that the larger the magnetic moment of the metal ion and the closer it is to the proton, the greater will be its effect. Thus Gd^{3+} with 7 unpaired electrons would be expected to be better than high spin Mn^{2+} or Fe^{3+}, each having 5 unpaired electrons, and these ions in turn should be better than Cr^{3+} with only 3 unpaired electrons. It might also be expected that a complex in which a water molecule is directly bonded to the metal ion would give the shortest metal to proton distance. Such a complex should have a greater effect than one in which other ligands saturate the metal ion coordination sphere and prevent any direct water metal interaction. In this context it is important to distinguish between inner sphere water molecules, which are directly bonded to the metal ion, and outer sphere water molecules which are adjacent to the metal complex but not bound to it. There is also the possibility that such outer sphere water molecules may hydrogen bond to a ligand donor atom bound to the metal and so, in effect, become inner sphere (Figure 6). In lanthanide ion complexes water or carboxylate ligands are particularly important in this respect. The situation is further complicated by the effects of

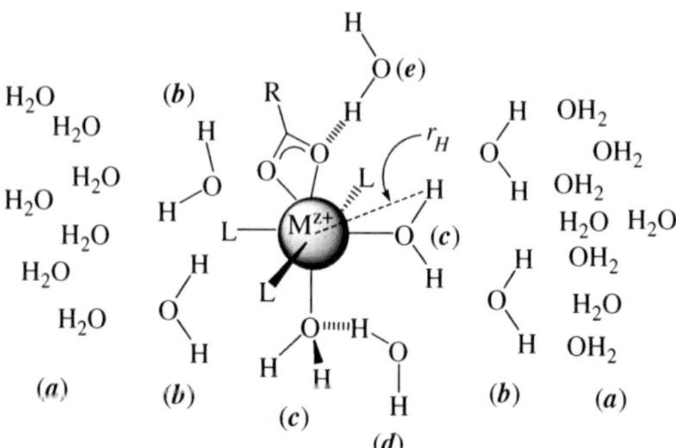

Figure 6 *A schematic representation of the different types of water molecule around a metal ion in a complex $[ML_3(RCO_2)(H_2O)_2]^{z+}$ in aqueous media (a) bulk water; (b) 'outer sphere' water; (c) 'inner sphere' water. The distance rH is from the metal ion to a proton in an inner sphere water molecule; (d) an 'outer sphere' water molecule hydrogen bonded to an inner sphere water molecule so that one proton becomes essentially inner sphere; (e) an 'outer sphere' water molecule hydrogen bonded to an inner sphere ligand carboxylate oxygen so that one proton becomes essentially inner sphere*

exchange between the different types of water molecule present. If a water bound to a paramagnetic metal ion exchanges with bulk water in the liquid at a very slow rate compared to the proton relaxation rate, communication of the effect of the paramagnetic ion to the bulk water is very poor. An effective MRI contrast agent will need to show fast exchange of the bound water relative to the relaxation rates. Thus kinetically inert ions such as Cr^{3+} are ineffective compared to labile ions such as Mn^{2+}. The interplay between the proton relaxation times and dynamic processes involving the water molecules in the region of the metal ion is an important contributor to the effectiveness of a metal complex as an MRI contrast agent. In designing metal complexes for use as MRI contrast agents it is necessary to select ligands which not only confer good *in vivo* properties on the complex, but which also optimise the dynamics of the metal water interaction with respect to relaxation times.

3.2.1.3 *Relaxivity*

The effect of a metal complex $\{ML\}$ on T_1 or T_2 can be conveniently summarised in a parameter known as the relaxivity, r_i ($i = 1$ or 2; r_1 and r_2 being respectively associated with T_1 and T_2). The relaxivity, r_i, is related to the observed reciprocal relaxation time, or relaxation rate, $1/T_{iO}$ ($i = 1$ or 2) and the concentration of the metal complex, $[\{ML\}]$. However, the observed relaxation rate is made up of two components, $1/T_{i\mathrm{P}}$ which is the relaxation rate in the presence of $\{ML\}$ and $1/T_{i\mathrm{D}}$ which is the relaxation rate in the absence of $\{ML\}$, that is the diamagnetic contribution to the relaxation rate, as shown in Equation (2). The relaxivity associated with the complex can be determined from measurements of the observed relaxation rate $1/T_{iO}$ at different concentrations of the metal complex. Since only $1/T_{i\mathrm{P}}$ is dependent on $[\{ML\}]$ a plot of $1/T_{iO}$ against $[\{ML\}]$ gives a line of slope r_i with an intercept of $1/T_{i\mathrm{D}}$ according to Equation (3).

$$1/T_{iO} = (1/T_{i\mathrm{P}}) - (1/T_{i\mathrm{D}}) \quad (i = 1 \text{ or } 2) \tag{2}$$

$$r_i\,[\{ML\}] + (1/T_{i\mathrm{D}}) = 1/T_{iO} \quad (i = 1 \text{ or } 2) \tag{3}$$

The units of $[\{ML\}]$ are mmol l^{-1} for solutions but should be expressed in mmol kg^{-1} for soft tissues which contain significantly less than 100% water. The units of r_i are thus l $\mathrm{mmol}^{-1}\ \mathrm{s}^{-1}$ for solutions and kg $\mathrm{mmol}^{-1}\ \mathrm{s}^{-1}$ for soft tissues.

Since the relaxivity, r_i associated with a metal ion depends upon $T_{i\mathrm{P}}$ it will be affected by the various factors which contribute to the value of $T_{i\mathrm{P}}$. These include the temperature, the viscosity of the medium, the magnetic field and so the radiofrequency used, the magnetic moment of the metal ion, the number of water molecules attached to the metal ion (q), the distance between the metal ion and the proton of the water molecule (r) and the dynamic behaviour of the system. The dynamics of the system can be described by a group of correlation times which relate to the rotation of the metal complex (τ_R), the residence time for a water molecule bound to the metal (τ_M) and the effect of solute-solvent collisions (τ_v). A further correlation time has to be taken into account which is associated with the unpaired electrons. An unpaired electron in a magnetic field behaves in a similar manner to a proton and has its own Larmor precession frequency, which is somewhat higher than that of the proton in the same magnetic field. As with the excited state proton, longitudinal and transverse relaxation times are associated with an excited paramagnetic state electron and can influence the magnitude of T_{1P}. Correlation times can be influenced by the structure of the metal complex and so by the nature of the ligands bound to the metal ion.

A theoretical derivation of T_{iP} for a water molecule bound to a metal ion can be made using the Solomon–Bloembergen–Morgan equations (SBM theory). These relate the Larmor frequency of the proton to the correlation times and magnetic properties of the system. The application of SBM theory to low symmetry systems with more than one unpaired electron is still developing and a detailed description lies beyond the scope of this discussion. However, the theory does provide a basis for identifying and understanding the factors which will be important for producing an effective contrast agent. The expression for $1/T_{iP}$ is made up of a dipolar through space component and a through bond or scalar component. The dipolar component is inversely related to r^6 and so falls off very rapidly with increasing metal ion to proton distances. The dipolar term also depends upon correlation times which reflect the rotation rate of the molecule, the longitudinal electronic relaxation time and the rate at which water, or the proton therein, exchanges between the bulk water and the metal ion bound state. The scalar component is dependent on the longitudinal electronic relaxation time and the electron-nuclear hyperfine coupling constant.

The SBM model takes account of the interplay between the correlation times and the Larmor frequencies of the proton and electron, which are determined by the applied field, and allows a simulation of the variation in relaxivity with frequency. The form this variation takes for a given system depends upon the rotation rate of the molecule. A correlation time for molecular rotation of 10^{-10} s might be typical of a metal complex in aqueous solution. This leads to a relaxivity which remains fairly constant from 0.01 to about 5 MHz, then decreases between *ca.* 5 and 10 MHz before levelling off again (Figure 7). A metal ion attached to more slowly rotating macromolecule such as a protein might exhibit a longer correlation time of *ca.* 10^{-9} s, provided the local dynamics of the metal binding site reflect the slower tumbling of the macromolecule as a whole. This leads to a pronounced maximum in the plot of relaxivity against frequency at *ca.* 100 MHz (Figure 7).

The number of water molecules bound to a metal ion (q) and the water exchange rate are factors affecting T_{1P}, which can be readily modified through the selection of the metal ion and its ligands. Although it may offer higher complex stability, a ligand set which can fully saturate the metal coordination sphere will lead to $q = 0$ and have an undesirable effect on relaxation times. Similarly if the residence time of the water molecule on the metal ion (τ_M) is much larger than T_{1P} a low relaxivity will result. This limits the utility of kinetically inert paramagnetic metal ions such as Cr^{3+} unless rapid proton

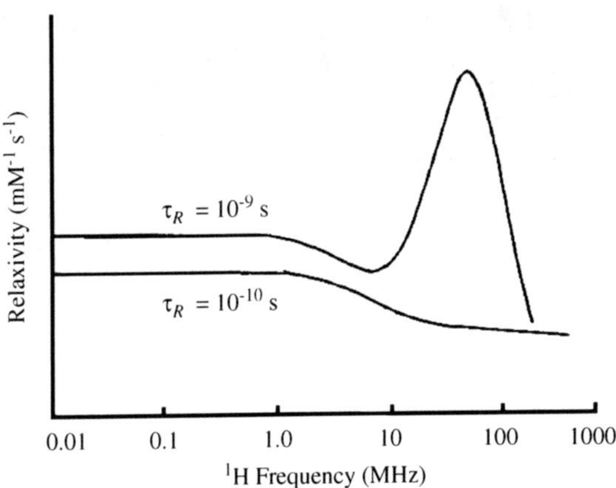

Figure 7 *Plots showing the qualitative form of the calculated relaxivity (r_1) against ^1H frequency for molecular rotation correlation times of 100 ps (top trace) and 1 ns (bottom trace) based on a theoretical model*

Table 1 *Some examples of factors affecting r_1*

	Gd(dtpa)$^{2-}$	*Mn(edta)*$^{2-}$	*Cr(edta)*$^{-}$
μ_{eff}^2	63	35	15
k_{ex} (s^{-1})	4.1×10^6	4.4×10^8	1.0
r_H (pm)	249	216	200
r_1 (mM^{-1} s^{-1})[a]	3.8	2.9	0.2

[a] Measured at 20°C.

exchange as shown in Equation (4) offers an alternative mechanism for the exchange of protons between metal-bound and bulk water.

$$\{CrL(OH_2)\} + H_2O \leftrightarrow \{CrL(OH)\}^- + H_3O^+ \quad (4)$$

Complexes of the metal ions Mn^{2+} and Gd^{3+} have no crystal field stabilisation energy and show much more labile behaviour so that $\tau_M << T_{1P}$. The water exchange rate (k_{ex}) for $[Gd(H_2O)_8]^{3+}$ in water is 8.3×10^8 s^{-1}, giving τ_M *ca.* 10^{-9}s, compared to a k_{ex} of *ca.*10^{-6} s^{-1} for Cr^{3+}. Complexation of the metal ion by a polydentate ligand can significantly reduce k_{ex}. In the case of Gd^{3+} binding to pdta^{4-} (**2**) or dtpa^{5-} (**3a**) to give $[Gd(pdta)(H_2O)]^-$ or $[Gd(dtpa)(H_2O)]^{2-}$, respectively results in respective k_{ex} values of 1.0×10^8 and 4.1×10^6 s^{-1}. The latter corresponds with τ_M in the range 10^{-6}–10^{-7} s which is only just sufficient for MRI applications. The Mn^{2+} complex $[Mn(edta)(H_2O)]^{2-}$ (edtaH_4 = **4**) shows a higher exchange rate and shorter metal ion-proton distance than the Gd^{3+} complexes (Table 1) but these benefits are offset by the lower magnetic moment of the d^5 first row transition metal ion compared to the f^7 lanthanide ion. The effectiveness of the Cr^{3+} complex is limited by both the lower magnetic moment of Cr^{3+} and its kinetically inert behaviour.

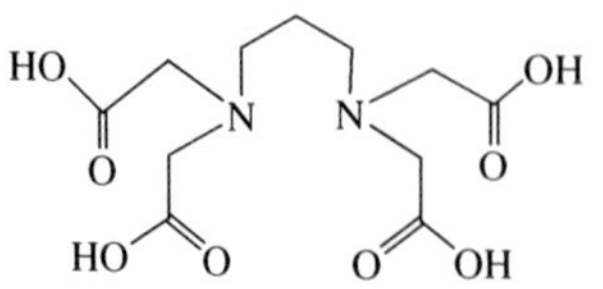

2
$pdtaH_4$

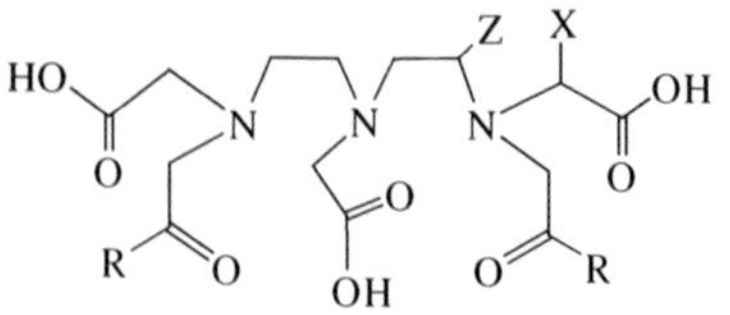

3

a	$dtpaH_5$:	R = OH, X = Z = H
b	$boptaH_5$:	R = OH, X = $CH_2OCH_2C_6H_5$, Z = H
c	dtpa-$eobH_5$:	R = OH, X = H, Z = $CH_2C_6H_4$-4-OC_2H_5
d	dtpa-$bmaH_3$:	R = $NHCH_3$, X = Z = H
e	dtpa-bmea H_3:	R = $N(CH_3)CH_2CH_2OCH_3$, X = Z = H

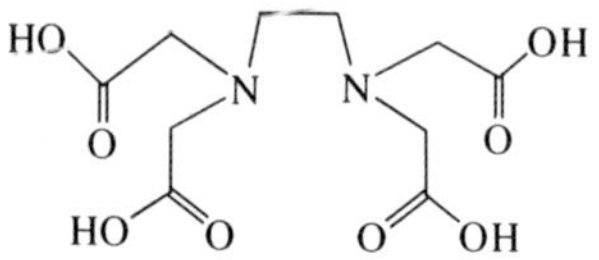

4
$edtaH_4$

In addition to the contribution from protons linked to the metal ion through bonding interactions with a donor atom, the relaxivity of a metal complex also contains a contribution from protons in outer sphere water molecules. In a complex where the ligands fully saturate the metal coordination sphere, so that water cannot bind directly to the metal ion, the outer sphere contribution to relaxation times dominates the relaxivity. The contribution of outer sphere water to the magnitude of T_1 depends upon the diffusion coefficient of water molecules, the diffusion coefficient of the complex, the concentration of metal ions and the closest approach distance of an outer sphere water molecule to the metal ion. Calculations indicate that the outer sphere contribution to the relaxivity of aquated Gd^{3+} is about 10% or *ca.* 1 mM^{-1} s^{-1} at 40°C at 20 MHz. The charge on the complex can also be important and the relaxivity of coordinatively saturated cationic Mn^{2+} complexes has been found to be lower than for coordinatively saturated anionic Mn^{2+} complexes (the term coordinatively saturated means that the metal first coordination sphere is fully occupied by the ligands). This reflects the closer approach of positively polarised hydrogen atoms in outer sphere water to anionic complexes compared to their cationic counterparts.

3.2.1.4 T_2 Agents

Metal complexes are usually used to increase relaxivity through their shortening T_1 and so increasing signal intensity. However, it is also possible to use paramagnetic metal compounds to reduce T_2 and so reduce signal intensity. Iron oxide particles can be very effective in this respect but may present problems with tolerance because of the relatively high doses required. The possibility that soluble coordination complexes might be used to decrease signal intensity through reducing T_2 has been investigated using dysprosium. The

Dy^{3+} ion has 5 unpaired electrons but because of a large orbital contribution its effective magnetic moment is 10.5 BM, substantially higher than the spin only value of 5.9 BM. The very short electronic relaxation time of Dy^{3+} means that it has little effect on T_1 but can act as an effective T_2 agent. Advantages over iron oxide particles include improved tolerance and more rapid injection with the prospect of differentiating ischemic tissue from that with normal perfusion. However, similar diagnostic information might be obtained using cheaper, more well established Gd^{3+} complexes.

3.2.1.5 Coordination Compounds as Paramagnetic MRI Contrast Agents

The design of coordination compounds for use as paramagnetic MRI contrast agents presents a particular challenge to the coordination chemist. In order to develop a successful agent it is necessary to reconcile several conflicting requirements. The biological requirement is for a compound with low toxicity combined with suitable biodistribution and pharmokinetic behaviour. Toxicity is a particular issue since relatively large doses are required for an MRI contrast agent to be effective, typically in the range 0.1–0.3 mmol kg^{-1}. This makes it particularly important that the metal is not readily released from the ligand as this would allow binding to serum proteins or other *in vivo* complexing agents. Transfer of the metal in this way would result in a loss of control over metal ion toxicity and biodistribution. Rather the complex should remain intact and be excreted completely and rapidly following the imaging procedure. This requires the complex to be thermodynamically stable and kinetically inert. Suitable modification of ligand structure might then be used to optimise biodistribution and pharmokinetic behaviour.

The basic requirements for an effective contrast agent are that it should have a high magnetic moment, accommodate at least one water molecule in the first coordination sphere of the metal ion and that the coordinated water should undergo rapid exchange with bulk water. In order to attain a high magnetic moment it is necessary to choose high spin metal ions with a large number of unpaired electrons. Such complexes have little or no crystal field stabilisation energy, a property typically resulting in labile behaviour. This labile behaviour is useful in being associated with rapid water exchange. However, the presence of a vacant binding site for water in the complex provides a potential pathway for initiating ligand dissociation. Complexes in which the metal coordination sphere is completely saturated by the ligand so that aquation is blocked would be more inert. Thus the requirements for an effective contrast agent appear to be at odds with those of a complex with suitable *in vivo* properties. Since complex stability *in vivo* cannot be attained by using kinetically inert metal ions, stability must be attained largely through the thermodynamics of complex formation. Complexes with very high stability constants are needed. This in turn implies polydentate ligands of suitable structure and with donor atom types chemically well matched to the metal ion. In the case of lanthanide ions such as Gd^{3+}, for example, hard oxygen donor atoms would be preferred for compatibility with the hard metal centre. The ligand also needs to be anionic as this will promote binding to the metal cation and, if an anionic complex is formed, interactions with outer sphere water will be increased, improving relaxivity. However, charged complexes may be less acceptable in terms of their *in vivo* effects as mentioned in Section 3.2.1.9. Careful selection of the ligand structure can allow the stability of the metal ligand complex to be improved without reducing the lability of coordinated water. As an example a relatively rigid backbone, which requires the ligand donor atoms to occupy locations close to those which will be occupied in the complex, should give a more stable complex than a corresponding ligand with a non-rigid backbone. The more rigid ligand is said to be 'preorganised' in that the structural arrangement of the ligand atoms needs to change little in forming the complex (Figure 8). This reduces the energy penalty associated with complex formation which arises from necessary changes in ligand structure.

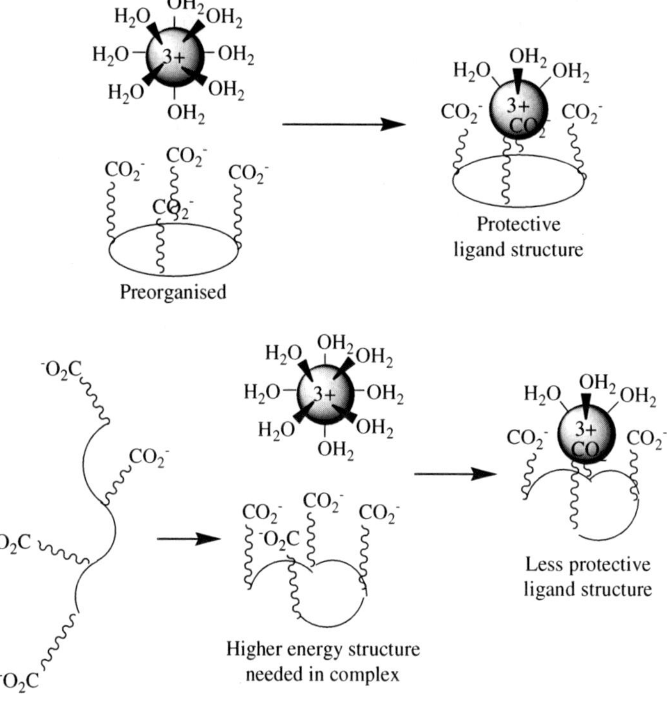

Figure 8 *A schematic representation of how a preorganised ligand can offer higher complex stability compared to a non-preorganised counterpart. Both complex types contain labile water. (a) Little structural change is needed in the ligand during complex formation and the rigid backbone offers protection of one side of the metal ion from attack by other ligands. (b) Energy is required to change the ligand structure to that present in the complex making complex formation less thermodynamically favourable. The ligand structure is less well suited to preventing attack by competitor ligands*

3.2.1.6 Choice of Metal

As indicated in the preceding discussion the metal ions of particular interest for MRI contrast agent applications include high spin d^5 Fe^{3+}, high spin d^5 Mn^{2+}, f^7 Gd^{3+} and f^5 Dy^{3+} (Table 1). The high magnetic moment and long electronic relaxation time of Gd^{3+} make it the most widely used ion in MRI applications. Compared to Gd^{3+} the smaller ionic radius of Mn^{2+} leads to a shorter metal-proton distance for coordinated water which to some extent offsets the smaller magnetic moment of Mn^{2+}. The long electronic relaxation time of Mn^{2+} is also an attractive feature but this metal ion shows cardiovascular toxicity at low doses limiting its utility in MRI applications. Iron, in the form of Fe^{3+}, is a d^5 ion like Mn^{2+} but without the cardiovascular toxicity. In MRI applications Fe^{3+} is typically used in the form of insoluble iron oxide particles which are neither coordination compounds in the usual sense nor solution species. There has been some interest in Dy^{3+} as a T_2 agent since its very short electronic relaxation time leads to its having a negligible effect on T_1 but the high magnetic moment is effective in reducing signal intensity from its effect on T_2. A Dy^{3+} complex has been tested in humans for identifying ischemic regions in heart and kidney. However, although the use of Dy^{3+} complexes offers interesting possibilities, in general they may not prove competitive with more familiar and medically established Gd^{3+} agents.

3.2.1.7 Choice of Ligand

The ability to provide very high thermodynamic stability, access of water to an inner sphere binding site and allow rapid exchange without compromising stability are important factors in the choice of ligand for MRI applications. Since lanthanide ions, particularly Gd^{3+}, are the metal ions most commonly used in MRI applications anionic hard donor ligands represent an obvious choice. In particular the polyaminepolycarboxylic acid proligand diethylene-triamine pentaacetic acid (**3a**, $dtpaH_5$ or DTPA[1]) and its derivatives (**3b–e**) show strong binding to Ln^{3+} ions through having 'hard' oxygen donor atoms, being polydentate (so stabilising the complexes through the chelate effect) and being able to achieve the high coordination numbers required by Ln^{3+} ions in a 1:1 ligand/metal complex. The well-known proligand $edtaH_4$ (**4**) is less well suited than $dtpaH_5$ for Ln^{3+} binding since it offers a maximum coordination number of only 6 in a 1:1 ligand/metal complex. Thus $edta^{4-}$ alone cannot saturate the coordination sphere of a lanthanide ion which might typically

[1] Very often ligand abbreviations are written in upper case ignoring acidic hydrogen atoms so that ethylenediamine tetra-acetic acid is often referred to as EDTA. However, this can create problems in accurately writing formulae. Under this definition, to be accurate, the formula of the dicalcium salt of ethylenediamine tetra-acetic acid has to be written $Ca_2(EDTA\text{-}4H)$ because four H^+ ions have been lost in forming the complex. To avoid this situation, in this book a lower case text abbreviation is also used which includes ionisable hydrogen in the formula of the proligand, hence $edtaH_4$ and $Ca_2(edta)$ for the compound containing $2Ca^{2+}$ and $edta^{4-}$. The abbreviation edta simply represents the core of the molecule with hydrogen removed as neutral H atoms. The upper case acronyms are also used in the usual less formal manner often seen in literature, but not in the construction of formulae representing the actual composition of compounds.

exhibit 8- or 9-coordination. Another structural motif used in lanthanide ion sequestering agents involves a cyclic polyamine core as found in dotaH_4 (**5a**) and its derivatives (**5b,c**). Here additional stability is achieved as a result of the rigid structure of the cyclic polyamine core which preorganises the carboxylate groups into an arrangement better adapted to coordinate to the metal ion (Figure 8). The cyclic polyamine structure also provides an element of steric protection inhibiting the approach of competitor ligands to the metal and slowing down any ligand dissociation.

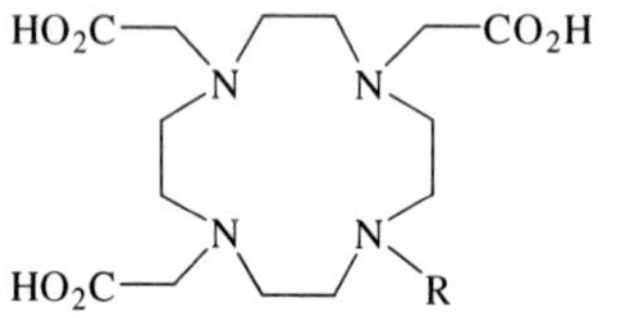

HO_2C CO_2H N N N N HO_2C R

5

a dotaH_4: R = CH_2CO_2H
b hp-do3aH_3: R = $CH_2CH(CH_3)OH$
c do3a-butrolH_3: R= HO OH OH

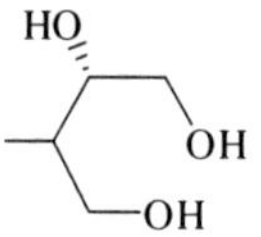

Stability constant data can provide a basis for comparing complexes and assessing the effect on their properties of structural changes in their ligands. However, it should be noted that the quoted values for the stability constants of a particular complex can show some variability. In part this may reflect differences in the conditions of measurement since the values obtained will often depend upon factors such as solution pH and ionic strength. Provided the stability constant data is obtained under similar conditions they provide a useful means of comparing complexes. The stability constants of some complexes of polyamine-carboxylate ligands are shown in Table 2. These reveal increases in the log K values for Gd^{3+} complexes in going from $edta^{4-}$ to

Table 2 *Stability constants for some selected 1:1 polyminecarboxylate complexes*

Complex[a]	*{Gd(edta)}*$^-$	*{Gd(dtpa)}*$^{2-}$	*{Gd(dota)}*$^-$	*{Gd(dtpa-bma)}*	*{Gd(hp-do3a)}*
log$K$	17.3	22.4	25.8	16.9	21.8
K_{obs} (10^3 s^{-1})[b]	14 × 10^3	1.2	0.021	>20	0.064
Complex[c]	*{Yb(dtpa)}*$^{2-}$	*{Pb(edta)}*$^{2-}$	*{Bi(dtpa)}*$^{2-}$	*{Gd(do3a)-butrol}*	
log$K$	22.6	18.1	27.8	23.8	

[a] Cited with *N*-methylglucammonium (NMG^+) as the counterion for log$K$ values.
[b] Observed rate constant for acid dissociation in 0.1 M acid.
[c] Cited with Na^+ as the counterion.

$dtpa^{5-}$ then $dota^{4-}$ as might be expected from the discussion above. The values also reveal how changes to the ligand structure can significantly affect the stability of the complex. The stability constant of [Gd(hp-do3a)] is 10,000 times smaller than that of $[Gd(dota)]^-$ yet, despite the general similarities between the hp-do3a and do3a-butrol ligands, in the case of [Gd(do3a-butrol)] the reduction is only 100-fold compared to $[Gd(dota)]^-$. In part this reduction in stability constant for the substituted $dota^{4-}$ ligands reflects the removal of a coordinating carboxylate group from the ligand structure. However the nature of the substituent is also important as shown by 100-fold difference in stability constants between [Gd(hp-do3a)] and [Gd(do3a-butrol)]. Depending upon its structure a substituent might inhibit ligand binding or, conversely, it might improve stability through inhibiting the approach of competitor ligands. If the substituent contains donor atoms in suitable locations for binding to the metal additional stability may be conferred on the complex through chelate formation.

The relative effects of different donor atom types on lanthanide ion complex stability have been measured and the results expressed in terms of a $\Delta\log K$ value. This value is the increase in $\log K$ resulting from the reaction in Scheme 1 in which the hydrogen of the amine dicarboxylate ligand NH group is replaced by an additional chelating group. The results obtained are summarised in Table 3.

The thermodynamic stability of the complex does not provide the complete picture. The rate of ligand dissociation is also important. A kinetically inert complex of low stability might be less prone to release the bound metal ion than a labile complex of high stability. One measure of ligand lability is provided by acid dissociation rates and some examples are provided in Table 2. The $edta^{4-}$ ligand, which is unable to saturate the Gd^{3+} coordination sphere, shows a high dissociation rate. Complexes of $dtpa^{5-}$ and its derivatives show much smaller dissociation rates but within this group a comparison of $\log K$ values and k_{obs} show that there is no good correlation between kinetic behaviour and

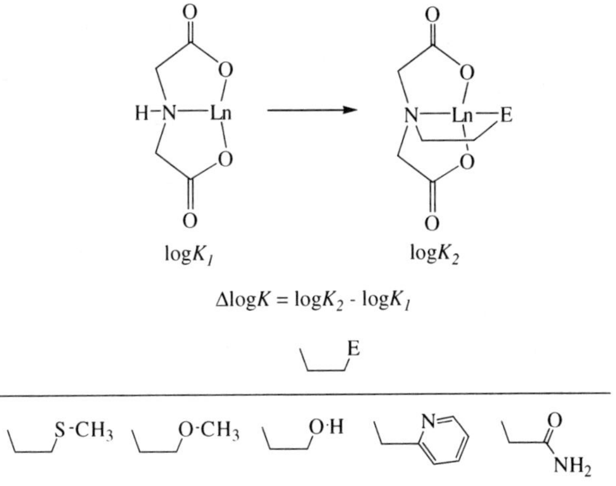

$\log K_1$ $\log K_2$

$$\Delta\log K = \log K_2 - \log K_1$$

Scheme 1

Table 3 *Variation in stability constants for chelating groups in Scheme 1 complexes*

logK	 E
0	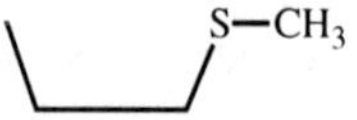 S—CH_3
1.6	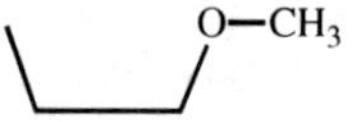 O—CH_3
2.9	 O-H
3.3	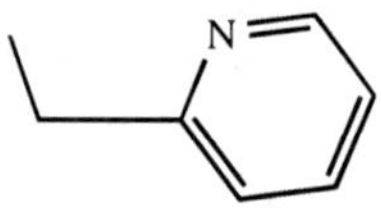 N
3.38	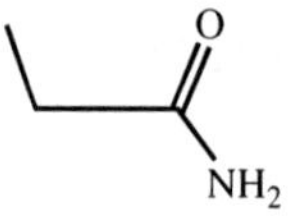 O NH_2

thermodynamic stability. The modification of a ligand structure through substitution must be expected to change the stability and kinetic behaviour of its complexes but it cannot be assumed that any particular change will be beneficial.

The nature of the metal ion is also important in determining the stability and kinetic behaviour of the complex. However, within the lanthanide series these effects may be small. The absence of significant crystal field effects means that ionic radius is an important parameter yet, despite the difference in ionic radii, there is little difference in log$K$ between the closely related lanthanide ions Gd^{3+} (ionic radius 93.8 pm) and Yb^{3+} (ionic radius 86.8 pm) for complexes with $dtpa^{5-}$ (Table 2). Typically it might be expected that the ion with the smaller radius would show the higher binding constant. However, with large polydentate ligands such as dtpa the structure and flexibility of the ligand also plays an important role in determining complex stability since this also reflects the 'goodness of fit' between the metal ion and the structural arrangement adopted by the ligand. In the case of the larger *p*-block ion Bi^{3+} (ionic radius 103 pm) a significantly higher stability constant is found with $dtpa^{5-}$.

3.2.1.8 Structures of Lanthanide Polyaminecarboxylate Complexes

The solid state structures of $[Gd(dtpa)(H_2O)]^{2-}$, $[Gd(bopta)(H_2O)]^{2-}$, $[Gd(do3a\text{-}butrol)(H_2O)]$, $[Gd(hp\text{-}do3a)(H_2O)]^{2-}$ and numerous other lanthanide polyaminecarboxylate complexes have been determined. The complexes typically contain 9-coordinate Gd^{3+} ions bonded to an octadentate polyamine carboxylate ligand, the ninth coordination site being occupied by the water. In the case of $dtpa^{5-}$, or its derivatives, the metal ion is bonded to three nitrogen and five carboxylate oxygen atoms, whereas in the complexes of $dota^{4-}$, or its derivatives, the metal ion is bonded to four nitrogen and four carboxylate oxygen atoms. The ionic character of lanthanide-ligand bonding and the absence of significant crystal field stabilisation effects with the f-block metal mean that the coordination geometries of the complexes are quite dynamic and strongly influenced by the structural requirements of the ligand. The most energetically favoured idealised geometric structure for a 9-cooordinate complex, ML_9 where L is a monodentate ligand, is the tricapped trigonal prism (TTP, Figure 9). However, the structural and energetic differences between this

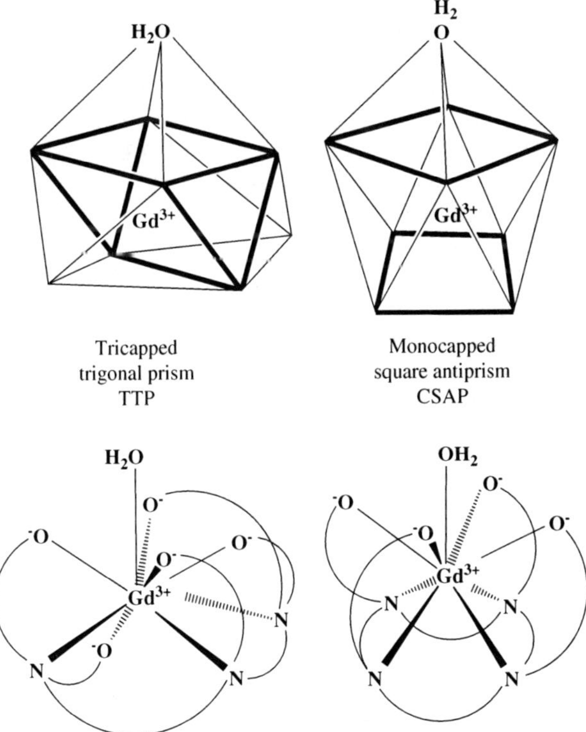

Figure 9 *Schematic representations of the metal coordination geometries (top) and ligand connectivities in 9-coordinate Gd^{3+} complexes of $dtpa^{5-}$ (TTP) and $dota^{4-}$ (CSAP) ligands*

and the other idealised geometry for 9-coordination, the monocapped square antiprism (CSAP), are small. In reality the requirements of the ligand structure will distort the structures from ideality, $dtpa^{5-}$ forming seven, and $dota^{4-}$ eight, chelate rings. In general structures approximating to TTP are more usually associated with $dtpa^{5-}$ and its derivatives while the cyclic backbone of the $dota^{4-}$ ligand predisposes it to the formation of CSAP structures. Compared to $\{Gd(dtpa)\}^{2-}$, the less open arrangement of the $\{Gd(dota)\}^{-}$ unit improves its stability and inhibits the approach of competitor ligands (Table 2).

Although the solid-state structures reveal much about how the polyamine carboxylate ligands can bind to lanthanide ions, they do not necessarily correspond with the solution state structures. Measurements on aqueous solutions are necessary to confirm or refute the persistence of the solid-state structure in solution. Information about the number of water molecules bound to a lanthanide in solution can be obtained from luminescence studies. The luminescence of lanthanide ions is quenched by coupling to the O–H bond vibrations of water. By comparing the fluorescence decay constants for solutions in H_2O and D_2O, which shows less effective coupling of the O–D vibrations, it is possible to estimate the number of water molecules, q, bound to the lanthanide ion. Unfortunately the electronic properties of Gd^{3+} make it unsuitable for such studies. However, both Tb^{3+} and Eu^{3+}, which lie either side of Gd^{3+} in the Periodic Table, exhibit strong luminescence. Therefore studies of Tb^{3+} and Eu^{3+} complexes give a good indication of the properties of the corresponding Gd^{3+} complex. The hydration numbers found for the $dtpa^{5-}$ and $dota^{4-}$ complexes and some derivatives typically lie in the range 1.0–1.3. These findings are further supported by NMR measurements of lanthanide ion complexes.

NMR studies also provide information about the dynamic behaviour of complexes in solution. Studies of $dtpa^{5-}$ complexes suggest that their structures in solution are, on average, similar to those found in the solid state. However, in solution dynamic processes are possible in which carboxylate groups in the bound ligand exchange between coordination sites. The rate of exchange in $\{Eu(dtpa\}^{2-}$ was found to be 360 s^{-1} and Gd^{3+} having a similar ionic radius to Eu^{3+} might be expected to show similar behaviour. The interconversion of structural isomers of $dota^{4-}$ complexes seems to be around ten times slower than for the $dtpa^{5-}$ complexes. The rates of ligand rearrangements in complexes of $hp\text{-}do3a^{3-}$ complexes fall in a similar range of 20–120 s^{-1}. Thus it appears that ligand rearrangements are substantially slower than water exchange in these lanthanide polyamine carboxylate complexes.

3.2.1.9 *In vivo* Behaviour

In addition to having suitable magnetic properties and satisfying the chemical criteria of stability and rapid water exchange, a metal complex must exhibit suitable *in vivo* behaviour if it is to be of use as an MRI contrast agent. The relatively large doses required for MRI applications mean that the viscosity and osmotic strength of the solution must be acceptable for injection. In the case of a Gd^{3+} agent a dose of *ca.* 0.1–0.3 mmol kg^{-1} would constitute 7.5–22.5 mmol or *ca.* 1–4 g of Gd^{3+} for a 75 kg subject. Administration of a 0.5 M solution would thus involve the injection of 15–45 ml of solution. If the osmotic strength of the solution is substantially different from that of the biological fluids in the region of the injection the 'osmotic shock' produced could have unacceptable consequences. The osmotic strength of a solution can be expressed in terms of osmolality;[2] a measure of the total number of osmotically active particles in a solution. The osmolality of blood and body fluids is *ca.* 0.3 mmol kg^{-1} and, ideally, an MRI contrast agent should have a similar osmolality.

In order to control toxicity it is also important that the metal ion is not readily exchanged with other species present *in vivo*. The complex needs to be stable for long enough that, prior to excretion, significant quantities of metal ion are not released to accumulate in the body. There may be competition for the paramagnetic metal ion between the ligand of the contrast agent and potential biological ligands such as proteins and natural metal sequestering agents. There may also be competition for the ligand between the paramagnetic metal ion and other metals such as Cu^{2+} or Zn^{2+} present in the body fluids.

The first clinically approved MRI contrast agents have been extracellular in type. That is they do not enter cells but reside in the extracellular fluids of the body. Such agents are prone to renal excretion offering an effective pathway for removal from the body. In order to extend the utility and range of application of MRI agents, complexes which can enter specific cell types are needed. An early target would be hepatocytes to allow liver imaging with excretion *via* the hepatobiliary system. Later targets might be reticuloendothelial cells associated with lymph nodes, spleen, lung and bone marrow. Excretion pathways might then involve decomposition of the complexes by macrophages with the associated challenge of preventing long-term retention of liberated metal ions as this could have unacceptable toxic effects. A number of complexes have been identified which meet the criteria for medicinal use as contrast agents to a sufficient extent to find application. However, there is continuing interest in new agents with improved properties and a capability to extend the established range of clinical applications.

3.2.1.10 Commercial MRI Contrast Agents

Several extracellular MRI contrast agents based on Gd^{3+} polyaminecarboxylate complexes are approved for clinical use or are in clinical trials and these are included in Table 4.

Magnevist®, ProHance® and Gadovist® show similar *in vivo* behaviour. After the injection they are rapidly and completely distributed from the

[2] Osmolality is given by the sum of the molalities of all the solutes present in that solution. The molality of a solute is a measure of its concentration expressed in moles per kilogram of solvent. The molality and molarity of aqueous solutions are usually almost identical since the density of water is 1 kg l^{-1}. However, in biological fluids there is a slight difference between molality and molarity because of the other components present, for example proteins and lipids. In human serum these make up about 6% of the total volume so that the molality of a substance in serum is about 6% higher than its molarity. The osmolality of human serum is affected mainly by sodium, potassium, chloride and bicarbonate. Urea and glucose may also be present in high enough concentrations to affect the osmolality. Together these solutes may account for *ca.* 95% of serum osmolality.

Table 4 *Soluble MRI contrast agents*

Names[a]	*Manufacturer*[a]	*Formula*	*Proligand/ structure*
AngioMARK® (renamed Vasovist®) Gadophostriamine trisodium	Mallinckrodt/Tyco Healthcare	$Na_3[Gd(MS\text{-}325)(H_2O)]$	**7**
Dotarem® Gadoterate meglumide	Guerbet	$(NMG)[Gd(dota)(H_2O)]$	**5a**
Eovist® (renamed Primovist®) Gadoxetic acid disodium	Schering	$Na_2[Gd(dtpa\text{-}eob)(H_2O)]$	**3c**
Gadovist® Gadobutrol	Schering	$[Gd(do3a\text{-}butrol)(H_2O)]$	**5c**
Magnavist® Gadopentetate dimeglumide	Schering	$(NMG)_2[Gd(dtpa)(H_2O)]$	**3a**
Multihance® Gadobenate dimeglumide	Bracco	$(NMG)_2[Gd(bopta)(H_2O)]$	**3b**
Omniscan® Gadodiamide	GE Healthcare (previously Nycomed-Amersham)	$[Gd(dtpa\text{-}bma)(H_2O)]$	**3d**
OptiMARK® Gadoversetamide	Mallinckrodt/Tyco Healthcare	$[Gd(dtpa\text{-}bmea)(H_2O)]$	**3e**
Prohance® Gadoteridol	Bracco	$[Gd(hp\text{-}do3a)(H_2O)]$	**5b**
Lumenhance®	Bracco	$MnCl_2$	
Teslascan®	GE Healtcare (previously Nycomed-Amersham)	$Na_3[Mn(Hdpdp)]$	**9**

[a] Some product names and manufacturers have changed since the product was first introduced and may do so again so that those cited here may not represent the current product name and manufacturer.

vascular compartment into the extracellular fluid. The plasma half-life is about 15 min and rapid excretion occurs *via* the kidneys with an overall elimination half-life of about 90 min. In each case >90% of the injected dose is recovered in the urine within 24 h. The biodistribution profile and pharmokinetics of Omniscan® are similar to those of Magnavist® but with an elimination half-life of about 70 min.

The properties of four of these complexes are compared in Table 5. All show high stability constants and appear to carry one coordinated water molecule. The metal to coordinated water hydrogen distances are similar and the rates of water exchange, although near the limit, are sufficient for MRI applications. The osmolalities of the neutral complexes are lower than those of the anionic complexes and more compatible with biological fluids, although the LD_{50} values are broadly similar, $[Gd(dtpa)]^{2-}$ being the most toxic to rodents. The half times for dissociation of the non-cyclic ligands $dtpa^{5-}$ and $dtpa\text{-}bma^{3-}$ are substantially shorter than those of the more rigid preorganised ligands with cyclic backbones, $dota^{4-}$ and $hp\text{-}do3a^{3-}$. This is also reflected in the greater extent of reaction with Cu^{2+} and Zn^{2+} ions and the greater long-term whole

Table 5 *Comparison of data for four MRI contrast agents*

Complex[a]	*Gd(hp-do3a)*	*Gd(dtpa-bma)*	$Gd(dtpa)^{2-}$	$Gd(dota)^{-}$
log *K*	23.8	17.1	22.2	25.3
log K^{*b}	17.1	14.9	17.8	18.3
r (Gd--H) (pm)	250	242	249	246
Rotation time[c] τ_R (s^{-1})	57	53	55	63
Water exchange rate (s^{-1})	2.86×10^6	0.45×10^6	3.3×10^6	4.10×10^6
Relaxivity r_1 (mM^{-1} s^{-1})	3.7	3.8	3.8	3.5
Osmolality (37°C)[d]				
0.5 M solution (Osmol kg^{-1})	0.63	0.65	1.96	1.35
1.0 M solution (Osmol kg^{-1})	1.91	1.90	5.85	4.02
Viscosity (37°C)				
0.5 M solution (cP)	1.3	1.4	2.9	2.0
1.0 M solution (cP)	3.9	3.9	>30	11.3
Dissociation[e] $t_{1/2}$ (min)	*ca.* 180	*ca.* 0.5	10	$> 4 \times 10^5$
LD_{50}[f] (mmol kg^{-1})	12	15	6	11
Reaction[g] Cu^{2+}	< 1%	35%	25%	< 1%
Reaction[g] Zn^{2+}	< 1%	25%	21%	< 1%
Whole body Gd at 1 day[h]	2	2	2	2
Whole body Gd at 7 days[h]	0.05%	1%	0.2%	0.1%
Whole body Gd at 14 days[h]	0.03%	1%	0.1%	0.05%

[a] Each complex is thought also to coordinate 1 water molecule not shown in the formula.
[b] Conditional stability constant at pH 7.4.
[c] Correlation time for molecular rotation.
[d] Osmolality represents the sum of the molalities of the osmotically active solutes present.
[e] Approximate half life for dissociation at pH 1.
[f] Values for rodents.
[g] The percentage of free Gd^{3+} released over 10 min at 22°C when the complex is challenged with 25 mM Cu^{2+} or 25 mM Zn^{2+} ions in the presence of 66 mM phosphate at pH 7.
[h] Residual whole body ^{153}Gd (%) in mice after intravenous injection of radiolabelled complex at 0.4 mmol/kg^{-1} body weight.

body retention of Gd^{3+} when the acyclic ligands are used. Except for Eovist® (now called Primovist®) and AngioMARK® (now called Vasovist®) all of these compounds are extracellular agents. Multihance® also has some hepato-biliary application, as does Eovist®, while AngioMARK® was developed for bloodpool imaging.

The first generation of soluble MRI contrast agents has been used primarily for brain imaging in order to diagnose tumors or other pathologies which interfere with the normal ability of the brain to exclude potentially damaging dissolved species present in the blood. Paramagnetic complexes dispersed in extracellular fluids can be effective in this application but have limited utility in other contexts where diffusion from blood into extracellular space in tissue is not suppressed. Potential applications such as angiography and imaging the hepatobiliary system require agents which show selective cellular uptake. Suitable compounds for this are being developed. Such tissue specific contrast agents present an important research topic for extending the range of applications of MRI contrast agents.

3.2.1.11 Brain MRI Contrast Agents and the Blood Brain Barrier

The use of extracellular MRI contrast agents in brain imaging exploits the specialised ability of the capillaries supplying blood to the brain to retain otherwise freely diffusing species in the blood and exclude them from the neural tissue. This 'blood brain barrier (BBB)' prevents molecular species, such as metal complexes, which have no specific membrane transport mechanism, from entering the highly regulated tissues in the brain. In order to obtain an adequate blood supply brain tumors need to stimulate the growth of new capillaries from the existing network. Fortunately the structure of these new capillaries does not give rise to an efficient BBB and so allows species such as MRI contrast agents into the tumor tissue providing contrast compared to normal brain tissue.

Often infections and other diseases of the brain may also disrupt the BBB, though in less subtle ways than tumors, allowing MRI imaging of diseased regions through the use of soluble contrast agents. A result of these effects, brain imaging dominated earlier applications of soluble MRI contrast agents such as Omniscan®, Magnavist®, ProHance® and OptiMARK®. Soluble extracellular Gd^{3+} contrast agents might be used as diagnostic aids for a variety of brain disorders. These include primary brain tumors, metastatic brain tumors, subdural hematoma or blood clots, aneurysms and abnormalities of blood flow, infarction or other tissue destruction, brain abscesses as well as structural abnormalities of the brain. An example of a brain scan obtained using ProHance® to enhance the image of a brain tumor is shown in Figure 10.

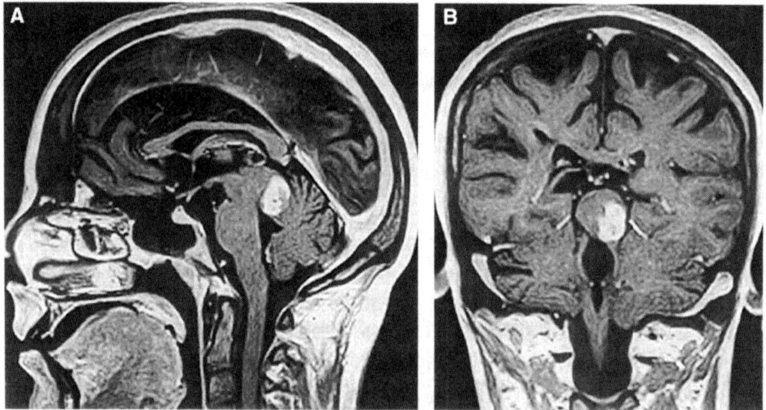

Figure 10 *A T_1 weighted image obtained after intravenous injection of ProHance® (0.1 mmol kg^{-1}) showing a Tectal glioma as the white region just to the right of centre in the Saggital image A and the coronal image B*
(Reproduced from M.F. Tweedle and K. Kumar, Magnetic Resonance Imaging (MRI) Contrast Agents. Figure 2 page 6 in Metallopharmaceuticals II: Diagnosis and Therapy, M.J. Clarke, P.J. Sadler (eds), Springer, Berlin, 1999 with permission from Drs. Michael F. Tweedle, Krishan Kumar and Dr. Val Runge, from whose lab the image originated, and from Springer Science and Business Media).

3.2.1.12 Bloodpool MRI Contrast Agents

Bloodpool imaging has two important applications, angiography which is used to study larger blood vessels and perfusion imaging which is used to study capillary bloodflow. Angiography investigations require image data to be collected over a period of several minutes during which time it is important that a contrast agent remains in the blood stream and does not percolate into extracellular spaces. It is also important that the contrast agent is rapidly and completely excreted after the imaging procedure has been carried out. Soluble metal complexes are typically freely diffusing species small enough in size to enter extracellular spaces quite readily and so these are generally unsuited for bloodpool imaging applications. However, large macromolecular or particulate species such as serum albumin, liposomes or dextrans tend to be excluded from extracellular spaces. If paramagnetic derivatives of these species could be prepared they might offer one possible approach to bloodpool imaging by providing contrast agents for MRI. Thus attaching a binding site for a paramagnetic metal ion such as Gd^{3+} to a species such as albumin could potentially provide a contrast agent suitable for bloodpool imaging. There is a particular advantage with this 'labelled particle' approach in that the slower molecular rotation rate of the particle can lead to enhanced relaxivity, provided the metal binding site has limited freedom to move independently of the particle.

The approaches which have been used to develop bloodpool agents based on Gd^{3+} linked to particles involve the exploitation of both covalent and non-covalent interactions between components of metal complex and macromolecules. A widely used approach to covalently linking a lanthanide ion binding site to a macromolecule involves the reaction of amine groups in the macromolecule with dtpa-dianhydride, (**6**, Scheme 2), in buffered aqueous media. This leads to the formation of amide links between the macromolecule and dtpa-dianhydride as shown in Scheme 2. In the case of Human Serum Albumin (HSA) up to about 30 dtpa amide groups can be attached in this way and subsequent treatment with Gd^{3+} ions affords the paramagnetic contrast agent. A chemically more complicated approach has been used to produce Gd^{3+} bound to dextran. Chloroacetic acid was used to introduce a carboxylate group which was then linked to a diaminoethane group leaving a free amine terminus which could be linked to $dota^{4-}$ which could then bind Gd^{3+} as shown in Scheme 3. One disadvantage of these examples is that the formation of the amide link modifies one of the carboxylate groups of the dtpa ligand potentially reducing the strength of its binding to Gd^{3+}. This could be avoided by attaching the linking group through the ligand backbone rather than one of the carboxylate groups so as not to reduce the number of metal binding sites.

Use of a covalent link between the particle and the Gd^{3+} binding site presents a problem in that the fate, *in vivo*, of the Gd^{3+} injected becomes coupled to that of the macromolecule. This could lead to the accumulation of Gd^{3+} in undesirable locations. An alternative approach involving non-covalent linking of the Gd^{3+} binding site to the particulate species offers a means of avoiding this difficulty. A good example is provided by the proligand MS-325, ($ms325H_6$, **7**,

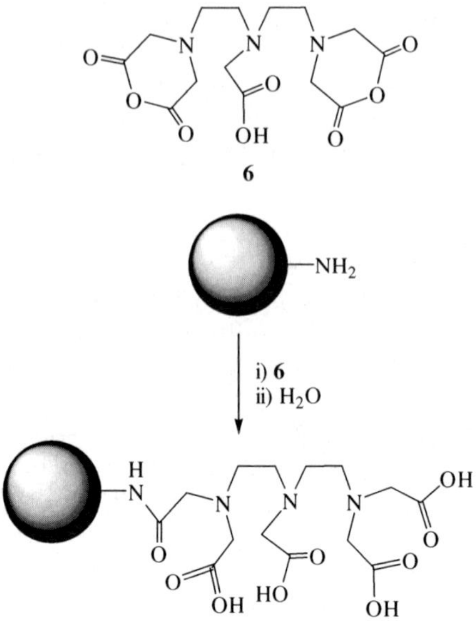

6

i) 6
ii) H_2O

Scheme 2

Figure 11), a derivative of $dtpaH_5$ in which the backbone has been modified to include a lipophilic group. Interactions of the lipophilic group with hydrophobic regions of HSA provide a non-covalent link between HSA and the complex to form a Gd-MS-325-HSA conjugate species (Figure 11). However, the Gd-MS-325 interaction with HSA is weaker than for a covalently bound Gd-complex and an equilibrium exists leading to a distribution of the Gd-MS-325 complex between the HSA bound and free states. The free Gd-MS-325 is subject to renal excretion so the Gd-MS-325 is gradually excreted and the proportion retained in the bloodpool due to binding to HSA decreases over time.

In this system the effect of molecular rotation rate on relaxivity can be used to good effect. The rotation rate of free Gd-MS-325 is *ca.* 10^{10} s^{-1} leading to a relaxivity of 6.6 mM^{-1} s^{-1}. In contrast the rotation rate of the Gd-MS-325-HSA macromolecule is *ca.* 10^{8} s^{-1} leading to a relaxivity of between 30 and 50 mM^{-1} s^{-1}. As a result the free Gd-MS-325 makes a far smaller contribution to image contrast than the HSA bound Gd-MS-325 so that signal enhancement is concentrated in regions with the highest albumin content thus favouring the bloodpool over extracellular spaces in terms of contrast enhancement. This approach has been called Receptor Induced Magnetisation Enhancement (RIME) and is unique to MRI applications. The method has advantages over those using covalently linked Gd^{3+} binding sites in that chemical modification of the macromolecule is not required and a renal excretion pathway is available for the Gd-complex. Phase III clinical trials have been carried out and marketing approval for the agent under the trade name Vasovist® (formerly AngioMARK®) was expected in 2005. However, this was delayed by an FDA request for further data.

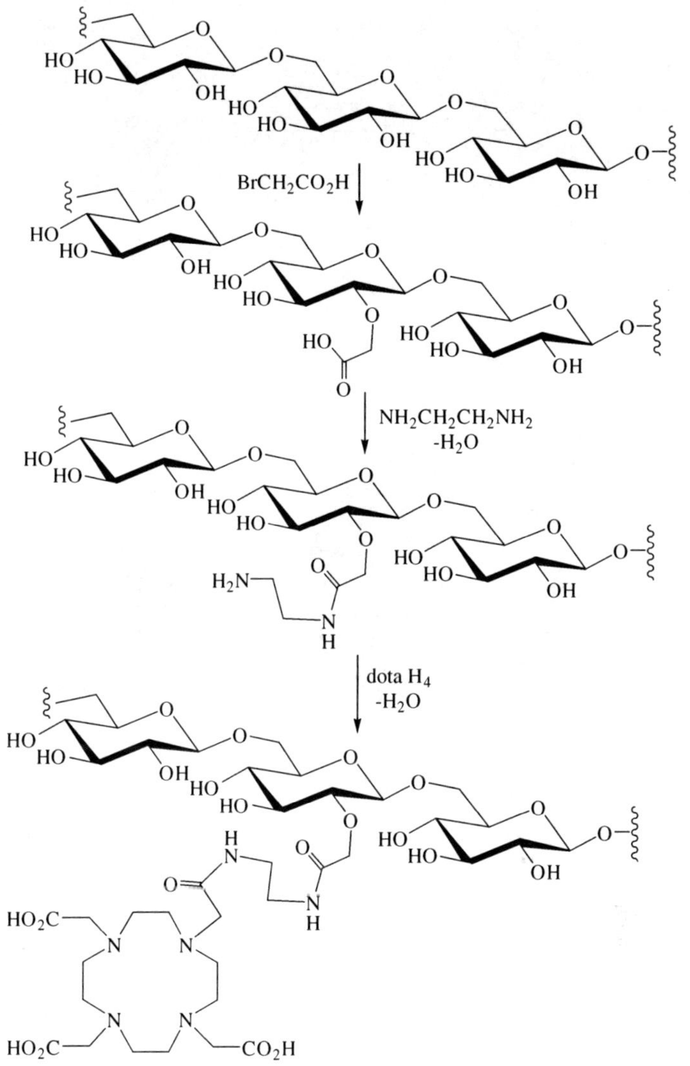

$BrCH_2CO_2H$

$NH_2CH_2CH_2NH_2$
$-H_2O$

dota H_4
$-H_2O$

Scheme 3

The use of imaging to estimate the amount of blood perfusing tissues requires that in one pass the perfusion agent is almost completely absorbed by the tissue or almost entirely retained in the bloodstream. Typically this type of investigation is carried out using radiopharmaceutical agents but MRI offers the prospect of providing better spacial resolution. However, the use of MRI contrast agents in this context is challenging and requires that the signal intensity varies in proportion to the amount of blood perfusing the tissues. The rapid bolus injection of a contrast agent coupled with fast imaging offers a possible approach to perfusion imaging. Analogues of extracellular contrast

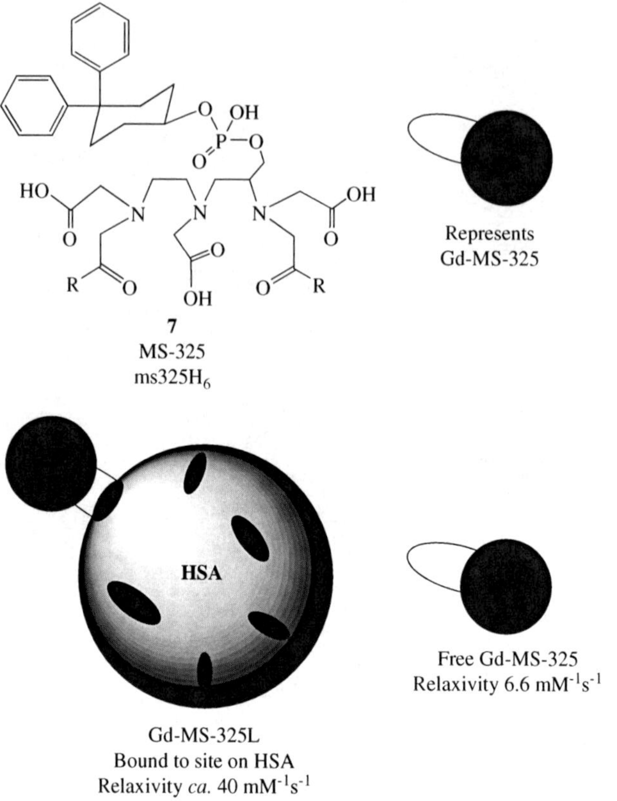

Figure 11 *A schematic showing the non-covalent binding of Gd-MS-325- to HSA*

agents in which Gd^{3+} has been replaced by Dy^{3+} have been investigated on the presumption that they will act as perfusion agents, which lead to signal loss. These agents rapidly diffuse into extracellular spaces and are around one third extracted so are far from ideal. However, the use of water soluble non-ionic complexes of this type at higher (*ca.* 1 M) concentrations might provide a means of perfusion imaging.

3.2.1.13 Gastrointestinal MRI Contrast Agents

Orally administered contrast agents for GI investigations must withstand conditions in the GI tract and, in particular, the acidic conditions (pH < 2) in the stomach. Superparamagnetic iron particles have found use in this application as has ferric ammonium citrate (under the tradename Ferriseltz®). Orally administered $MnCl_2$ produces a rapid decrease in the T_1 relaxation time causing a bright signal on T_1 weighted images and also shows negative enhancement on T_2 weighted images. Phase III clinical trials have been conducted on an $MnCl_2$ preparation developed under the tradename Lumenhance®. There is also interest in Gd^{3+} complexes for GI imaging and

Magnevist® enteral is a $\{Gd\text{-}dtpa\}^{2-}$ preparation formulated specifically for use in GI imaging. More recently $\{Gd\text{-}dota\}^{-}$ and {Gd-(hp-do3a)} have aroused interest as possible oral GI contrast agents due to their slower acid dissociation kinetics compared to $\{Gd\text{-}dtpa\}^{2-}$.

3.2.1.14 Hepatobiliary MRI Contrast Agents

MRI imaging of the liver might use suitable contrast agents to provide high-resolution images of the gall bladder and bile duct, to selectively enhance signals from normally functioning liver tissue so revealing focal liver disease, or to identify poor liver function resulting from diffuse disease *e.g.* cirrhosis. Water soluble hydrophilic complexes such as $\{Gd(dtpa)\}^{2-}$ are excreted though the renal pathway and, in order to promote uptake in hepatocytes and excretion into the intestine *via* the hepatobiliary system, complexes with some lipophilic character are usually required. However, if the complex is too lipophilic in character it may be retained in the reticuloendothelial cells of the liver and spleen leading to unacceptable toxicity. Thus in a hepatobiliary contrast agent a balance needs to be struck between promoting hepatocyte uptake and suppressing sufficiently rapid excretion.

The first Gd^{3+} complex to obtain approval in Europe for use in liver imaging was $[Gd(bopta)(H_2O)]^{2-}$ (boptaH_5 = **3b**) marketed as its dimeglumine salt Multihance®. FDA approval for marketing in the USA was granted in 2004. The value of log$K$ for $\{Gd(bopta)\}^{2-}$ has been quoted as 22.6 and the LD_{50} in mice is similar to that of $\{Gd(dtpa)\}^{2-}$. $[Gd(bopta)(H_2O)]^{2-}$ exploits the ability of hepatocytes to absorb lipophilic anions and is both absorbed in hepatocytes and renally excreted. Initially the complex is distributed in the blood and diffuses into extracellular spaces from which it ultimately clears as a result of renal excretion. A proportion of the agent is absorbed by hepatocytes and is excreted more slowly *via* the hepatobiliary system. Metastases in the liver lack the hepatocyte capability for taking up the contrast agent. Initially the signal from both the normal liver tissue and the metastases is enhanced through the presence of the agent in the extracellular spaces. However, subsequent renal clearance of the extracellular agent leaves an excess in the hepatocytes compared to the metastases so that any lesion appears darker in the MRI image.

The related complex $[Gd(dtpa\text{-}eob)(H_2O)]^{2-}$ (dtpa-eobH^5 = **3c**) shows similar physical properties to $[Gd(bopta)(H_2O)]^{2-}$ and similar signal enhancement in liver tissue a few minutes post injection. This contrast agent was granted EU approval in 2004 under the name Primovist®. In rats the signal enhancement produced by $[Gd(dtpa\text{-}eob)(H_2O)]^{2-}$ declines after 20 min and is substantially lost after 1 h while significant enhancement persists for at least 2 h with $[Gd(bopta)(H_2O)]^{2-}$. These observations show how quite small changes in ligand structure can have a substantial effect on the suitability of a complex for use in MRI applications. This suggests that there are good opportunities for contrast agent optimisation through careful ligand design and this is stimulating continuing research. One example of a new experimental ligand design is found in dota-tppH_4 (**8**). The Gd^{3+} complex of this proligand gave good biliary images in rats even at doses as low as 5×10^{-6} mol kg^{-1}. Tumors could be visualised up to 8 h post injection and the complex was 99.99% cleared after 6 days.

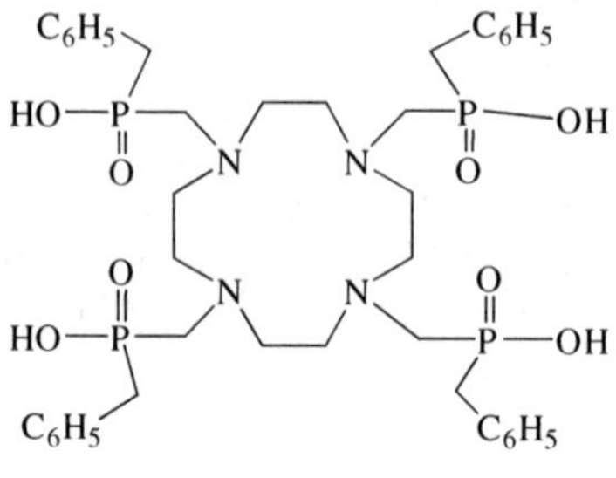

8
dota-tppH_4

The Mn^{2+} complex $[Mn(Hdpdp)]^{3-}$ (**9**, marketed as its trisodium salt Teslascan®) also shows liver uptake but, having no coordinated water has lower relaxivities than $[Gd(bopta)(H_2O)]^{2-}$ (respective values at 40°C and 20 MHz being 2.8 and 4.4 mM^{-1} s^{-1} for T_1 and 3.7 and 5.6 mM^{-1} s^{-1} for T_2). $[Mn(Hdpdp)]^{3-}$ is also less stable than $[Gd(bopta)(H_2O)]^{2-}$ having $\log K = 15.1$ and, at physiological pH, $\log K' = 9.4$. Nonetheless, the dpdp^{5-} ligand appears to suppress acute Mn^{2+} toxicity at low doses and the reported LD_{50} value of the complex appears to be less than that of $[Gd(bopta)(H_2O)]^{2-}$. Ascorbic acid is also used in the formulation of the contrast agent and may be present to suppress oxidation of the Mn^{2+}. The liver uptake mechanism appears to involve both the complex itself and, through natural sequestering, the free Mn^{2+} liberated through dissociation.

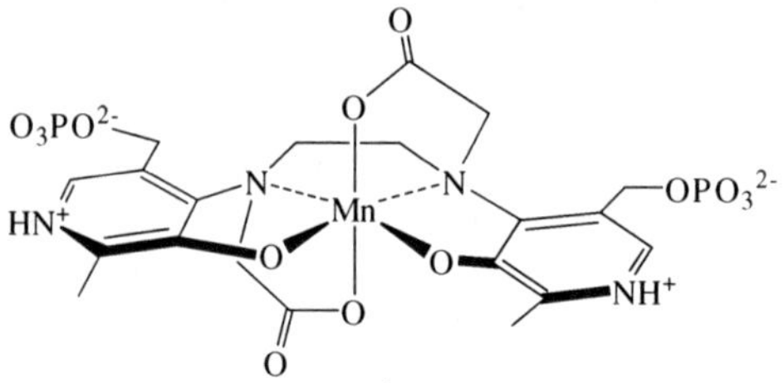

9
$Mn(dpdp)^{4-}$

3.2.2 X-ray Contrast Agents

3.2.2.1 Historical Development

In the past a variety of 'heavy' metals, or their oxides, have appeared in the many different preparations tested for use as X-ray contrast agents. Applications included bronchography and angiography, as well as imaging of the liver, urethra and bladder. An example is provided by the once commonly used commercial agent, Thorotrast. This contained 20 weight % colloidal ThO_2 with dextrin and a preservative. Thorium agents produced little acute toxicity or discomfort and between 1928 and 1950 several million people were exposed to thorium during diagnostic procedures. However, the high retention and associated long-term radiotoxicity of thorium contrast agents make them quite unacceptable for modern day use. Preparations containing other 'heavy metals' such as silver, tin, bismuth, tantalum and tungsten have been investigated but have not found significant clinical use, although they may have some applications, for example, in improving the opacity to X-rays of dental implants or catheters. One notable exception to the use of toxic metals as X-ray contrast agents is provided by barium sulfate. Soluble barium compounds are extremely toxic and quite unsuitable for clinical use. However, the sulfate of barium, $BaSO_4$, is highly insoluble and so can be used in preparations such as a 'barium meal' without presenting a significant toxic hazard. A dispersion of $BaSO_4$ in the GI tract shows up in an X-ray image as an opaque region which can reveal abnormalities in structure through, for example, coating the stomach wall and revealing ulcerated areas.

More recently there has been some interest in the potential use of coordination compounds as X-ray contrast agents. These offer the prospect that the properties of the complex administered to the patient might be optimised by a suitable choice of ligand extending the range of applications for X-ray tomography. Soluble complexes might be quite readily excreted preventing heavy metal retention and any associated chronic toxicity effects. Two broad classes of compound have attracted attention. The first consists of coordination complexes of lanthanides with polyaminecarboxylate ligands and has developed from the use of such compounds as MRI contrast agents. The second consists of polymetallic compounds containing two or more atoms of a heavy metal such as tungsten. These have sometimes been described as cluster compounds, although in strict chemical terminology they are not metal clusters.

3.2.2.2 Lanthanide Complexes

The use of $\{Gd(dtpa)\}^{2-}$ as an MRI contrast agent (Magnevist®, Section 3.2.1.10) followed by X-ray investigations led to the discovery of its potential use as an X-ray contrast agent. Animal studies of both $\{Gd(dtpa)\}^{2-}$ and $\{Yb(dtpa)\}^{2-}$ have shown them to give superior contrast to iodinated agents when higher X-ray tube voltages (*ca.* 125 kVp), and hence higher X-ray photon energies, are used. Higher energy X-rays give lower patient radiation doses than lower energy X-rays, which are scattered and absorbed more efficiently. However, in conventional radiography higher energy X-rays show much poorer contrast between bone, muscle and fatty tissue and so are of limited use. A different situation arises when the visualisation of a contrast agent location is required. In this case it is the contrast between the contrast agent and surrounding tissue which is important and a different optimum X-ray photon energy will apply. Thus in applications where patients show allergic reactions to the iodine-based agents, or it is important to minimise patient radiation doses, lanthanide-based contrast agents may be advantageous.

Early investigations of the effect of Magnevist on X-ray images from humans used the standard MRI dose of 0.1 mmol kg^{-1} and gave unpredictable results. Subsequently satisfactory vascular enhancement was obtained in a cranial CT scan using a dose of 0.5 mmol kg^{-1}. Time-delayed images of kidney and bladder using $(NMG)_2$\{Gd(dtpa)\} (NMG^+ = *N*-methylglucammonium) also showed good contrast, although less marked than with iodinated agents. Gadolinium polyamine carboxylate complexes have also been used in Digital Subtraction Angiography (DSA) with some success. However, there has been some debate about the relative toxicities of the gadolinium and iodinated contrast agents.

The neutral complex \{Gd(dtpa-bma)\} (Omniscan®) shows broadly similar *in vivo* behaviour to $(NMG)_2$\{Gd(dtpa)\} as do \{Gd(hp-do3a)\} (ProHance®) and \{Gd(do3a-butrol)\} (Gadovist®). Animal studies suggest that ProHance® and Gadovist® have reduced acute toxicity compared to Magnevist® and both were tolerated well in healthy volunteers, even up to a dose of 0.5 mmol kg^{-1} in the case of Gadovist®. The doses of gadolinium required for X-ray contrast effects are higher than those typically used in MRI and, although the complexes appear to be well tolerated when administered intra-arterially in doses of less than 0.3–0.4 mmol kg^{-1}, more data are needed on patients with renal insufficiency. According to some researchers the incidence of contrast medium–induced nephropathy is decreased with gadolinium compared with iodinated contrast agents. However, others have expressed concern that gadolinium-based agents are hypertonic and this can be a pathogenic factor in nephropathy following renal angiography. At X-ray tube peak voltages of 70 kVp equally attenuating doses of iodinated agents were thought to be less toxic than the gadolinium agents. Thus while it appears feasible to use the gadolinium or ytterbium polyamine carboxylates as X-ray contrast agents the issues of toxicity compared to iodinated agents need to be fully resolved. The gadolinium agents may offer a suitable alternative in cases where iodinated agents produce adverse reactions, particularly if higher X-ray tube voltages can be used. The use of Pb^{2+} or Bi^{3+} polyaminecarboxylate complexes as X-ray contrast agents appears to be limited by their toxicity since, at useful doses, $\{Pb(edta)\}^{2-}$, $\{Pb(dtpa)\}^{3-}$ and $\{Bi(dtpa)\}^{2-}$ were found to be too toxic for use in humans.

3.2.2.3 Polymetallic Complexes

Two types of polymetallic tungsten complex have been investigated as potential X-ray contrast agents. The first involves complexes of polyaminecarboxylate ligands with polytungstate groups, a simple example of which is provided by $[W_2O_2S_2(edta)]^{2-}$ (**10**). Other examples include $[W_3S_4(ttha)]^{2-}$ ($tthaH_6$ = **11**) and $[(W_3S_4)_2(egta)_3]^{2-}$ ($egtaH_4$ = **12**). The second type involves the well-known heteropolytungstate ions exemplified by $[PW_{12}O_{40}]^{3-}$ and $[P_2W_{18}O_{62}]^{6-}$. CT phantom studies showed a factor of 2 higher X-ray attenuation for a tritungsten compound as compared to a tri-iodinated compound when using a peak X-ray tube voltage of 125 kVp. Animal studies (rat) with $Na_2[W_3S_4(ttha)]$ were in accord with the CT phantom study and good angiograms of rat paw and forearm vasculature were obtained using $Na_6[PW_{12}O_{40}]$. However, the toxicity of the polytungsten compounds is higher than the iodinated compounds and may limit the development of these types of tungsten complex for clinical use. The results obtained so far nonetheless suggest that it may be possible to develop polymetallic tungsten complexes for use as X-ray contrast agents provided that suitable toxicity, biodistribution and pharmokinetic properties can be attained.

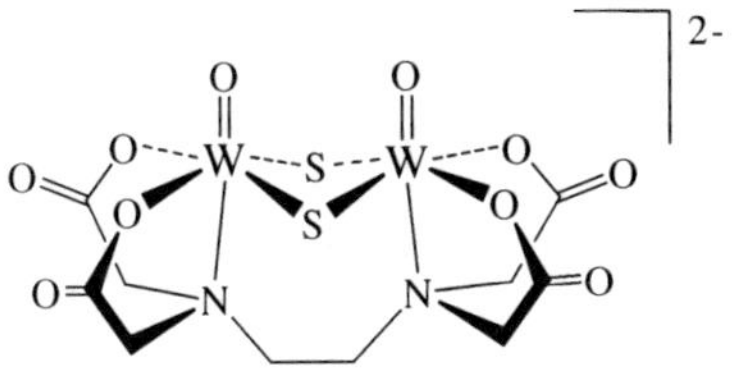

10
$[W_2O_2S_2(edta)]^{2-}$

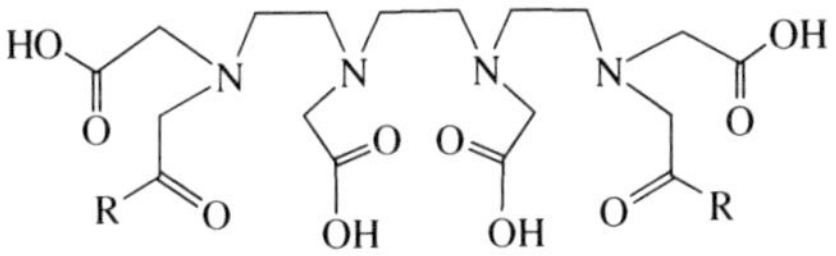

11
$tthaH_6$

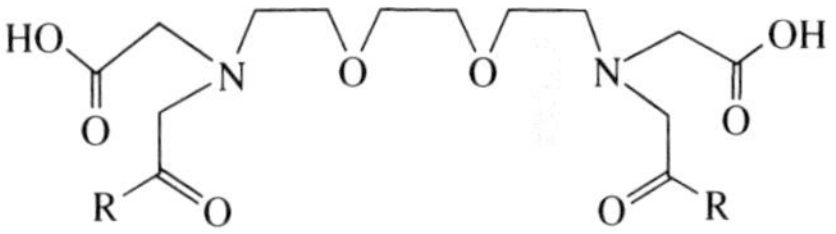

12
$egtaH_4$

3.3 Functional Imaging

3.3.1 Radiopharmaceuticals for Functional Imaging

Non-invasive imaging is a most important area of diagnostic medicine in which coordination chemistry plays an important part. More than 10 million nuclear medicine procedures are carried out each year in the USA alone and, of these, at least 85% involve imaging procedures using the radioactive nuclide ^{99m}Tc. The nuclear properties and chemistry of this man made element make it almost ideal for the purpose. Furthermore the diverse chemistry of technetium allows a wide variety of compound types to be prepared offering an extensive array of applications. These features have resulted in ^{99m}Tc becoming the radionuclide of choice for most diagnostic imaging radiopharmaceuticals and it is replacing several other radionuclides previously used for this purpose. The crucial feature of most ^{99m}Tc-radiopharmaceuticals used in diagnostic imaging today is that they reveal structure through function. That is the *in vivo* biological processing of the compound administered to the patient determines its biodistrbution and hence the appearance of the image produced.

Two main types of ^{99m}Tc-radiopharmaceutical can be distinguished, technetium essential agents and bifunctional agents. The technetium essential agents are complexes formed from a proligand with no specific biological distribution and ^{99m}Tc. The biodistribution of the radiopharmaceutical is determined by the nature of the complex formed and requires the presence of both ^{99m}Tc and its ligand set. Because ^{99m}Tc can be incorporated in a variety of different coordination compound structures, each with its own particular biodistribution, many different technetium essential agents can be formulated for different diagnostic purposes. The bifunctional agents consist of a conjugate molecule containing a biologically active carrier part and a metal ion binding part. The carrier part of the molecule is selectively taken up at specific receptor sites and so controls the biodistribution of the radiopharmaceutical and it is important that the incorporation of the radionuclide into the metal binding site does not adversely affect the function of the carrier part.

Apart from ^{99m}Tc several other metallic radionuclides have been, and still are, used for diagnostic imaging; in particular these include ^{201}Tl, ^{67}Ga, ^{68}Ga, and ^{111}In. However, none offers the highly varied chemical behaviour of technetium and this, together with less ideal nuclear properties, limits their utility to some degree. The utilisation of radionuclides in diagnostic or therapeutic medicine falls within the specialised discipline of nuclear medicine and the radioactive drugs involved are known as radiopharmaceuticals.

3.3.2 Nuclear Medicine

3.3.2.1 Ionising Radiation

Nuclear medicine exploits the special properties of radioactive elements, *i.e.* radionuclides, for the diagnosis or treatment of disease. Radionuclides emit ionising radiation which has sufficient energy to cause chemical changes in the material though which it passes. X-rays represent one well-known type of ionising radiation but radionuclides emit one or more of four other types of ionising radiation designated α, β^-, γ and β^+, each of which can have a specific medical application.

The first type involves α-particles which contain two protons and two neutrons so they are, in effect, helium nuclei. They are energetic, dipositive ions typically having energies in the range of about 4–8 MeV (1 MeV $\approx 10^8$ kJ mol^{-1}). Their high charge means that they are strongly absorbed in tissue, usually within 0.1 mm, and the deposition of this much energy in such a short distance will cause the death of any cell through which an α-particle passes. As result, α-emitters are highly radiotoxic and unsuitable for diagnostic use, although it is possible that they may find use in radiotherapy. An alternative approach to the use of α-particles in therapy, known as Boron Neutron Capture Therapy (BNCT), is becoming important and involves the creation of the α-particle from a non-radioactive element *in vivo*. This is described briefly in Section 4.7.2.

The second type involves β^-–particles which are high energy electrons (mostly in the range 0.5–5 MeV) emitted from nuclei. Being lighter than α-particles, and having half the charge, β^-–particles have a longer range in tissue, up to 10^{-2} m. This is still insufficient for imaging studies using external radiation detectors but is suitable for some therapeutic applications where the energy deposited can cause cell death.

The third type of radiation in the list, γ, is a high-energy form of electromagnetic radiation. Typical γ -ray energies range from a few tens of keV up to several MeV, compared to around 50 keV for the typical X-ray energy used in anatomical imaging. Since γ -rays interact with tissue rather weakly they can be quite penetrating with ranges of up to several metres. This makes internally produced γ -rays of suitable energy for detection by external devices particularly well suited for imaging purposes. Radionuclides, such as ^{99m}Tc, which emit γ–rays as a result of internal nuclear transitions, are of particular use for diagnostic imaging since the γ–rays are not associated with other radioactive decay processes, which would deliver an unwanted radiation dose to the subject. The emitted γ–rays are detected by a γ–camera consisting of a multihole lead collimator, a scintillating medium and an array of photomultipliers. The data from the detection system is fed to a computer system where it may be used to create an image of the distribution of the γ–emitter within a region of the patient's body (Figure 12a). This imaging method is known as Single Photon Emission Computed Tomography (SPECT). A combined SPECT-CT

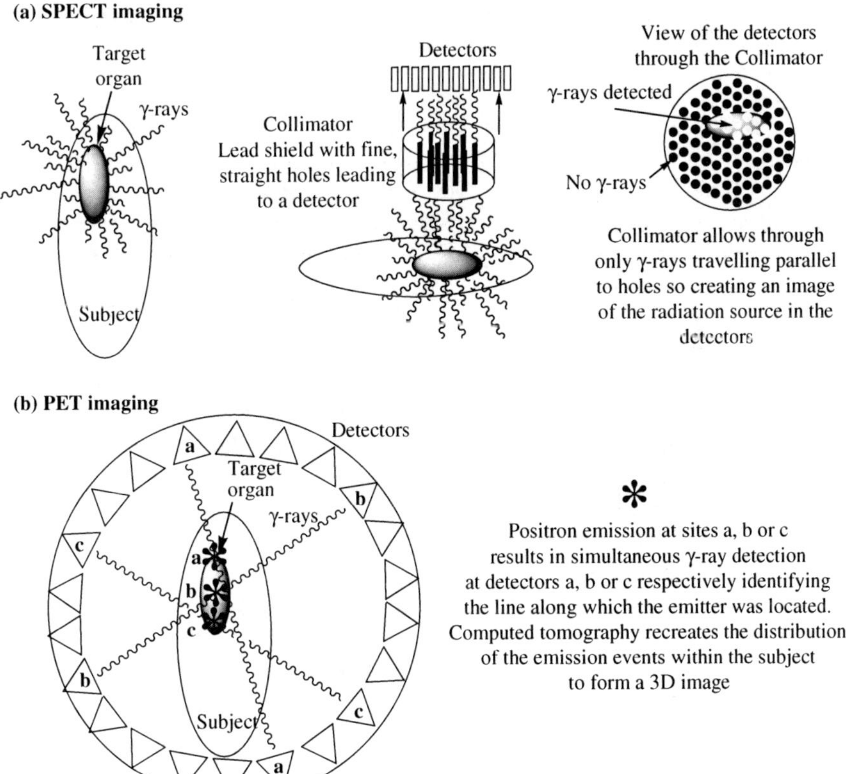

Figure 12 *A schematic representation of the means of creating an image using (a) single photon emission computed tomography (SPECT) and (b) positron emission tomography (PET)*

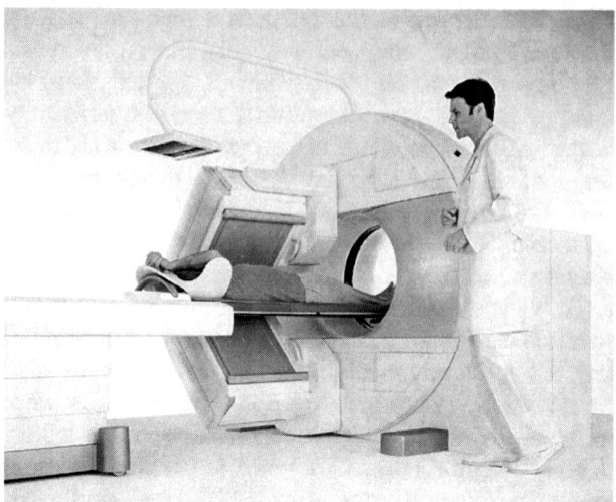

Figure 13 *A scanner used for combined SPECT and CT imaging. The two rectangular units above and below the subject are γ–cameras used for SPECT imaging and the circular structure at the subject's feet is the CT scanner (photo courtesy of Siemens Medical Solutions)*

scanner system is shown in Figure 13. This allows anatomical images from the X-ray CT scanner section to be combined with functional images from the SPECT scanner offering a very powerful diagnostic imaging tool.

The fourth type of radiation, β^+ (positron or positively charged electron), is of diagnostic interest since positron–electron annihilation processes occur close to the β^+ source and produce two 511 keV γ–rays oriented at 180° to one another. Simultaneous detection of these two γ–rays defines a line through the subject in which the decayed nucleus lay (Figure 12b). The observation of multiple decay events allows a detailed image of the distribution of the β^+ emitter to be computed by the technique of Positron Emission Tomography, (PET). Historically radionuclides such as ^{18}F have been incorporated into radiopharmaceuticals for this purpose. However, the short half-life (110 min) of ^{18}F and the chemical procedures required for its incorporation into the radiopharmaceutical are limiting factors on its utility. The metallic β^+ emitter ^{68}Ga is also finding use in diagnostic procedures and, despite a short half-life of 68 min, complex formation offers a rapid and convenient means of incorporating $^{68}Ga^{3+}$ into a radiopharmaceutical formulation. It is in the use of metallic radionuclides such as ^{99m}Tc and ^{68}Ga that coordination chemistry is finding its greatest level of application in providing new radiopharmaceutical agents for non-invasive diagnostic imaging based on γ–ray emission.

3.3.2.2 Radionuclide Production

In order for a radionuclide to be useful in the context of nuclear medicine it must be readily available in a sufficiently pure form, have a suitable half-life, emit radiation of usable energy and have a chemistry which allows the facile synthesis of the active radiopharmaceutical under clinical conditions. The availability of a particular radionuclide is an important factor in determining its general clinical utility. After a radionuclide is produced it must be transported from the production site to the nuclear medicine facility for use. This can be problematic if the radionuclide has a short half-life and must be transported a considerable distance. Fortunately some radionuclides (*e.g.* ^{99m}Tc and ^{68}Ga) can be conveniently obtained from the radioactive decay of a longer lived and more conveniently transported precursor. This makes it possible to produce the medically useful radionuclide at the nuclear medicine facility itself through the decay of the precursor in what is known as a radionuclide generator. These are convenient to use and readily transported to clinical laboratories remote from the precursor radionuclide production site. This can make an important contribution to the acceptance of a particular radionuclide for clinical use. There are two main sources of radionuclides, nuclear reactors and particle accelerators (cyclotrons).

In a nuclear reactor the neutron flux may be used to irradiate a target in which neutron capture converts part of the target material to the required radionuclide, or a precursor which decays to it. Chemical treatment can be used to obtain a solution containing the required radionuclide or its precursor. As an example irradiation of molybdenum oxide isotopically enriched in ^{98}Mo converts some of the ^{98}Mo to ^{99}Mo. If the irradiated target material is dissolved in alkali, a solution of $(NH_4)_2[MoO_4]$ is produced containing a mixture of ^{98}Mo and ^{99}Mo. A major disadvantage of this neutron capture approach to ^{99}Mo production is that the active $[^{99}MoO_4]^{2-}$ is diluted by inactive $[^{98}MoO_4]^{2-}$ so that it is said to be 'not carrier free'. This means that the desired ^{99}Mo only constitutes a part of the Mo present so that larger amounts of material are required to obtain the effect of a given amount of ^{99}Mo. Fortunately, in this case the problem can be avoided by using nuclear fission in the reactor, rather than neutron capture, to produce ^{99}Mo. The fission of ^{235}U produces a mixture of fission products including ^{99}Mo. Since the other fission products are chemically different from Mo, a chemical separation can be carried out using ion exchange chromatography to obtain 'carrier free' $[^{99}MoO_4]^{2-}$ which is free of other Mo isotopes. This can be adsorbed onto an activated alumina column where the ^{99}Mo decays, with a half-life of 66 h, to ^{99m}Tc which in turn decays with a 6 h half-life to ^{99}Tc. Elution of the column with saline solution displaces the $[TcO_4]^-$ daughters by exchange with chloride, but leaves the more highly charged, undecayed $[^{99}MoO_4]^{2-}$ absorbed on the column. This system provides the basis for a '^{99m}Tc generator' from which a solution containing $[^{99m}TcO_4]^-$ can be eluted as long as sufficient undecayed $[^{99}MoO_4]^{2-}$ remains (Figure 14).

Although ^{99m}Tc is being continually produced from ^{99}Mo, once formed it is continually decaying to ^{99}Tc, thus the eluted $[^{99m}TcO_4]^-$ is not completely carrier free. Depending on the time since the column was last eluted, some ^{99}Tc will also be present. Consequently when the radiopharmaceutical is prepared a part will contain the ^{99}Tc label, which is no use for imaging. Fortunately this

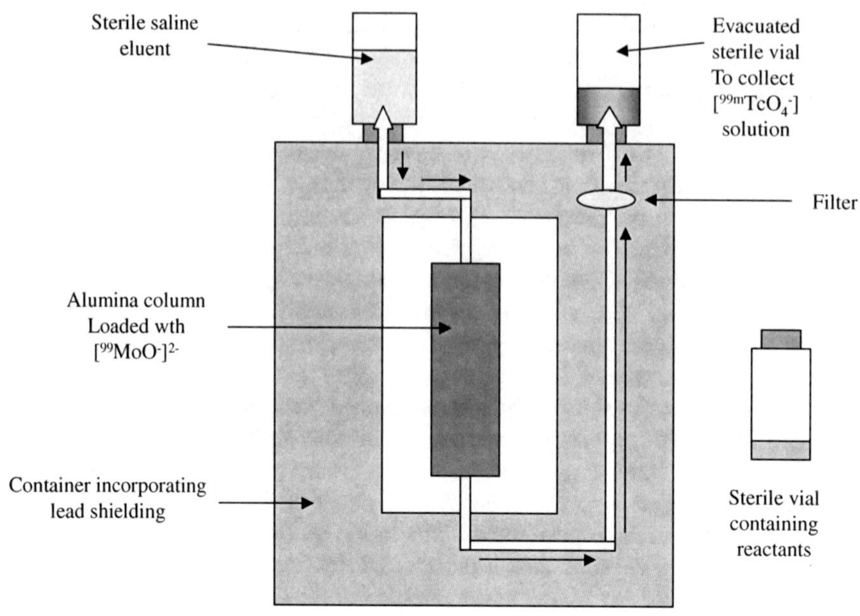

Figure 14 *A schematic representation of a ^{99m}Tc generator producing a saline solution of $[^{99m}TcO_4]^-$*

problem is easily managed and the optimum time between elutions is about 24 h to obtain the maximum ^{99m}Tc activity with minimal ^{99}Tc formation. This can be particularly important when bifunctional agents are involved since the ^{99}Tc-labelled agent will compete with the ^{99m}Tc-labelled agent for receptor sites and reduce the efficacy of the procedure. Once the ^{99}Mo is largely decayed the generator can be returned to the production site and recharged with $[^{99}MoO_4]^{2-}$ for reuse and recycle. Typically a nuclear medicine facility might obtain a new ^{99m}Tc generator on a weekly basis.

The other means of preparing radionuclides involves irradiating a suitable target with ions accelerated in a cyclotron. As an example proton irradiation of ^{69}Ga produces ^{68}Ge in a (p, 2n) reaction whereby a proton is absorbed by the ^{69}Ga nucleus and two neutrons are subsequently lost forming ^{68}Ge. The target material can be dissolved in HCl to give a solution containing $^{68}Ge^{2+}$ which can be absorbed on an alumina column which has been pre-treated with ethylenediaminetetra acetic acid (edtaH_4, **4**). The ^{68}Ge decays with a 280 day half-life to the positron emitter ^{68}Ga. This ^{68}Ga daughter may then be eluted from the system providing the basis of a $^{68}Ga^{3+}$ generator. Cyclotron production of radionuclides is expensive compared with reactor irradiations, but higher specific activities are possible than with the neutron capture process. Some radionuclides, which cannot be obtained from a reactor, may be prepared by cyclotron irradiation. Examples of the more important radionuclides used in diagnostic imaging are presented in Table 6.

Table 6 *Some radionuclides used in diagnostic nuclear medicine*

Radionuclide	$T_{1/2}$[a]	γ *energy (keV)*	*Source or Generator*[b] ($T_{1/2}$)	*Applications*
^{51}Cr	27.8 d	322	Reactor ^{50}Cr (n, γ)	Red blood cell labeling
^{57}Co	267 d	120	Cyclotron ^{60}Ni (p, α)	Pernicious anaemia studies
^{62}Cu (PET)	10 min	511	Generator ^{62}Zn (9 h)	Pre-clinical studies for myocardial and cerebral perfusion, hypoxia
^{67}Ga	78 h	300, 180	Cyclotron ^{65}Cu (α, 2n)	Abcesses and neoplasms
^{68}Ga (PET)	1.13 h	511	Generator ^{98}Ge (288 d)	Tumor imaging
^{99m}Tc	6 h	140	Generator ^{99}Mo (6 d)	Many and varied
^{111}In	2.8 d	250, 170	Cyclotron ^{109}Ag (α, 2n)	White and red blood cells, antibody and peptide labelling for cancer diagnostics
^{201}Tl	74 h	0	Cyclotron ^{203}Tl (p, 3n)^{201}Pb(EC)	Myocardial perfusion, parathyroid and tumor

[a] Half life, d = days, h = hours, min = minutes
[b] EC = electron capture, n = neutron, p = proton

3.3.3 Radiopharmaceuticals

In order for a radiopharmaceutical to be effective in diagnostic imaging it must meet a number of criteria

(i) It must have low radioactive and chemical toxicity.
(ii) In a clinical environment it must be easily prepared from the radionuclide supplied in adequate purity and yield using simple procedures. The preparative timescale must be acceptable in comparison with the half-life of the radionuclide involved.
(iii) It must be sufficiently stable to reach the target tissue intact.
(iv) It must show sufficiently selective uptake in the target tissue and be cleared rapidly enough from non-target tissue to give a sufficiently high target to non-target concentration ratio that useful imaging data can be obtained.
(v) It must remain in the target tissue long enough for the imaging procedure to be completed and not redistribute significantly during this time.

Technetium can adopt oxidation states ranging from (−1) to (+7). All but the d^0 (+7) ions are partially occupied electron shell systems and their bonding to ligands may be stabilised by ligand to metal (high oxidation states) or metal to ligand (low oxidation states) π-electron donation. This together with Crystal Field Stabilisation Energy effects can confer a degree of kinetically inert character on the complexes. In the lower or intermediate oxidation states the metal is present in complexes as simple ions, *i.e.* Tc^+, Tc^{2+}, Tc^{3+} and Tc^{4+} but in the higher oxidation states oxo- or nitrido-ions are present, *i.e.* TcO^{3+}, $TcO_2{}^+$, TcN^{2+}, TcN^{3+} and $[TcO_4]^-$. This offers a variety of ways in which ^{99m}Tc might be incorporated into a Tc-essential radiopharmaceutical. Although $[^{99m}TcO_4]$– itself can be used as a radiopharmaceutical, for example to image the thyroid, it is usually necessary to reduce the $[^{99m}TcO_4]^-$ to a lower oxidation state form in order that it can be incorporated in the radiopharmaceutical agent. The most common reducing agent used for this purpose is $SnCl_2.2H_2O$, which typically effects a 2-electron reduction from Tc(+7) to Tc(+5). However, various other reduction reactions are possible providing access to the different oxidation state species. The lower oxidation state Tc ions are soft in character and the high oxidation states hard. However, Tc(+3) and Tc(+5) can show rather mixed behaviour readily forming complexes with ligands containing hard O, intermediate N and soft S or P donor atoms. A very brief overview of the chemistry of Tc follows.

3.3.4.1 *Tc(+7) and Tc(+6)*

The highest Tc oxidation state, Tc(+7), is found in $[TcO_4]^-$, the form in which ^{99m}Tc is eluted from a generator, and in the oxide Tc_2O_7. Unlike $[MnO_4]^-$, $[TcO_4]^-$ is not a strong oxidising agent but it is easily reduced. In the absence of complexing agents aqueous reduction of $[TcO_4]^-$, leads to insoluble TcO_2 and Tc_2O_3. In non-aqueous solvents one-electron reduction to $[TcO_4]^{2-}$ is possible but this Tc(+6) complex is not stable in water. Some Tc(+7), d^0, complexes are known, for example 6-coordinate $[TcO_3Cl(L–L)]$ (L–L = 2,2′-bipyridyl, **13** or 1,10-phenanthroline, **14**). Reaction of $[TcO_4]^-$ with azide, $N_3{}^-$, affords the Tc(+6) complex $[TcNCl_4]^-$ which undergoes chloride substitution reactions with proligands such as cysteine or polyamine carboxylates.

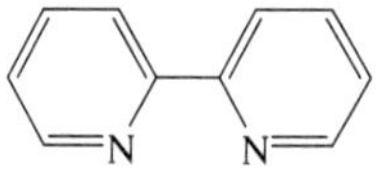

13 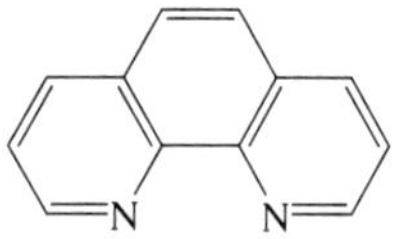 **14** 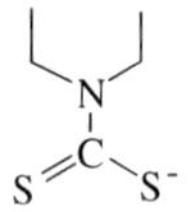 **15**

In the presence of complexing agents reduction of $[TcO_4]^-$ leads to the formation of various lower oxidation state complexes, the nature of which depends on the reductant and complexing agent used. Different reductants can lead to different products even when the same complexing agent is used. For example the reduction of $[TcO_4]^-$ in the presence of diethyldithiocarbamate (edt^-, **15**) affords $[\{TcO(edt)_2\}_2(\mu\text{-}O)]$ [**16** containing Tc(+5)] with dithionite,

$S_2O_4^{2-}$, $[TcN(edt)_2]$ [**17** containing Tc(+5)] with hydrazine, NH_2NH_2, and $[Tc(edt)_3]$ [**18** containing Tc(+3)] with formamidine sulfinic acid, $NH_2C(=NH)SO_2H$. Sometimes the proligand itself can act as both reductant and complexing agent so that $[TcO_4]^-$ reacts with excess thiourea, $S = C(NH_2)_2$ (tu), to give the octahedral Tc(+3) complex $[Tc(tu)_6]^{3+}$ and with excess $P(OCH_3)_3$, under more forcing conditions, to give the octahedral Tc(+1) complex $[Tc\{P(OCH_3)_3\}_6]^+$. The usual product of aqueous $[TcO_4]^-$ reduction by $SnCl_2$ in the presence of a complexing agent is a complex containing the TcO^{3+} core and, historically, this has been the usual target in ^{99m}Tc-radiopharmaceutical synthesis. However, in recent times complexes containing Tc^{3+} and more particularly, Tc^+, have been of interest. The complicated nature of the $[TcO_4]^-$ reduction reactions can present problems in the formulation of new ^{99m}Tc-radiopharmaceuticals so careful chemical studies are needed to underpin the development of imaging agents. These also need to take account of differences in chemical behaviour between the conditions typically used to study new compounds in a chemical laboratory and the very low concentrations used clinically.

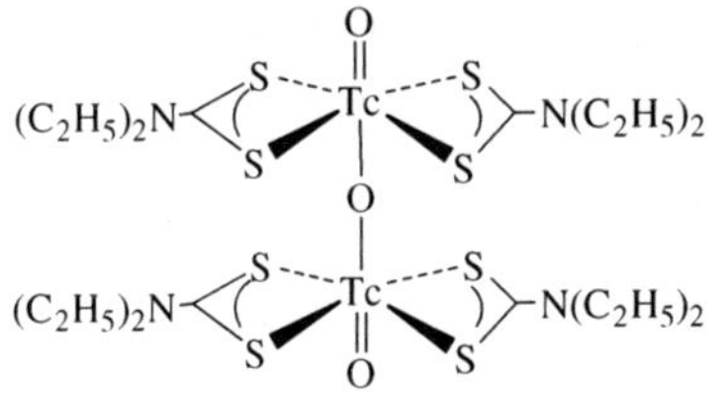

$(C_2H_5)_2N$ S S Tc O S S $N(C_2H_5)_2$ O $(C_2H_5)_2N$ S S Tc O S S $N(C_2H_5)_2$

16

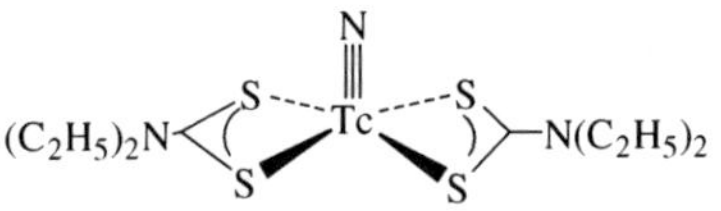

N $(C_2H_5)_2N$ S S Tc S S $N(C_2H_5)_2$

17

$(C_2H_5)_2N$ S S S Tc S S S $N(C_2H_5)_2$ $(C_2H_5)_2N$

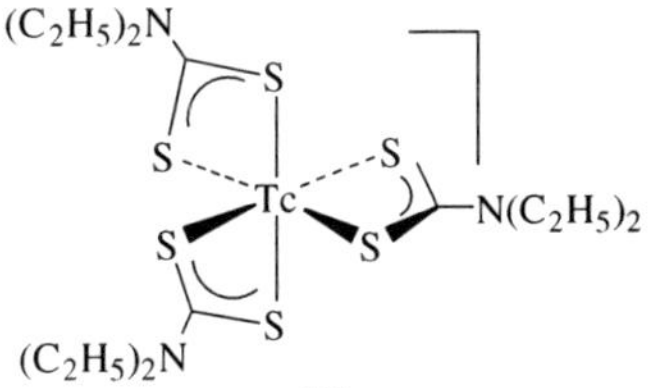

18

3.3.4.2 *Tc(+5)*

Tc(+5) is perhaps the most versatile oxidation state in Tc chemistry since it can appear in complexes containing the TcO_2^+, TcO^{3+}, TcN^{2+} and Tc^{5+} cores. The TcO_2^+ core forms on reduction in the presence of neutral amine ligands such as pyridine, which gives $[TcO_2(NC_5H_5)_4]^+$, or cyclam, which gives $[TcO_2(cyclam)]^+$, **19**. Both complexes being octahedral with *trans*-dioxo groups in the TcO_2^+ core. Organophosphine ligands such as $(CH_3)_2PCH_2CH_2P(CH_3)_2$ (dmpe) can also form complexes containing TcO_2^+, (*e.g.* **20**, Scheme 4), but under more forcing conditions and with excess phosphine these can undergo further reduction to Tc(+3) complexes, *e.g.* **21**, and ultimately to Tc (+1) complexes (*e.g.* **22**). The Tc(+3) complex **21** is easily reduced to its Tc(+2) counterpart $[TcCl_2(Me_2PCH_2CH_2PMe_2)_2]$, **23**, (Me = CH_3). Organophosphines,

R_3P, can act as reducing agents through the formation of phosphine oxides, $R_3P = O$ as shown in Equations (5) and (6).

$$[TcO_4]^- + R_3P + 2H^+ \rightarrow TcO_2^+ + R_3P = O + H_2O \qquad (5)$$

$$TcO_2^+ + R_3P + 2H^+ \rightarrow Tc^{3+} + R_3P = O + H_2O \qquad (6)$$

The complexes **21**, **22** and **23** demonstrate the stability of the low spin octahedral geometry found for the d^4 Tc(+3), d^5 Tc(+2) and d^6 Tc(+1) centres, all of which have large crystal field stabilisation energies. Reduction of $[TcO_4]^-$ in the presence of diamine dioxime proligands containing four nitrogen donor atoms leads to complexes containing the TcO^{3+} core exemplified by **24**. Here the ligand has a charge of (3−) resulting from the loss of $3H^+$ from the diamine dioxide proligand. Compared to **19**, in these complexes the negative charge on the ligand increases the electron density at the Tc(+5) core, reducing the capacity of the metal to act as a π-electron acceptor towards the O ligands. This is reflected in the Tc(+5) centre supporting only one Tc=O interaction rather than two as with the neutral ligand cyclam. The cyanide ligand is somewhat ambivalent towards Tc(+5) in being anionic but also having some π-acceptor character. In this case complexes containing the TcO_2^+ or TcO^{3+} cores can be isolated depending on the reaction conditions. In water $[Tc(=O)(CN)_5]^{2-}$ converts to $[Tc(=O)_2(CN)_4]^{3-}$ as shown in Equation (7).

$$[Tc(=O)(CN)_5]^{2-} + H_2O \rightarrow [Tc(=O)_2(CN)_4]^{3-} + HCN + H^+ \qquad (7)$$

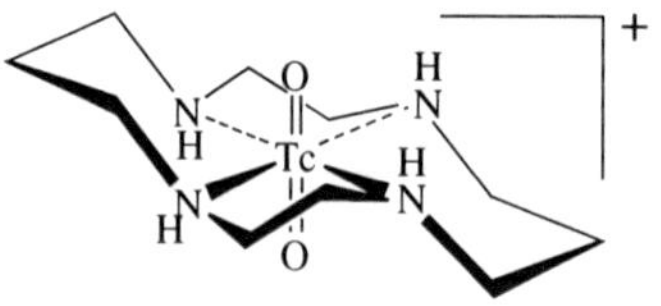

19

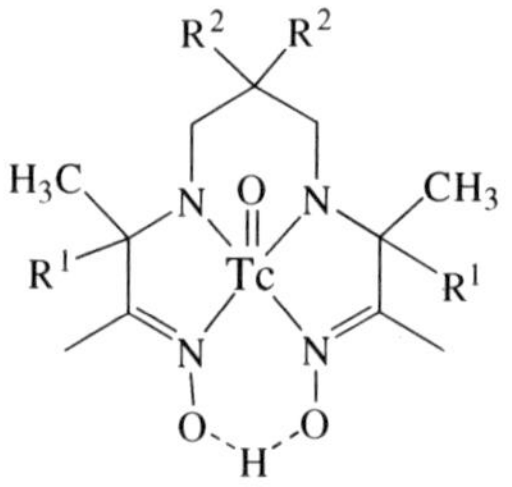

24

$R^1 = H, R^2 = CH_3$
$R^1 = CH_3, R^2 = H$

In hydrochloric acid $[^{99}TcO_4]^-$ can be reduced to $[^{99}TcOCl_4]^-$ which provides a useful precursor in macroscopic chemical studies of complexes containing the $^{99}TcO^{3+}$ core. Reduction of $[TcO_4]^-$ by $SnCl_2$, the reductant commonly used in radiopharmaceutical preparations, typically affords Tc(+5) complexes but can also lead to Tc(+4) and even Tc(+3) if present in sufficient excess. The products of the reduction depend on the nature of the proligand, and the pH of the solution. Often more than one product is formed and polynuclear species may be present so that reaction conditions may need to be optimised to favour a particular desired product. As examples reactions involving polyamine carboxylates have produced several different types of complex. The

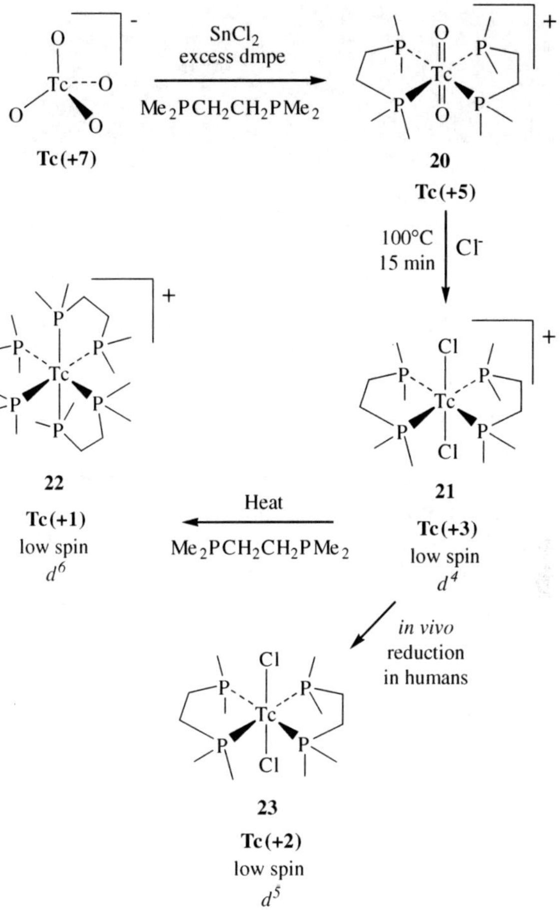

Tc(+7)

$SnCl_2$
excess dmpe
$Me_2PCH_2CH_2PMe_2$

20
Tc(+5)

100°C
15 min
Cl^-

21
Tc(+3)
low spin
d^4

Heat
$Me_2PCH_2CH_2PMe_2$

22
Tc(+1)
low spin
d^6

in vivo
reduction
in humans

23
Tc(+2)
low spin
d^5

Scheme 4

7-coordinate mononuclear Tc(+5) complex $[TcO(edta)]^-$, **25**, or the 6-coordinate binuclear Tc(+4) complex $[\{TcO(edtaH_2)\}(\mu\text{-}O)_2]^{2-}$, **26**, have been obtained from reactions involving edtaH_4, **4**. A 6-coordinate binuclear complex $[\{Tc(nta)\}_2(\mu\text{-}O)_2]$, **27**, was obtained from a reaction with nitrilotriacetic acid, ntaH_3, and it is said that octahedral Tc(+3) complexes $[Tc(ntaAr)_2]^-$, **28**, can form with monoamide derivatives of ntaH_3, ntaArH_2.

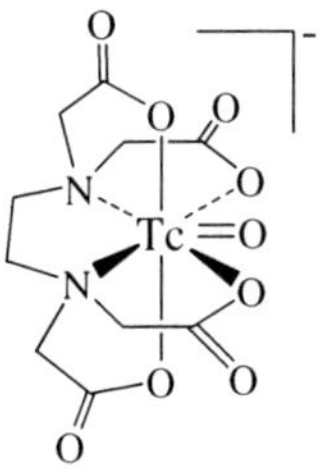

25 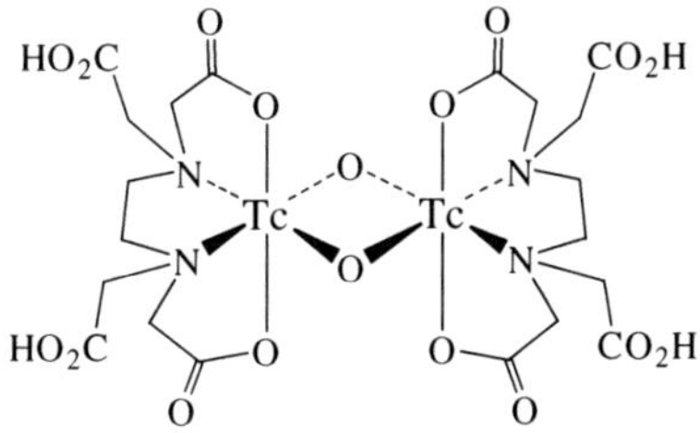 **26**

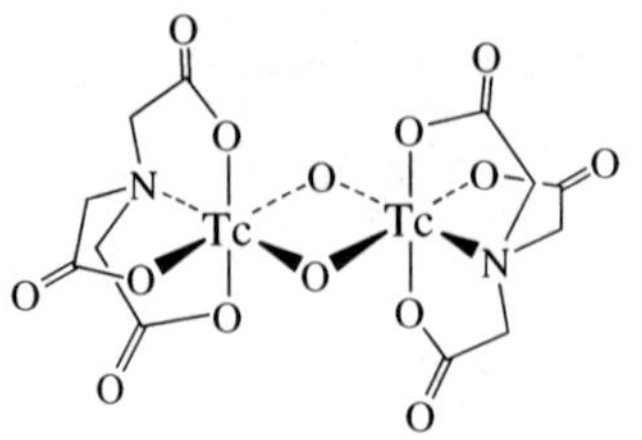

27

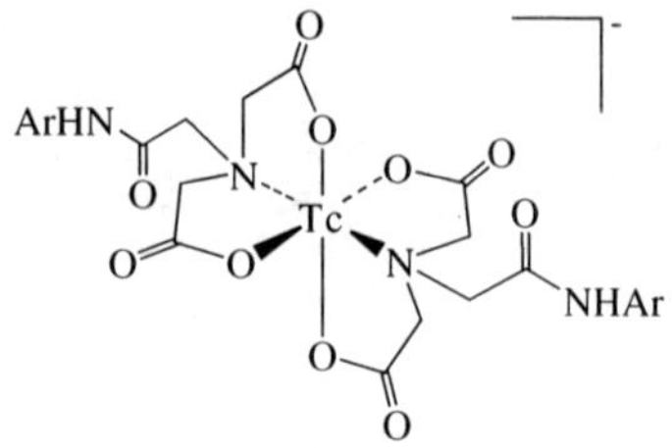

28
Ar = aryl

The TcO^{3+} core is especially well known in complexes with coligands containing all S or a mixture of S and N donor atoms. When four large S donor atoms are present, pyramidal structures are typical as in **29** formed with thiomercaptoacetate. This structure is also typically found for TcO^{3+} complexes with ligands containing both N and S donor atoms although a 6-coordinate octahedral complex, **30**, has been obtained with penicillamine. Such complexes show good chemical stability and this may be compared with the rather labile behaviour of the all oxygen donor atom glycolate complex **31** (E = O). This is a useful precursor to other Tc(+5) complexes and undergoes facile ligand substitution reactions, for example with $HSCH_2CH_2SH$ to give more stable **31** (E = S).

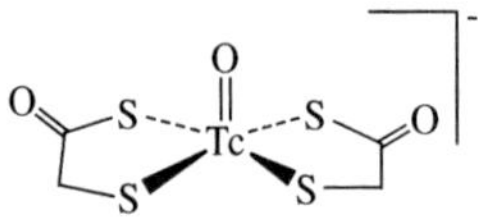

29

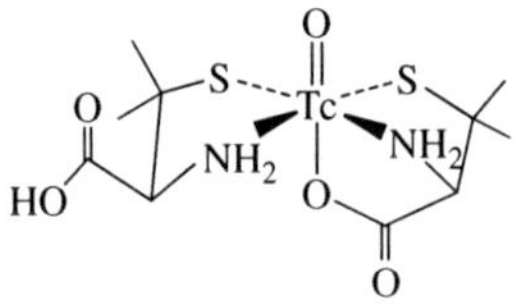

30

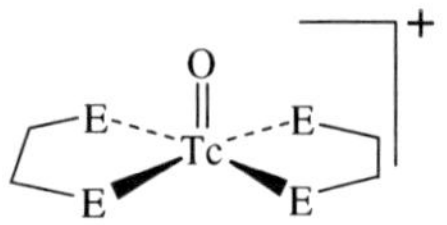

31
E = O, S

A particularly important type of ligand used in the production of ^{99m}Tc radiopharmaceuticals is represented in **32**. This structure contains three chelate rings offering high complex stabilities and is very versatile from a synthetic point of view. The sizes of the chelate rings can be varied, but 5- and occasionally 6-membered rings are the norm. The atoms E could be neutral thioether S, neutral amine >NR (R=hydrocarbyl or other substituent), negative amide $>N^-$ or neutral phosphine >PR. The atoms X could be any of these types as well as thiolate $-S^-$ or, less likely, $-O^-$. Ligands of this type can usually be synthesised relatively easily from two or three reactants offering a huge variety of ligand structures and donor atom types. The substitution pattern on the chelate rings can be varied to modify the lipophilicity/hydrophilicity of the complex. The inclusion of amide links in the chelate rings promotes the ionisation of amide hydrogen in -NH-C(=O)- to give $-N^-$-C(=O)-, offering a means of reducing positive charge on the complex. Also substituents on the chelate ring can be used to connect the metal binding group to other biologically active molecules. The possibilities are illustrated by the examples of Tc(+5) complexes described in subsequent sections on the applications of Tc coordination chemistry in diagnostic imaging.

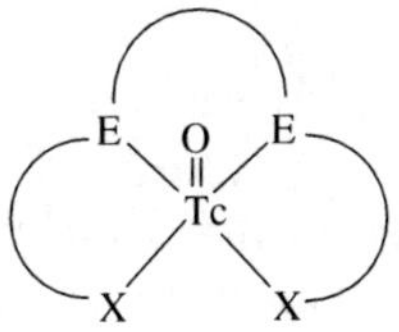

32

The reduction of $[^{99}TcO_4]^-$ by hydrazine in the presence of suitable pro-ligands affords Tc(+5) complexes containing the TcN^{2+} core. Octahedral structures are found for nitrogen donor ligands such as $NH_2CH_2CH_2NH_2$ (en), which forms $[TcN(en)_2Cl]^+$, **33**, and the complex **34** containing a macrocyclic ligand. The presence of larger S donor atoms promotes 5-coordinate square pyramidal structures as found in **35**, **36** and **37**, for example. The TcN^{2+} core offers the possibility of a parallel chemistry to that of the TcO^{3+} core but with the positive charge reduced by one unit.

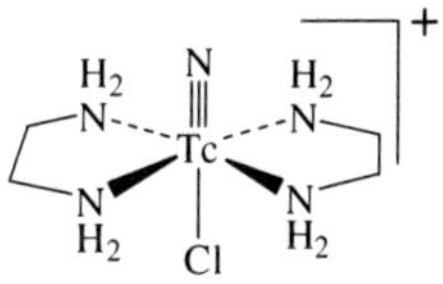

33

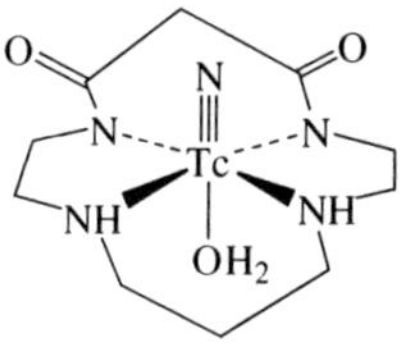

34

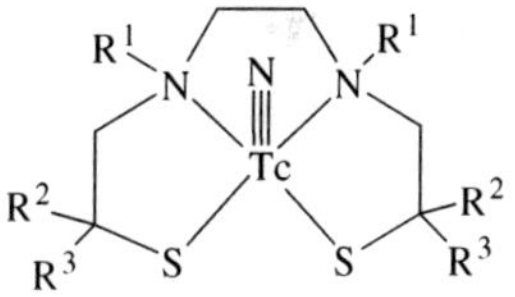

35

$R^1, R^2, R^3 = H$ or hydrocarbyl

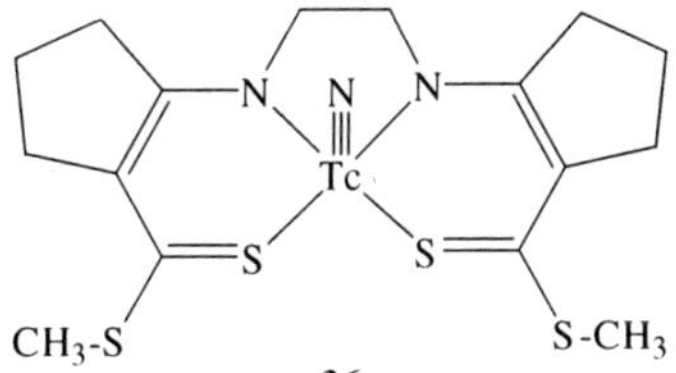

36

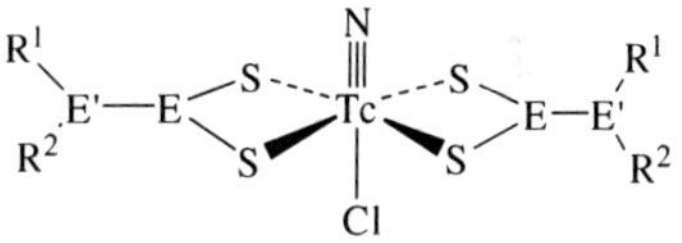

37

$R^1, R^2 = H$ or hydrocarbyl

E-E' = C-N or P

Complexes of the simple Tc^{5+} core are rare but 8-coordinate $[Tc(diars)_2Cl_4]^+$ (diars = **38**), having a dodecahedral structure, provides one example.

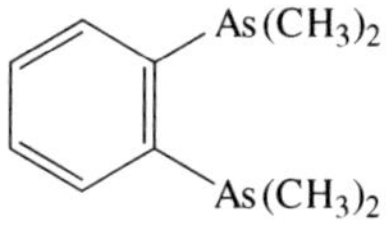

38

3.3.4.3 Tc(+4) and Tc(+3)

The d^4 Tc^{3+} and d^3 Tc^{4+} ions are typically 6-coordinate forming octahedral complexes. The higher charges on these ions might be expected to result in higher crystal field splittings compared to Tc^+ and Tc^{2+} but this must be set against the smaller number of d-electrons and reduced π–backbonding capability for forming stable complexes with π–acceptor ligands. Boiling hydrochloric acid can effect the reduction of Tc(+7) in $[TcO_4]^-$ to Tc(+4) in the octahedral complex $[TcCl_6]^{2-}$. This complex is rather kinetically inert and a less reactive precursor to other Tc complexes than $[TcOCl_4]^-$. Some Tc(+4) complexes with polyamine carboxylates are binuclear with bridging O^{2-} ligands (*e.g.* **26**) and Tc(+4) is thought to be present in diphosphonate complexes used in bone imaging.

Proligands such as phosphines or thiourea, which contain soft donor atoms and are mild reducing agents, can react with $[TcO_4]^-$ to form Tc(+3) complexes directly. A particular example is provided by $[Tc\{S = C(NH_2)_2\}_6]^{3+}$, formed in the reaction of $[TcO_4]^-$ with thiourea and in which the thiourea is bonded to Tc^{3+} through the S atoms. This complex reacts further with isonitriles, (C = NR) in the presence of a reductant to give $[Tc(CNR)_6]^+$ and is a potential precursor to many other complexes. The positively charged Tc(+3) phosphine complexes $[TcCl_2\{P(CH_3)_2CH_2CH_2P(CH_3)_2\}_2]^+$, **20** (Scheme 4), can also be formed directly from $[TcO_4]^-$ and were first investigated as potential heart agents, but failed in this application through their *in vivo* reduction to the neutral Tc(+2) complexes $[TcCl_2\{P(CH_3)_2CH_2CH_2P(CH_3)_2\}_2]$ (Section 3.3.9.1).

One topic in the chemistry of Tc(+3) which has aroused interest as an efficient means of attaching a Tc^{3+} ion to an organic molecule is use of hydrazino nicotinic acid (HYNIC), **39**, derivatives. A hydrazine derivative, $RNHNH_2$ (R = *e.g.* pyridyl or substituted pyridyl) can bind to a metal ion in any of several ways (Scheme 5). For example the neutral molecule might bind as a monodentate ligand in a similar manner to NH_3 but it is also possible that one hydrogen might ionise to give monodentate hydrazido(1−) donor group [*cf.* the amide donor group -C(=O)N-$^-$] [Scheme 5(a)]. It is also possible that two hydrogens might ionise to give a hydrazido(2-) complex. In addition to being proligands hydrazine derivatives can also act as reducing agents losing dihydrogen, or the equivalent $2H^+ + 2e^-$ (Scheme 5b). The oxidation of a hydrazine in this way produces a diazene and again this may coordinate as a neutral diazine or isodiazine ligand, or lose H^+ to bind as a uninegative diazenido(1-) ligand [Scheme 5(c)]. If group R is pyridyl chelation becomes possible so that in addition to monodentate structures [Scheme 5(c)] didentate structures have also been found [Scheme 5(d and e)]. The oxidation state of Tc in the Tc complexes of HYNIC derivatives has been assigned as (+3), although there often tends to be some uncertainty about the exact nature of Tc–N–N–R bonding when dealing with ligands such as diazenes. The complexes formed appear to have 6-coordinate octahedral structures of the type shown in **40** and some examples are described in Section 3.3.12.3.

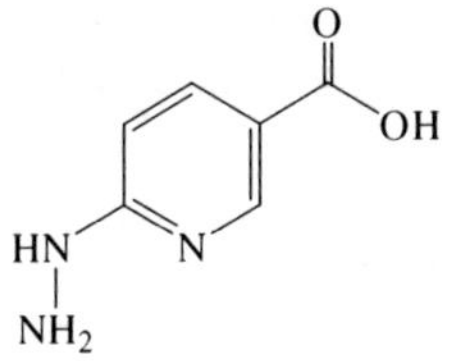

39

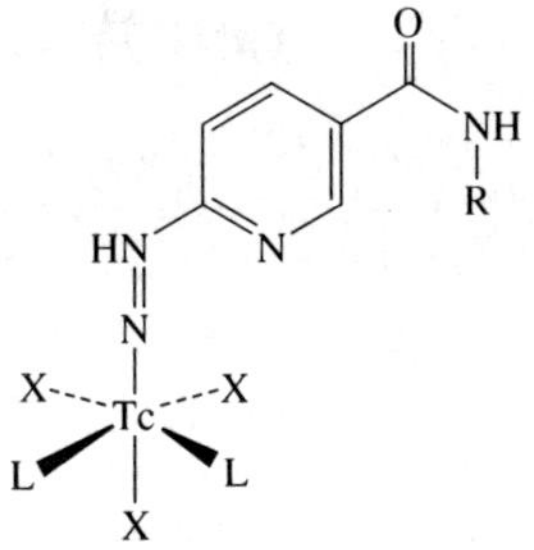

40

L = neutral donor atom
X = anionic donor atom

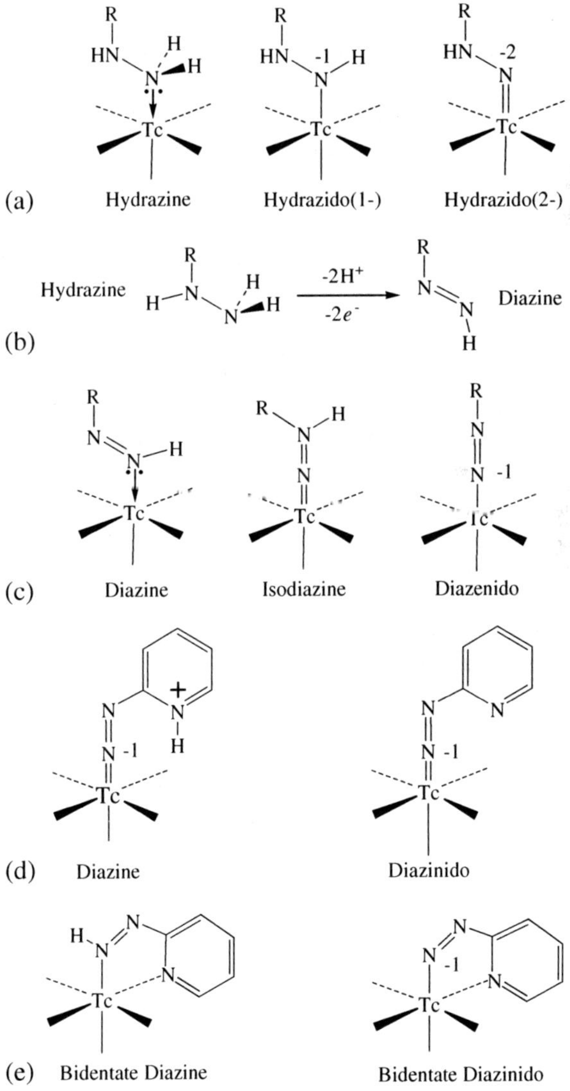

(a) Hydrazine Hydrazido(1-) Hydrazido(2-)

(b) Hydrazine $\xrightarrow[-2e^-]{-2H^+}$ Diazine

(c) Diazine Isodiazine Diazenido

(d) Diazine Diazinido

(e) Bidentate Diazine Bidentate Diazinido

Scheme 5

3.3.4.4 Tc(+2) and Tc(+1)

Oxidation numbers (+1) and (+2) must be considered rather low for Tc and Tc^+ or Tc^{2+} are only stable when bound to π–acceptor ligands such as CO, CNR (R = hydrocarbyl), PR_3 or $P(OR)_3$ (Section 2.5.2). The simple d^6 Tc^+ and d^5 Tc^{2+} ions are typically 6-coordinate forming octahedral complexes, which offer the advantage of large crystal field stabilisation energies in the presence of strong field ligands. This can contribute to the kinetic as well as the thermodynamic stability of complexes having these oxidation numbers. The chemistry of Tc(+2) is not so well developed, though octahedral complexes of the type $[TcCl_2\{E(R_2)\text{-}Z\text{-}Z\text{-}E(R_2)\}_2]$ (**41**, E = P, Z–Z = CH_2CH_2; E = As, Z–Z = 1, 2-C_6H_4) are well known. In early studies of heart imaging agents the redox properties of the $[TcCl_2\{P(CH_3)_2CH_2CH_2P(CH_3)_2\}_2]^{+1/0}$ system were problematic in the formulation of a radiopharmaceutical (Section 3.3.9.1).

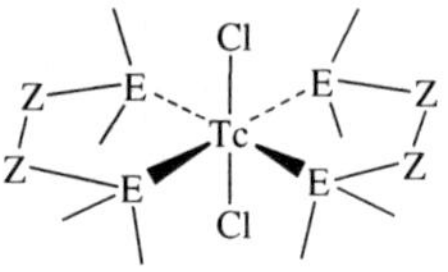

41
E = P, Z-Z = CH_2CH_2;
E = As, Z-Z = 1,2-C_6H_4

It is possible to obtain the octahedral Tc(+1) complex $[Tc\{P(OCH_3)_3\}_6]^+$ directly from the reaction between $[TcO_4]^-$ and $P(OCH_3)_3$ under forcing conditions *i.e.* strong heating. The related isonitrile complexes $[Tc(CNR)_6]^+$ were originally prepared from $[Tc\{S = C(NH_2)_2\}_6]^{3+}$ and CNR using sodium dithionite as the reductant. However, it proved possible to prepare $[Tc\{CNC(CH_3)_3\}_6]^+$ directly from $[^{99m}TcO_4]^-$, $SnCl_2$ and excess $CNC(CH_3)_3$. These complexes are related to the hexacarbonyl cation $[Tc(CO)_6]^+$ which is less easily prepared. However, recently it has been found that the reduction of $[TcO_4]^-$ by borohydride (BH_4^-) under a CO atmosphere produces the water soluble complex *fac*-$[Tc(CO)_3(H_2O)_3]^+$ (**42**). This complex is stable in aqueous media and offers a versatile reagent for the synthesis of a variety of Tc(+1) complexes. Unlike the CO ligands, the water ligands are not π-acceptors and so their binding to the Tc^+ centre is not stabilised by synergic bonding (Section 2.5.2). As a result the water ligands are labile and can undergo facile ligand exchange reactions with a variety of proligands (Scheme 6). Ligating groups with soft donor atoms such as S or P would seem best suited to binding the low oxidation state Tc(+1) centre but compounds containing macrocyclic N donor ligands are also formed.

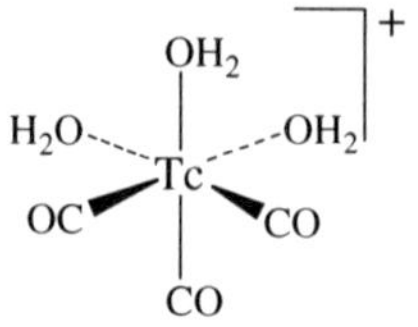

42

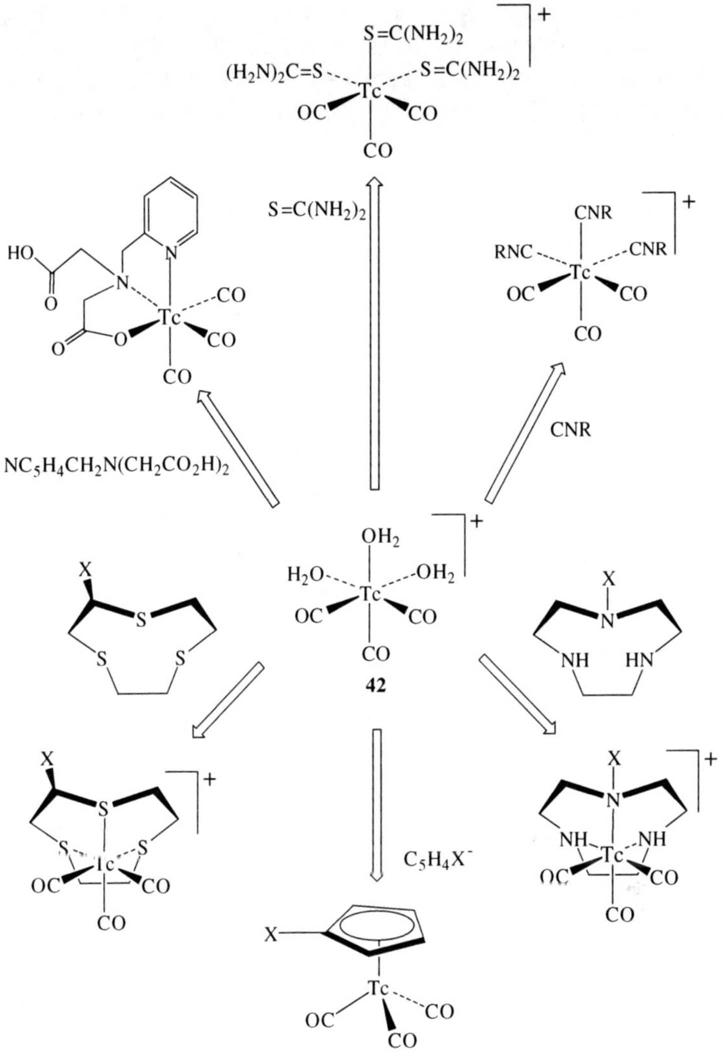

R = Unreactive substituent,
X = H, an unreactive substituent or a reactive substituent offering a possible point of attachment to a carrier function

Scheme 6

3.3.4.5 Tc(0) and Tc(-1)

The lowest Tc oxidation states are found in the carbonyl complexes $[Tc_2(CO)_{10}]$, **43** containing Tc(0), and $[Tc(CO)_5]^-$, **44** containing Tc(-1). Unlike isonitrile derivatives, the CO ligands in these complexes offer no opportunity for derivatisation, ligand substitution reactions would be necessary to functionalise the complexes and vary their properties. This, coupled with the difficulties in synthesising such complexes, the ease of oxidation of $[Tc(CO)_5]^-$ and the relatively substitution inert nature of $[Tc_2(CO)_{10}]$, suggests that these compounds are far from ideal as candidates for radiopharmaceutical preparation. The Tc(+1) complex **42** offers a far more promising starting point.

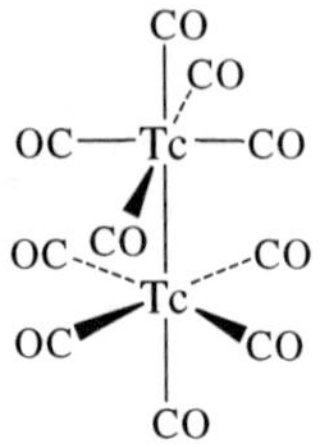

CO CO OC—Tc—CO OC CO CO Tc OC CO CO **43**

CO OC Tc—CO OC CO **44**

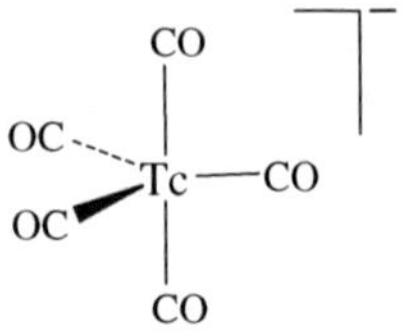

3.3.5 Blood-pool Imaging and Labelled Blood Cells

Radiolabelled macromolecular or particulate species have a variety of applications in diagnostic nuclear medicine, although the exact role of coordination chemistry in such formulations may not always be well defined. Blood-pool imaging, the evaluation of heart function and the detection of GI haemorrhages are possible using radiolabelled red blood cells. If the cells are denatured before use red blood cell sequestration by the spleen can also be studied. ^{99m}Tc-labelled HSA may be used to study cardiac output, blood volume and, in aggregated particulate form, for lung perfusion imaging. The reduction of $[^{99m}TcO_4]^-$ by dithionite ions forms a colloid useful for liver and spleen imaging. This formulation is thought to contain Tc_2S_7 associated with colloidal sulfur and some 80–85% accumulates by phagocytosis in the Kupffer cells of the liver.

There are two basic approaches to the preparation of radiolabelled red blood cells depending on whether the cells are labelled *in vitro* or *in vivo*. The *in vitro* approach involves the removal of blood from the patient, addition of the labelling agent and re–injection into the patient. A blood purification procedure and/or chemical pretreatment may then be required to facilitate binding of the labelling agent and re–injection into the patient. In the cases of $^{111}In^{3+}$ or $^{68}Ga^{3+}$ such a process can be inefficient because of the binding of the In^{3+} or Ga^{3+} ions to serum proteins, especially to transferrin. However, if these ions are first complexed by 8–hydroxyquinoline (oxineH) to produce lipophilic species, presumed to be the neutral tris complexes $[M(oxine)_3]$ (M = Ga, In), efficient labelling of red blood cells, platelets or leukocytes is possible. It has also been shown that 2–mercaptopyridine *N*–oxide may be used to label platelets and leukocytes with ^{111}In. Labelling of red blood cells with ^{99m}Tc *in vitro* involves the treatment of the cells with $SnCl_2.2H_2O$ which enters the cells where it can act as a reducing agent. Residual extracellular Sn(+2) is oxidised with hypochlorite to prevent reduction of $[^{99m}TcO_4]^-$ outside the cells, and when $[^{99m}TcO_4]^-$ is subsequently added it diffuses into the cells where it is reduced and trapped by binding to the β-chain in the globin unit of hemoglobin. Labelling yields of 95% are possible and a kit for performing this procedure is available under the name Ultratag® (Mallinckrodt/Tyco Healthcare).[3]

In vivo labelling of red blood cells by ^{99m}Tc also involves the intracellular reduction of $[^{99m}TcO_4]^-$ but in this case Sn(+2)-pyrophosphate [*a.k.a.* Phosphotec (Bracco), Technescan PYP® (Mallinckrodt/Tyco Healthcare), Amersham PYP (GE Healthcare, was Nycomed-Amersham)] is given intraveously. After 30 min the blood is largely free of extracellular Sn(+2) and a solution of $[^{99m}TcO_4]^-$ can be injected. As in the *in vitro* approach $[^{99m}TcO_4]^-$ enters the red blood cells where it is reduced and trapped. Again labelling yields of 95% are found.

Human plasma proteins such as serum albumin, fibrinogen or immune γ–globulin can also be radiolabelled using $[^{99m}TcO_4]^-$ and a reducing agent, usually Sn(+2). The preparation may be freed of unbound $[^{99m}TcO_4]^-$ by ion exchange procedures and the ^{99m}Tc label is sufficiently stable *in vivo* for imaging purposes. In the case of albumin reduction of disulfide (–S–S–) links to dithiol (–SH HS–) would create a chelating moiety of a type which is known to form robust complexes with TcO^{3+}. Kits containing HSA treated with Sn(+2) are available for studying pulmonary perfusion [*e.g.* Macrotec (Bracco), Technescan® MAA (Mallinckrodt/Tyco Healthcare), Amersham MAA (Nycomed-Amersham then GE Healthcare)] and are reconstituted with $[^{99m}TcO_4]^-$ prior to injection.

3.3.6 ^{99m}Tc SPECT Imaging Agents for Bone

An important application for bone imaging agents is in the diagnosis of occult metastatic bone disease in cancer patients. Functional radiopharmaceutical agents are selectively taken up in regions of abnormal osteogenesis even though the anatomical structure of the bone may appear normal. In this way they can reveal bone or joint disease undetected by anatomical imaging methods (X-ray CT scans or MRI). The strontium isotopes ^{87m}Sr ($T_{1/2} = 2.83$ h, γ390, keV) and ^{85}Sr ($T_{1/2} = 64$ d, γ510, keV) were used in some early investigations of bone lesions but neither is available from a generator and both have major disadvantages. ^{87m}Sr was also taken up in soft tissue giving a high background and, although better target/background radiation ratios could be obtained with ^{85}Sr, use of this radionuclide resulted in high patient radiation doses. These limitations can be overcome by using less radiotoxic ^{99m}Tc in the formulation of bone imaging agents.

Since phosphate is an important component of bone, and capable of acting as a ligand for Tc ions, early bone scanning studies using ^{99m}Tc employed complexes with pyrophosphate (**45a**) or polyphosphate ligands. These were found to be effective but the proligands were prone to hydrolysis and

[3] The names of some of the products and their manufacturers have changed since the product was first introduced. Both the original and the new names and manufacturers are cited where known but may not represent the current product name and manufacturer.

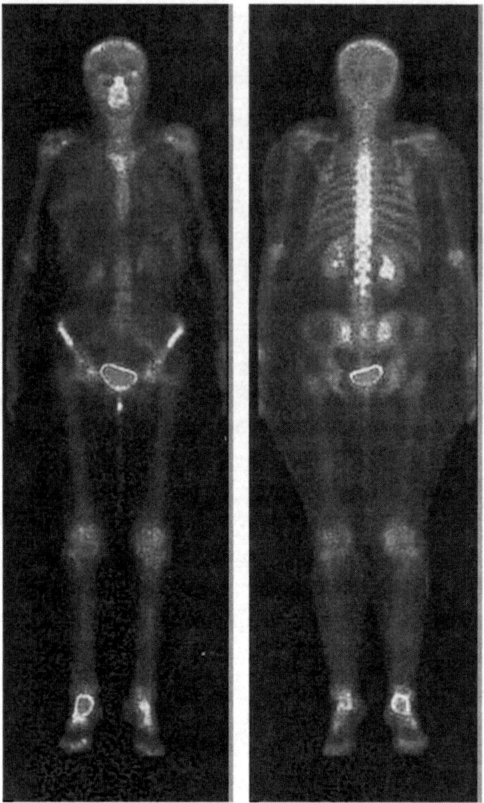

Figure 15 *SPECT images of the skeleton obtained using the bone imaging agent ^{99m}Tc-MDP. In addition to skeletal structures activity can be seen accumulating in arthritic regions of the right ankle and in the bladder*
(Reproduced from J.R. Dilworth and S.J. Parrott, *The Biomedical Chemistry of Technetium and Rhenium, Chem. Soc. Rev.*, 1998, **27**, 43 with permission of Prof S.J. Mather, St Batholomews Hospital, London, and the Royal Society of Chemistry, Copyright 1998).

so were superseded by more robust diphosphonate ligands (**45b–d**) among which ^{99m}Tc- MDP (*a.k.a.* Tc-99m Medronate or Osteolite) gave superior images because of its rapid renal clearance from the blood (Figure 15). Blood serum proteins may compete with bone for the ^{99m}Tc complexes and reduce the efficiency of the radiopharmaceutical. Thus the proportion of an administered dose of ^{99m}Tc taken up in bone was found to increase linearly with the fraction of ^{99m}Tc in the plasma which was not protein bound. In this respect the ^{99m}Tc-diphosphonate formulations showed better performance than ^{99m}Tc-PYP or ^{99m}Tc-tripolyphosphate. Two hours after administration the proportions of plasma ^{99m}Tc, which were protein bound, were found to be 16, 31, 54 and 85% for HEDP, MDP, PYP and tripolyphosphate, respectively.

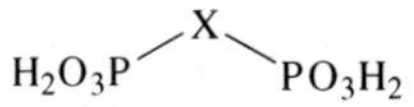

H_2O_3P–X–PO_3H_2

45

a X = O; PYP, $pypH_4$
b X = CH_2; MDP, $mdpH_4$
c X = CH(OH); HMDP, $hmdpH_4$
d X = $C(CH_3)(OH)$; HEDP, $hedpH_4$

The relationship between ligand structure and biodistribution has been investigated for the ^{99m}Tc-phosphonate systems. It appears that only one phosphonate group is necessary for bone uptake to occur and, in the series of ligands $X\{P(=O)O_2^{2-}\}_2$, those in which the group X, together with a P–O moiety, could form a 5- or 6-membered chelate ring on technetium showed the highest degree of bone localisation. However, the exact nature of the Tc-diphosphonate complex or complexes involved in the bone uptake process are not known with certainty. A structural study of oligomeric $\{^{99}Tc(OH)MDP\}_n$ (**46**) formed from $[TcBr_6]^{2-}$ and MDP has revealed a polymeric structure involving Tc(+4) centres and bridging OH groups. Intuitively this structure would seem to provide a realistic model for the structure of the ^{99m}Tc-MDP radiopharmaceutical but the complex was obtained from solutions of *ca.* millimolar concentration compared to *ca.* nanomolar for the radiopharmaceutical. This concentration difference is important since anion exchange high pressure liquid chromatography (HPLC) studies of ^{99m}Tc-$NaBH_4$-HEDP formulations prepared under carrier-added conditions (3×10^{-3} M ^{99}Tc concentration) showed that oligomerisation occurred giving rise to at least seven different complexes. In contrast oligomer formation was not favoured in the carrier-free ^{99m}Tc system where one product predominated. The situation is further complicated by the choice of reducing agent. The use of Sn(+2) as a reducing agent for $[^{99m}TcO_4]^-$ may complicate the product distributions since Sn(+4) might replace Tc(+4) in some sites in the polynuclear structure, or bind to peripheral phosphate oxygens. Electrophoretic studies of a ^{99m}Tc-MDP-$SnCl_2$ formulation showed that at least four components were present. A study of the biodistributions of individual ^{99m}Tc-$NaBH_4$-HEDP solution components formed under carrier-added conditions has shown that the complexes of lowest retention time, and thus lowest negative charge density, give the largest bone uptake, lowest kidney uptake and fastest blood clearance.

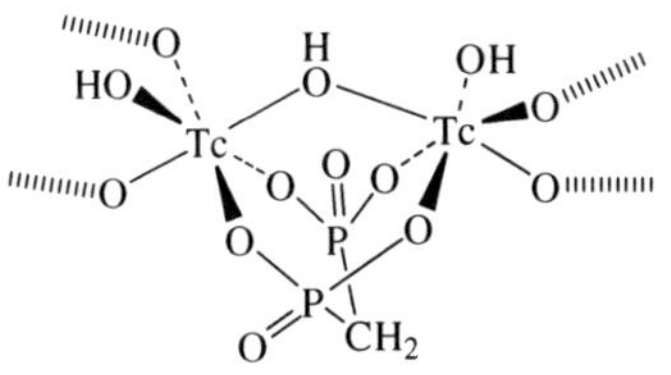

46
Tc(OH)-MDP

The exact mechanisms by which ^{99m}Tc-diphosphonate agents selectively accumulate at sites of newly forming or cancerous bone are not fully understood.

Adherence to denatured protein, binding to iron deposits in soft tissue, extracellular metastatic calcification during hypercalcemia and binding to calcium complexes within the cells of infarcted tissues have all been proposed. An important chemical basis for the uptake by bone of diphosphonate agents must be their high affinity for calcium. This behaviour is demonstrated by the accumulation in bone of free HEDP and its complexes with tin or rhenium, and by the ability of cobalt diphosphonate complexes to bind Ca^{2+} ions. It has been suggested that the selective accumulation of HMDP and HEDP complexes of ^{99m}Tc at sites of new bone formation is due to preferential binding to the (001) faces of the hydroxyapatite crystals in the bone. Since these crystals grow most rapidly along the c–axis, the (001) face containing this axis contributes more to the surface area of newly forming bone mineral than to that of old mineral. This would lead to a higher diphosphonate/calcium ratio on newly forming as opposed to non-growing bone tissue. In addition stressed bone has higher concentrations of Ca^{2+} ions which will increase accumulations of Ca^{2+} seeking agents. The exact nature of the ^{99m}Tc complex bound to the bone remains uncertain but the dissolution of bone treated with ^{99m}Tc-MDP in aqueous EDTA afforded a ^{99m}Tc product which could not be distinguished from ^{99m}Tc-MDP suggesting that Tc transfer into the bone itself is not an important feature of the bone uptake.

^{99m}Tc- MDP has been marketed for skeletal imaging to reveal regions of altered osteogenesis by several companies (Osteolite, Dupont-NEN then Bristol-Meyers Squibb; Amerscan-MDP, Amersham International then GE Healthcare; Technescan-MDP. Mallinckrodt/Tyco Healthcare; MDP-Squibb, Bristol-Meyers Squibb, AN-MDP-Syncor International; MPI MDP Kit, Medi-Physics). The ^{99m}Tc complexes of the substituted diphosphonates HMDP and HEDP (**45c** and **d**) have also been studied and show greater uptake in metabolically active bone than ^{99m}Tc-MDP. ^{99m}Tc-HMDP also shows faster blood clearance and has found clinical application (Osteoscan-HDP, Mallinckrodt).

3.3.7 Kidney Imaging

Imaging agents for the kidney are used to provide clinical information regarding renal bloodflow, cortical structure and renal function in terms of both glomerular filtration and tubular secretion. Near-ideal ^{99m}Tc kidney imaging agents are now available and have come to dominate this area of application. Renal agents may be divided into three types: those that are filtered, those that are secreted and those that exhibit mixed behaviour. Complexes which exhibit mixed behaviour may be useful in the study of kidney morphology.

Historically ^{99m}Tc-glucoheptonate (GlucoScan, DuPont then Bristol-Meyers Squibb; Technescan®, Mallinckrodt/Tyco Healthcare), which is cleared from the blood by both glomerular filtration and tubular secretion, was used for renal perfusion imaging. The structure of the complex is not known with certainty but it is thought to contain a 5-coordinate TcO^{3+} core (**47**). The complex is not very stable *in vivo* and a decomposition product binds to the tubules in the renal cortex leading to *ca.* 15% retention of the injected dose in the kidney 4 h post injection. A ^{99m}Tc-DMSA formulation (^{99m}Tc-succimer, Nycomed-Amersham then GE Healthcare; DMSA is dimercaptosuccinic acid or dmsaH_4) offers a more stable radiopharmaceutical and this has replaced ^{99m}Tc-glucoheptonate which is no longer routinely used for renal imaging. ^{99m}Tc-DMSA is prepared from $[^{99m}TcO_4]^-$, *meso*-dmsaH_4 and $SnCl_2.2H_2O$ buffered to pH 4.5. The agent clears from the blood relatively slowly with a half time of 56 min, presumably as a result of protein binding. At 4 h post injection total renal uptake of over 40% has been found and the agent localises in the proximal convoluted tubule. The structure of the complex formed from Tc and DMSA at pH 4.5 is not known but an octahedral structure containing a Tc(+3) core and two H_4DMSA ligands, each binding through two S and one carboxylate O, has been proposed. However, the square pyramidal Tc(+5) complex, $[Tc(=O)\{S_2(CHCO_2H)\}_2]^-$, **48**, and its Re analogue are known structures and are typical of TcO^{3+} dithiolates. The ^{99m}Tc(+5) complex can be obtained by modification of the standard ^{99m}Tc-DMSA kit and raising the reaction pH to 7.5–8.0. The higher pH favouring the higher oxidation state. The kidney uptake of ^{99m}Tc-DMSA can be affected by the conditions of $[^{99m}TcO_4]^-$ reduction and the use of Sn(+2) has been found to increase ^{99m}Tc retention in some cases. Labelling studies using ^{117m}Sn, ^{14}C and ^{99m}Tc revealed that, at an Sn- DMSA:Tc ratio in excess of 10^4, one ^{99m}Tc species was formed, but at lower ratios, four components were present. Two Sn-DMSA compounds were also identified using reversed-phase HPLC and ion exchange methods. An improved protocol for preparing the ^{99m}Tc-DMSA agent was developed to overcome the variability in performance observed initially. Binding to protein with the displacement of one DMSA ligand has been proposed as an uptake mechanism for ^{99m}Tc from the ^{99m}Tc-DMSA complex.

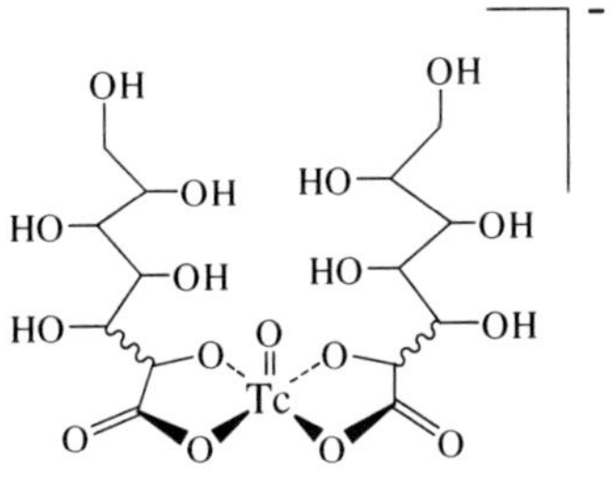

47 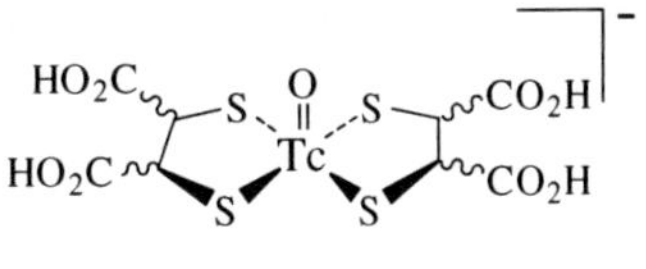 **48**

The hydrophilic complexes formed between ^{99m}Tc and $edta^{4-}$ (edtaH_4 = **4**) or $dtpa^{5-}$ dtpaH_5 = **3a**) are cleared from the blood by glomerular filtration and this has led to the use of ^{99m}Tc-DTPA (MPI-DTPA, Nycomed-Amersham then GE Healthcare, DTPA, DuPont/Syncor International then Bristol-Myers Squibb; Techniplex, Bracco Diagnostics) as an agent for investigating kidney function. The complex is also known as ^{99m}Tc-Pentetate and is formed from $SnCl_2$, dtpaH_5 and $[^{99m}TcO_4]^-$. ^{99m}Tc-Pentetate appears to be reasonably stable *in vivo* and is excreted unchanged in the urine. The radiopharmaceutical circulates freely in the bloodstream until it is filtered out through the glomerulus to the extent of about 20% for each passage of the bloodpool through the kidneys. Binding to blood serum protein may affect the filtration rate of ^{99m}Tc-DTPA and some allowance may need to be made for this when evaluating renal function with this agent. However, alternative agents such as ^{51}Cr- EDTA or ^{169}Yb-DTPA do not appear to offer any advantages over ^{99m}Tc-DTPA. The structure of the ^{99m}Tc-DTPA complex is not known and structures containing either Tc(+4), **49a**, or Tc(+5), **49b**, have been proposed. Various oxo- and hydroxo-complexes of ^{99}Tc with polycarboxylate ligands are known from work at much higher concentrations than for the radiopharmaceutical and Tc(+5) seems a more typical product of Sn(+2) reduction (Section 3.3.4.2). In a study of the kidney uptake of various Tc(+5) and Tc(+4) carboxylic acid complexes the retention of Tc(+5) complexes was found to be higher than for Tc(+4) complexes with citrate, tartrate, malate and hydroxyisobutanoate.

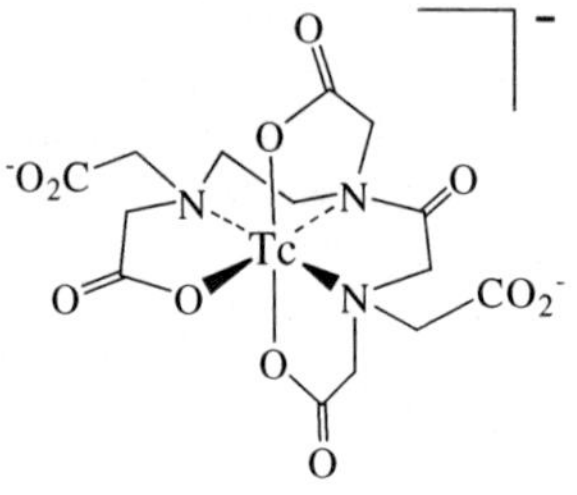

49a
Tc(+4)

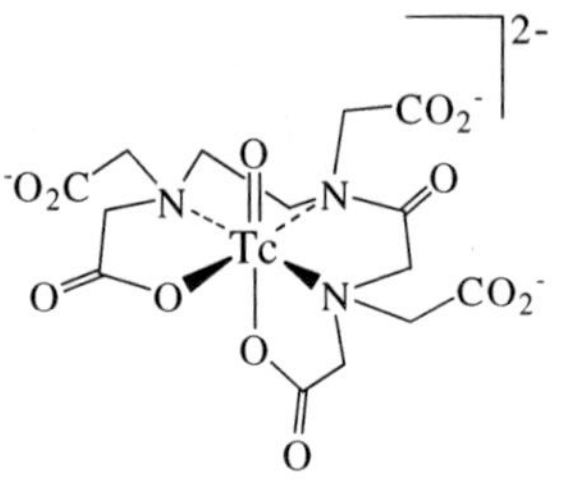

49b
Tc(+5)

Hippuric acid, $C_6H_5C(=O)NHCH_2CO_2H$, labelled in the *ortho*-position with ^{131}I (^{131}I-hippuran), was once the principal agent used for evaluating renal tubular secretory function, but the nuclear properties of ^{131}I or are far from ideal for this purpose. This stimulated the search for a ^{99m}Tc agent suitable for studying renal secretion and measuring effective renal plasma flow. The presence of a free carboxylic acid group in hippuric acid provides hydrophilic character important in kidney agents. The other requirement for a renal ^{99m}Tc agent to replace ^{131}I-hippuran would be the presence of ^{99m}Tc bound in a ligating group capable of forming a stable complex. In early investigations promising results were obtained with a group of complexes (**50**) in which a TcO^{3+} core is complexed to a ligand with an N_2S_2 donor atom set. Rapid renal secretion was observed for the **50** system and a series of closely related compounds containing chelate rings of varying size was investigated. It was found that the complexes containing three 5-membered chelate rings were most readily formed, stable and rapidly excreted. Unfortunately, with these agents biliary excretion also occurs and some depression of renal function was found. Various other potential renal agents of this type were prepared and some gave satisfactory renal images but could only mimic the distribution behaviour of ^{131}I-hippuran in rabbits, not in man.

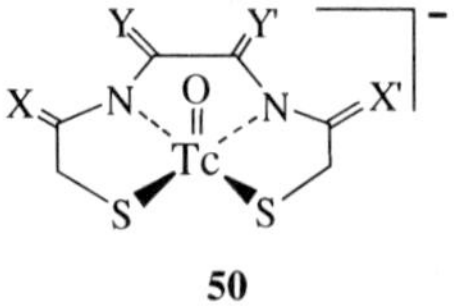

50
(i) X = X' = O, Y = Y' = H_2
(ii) X = X' = H_2, Y = Y' = O
(iii) X = O, X' = H_2, Y = H_2, Y' = O

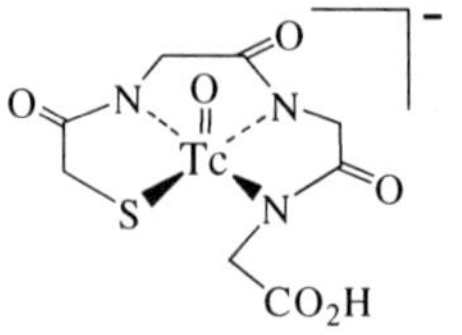

Tc-MAG$_3$
51

The compound which finally emerged as a successful renal imaging agent for the study of tubular secretion was ^{99m}Tc-MAG$_3$ (**51**; TechneScan MAG$_3$, Mallinckrodt/Tyco Healthcare) in which the mercaptoacetylglycylglycylglycinato ligand contained a carboxylate group appended to a terminal coordinated N-donor atom in the SN$_3$ donor atom set. The absence of chiral centres in the ligand means that there are no isomer issues with this compound. However, to prevent the oxidation of the thiol to disulfide the proligand contains a protective benzoyl substituent on the sulfur. This can be removed by heating after addition of the $[^{99m}TcO_4]^-$ and to avoid any reduction to Tc(+3) as a result of heating with a thiol, a low $SnCl_2$ content is used in the kit and the formulation is exposed to air before use to oxidise excess Sn(+2).

A few minutes after injection some 1–2% of the ^{99m}Tc is taken up in the kidneys. The carboxylic acid group is ionised at physiological pH so that the complex is able to act as a substrate for the anionic receptors in the renal tubules and be excreted actively in a similar manner to ^{131}I-hippuran. Approximately 50% of the radioactivity in the blood is actively excreted into urine by the renal tubules at each pass, despite the complex showing significant protein binding. ^{99m}Tc-MAG$_3$ has now become the agent of choice for measuring effective renal plasma flow even though its plasma clearance is only 60% that of ^{131}I-hippuran.

Despite the success of ^{99m}Tc-MAG$_3$ for assessing renal function work continues on developing new agents for this purpose. One approach involves the replacement of glycine in MAG$_3$ with alanine and another involves the hydrolysis product (**52**, ^{99m}Tc-*l, l*-ECD) of the brain imaging agent Neurolite (Section 3.3.10.1). This complex is actively secreted through the renal tubules and shows a 39% higher plasma clearance and lower liver accumulation than ^{99m}Tc-MAG$_3$. One of the amine groups is deprotonated so that with the two thiolate donors and two ionised carboxylate substituents, the complex carries a 2- charge. An even higher negative charge is associated with the Tc(+1) complex $[Tc(C{=}NCH_2CO_2)_6]^{5-}$, the hydrolysis product of the potential heart imaging agent $[Tc(C{=}NCH_2CO_2CH_3)_6]^+$. The ^{99m}Tc form of this highly charged hydrophilic complex has shown some promise for effective renal plasma flow measurements.

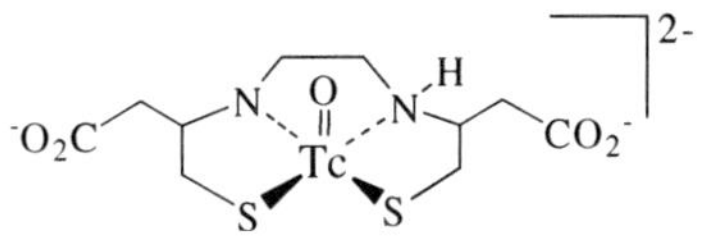

^-O_2C … N … O … N–H … CO_2^- … Tc … S … S … 2-

Tc-*l,l*-ECD

52

3.3.8 Liver and Hepatobiliary System

The liver contains two main cell types, parenchymal and reticuloendothelial. The parenchymal (polygonal) cells make up some 85% of the tissue and are the primary metabolic cells of the liver. The reticuloendothelial (Kupffer) cells comprise some 15% of the liver and are associated with phagocytosis of foreign particles. A colloid produced by dithionite reduction of $[^{99m}TcO_4]^-$, and presumed to be $^{99m}Tc_2S_7$, is taken up by the Kupffer cells and can be used for imaging the reticuloendothelial system. More recently there has been interest in the use of ^{99m}Tc-labelled HSA and other labelled particles for this purpose. However, in many cases ultrasound and CT scans can more conveniently provide information about the morphology of the reticuloendothelial system.

A different situation arises with non-reticuloendothelial diseases, such as obstruction of the bile duct and choleocystitis, and these are usually best investigated using nuclear medicine procedures. In particular, hepatobiliary agents provide real time information on the passage of bile from the liver through the gall bladder into the small intestine. Originally, ^{131}I-labelled rose bengal was used to investigate liver function but in the 1970s ^{99m}Tc hepatobiliary agents became available and have since superseded ^{131}I-labelled rose bengal in this application. Three agents became available commercially, ^{99m}Tc-Lidofenin (Technescan HIDA, now discontinued), ^{99m}Tc-Mebrofenin (Choletc, Bristol-Meyers Squibb) and ^{99m}Tc-Disofenin (Hepatolite, DuPont-NEN then Bristol-Meyers Squibb), **53**, all of which contain ^{99m}Tc bound to the iminodiacetate moiety [ida^{2-}, $NH(CH_2CO_2^-)_2$]. All three agents show similar behaviour *in vivo* with maximum liver uptake some 10 to 15 min post injection and *ca.* 13% of the injected dose being excreted in urine within 2 h of injection. These hepatobiliary agents are removed from the blood by hepatocytes in the liver using an active transport mechanism. They are cleared from the plasma as anions by the same general mechanism as bilirubin so that high bilirubin levels can competitively inhibit the clearance of ^{99m}Tc-IDA agents through the liver. Following injection the liver can be visualised after *ca.* 5 min, the hepatic duct and gall bladder between 10 and 40 min and the intestine between 15 and 60 min. Increased urinary excretion, lack of gall bladder visualisation and slow liver clearance all provide indications of hepatobiliary dysfunction.

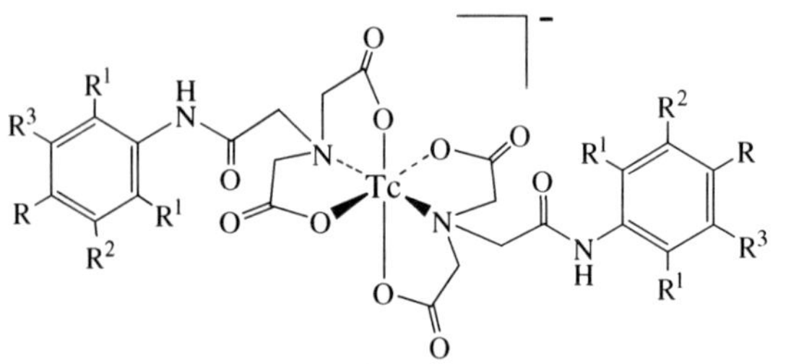

53

a Mebrofenin $R = R_1 = CH_3$, $R^2 = H$, $R^3 = Br$
b Disofenin $R_1 = (CH_3)_2CH$ $R = R^2 = R^3 = H$
c Lidofenin $R_1 = CH_3$, $R = R^2 = R^3 = H$

The ability of ^{99m}Tc complexes of IDA derivatives to undergo rapid hepatobiliary excretion was discovered accidentally by Loberg and co-workers. The free ligand was found to be rapidly excreted through the kidneys while the ^{99m}Tc complex was excreted through the liver and bile duct providing a clear-cut example of a technetium essential radiopharmaceutical. Studies of the relationship between structure and biodistribution have been carried out on the ^{99m}Tc-IDA derivatives and related compounds. The ^{99m}Tc complexes of more hydrophilic ligands, such as ^{99m}Tc-methyliminodiacetic acid, are excreted primarily through the kidneys. However, as the lipophilicity of the ligand increases, hepatobiliary excretion of its ^{99m}Tc complex becomes more important. Increasing alkyl substitution of the aromatic ring, other than at the R^1-positions (**53**), increases the ratio of hepatobiliary to renal excretion. Substitution of the R^1-positions promoted urinary as opposed to biliary excretion. The radiochemical purity of the ^{99m}Tc formulations was found to be dependent upon the preparative conditions used; the pH of the preparative medium and the *pKa* of the complexing agent appear to be important factors. At least two components were identified in formulations with **53** ($R = R^2 = R^3 = H$, $R^1 = C_2H_5$).

The structure of the ^{99m}Tc-IDA agents is not known with certainty and an octahedral structure involving a Tc(III) complex has been proposed as shown in **53**. Certainly evidence has been obtained for the 2:1 ligand/Tc ratio and the -1 charge. However, it is difficult to distinguish between complexes containing Tc^{3+} or TcO^{3+} cores. The Tc(+5) formulation would be consistent with other known Tc-iminoacetate ligand complexes isolated from chemical reactions (Section 3.3.4.2). Under the high dilution conditions of a radiopharmaceutical kit it is not obvious why Tc(+3) might be preferred over Tc(+5) in the absence of a 'soft' donor ligand such as isonitrile or a sulfur compound.

Other nuclides which have been evaluated for use in liver function studies include ^{64}Cu and ^{61}Cu complexes of the tubercularstatic agent Myambutol (**54**). The Budd Chiari syndrome, occlusion of a vein in the liver, has been investigated by use of ^{67}Ga citrate but ^{99m}Tc formulations are also suitable for this purpose. An interesting alternative to ^{99m}Tc is provided by ^{97}Ru, which has a longer half-life and is thus suited for studies requiring delayed scintigraphy. Preparations of ^{103}Ru chloride with two types of IDA ligand containing Ru(+3) and Ru(+4) were found to exhibit similar biological behaviour to that of the ^{99m}Tc derivatives suggesting that ^{97}Ru agents might be developed for use in delayed scintigraphic studies of the liver, for which ^{99m}Tc might be unsuitable.

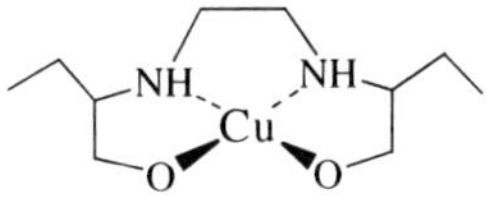

54

3.3.9 Heart SPECT and PET Imaging

3.3.9.1 SPECT Heart Agents

Important uses of heart imaging agents are to study the perfusion of cardiac tissues by blood, and so identify ischemic regions where the blood supply is impaired, or to reveal infarcts, areas of dead muscle tissue. Myocardial blood-flow imaging was originally carried out using the 74 keV γ-emitter ^{201}Tl ($T_{1/2}$ = 73 h). This was thought to behave like K^+ since Tl^+ has an ionic radius only slightly larger than K^+. ^{201}Tl is administered in the form of aqueous thallous chloride, ^{201}TlCl, and is efficiently taken up in myocardial tissue, probably through the action of Na^+/K^+-ATPase. It is cleared rapidly from the blood and may be used to reveal the effects of coronary artery obstruction and myocardial ischemia. However, release of Tl^+ from myocytes during contraction can lead to dispersion of the activity impairing image clarity. This, coupled with the low γ-photon energy of ^{201}Tl and the fact that it is not available from a generator, but must be obtained from a cyclotron, provided a strong incentive to develop a ^{99m}Tc-based heart imaging agent.

The pyrophosphate complex of ^{99m}Tc (Technescan PYP®, Mallinckrodt/Tyco Healthcare; Pyrolite, Dupont-NEN then Bristol-Meyers Squibb; Phosphatec, Bristol-Meyers Squibb) has been used for imaging myocardial infarcts and is thought to be taken up by the calcium deposits known to form in infarcted heart muscle. This affinity may be similar to the binding of diphosphonates to bone (Section 3.3.6) and ^{99m}Tc-MDP is also taken up in myocardial tissue but is not superior to the pyrophosphate in this application. Although they can accumulate in infarcted heart muscle, these compounds are not myocardial perfusion agents and a different design approach was needed to find agents suitable for imaging myocardial perfusion. It was surmised that a lipophilic cationic complex would be a good candidate for myocardial perfusion imaging and various attempts were made to exploit the chemistry of Tc to produce such an agent. Initially these focused on the use of neutral π-acceptor ligands to stabilise low oxidation states of technetium. Oxidation states of +3 and below may be considered low for an element like Tc which readily forms oxo-ions in oxidation states +5 and +7 and forms stable low spin d^3 octahedral complexes in oxidation state +4.

Deutsch *et al.* developed the Tc(+3) complex $[TcCl_2(Me_2PCH_2CH_2PMe_2)_2]^+$ (**21**, Scheme 4, Me = CH_3) which gave promising results in animal models but failed in humans due to *in vivo* reduction to its neutral Tc(+2) counterpart (**23**). Despite this rather unpromising start work continued on this class of compound and the complexes $[TcO_2(dmpe)_2]^+$, (**20**), and $[TcCl_2(dmpe)_2]^+$, **21**, and $[Tc(dmpe)_3]^+$, **22**, were obtained. The compound **22** is highly lipophilic and showed very good myocardial uptake in animal models. However, it proved unsuitable for use in human heart imaging because of its retention in blood leading to high background radiation levels which interfere with myocardial perfusion studies. A wide variety of other Tc(+1) phosphine or arene complexes were evaluated as myocardial imaging agents including $[Tc(PMe_3)_6]^+$, $[TcCl_2\{(MeO)_2PCH_2CH_2P(OMe)_2\}_3]^+$ and the arene complex $[Tc(Me_6C_6)_2]^+$ but these also proved unsuitable due to prolonged retention in blood through protein binding.

The first ^{99m}Tc compound to show high heart uptake was indeed a Tc(+1) complex but not one containing phosphine ligands. The chemistry underlying the development of this agent relates to the octahedral d^6 Tc(+1) hexacarbonyl complex $[Tc(CO)_6]^+$ which is strongly stabilised by crystal field effects and the strong π-acceptor properties of the carbon monoxide ligand coupled with the electron rich nature of the Tc(+1) centre. However, the carbonyl ligand is not lipophilic and does not offer opportunities for structural variations to optimise the properties of its complex. Davison *et al.* recognised that isonitriles (RNC, R = hydrocarbyl), which are known to form similar complexes to carbon monoxide and are stronger π-acceptors than phosphines, would similarly form cationic Tc(+1) complexes $[Tc(CNR)_6]^+$. Furthermore these would offer opportunities for modifications in ligand structure through the variation of the group R. Firstly, the highly lipophilic complex $[Tc(CNCMe_3)_6]^+$, **55a** was investigated. It was possible to prepare $[^{99m}Tc(CNCMe_3)_6]^+$ directly from $[^{99m}TcO_4]^-$, $SnCl_2$ and excess $CNCMe_3$. The complex proved to be an effective myocardial perfusion agent but showed prolonged retention in the lungs and liver. Washout from the lungs returning activity to the bloodstream made the complex less than ideal for imaging transient ischemic blood flow changes during exercise, while the high liver uptake obscured the apex of the heart. A further problem with this system is that isonitrile ligands are typically liquids, not indefinitely stable and have a disgusting smell which is not well accepted by patients. It was necessary to develop a system which was more acceptable to patients and which showed improved biodistribution with better uptake ratios for heart compared to non-target organs. The shelf life and acceptability of the isonitrile ligands could be improved by preparing solid Cu^+ or Zn^{2+} complexes which would undergo ligand exchange reactions to form the Tc(+1) complex. Modification of the R substituent in the ligand allowed complexes to be developed which have improved properties with respect to uptake and clearance. The complex of the ester ligand $CNC(CH_3)_2C(=O)OCH_3$, **55b**, was investigated and enzymatic hydrolysis of the ester in the liver to form a more hydrophilic carboxylic acid derivative led to faster clearance from the liver. Hydrolysis in the heart was much slower leading to good heart/liver uptake ratios. The neutral hydrolysis product $[^{99m}Tc^+(CNC(CH_3)_2C(=O)O^-)(CNC(CH_3)_2C(=O)OCH_3)_5]$ did not show heart uptake in animals supporting the supposition that a cationic complex was necessary for heart uptake.

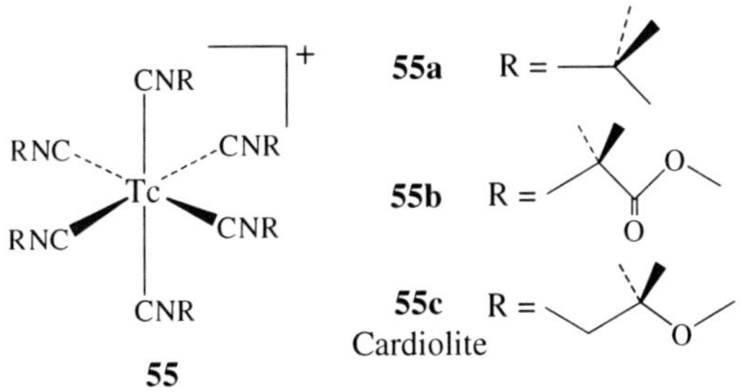

55a R =

55b R =

55c R =
Cardiolite

55
Tc(+1)

A further improvement in performance for this type of compound was obtained using the ether substituted ligand $CNCH_2C(CH_3)_2OCH_3$ to form $[^{99m}Tc^+\{CNCH_2C(CH_3)_2OCH_3\}_6]^+$, **55c**. This complex, known as Cardiolite or Tc-SESTAMIBI, shows reduced lung and liver uptake, increased kidney

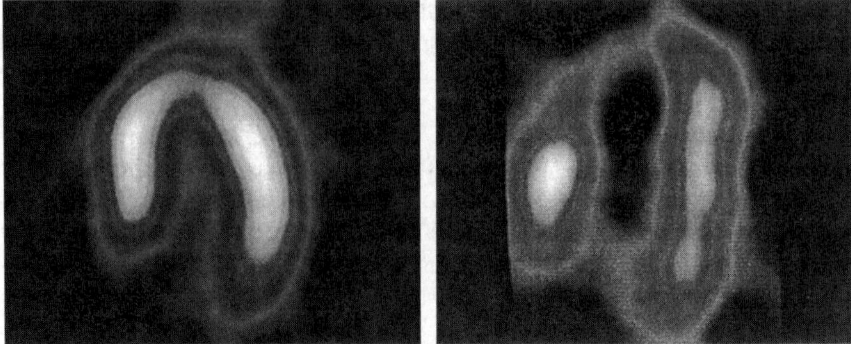

Figure 16 *SPECT images of the heart obtained using the cardiac perfusion imaging agent Myoview. The left hand image shows normal perfusion of the heart and the right image restricted perfusion in a diseased heart*
(Reproduced from J.R. Dilworth and S.J. Parrott, The biomedical chemistry of technetium and rhenium, *Chem. Soc. Rev.*, 1998, **27**, 43 with permission of GE Healthcare (Copyright www.gehealthcare.com)).

uptake and about 1.5% uptake in the heart at *ca.* 1 h after injection falling to *ca.* 1% after 4 h. The complex appears to be solvated in water, is not bound to membrane or cytosolic proteins and accumulates in myocardial tissue through a flow-dependent, diffusion mediated mechanism. The lipophilic character of the complex facilitates its passage through the sarcolemmal membrane and it is thought that its cationic charge promotes intracellular retention in association with mitochondria known to have negative membrane potentials.

Despite the early failures, the chemical approach of using chelating phosphine ligands to produce heart imaging agents finally met with success in the form of the Tc(+5) complex **56** known as Tc-Terofosmin or Myoview (Nycomed-Amersham then GE Healthcare). This was approved for use in the USA in 1996 (Figure 16). The compound is of similar structure to **21** but the methyl groups are replaced by polar ethoxyethyl groups which help the complex achieve a suitable balance between aqueous solubility and lipophilicity. ^{99m}Tc-Terofosmin/Myoview forms almost quantitatively at room temperature over a period of 15 min from the ligand $\{(C_2H_5OC_2H_4)_2PCH_2\}_2$, $SnCl_2$ and $[^{99m}TcO_4]^-$. Further reduction to Tc(+3) species is avoided by controlling the temperature and the ligand and $SnCl_2$ concentrations. The radiopharmaceutical shows membrane potential dependent retention in myocytes and is cleared rapidly from the blood through the hepatobiliary and renal systems. A resting heart accumulation of *ca.* 1.5% is possible in humans. The development of Myoview demonstrates how an initially unsuccessful approach can yield positive results through appropriate modification of the ligand system and careful control of the synthetic conditions used to prepare the radiopharmaceutical. Another approach using chelating phosphines exploited NO as a ligand to stabilise Tc(+1) in the $\{Tc(NO)Cl\}^+$ core of **57** (PL-37, Nycomed-Amersham then GE Healthcare). In such systems the nitric oxide ligand is regarded as a 2-electron donor NO+ group bound to a d^6 Tc(+1) centre to form a linear Tc-N=O moiety. Although **57** showed good heart uptake it gave higher relative background activity than other agents and was not developed commercially.

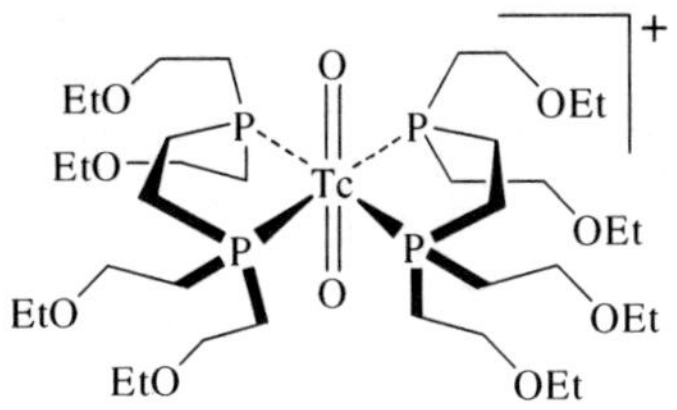

56
Myoview
Tc(+5)

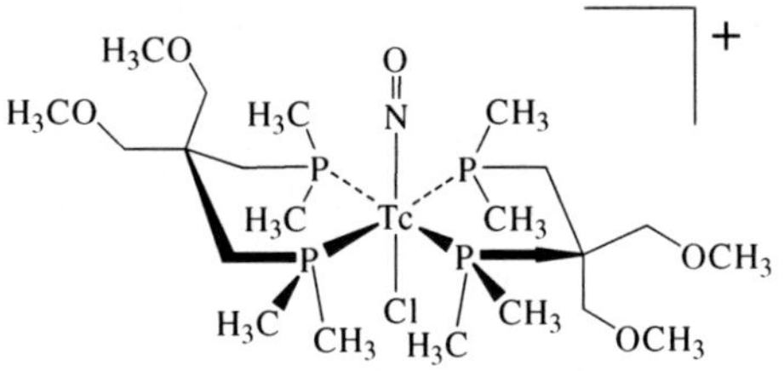

57
PL-37
Tc(+1)

Another versatile approach to the formation of cationic Tc complexes exploits the use of polydentate ligands with N and O donor atoms (Scheme 7). The tetradentate ligand **58** is reacted with $[^{99m}TcO_4]^-$ and $SnCl_2$ to produce the Tc(+5) oxo complex **59** which then reacts with phosphines such as $P(CH_2CH_2OCH_3)_3$ to form the cationic Tc(+3) complex **60**. The presence of the N and O donor atoms in the tetradentate ligand favour Tc(+3) over Tc(+2) so that the complex is less likely to be reduced to a neutral species. Structural variations in the phosphine and the tetradentate ligand allow the properties of the complex to be optimised. The furanyl units on the tetradentate ligand help stabilise the complex and add lipophilic character while the incorporation of ether groups in the phosphine improves hydrophilic character. Among the compounds tested **60** (Tc-Q12) seemed to have the best balance between hydrophilic and lipophilic character showing lower binding to serum proteins and more rapid clearance from the blood and lungs.

Although the initial focus of myocardial imaging agent design centred primarily on cationic ^{99m}Tc complexes, a neutral heart agent has also been developed. The complex is formed through a template reaction in which three cyclohexane dione dioxime ligands, **61**, are assembled around a Tc(+3) core. Methyl boronic acid condenses with the three oxime oxygens at one end of the molecule while two H^+ ions are bound to the other three. A chloride completes 7-coordination around the Tc(+3) centre leading to a distorted structure which does not allow reaction of a second methyl boronic acid with the remaining oxime oxygens. The resulting neutral complex, **62** known as Cardiotec (Bristol-Meyers Squibb), is highly lipophilic and is efficiently (*ca.* 90%) taken up by myocardial tissue. It is also rapidly cleared from the blood, mostly by the liver. However, washout from the heart is rapid compared to Cardiolite limiting the time over which imaging studies can be performed. This, combined with the high liver uptake, makes the complex uncompetitive with other heart agents and its production has now been discontinued.

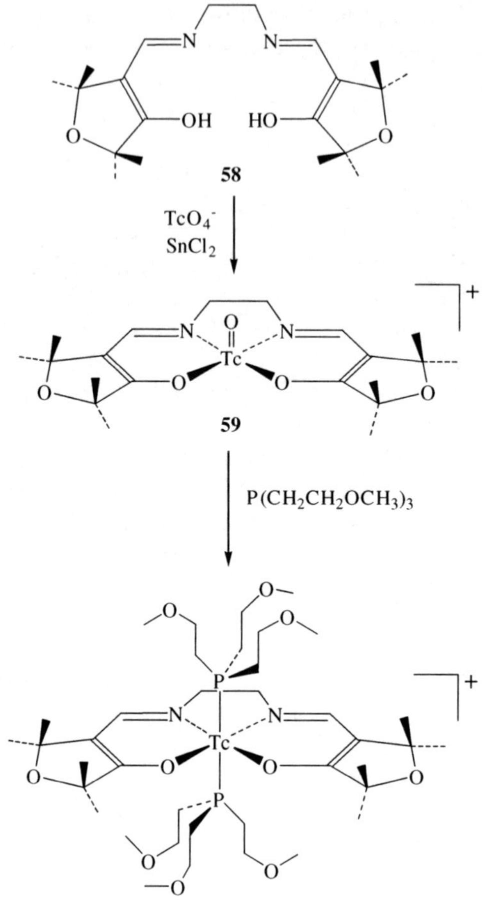

58

TcO_4^-
$SnCl_2$

59

$P(CH_2CH_2OCH_3)_3$

60

Scheme 7

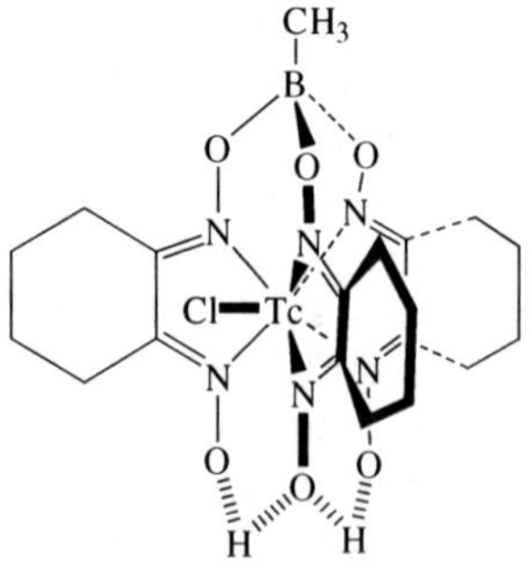

62
Cardiotec
Tc(+3)

derived from

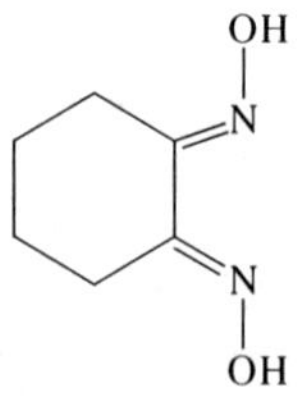

61

Nitrido complexes of technetium have been studied in some detail and can be made in reactions between pertechnetate and hydrazine. Unlike the Tc(+5) oxo unit which provides a tripositive TcO^{3+} core in complexes, the Tc(+5) nitrido unit gives a dipositive TcN^{2+} core. Attempts to find a heart agent based on this coordination entity have included the neutral dithiocarbamate derivative **63**. The ligand in this highly lipophilic compound contains ethyl and ethoxy groups exploiting similar ligand design principles to those seen in Cardiolite, Myoview and Tc-Q12. The compound localises in the myocardial cell membrane but, unlike the cationic heart agents, redistributes in the heart following initial uptake. High lung, and liver uptake together with slow blood clearance limit the utility of this agent but its development illustrates another coordination chemistry approach to the formation of lipophilic neutral Tc complexes.

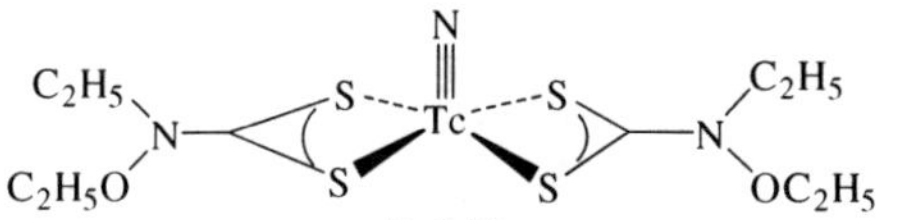

Tc(+5)

63

Another approach to cardiac imaging is to use antibody fragments to carry a radionuclide to a specific target tissue. An example of this approach is provided by Myoscint which was given market approval in the USA in 1996 for detecting the presence and location of myocardial infarction. Myoscint is a ^{111}In-DTPA labelled murine monoclonal antibody Fab fragment which binds with high affinity and specificity to human cardiac myosin which is exposed through damage to myocyte cell membranes. The modification and ^{111}In labelling of antibody fragments is discussed in more detail in Section 3.3.11 relating to tumor imaging.

3.3.9.2 PET Heart Imaging Agents

In a similar manner to $^{201}Tl^{+}$, the positron emitter $^{82}Rb^{+}$ (Cardiogen-82, Bracco Diagnostics, $T_{1/2} = 75$ s) may also be regarded as a K^{+} mimic and gives excellent PET images of myocardial perfusion. The radionuclide is available from a generator (^{82}Sr, $T_{1/2} = 25$ d) and was approved by the FDA in 1989. The very short half-life of ^{82}Rb is not compatible with exercise stress tests to reveal ischemia but does allow sequential images to be taken every few minutes. As an s-block metal Rb^{+} shows highly labile chemistry and does not offer opportunities to exploit coordination chemistry in controlling its biodistribution. Its function depends on its behaving in a similar manner to K^{+}.

In contrast to Rb, Cu is a d-block metal to which coordination chemistry can be usefully applied *in vivo*. In particular ^{62}Cu ($T_{1/2}$ = 9.7 min) is a positron emitter with the further advantage of being available from a generator (^{62}Zn, $T_{1/2} = 9.26$ h, Proportional Technologies, Houston). The Cu(+2) complex **64** (Cu-PTSM) has aroused considerable interest as a possible myocardial perfusion agent. Compounds of this type initially aroused interest over 30 years ago as possible anti-tumor agents. Although they proved ineffective in the field of therapy, their chemistry has now been revived in the context of diagnostic medicine. ^{62}Cu-PTSM is a lipophilic neutral agent which clears quite rapidly from the blood and experiments with rats show good myocardial uptake, although liver and lung uptake is also relatively high. Studies have shown that the complex is reduced by glutathione (GSH) to a ^{62}Cu(+1) derivative which decomposes rapidly leading to intracellular trapping of ^{62}Cu in the heart. The complex also decomposes rapidly in the blood. The addition of a further methyl substituent to the $NHCH_3$ groups in the complex converting them to $N(CH_3)_2$ makes the complex harder to reduce, owing to the electron releasing character of the methyl substituents. This prevents the reduction to Cu(+1) and so the $N(CH_3)_2$ substituted complex is not trapped in the heart but washes out. This provides another example of the way in which small changes in ligand structure can radically alter the biodistribution of a complex; in this case through a change in redox properties.

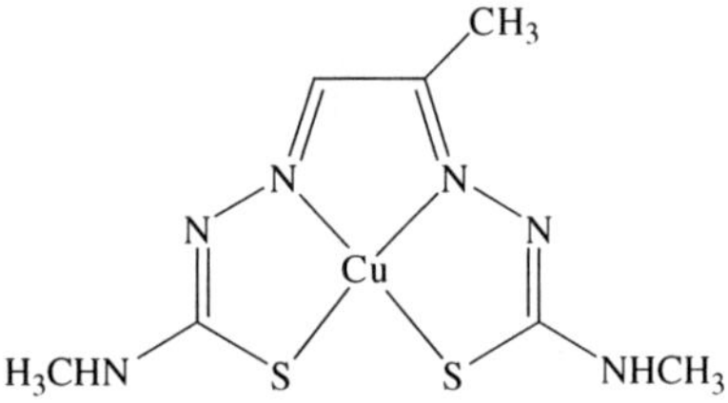

CH_3
N N
N Cu N
H_3CHN S S $NHCH_3$

64

The radionuclide ^{68}Ga ($T_{1/2}$ = 9.7 min) offers another positron emitter available from a generator (^{68}Ge, $T_{1/2}$ = 271 d). Gallium is a p-block element and typically exists in aqueous media as a tripositive ion, Ga^{3+}. Ions of the p-block elements contain closed electron subshells and so do not show crystal field stabilisation energy effects; consequently complexes of Ga^{3+} will be rather labile in comparison with those of Tc ions. There is also a strong similarity between Ga^{3+} (ionic radius 0.62 Å) and high spin d^5 Fe^{3+} (ionic radius 0.65 Å), which also has zero crystal field stabilisation energy. Both of these ions are highly charged and prone to hydrolysis in aqueous media. In the case of $^{68}Ga^{3+}$ hydrolysis occurs above pH 5.1 for solutions containing more than 3.7 10^8 Bq dm^{-3} of carrier free radionuclide. This means that a complexing agent must be added to solutions of $^{68}Ga^{3+}$ to suppress hydrolysis at physiological pH and citrate ($\log K_1 = 10.1$) has been used for this purpose. However, the lability of Ga^{3+} makes it available from citrate to stronger binding agents such as the Fe^{3+} carrier protein transferrin ($\log K = 20.3$ for Ga^{3+}). In order to prevent Ga^{3+} from binding to, and following, transferrin *in vivo* it is necessary to develop ligands which can retain Ga^{3+} *in vivo* and confer useful biodistribution characteristics on the complex. Several lipophilic $^{68}Ga^{3+}$ complexes have been tested in animals and the proligand **65** formed a $^{68}Ga^{3+}$ complex giving a heart/blood ratio of 11:1 60 min after injection into Sprague–Dawley rats. PET heart images were obtained from a dog but lung and liver uptake led to a high background. Generally, as was found with ^{99m}Tc complexes, neutral Ga complexes are poorly retained even though initial uptake may be acceptable. Better retention of the cationic complex **66** was found but the supposedly cationic complex **67** was poorly retained. In the latter case it is thought that, at physiological pH, an amine group in the ligand of **67** loses H^+ rendering the complex neutral. A further obstacle to the use of lipophilic $^{68}Ga^{3+}$ complexes as heart agents, which remains to be overcome, is their high liver uptake. At the time of writing there are no FDA-approved ^{68}Ga myocardial perfusion agents.

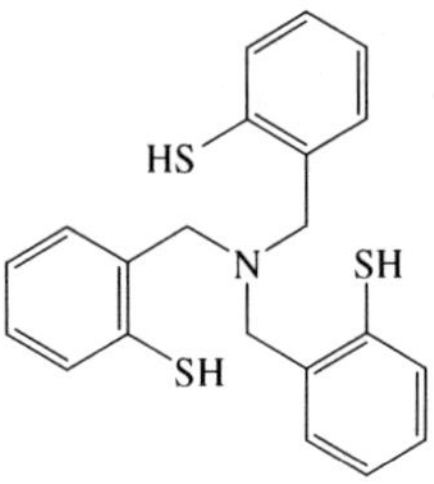

65 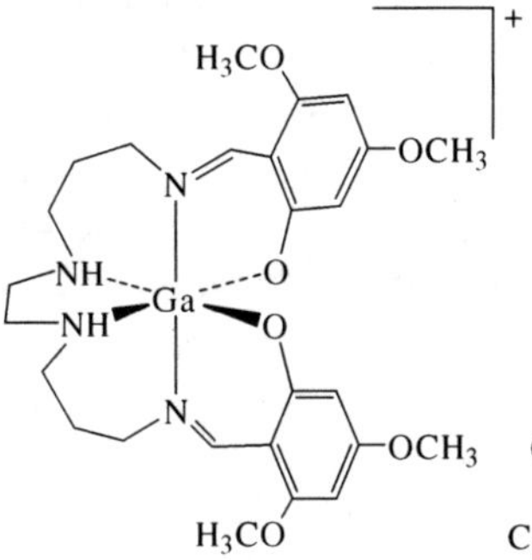 **66** 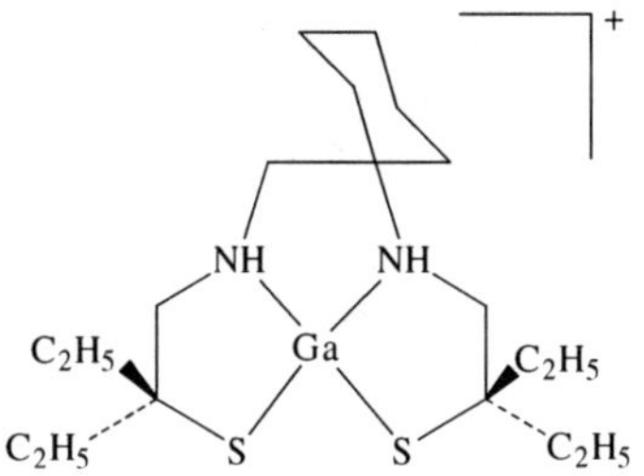 **67**

3.3.10 Brain Imaging

The brain presents a particular challenge for the development of regional cerebral blood flow (rCBF) agents because the unique structure of its capillary blood vessels incorporates a blood brain barrier (BBB) to exclude potentially harmful substances from the brain tissue. To be successful any rCBF agent must be able to traverse the BBB by passive diffusion. In order to achieve this a compound will need to be neutral in charge, of low molecular mass (<*ca.*600 Da) and exhibit a suitable balance between lipophilic and hydrophilic character. This latter requirement can be quantified in terms of an octanol/water partition coefficient P and a value of log P between 0.5 and 2.5 appears necessary. Values below this range indicate that the compound will be insufficiently lipophilic to cross the BBB while values above this range suggest that binding to proteins in the blood will compete with diffusion across the BBB and so render the compound ineffective. The first ^{99m}Tc brain imaging agents to be used did not in fact meet these criteria but instead depended for their uptake on the breakdown of the BBB in areas of the brain damaged by disease. In addition to simple $[^{99m}TcO_4]^-$, both the ^{99m}Tc-DTPA and ^{99m}Tc-glucoheptanoate complexes (Section 3.3.7) have been used to reveal areas damaged by stroke or tumors in the brain and both have been approved for brain imaging applications. However, an effective rCBF agent needs to show an uptake in the brain which is directly related to blood flow through the tissue.

3.3.10.1 Cerebral Perfusion Agents

Several ligand design approaches to the rCBF agent problem have been pursued with varying degrees of success. The PnAO tetradentate dioxime ligand system developed by Troutner and co-workers provided promising early results with the Tc(+5) complex **68** showing diffusion into the brain but poor retention. The PnAO ligand itself is trinegative through the loss of one oxime and two amine protons from the PnAO proligand. The four nitrogen donor atoms are compatible with a TcO^{3+} core and the trinegative charge on the ligand results in the desired neutral charge on the complex. The formation of a hydrogen bond between the oxime oxygens results in a macrocyclic 'cyclam-like' ligand structure which confers additional stability on the complex. The metal centre is further protected by the steric bulk of the methyl substituents associated with the 5-membered chelate rings. It was found that the PnAO proligand did not cross the BBB so that the presence of the metal is essential for brain uptake showing that ^{99m}TcO-PnAO is a truly 'technetium essential' agent.

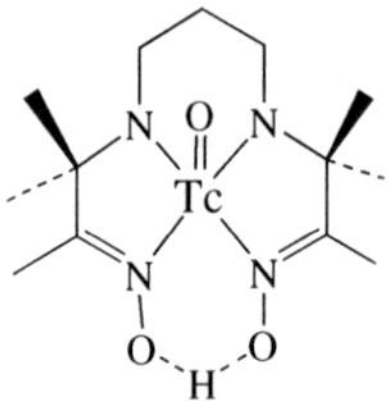

68
[TcO(PnAO)]
Tc(+5)

In order to identify a compound with improved cerebral retention a large number of derivatives similar in structure to ^{99m}TcO-PnAO were investigated and from this group an isomer of **68** emerged as the most effective one. The complex ^{99m}TcO-HMPAO, **69** (Ceretec, Nycomed-Amersham then GE Healthcare) differed from ^{99m}TcO-PnAO only in the disposition of its methyl substituents but was found to have sufficient retention in the brain for clinical use (Figure 17). ^{99m}TcO-HMPAO contains two chiral centres in the ligand and can form four isomers of the complex but the *meso*-isomer with the Tc=O group *syn* to the two methyl substituents has not been isolated. Although all isomers diffuse into the brain similarly the *meso* and *d, l* forms show differing chemical behaviour; the *meso*-isomer being the more stable and diffusing back out of the brain unchanged. Once in the brain the *d*- and *l*-isomers appear to react with GSH resulting in their conversion into hydrophilic derivatives which cannot traverse the BBB so that they become trapped. This gives rise to improved retention of the agent in the brain allowing its use in SPECT imaging. It is possible that the reduced steric protection of the core in ^{99m}TcO-*d,l*-HMPAO compared to ^{99m}TcO-PnAO contributes to its lower stability and improved retention. However, the relatively poor stability of ^{99m}TcO-HMPAO is a disadvantage as it leads to a decline in radiochemical purity to below 85% within 30 min, rendering the agent unusable after this time. The time for which

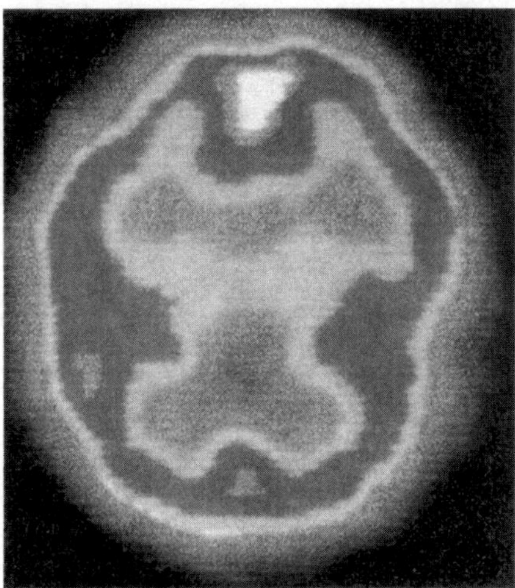

Figure 17 *A SPECT image of the brain obtained using the cerebral perfusion imaging agent Ceretec. The image shows normal perfusion in a healthy brain* (Reproduced from J.R. Dilworth and S.J. Parrott, The biomedical chemistry of technetium and rhenium, *Chem. Soc. Rev.*, 1998, **27**, 43 with permission of GE Healthcare (Copyright www.gehealthcare.com)).

^{99m}TcO-HMPAO can be used for SPECT imaging may be extended up to about 4 h by adding free radical scavengers to limit radiolytic damage which appears to be an important cause of instability. However, even the stabilised formulations are not well suited to applications which require the radiopharmaceutical to be available for extended periods. An important example is provided by the study of epilepsy where imaging during a seizure can reveal a probable seizure focus. To achieve this the agent must be prepared for use ahead of time in anticipation of a seizure and remain sufficiently pure for use when the time comes. The decomposition of Ceretec also contributes to an accumulating background in surrounding tissues which can limit the utility of this agent.

meso, anti-**69** *meso, anti*-[TcO(HMPAO)] not isolated

meso, syn-**69** *meso, syn*-[TcO(HMPAO)]

d,l-**69** *dl*-[TcO(HMPAO)]

Some of the shortcomings of Ceretec are addressed in Neurolite (Dupont-Merck then Bristol-Myers Squibb), **70** (*a.k.a.* ^{99m}Tc-ECD or ^{99m}Tc-bicisate) another neutral Tc(+5) complex containing a trinegative tetradentate ligand

Table 7 *A comparison between Ceretec and Neurolite*

Compound	*Accumulation in brain (%)*	*Blood clearance Agent remaining 20 min post-injection (%)*	*Half life in brain (hours)*
Ceretec	4.2 (20 min post-injection)	17	71
Neurolite	4.6 (5 min post-injection)	7–8	17

design, but in this case an N_2S_2 donor atom set is bound to the TcO^{3+} core. This type of compound was known to diffuse through the BBB but early examples showed poor retention in brain tissue. The incorporation of the $CO_2C_2H_5$ ester group, which is prone to enzymatic hydrolysis by intracellular esterases, leads to the complex being converted to hydrophilic derivatives which become trapped within the brain. As with Ceretec the process is steroselective and the *l,l* form of **70** is hydrolysed much more rapidly than the *d,d* isomer in monkey brain homogenate. Hydrolysis by esterases in the blood is slower than in brain tissue but sufficiently rapid, compared to Ceretec, to offer relatively rapid clearance of the hydrophilic products *via* the kidneys. Thus Neurolite shows similar brain uptake to Ceretec but faster blood clearance (Table 7). The shorter residence time of Neurolite in the brain is not of great clinical significance since it exceeds the half-life of ^{99m}Tc by almost a factor of three. The ^{99m}Tc-N_2S_2 core of Neurolite has been shown to be robust *in vivo* and to be excreted intact in the urine following hydrolysis.

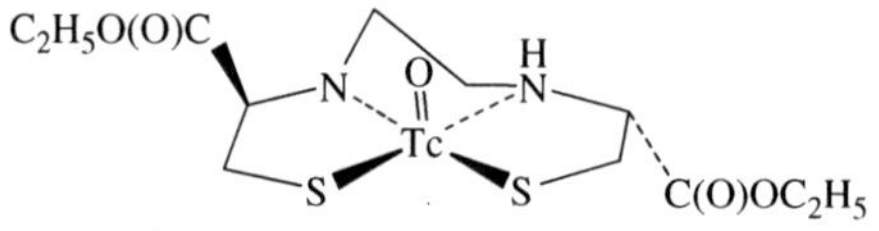

$C_2H_5O(O)C$ N O H N Tc S S $C(O)OC_2H_5$

70

Among the other complexes which have been evaluated as potential brain imaging agents, ^{99m}Tc-MRP20 (**71**) contained a trinegative ligand with an N_3O donor atom set. Although this agent showed brain uptake and the extent was similar to Ceretec or Neurolite, clearance from the blood was slow compared to Ceretec. A different structural approach is represented by the neutral 7-coordinate Tc(+3) dimethylglyoxime derivative **72** (^{99m}Tc-BATO-2MP) which was investigated at Squibb. This neutral complex is similar in conception to Cardiotec (**62**, Section 3.3.9) which was developed for myocardial imaging, both contain three dioxime binding groups linked at one end by a boroxime moiety and at the other by hydrogen bonds. Despite being taken up in the brain ^{9m}Tc-BATO-2MP clears too rapidly to compete with Ceretec or Neurolite as a potential imaging agent. The neutral technetium(+5) nitride complex $[Tc(N)\{S_2CNCH(CH_3)(C_2H_5)\}_2]$ (*cf.* **63**) has also been evaluated and biodistribution studies in mice have shown that the complex accumulated in the brain with good uptake and retention.

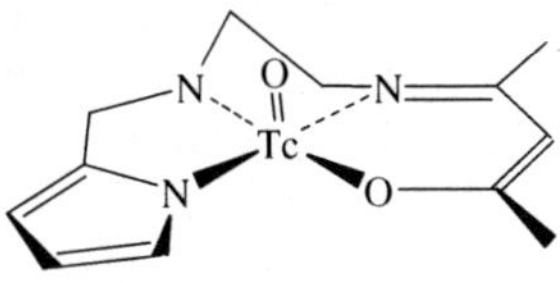

71
Tc-MRP20

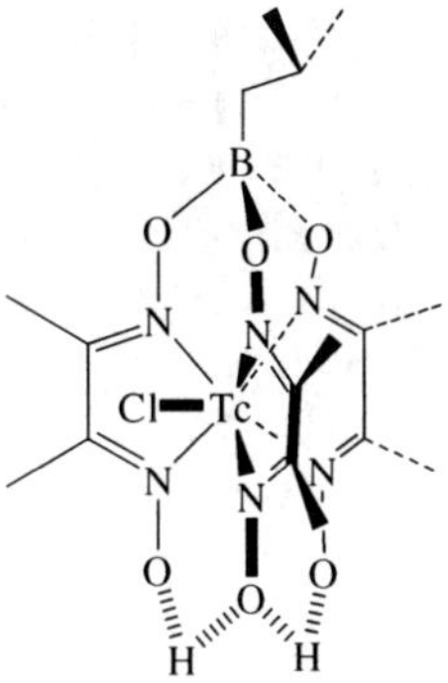

72
Tc-BATO-2MP
Tc(+3)

As yet no Ga-based rCBF PET agents have been developed. A ligand which prevents Ga^{3+} transfer to transferrin but allows transfer through the BBB followed by retention has yet to be identified. There has been interest in ^{62}Cu-PTSM as the complex is neutral and might be retained in the brain through a similar mechanism to myocardial uptake (Section 3.3.9).

3.3.10.2 Bifunctional Receptor Specific SPECT Brain Imaging Agents

Cerebral perfusion agents reveal disease through changes in regional perfusion and through metabolic abnormalities which can precede the appearance of any structural change detectable by anatomical imaging methods. Neurological disease which is not associated with such changes requires a different approach. Diseases and psychiatric conditions associated with changes in neurotransmitter receptor sites might be investigated using 'bifunctional' molecules (bioconjugates) which contain both a radiolabel and a receptor specific binding functionality. There is much interest in developing imaging agents of this type and PET imaging offers one approach but requires single atom labelling by isotopic substitution using ^{11}C or ^{13}N or halogenation using ^{18}F. This is extremely difficult due to the problems of obtaining and using short half-life positron emitters coupled with the complicated chemical manipulations involved. Similar objections apply to ^{123}I labelled SPECT imaging agents. A more attractive approach is to construct a 'bifunctional', or ''second generation', ^{99m}Tc imaging agent which contains both the receptor specific binding moiety and a chelating group capable of strongly complexing a ^{99m}Tc ion. The non-radioactive hybrid receptor binding proligand would be synthesised in advance then labelled with ^{99m}Tc immediately before use, using a 'kit' in a similar manner to Tc-essential imaging agents. However, such systems present a major design challenge since the presence of the metal centre, and its binding group, must not eliminate the receptor specific binding capability of the remainder of the molecule. As yet, no such agent has been approved for use but progress towards this goal is being made.

Target sites for Central Nervous System imaging agents include dopaminergic (Parkinson's disease), muscarinic and nicotinic (Alzheimer's disease), benzodiazapene (epilepsy) and serotoninergic (psychiatric conditions) receptors. In order to couple receptor binding agents for these sites to a γ-emitting ^{99m}Tc ion it is necessary to include a proligand structure which can form a strong, kinetically inert complex. The tetradentate N_2S_2 ligand system has featured prominently in the search for suitable bifunctional agents since examples are known to bind to Tc(+5) centres under conditions suitable for clinical use and to be excreted intact in urine after clearance from the body.

The dopamine transporter (DAT) is a receptor for cocaine, **73**, and several structurally related derivatives of this basic structure have been investigated. Simply connecting a Tc-N_2S_2 group to cocaine analogues *via* a trimethylene bridge afforded a series of complexes including Technepine, **74**. The presence of the C=O group in the Tc binding moiety facilitates deprotonation of the amide nitrogen to give a neutral -{Tc(O)[S(CH_2)$_2$NC(O)CH_2N(CH_2)$_2$S]} group. The complex exists as a mixture of diastereomers which were shown by studies of the non-radioactive Re(+5) analogue to have similar binding to the DAT (IC_{50} = 7.38 and 4.04 nM). Quite different binding to serotonin receptors was found (IC_{50} = 66.9 and 299 nM) giving selectivities for the DAT over serotonin receptor binding of 9 and 74, respectively. The mixture of diastereomers itself showed a selectivity of 21 which may be sufficient for clinical use. A different point of attachment for the Tc binding group in a cocaine analogue is used in the TRODAT system of which Tc-TRODAT-1, **75**, showed the highest brain uptake (0.43% in rats at 2 min post injection). *In vitro* studies of the Re analogue of Tc-TRODAT-1 showed a binding constant of 14.1 nM significantly higher than the value of 1.29 nM found for a high affinity DAT ligand, *N*-(3-iodopropen-2-yl)-2β-carbomethoxy-3β-(4-chlorophenyl)-tropane. This agent produced the first *in vivo* SPECT images of the D_2 DAT sites in man. Initial images showed normal rCBF behaviour clearing to show significant uptake in regions of the brain rich in DAT sites in subsequent scans at 60–80 min and 120–140 min. A disadvantage of the Tc-TRODAT system is the relatively demanding conditions for its preparation [125°C in an autoclave for 30 min, solvent extraction, HPLC] and the presence of a mixture of several isomers presumed to have different properties. A simpler preparative approach was possible using a '3+1' system in which an *N*-ethanethiol substituted cocaine analogue was treated with [$^{99m}TcO_4$]$^-$, tin(+2)-glucoheptanoate and $CH_3N(CH_2CH_2SH)_2$ at room temperature. The product Tc-NS2-Tropane, **76**, was purified by HPLC with a >95% radiochemical purity and the solid-state structure of the Re analogue of **76**, was determined. However, although the compound showed good selectivity for dopamine receptor sites, the affinity and uptake were much lower than for Tc-TRODAT-1 and unsuitable for use in human diagnostic studies.

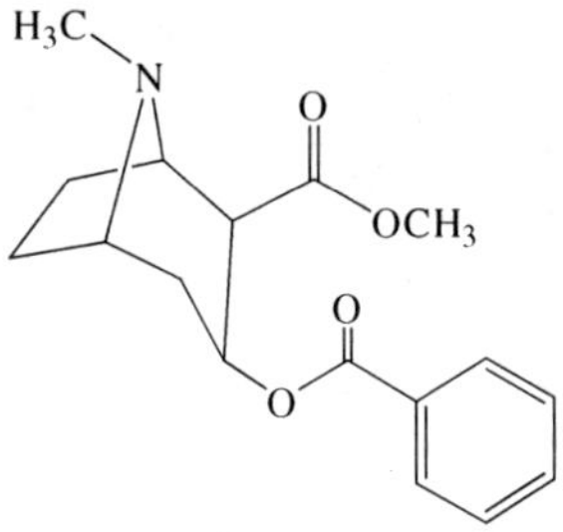

73
Cocaine

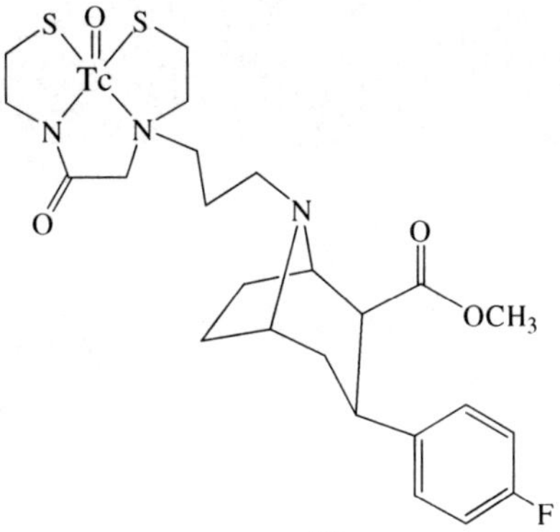

74
Technepine

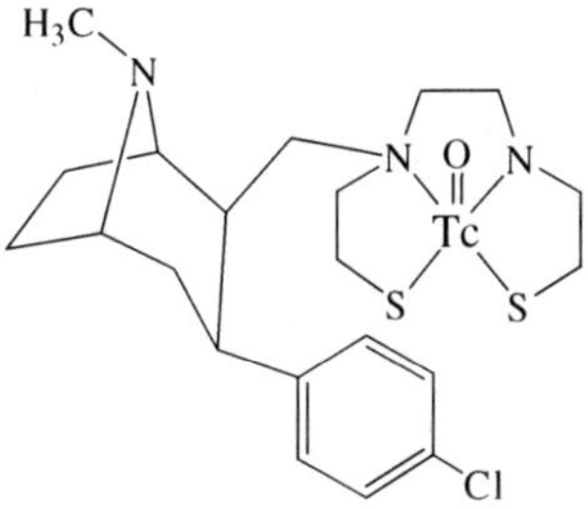

75
Tc-TRODAT-1

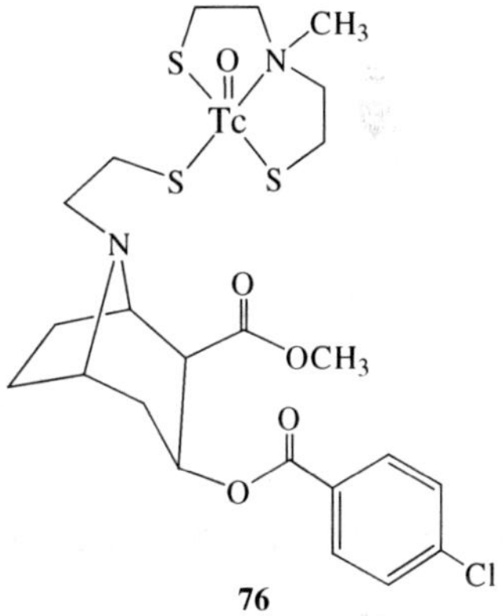

76
Tc-NS_2-Tropane

A similar '3+1' approach has been applied to targeting serotonin 5-HT receptor sites. Ketanserin, **77**, is a potent antagonist for serotonin binding at the 5-HT sites and complexes exemplified by **78** were investigated as ketanserin analogues. The complex with X=CN- and Y=O was found to have a high affinity for 5-HT receptor sites in rat brain homogenate. Attempts to develop a bifunctional imaging agent for the muscarinic acetylcholine receptor, associated with diagnosis of Alzheimer's disease, have focused on the replacement of the $(C_6H_5)_2C(OH)$- unit in quinuclidinyl benzylate, **79**, with a Tc complex. Use of a Tc-N_2S_2 binding motif as in **80**, led to a neuroreceptor binding specificity of only 15% and relatively low uptake in mouse brain (0.3% 5 min and 0.12% 1 h post injection). However, this agent did provide an early example of a Tc neuroreceptor binding agent which retained affinity for the receptor. Use of a BATO-2MP (**72**) type of Tc-complex substituent in place of the Tc-N_2S_2 unit gave an agent which showed no measurable specific binding. This emphasises the importance of selecting the appropriate structure, and hence properties, for the Tc binding site in bifunctional agents.

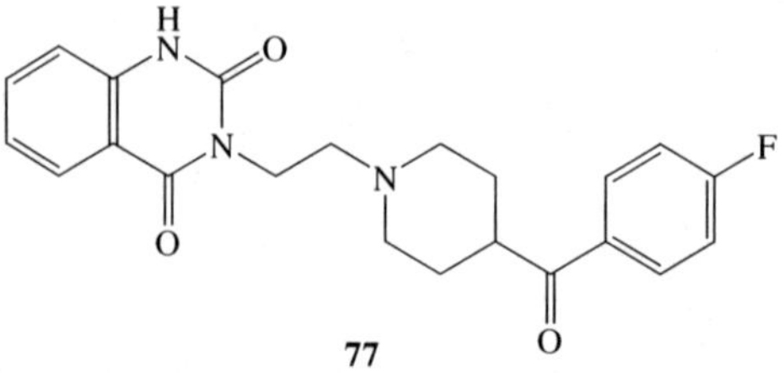

77
Ketanserin

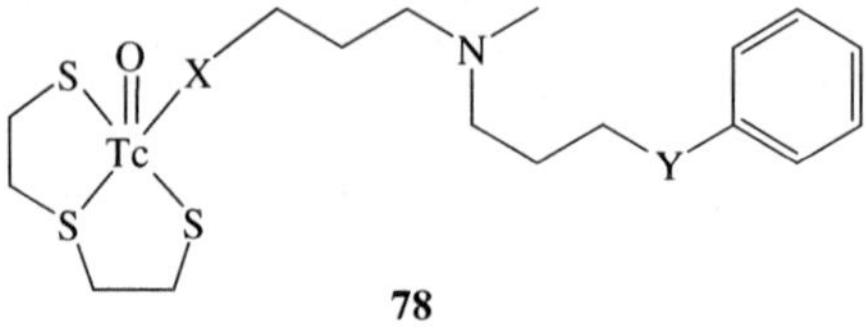

78
X = S, Y = nothing
X = CN-, Y = O

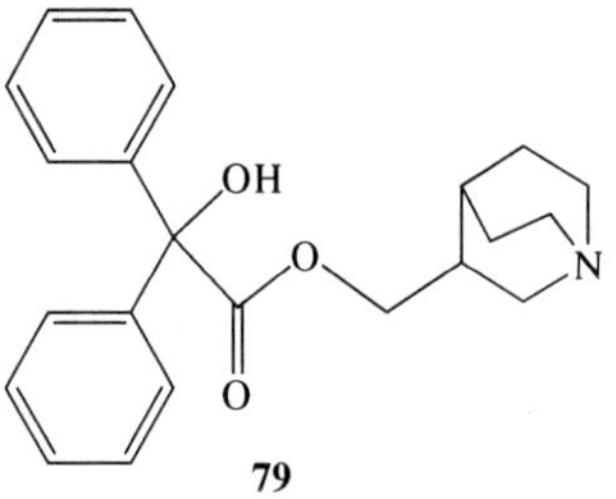

79

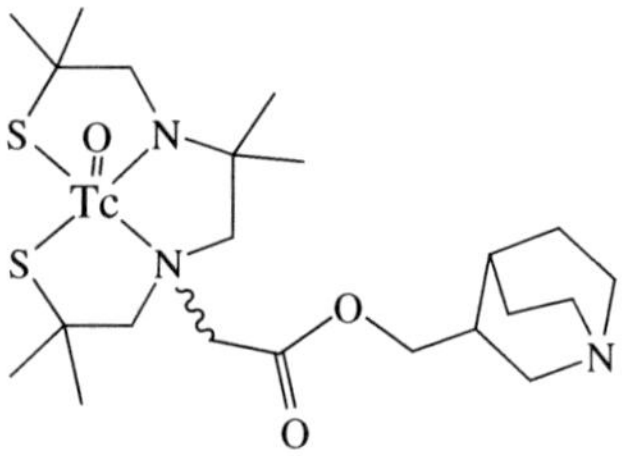

80

3.3.11 Tumors Abscesses and Hypoxia

The application of radiopharmaceutical imaging to diagnostic oncology offers the attractive possibility that malignant tissue might be identified by a functional imaging agent at an earlier stage than is possible using anatomical imaging methods. However, this presents several challenges. Cancer is not a single disease and malignant tissue may be any of various types, so no single agent can be expected to have universal application. Distinguishing malignant from normal tissue can be difficult and a successful functional imaging agent may need to exploit subtle differences in biochemistry. Furthermore, since early diagnosis of malignancy and its spread is so important, false negative results are particularly unacceptable. Historically ^{67}Ga has been the radionuclide most commonly used for imaging tumors and abscesses. However, the nuclear properties of this isotope are not ideal. The photon yield is low and the photon energies poorly matched to the optimum for imaging equipment. Furthermore ^{67}Ga is not available from a portable generator making it inconvenient to use, a problem also associated with ^{111}In, another radionuclide used in this context. As a result there is strong interest in developing ^{99m}Tc agents for imaging tumors. These have the advantage that ^{99m}Tc is conveniently available from a generator and has a near ideal photon energy with a high photon yield. Examples of tumor avid ^{99m}Tc complexes are known and there is continuing interest in what might be called 'bifunctional' agents (Section 3.3.12). These combine a radionuclide, which functions as the radiation source for imaging, with a biologically active carrier molecule which accumulates selectively in malignant tissues and functions as the target seeking part of the agent. As an example a nuclide such as ^{99m}Tc might be linked to a tumor specific antibody, which acts as a carrier to transport the radionuclide to the target site. It is, of course, important in developing radiopharmaceuticals of this type that the presence of the metal ion, and any associated metal binding groups, does not adversely affect the biodistribution of the carrier part of the molecule. Furthermore the metal ion complex itself must be sufficiently kinetically inert *in vivo* to prevent unacceptable losses of the metal ion from the carrier during the imaging procedure. Thermodynamic stability alone is insufficient, particularly when using hard metal ions such as In^{3+}, Y^{3+} or lanthanide(+3) ions.

3.3.11.1 Gallium and Indium Complexes for Diagnostic Oncology

A primary use of ^{67}Ga is during the treatment of Hodgkin's and non-Hodgkin's lymphoma to assess tumor viability. Under physiological conditions only oxidation number +3 is important in the chemistry of gallium and the ionic radius of Ga^{3+} (62 pm, CN6) is very similar to that of high-spin Fe^{3+} (65 pm, CN6). Since high-spin Fe^{3+} is a d^5 system, having no ligand field stabilisation energy in complexes with weak field ligands, its aqueous chemistry is rather like that of Ga^{3+}, which is a closed electron shell ion of a p-block element. The larger (82 pm) In^{3+} ion, below Ga in the Periodic Table, also shows some similarities to Ga^{3+} and Fe^{3+} in its behaviour. All three ions may be described as hard 'class a' acceptors which form more stable complexes with oxygen donor ligands. Their complexes are labile and the ions are prone to hydrolysis in aqueous media, forming insoluble precipitates of the metal hydroxides. In order to suppress this hydrolysis reaction $^{67}Ga^{3+}$ is administered as its citrate complex. Once in the blood stream $^{67}Ga^{3+}$ is scavenged from citrate by the iron transport protein transferrin (Tf) and equilibrium constants of $K_1 = 10^{23.7}$ and, more recently, $10^{20.3}$ mol^{-1} have been reported for the binding of Ga^{3+} to transferrin. The binding of Ga^{3+} to Tf is enhanced by bicarbonate and correlates well with the capacity of the serum to bind free iron so that Fe^{3+} and Ga^{3+} have been said to follow kinetically equivalent transport pathways. The binding of ^{67}Ga labelled Tf to cell surface receptor sites has been proposed to account for the tumor uptake of $^{67}Ga^{3+}$. Another iron binding protein, lactoferrin (Lf), can compete with Tf for Ga^{3+} and, since levels of Lf may be elevated in the region of tumors as part of an inflammatory response, it has been proposed that this serum protein may serve to localise ^{67}Ga in tumors. A good blood supply to the tumor is necessary for useful images to be obtained using $^{67}Ga^{3+}$ but activity remaining in the blood may mask the image of the target site. This may be countered by injecting iron to promote urinary excretion of the residual $^{67}Ga^{3+}$ in the blood by saturating the potential Ga^{3+} binding sites in normal tissue and blood.

A more targeted approach to the imaging of malignant tissue is provided by the ^{111}In radiopharmaceutical OctreoScan (Mallinckrodt/Tyco Healthcare) designed to mimic somostatin (**81**), a regulator of hormone secretion. Somostatin receptors are expressed in a number of malignant conditions including small-cell lung cancer, neuroblastoma, carcinoids, gastrinomas and paragangliomas. Since somostatin has a very short biological half-life of a few minutes, a more long-lived compound, Octreotide (**82**), with similar affinity for malignant tissue is used to act as a carrier for the radionuclide. Octreotide is a smaller molecule than somostatin but preserves the -Phe-Trp-Lys-Thr- part of the molecule which is recognised by somostatin receptors. In order to incorporate $^{111}In^{3+}$ into the molecule it is necessary to attach a suitable metal ion binding site and this can be done using dtpa-anhydride (**6**, Scheme 8) to form an amide link with Octreotide. The use of one dtpa carboxylate in amide formation prevents it from participating in metal binding but, since reasonably stable In^{3+} complexes with coordination number 6 or 7 are known, the remaining 3 nitrogen and four carboxylate donor groups appear sufficient to saturate the $^{111}In^{3+}$ coordination sphere. However, the larger Y^{3+} ion (ionic radius 90 pm, CN6), for which coordination number 8 is typical, forms a less stable complex with modified Octreotide and significant losses of $^{90}Y^{3+}$ from the radiopharmaceutical are found.

(NH_2) Ala–Gly—Cys—Lys—Asn—Phe—**Phe**—**Trp**—**Lys**—**Thr**—Phe—Thr—Ser—Cys (CO_2H)

81
Somostatin

(NH_2)*d*-Phe—Cys—**Phe**—***d*-Trp**—**Lys**—**Thr**—Cys—Thr(CO_2H)

82
Octreotide, OctreoScan

3.3.11.2 Technetium Complexes in Diagnostic Oncology

Some molecular complexes of ^{99m}Tc have shown promise in specific diagnostic oncology applications. The positively charged isonitrile complex **55c** (Cardiolite) has been found to accumulate in some tumors and has received FDA approval for use in mammography under the name Miraluma. A tumor-avid $^{99m}Tc(+5)$ complex can be obtained with *meso*-dimercaptosuccinic acid, *meso*-dmsaH_4 normally used for renal imaging, by raising the pH used in forming the radiopharmaceutical. The kit used to prepare a radiopharmaceutical for kidney imaging involves the reaction of *meso*-dmsaH_4 with $[^{99m}TcO_4]^-$ and Sn Cl_2 at pH 4.5 to form what is thought to be a Tc(+3) complex. Use of a higher pH (*ca* 7.5–8.0) favours the formation of a higher oxidation number Tc(+5) complex which can be used to evaluate medullary thyroid carcinoma, brain metastases and breast cancer. The complex is presumed to have the usual pyramidal structure found for tetrathiolate complexes of TcO^{3+} but three isomers are possible, *anti* **83**, *syn exo* **84**, or *syn endo* **85**.

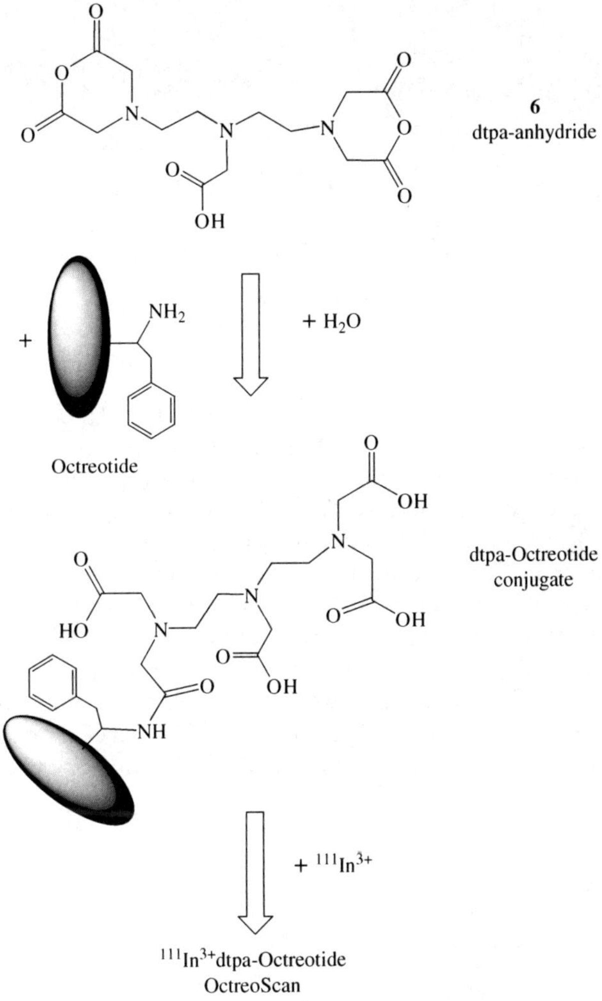

6
dtpa-anhydride

+ H_2O

Octreotide

dtpa-Octreotide conjugate

+ $^{111}In^{3+}$

$^{111}In^{3+}$dtpa-Octreotide
OctreoScan

Scheme 8

Although it may not be similar to that of the pharmacologically active Tc complex in Tc-DMSA, the solid state structure of the related Re(+5) complex $[(C_2H_5)_4N][Re(=O)(dmsaH_2)_2].1.5H_2O$ was found to have the *syn endo* structure. More recently ^{99m}Tc labelled transferrin has been prepared using a 'transfer ligand' approach in which a preformed ^{99m}Tc complex is used as an intermediate from which to prepare labelled transferrin. Labelling efficiencies of $>70\%$ are claimed using this approach offering a route to a ^{99m}Tc radiopharmaceutical with similar *in vivo* properties to ^{67}Ga-citrate, but with superior radiological properties. In trials ^{99m}Tc-labelled transferrin allowed detection of a 2 mm diameter tumor in a mouse.

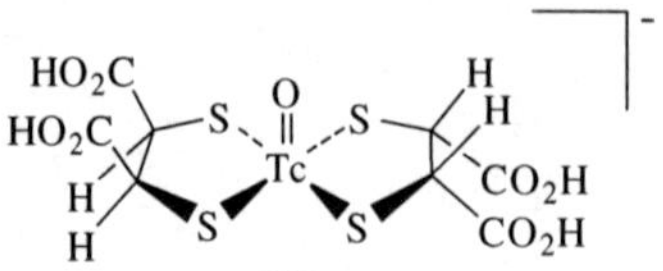

83
anti

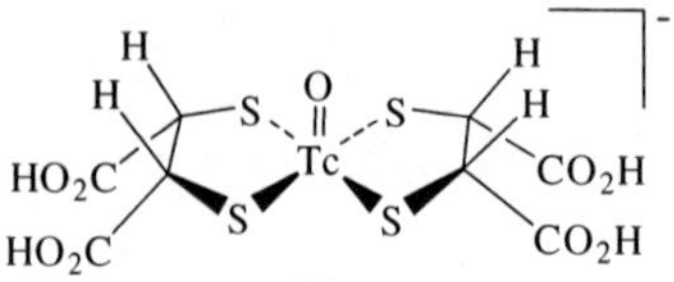

84
syn exo

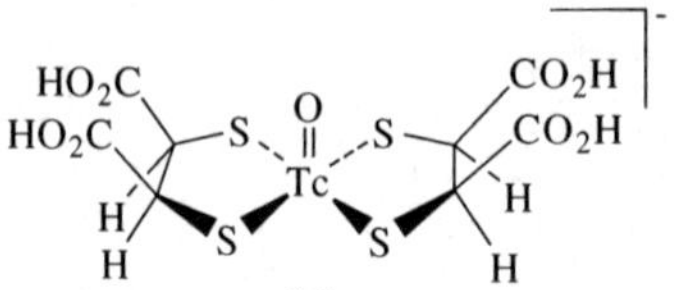

85
syn endo

Depreotide (NeoTect®, Diatide then Schering), a ^{99m}Tc counterpart to the ^{111}In agent Octreoscan received FDA approval in 1999 and can be used for imaging non-small cell lung cancer. It has also shown promising results in breast cancer, non-Hodgkin's lymphoma and Hodgkin's lymphoma patients. Like Octreoscan, Depreotide is a bifunctional agent but differs in the type of binding site used to retain the radionuclide. In the case of $^{111}In^{3+}$-Octreoscan an appended polyamine carboxylate group formed the ligand whereas in ^{99m}Tc-Depreotide an appended tripeptide chain provides a binding group for TcO^{3+} as shown in **86**. The biologically active part of the structure has been modified to eliminate the disulfide linkage of Octreotide which would be susceptible to reduction during the formation of TcO^{3+} from $[TcO_4]^-$. This allows ^{99m}Tc labelling under the usual reductive conditions without compromising the biological activity of the carrier.

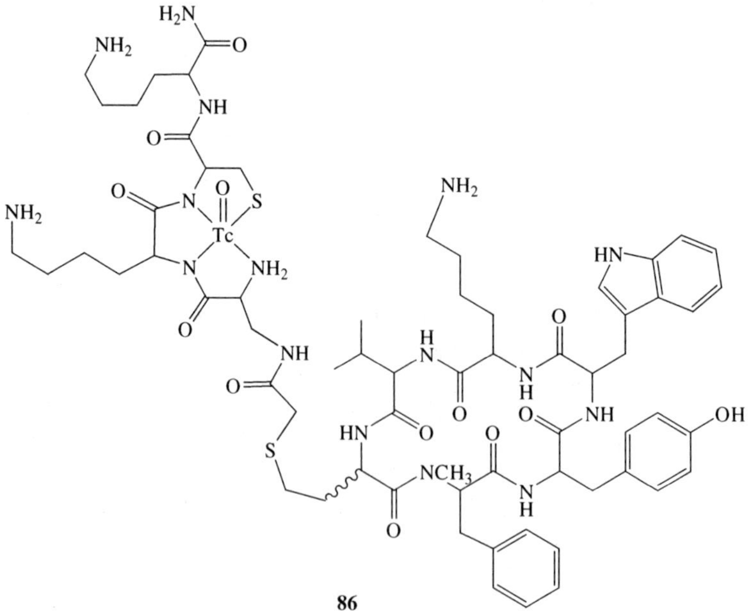

86
Tc-Depreotide (P829)

3.3.11.3 Antibody and Fragment Antibody Imaging Agents

Another directly targeted approach to tumor imaging is to use, as carriers for radionuclides, antibodies (Ab) or antibody fragments which are specifically bound by malignant tissue. Such tumor specific monoclonal antibodies can be prepared and modified chemically to provide suitable binding sites for metal ions. Since intact antibodies are cleared rather slowly from non-target tissue, at least a day must usually elapse before the background radiation falls sufficiently to allow successful imaging of the target tissue. This may be acceptable radiologically if a reasonably long-lived radionuclide such as ^{111}In with a half-life of 2.8 days is used, but is inconvenient for the patient. However, if ^{99m}Tc with a half-life of about 6 h is used, most of the injected radiation source will have decayed before imaging is feasible. This difficulty can be avoided by using antibody fragments which clear much more rapidly from non-target tissues.

The labelling of intact antibodies with $^{111}In^{3+}$ requires that a suitable metal ion binding site is linked to the antibody. Several methods are available to achieve this, dtpa anhydride, **6**, can be used to link to proteins through surface -NH_2 groups in a similar manner to the preparation of OctreoScan. The carbonic anhydride **87** offers an alternative reagent for forming the dtpa-antibody conjugate in a reaction involving amide formation with the evolution of CO_2. However, both of these approaches consume one carboxylate group of the dtpa so it is able to offer a maximum coordination number of 7 when binding to a metal ion. This affords sufficiently stable complexes with In^{3+} but complexes of larger ions such as lanthanide (+3) or Y^{3+} requiring 8-coordination show insufficient stability for radiopharmaceutical applications. This difficulty can be avoided by using a dtpa derivative functionalised with a protein binding group so that the dtpa in the conjugate retains its ability to offer a maximum of 8-coordination to a metal ion. An example of this is provided by the aryl thiocyanate derivative of $dtpaH_5$, **88** (Scheme 9). The application of ^{111}In labelled dtpa-antibody conjugates in oncology is exemplified by two radiopharmaceuticals. OncoScint (Cytogen), derived from the B72.3 murine antibody, was used for imaging colorectal and ovarian cancers but is no longer on the market and ProstaScint® (Cytogen), derived from the murine monoclonal antibody 7E11-C5.3 conjugated to the linker and chelator glycyl-tyrosyl-(*N*,-diethylenetriaminepentaacetic acid)-lysine hydrochloride, is used for preoperative staging of prostate cancer.

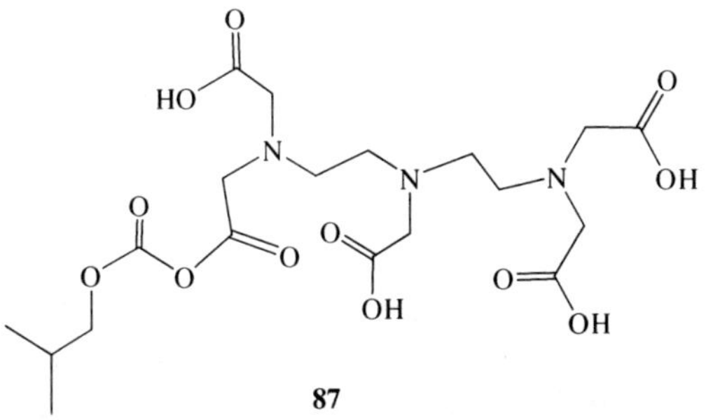

HO O N N N O OH O O O O O OH O OH

87

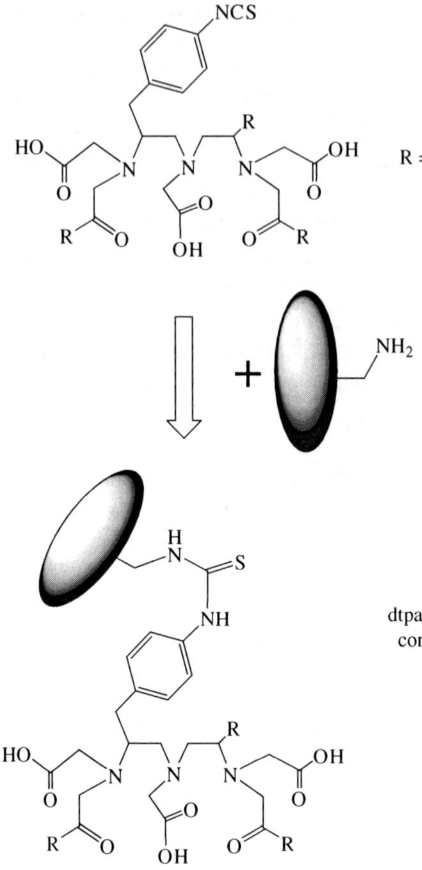

NCS

88
R =H, CH_3

NH_2
Protein

dtpa-protein conjugate

Scheme 9

The use of antibody fragments as carriers for tumor targeting allows shorter half-life ^{99m}Tc to be used in imaging procedures. In 1996 FDA advisory committees recommended market approval for CEA-scan (Immunomedics) to be used in the detection of recurrent and/or metastatic colorectal cancer. This radiopharmaceutical is made from a fragment of the murine monoclonal IMMU-4 antibody which has been reduced and labelled with ^{99m}Tc. This radiopharmaceutical associates with carcinoembryonic surface antigen (CEA), a tumor marker for colon and rectal cancer, providing information about the presence, location and extent of disease. In the USA there are approximately 134,000 new cases of colorectal cancer each year with an estimated death toll of over 55,000. The procedure for preparing the radiopharmaceutical exploits the affinity between thiolate groups and TcO^{3+}. Disulfide bridges (–S–S–) in the antibody fragment are reduced to thiols (–SH HS–) which can then act as donor atoms for technetium. Since the sulfur atoms originate from a disulfide link they must be able to adopt adjacent locations in the antibody fragment so are suitable for chelating to the metal ion (Scheme 10). Labelling is possible using

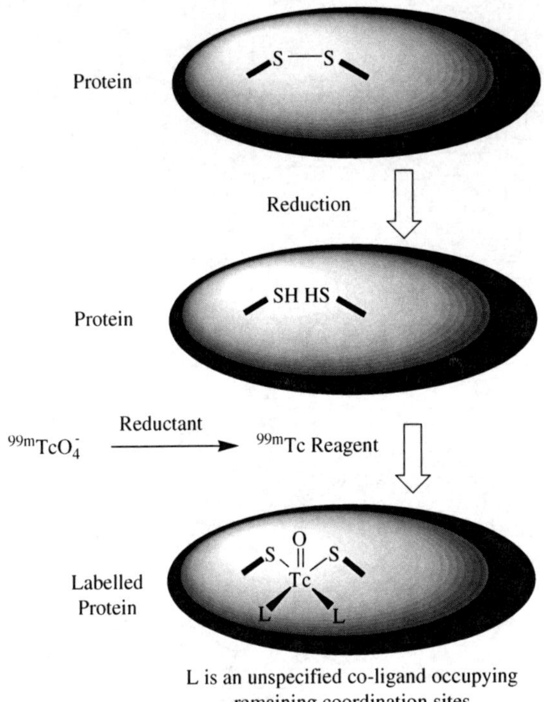

Scheme 10

$[^{99m}TcO_4]^-$ and a reducing agent in the conventional manner. CEA-scan shows rapid blood clearance and has a biological half-life of about 13 h giving improved target-to-background ratios. At 1, 5 and 24 h after infusion blood levels of 63, 23 and 7% of the injected dose are claimed, allowing imaging within 5 h of administration. Over 24 h 28% of the agent was excreted in the urine. Verluma (Dupont-Merck then Bristol-Myers Squibb) is another ^{99m}Tc labelled antibody fragment, which received marketing approval in 1996 for staging small cell lung cancer but is no longer on the market. In this case a preformed chelating group (**89**, Scheme 11) containing an N_2S_2 donor atom set and a reactive ester group was used to form a ^{99m}Tc complex. This was then reacted with the antibody fragment to form amide links with surface -NH_2 groups giving a bifunctional agent as shown in Scheme 11.

3.3.11.4 Hypoxia

Malignant tumors may grow at a rate which outstrips the development of a sufficient blood supply. This leads to hypoxic regions where the oxygen supply is inadequate. This offers another means of imaging tumors in that compounds which become trapped in hypoxic regions might also act as tumor specific carriers for radionuclides. A hypoxia seeking agent might have other applications since the condition is associated with myocardial ischemia and brain

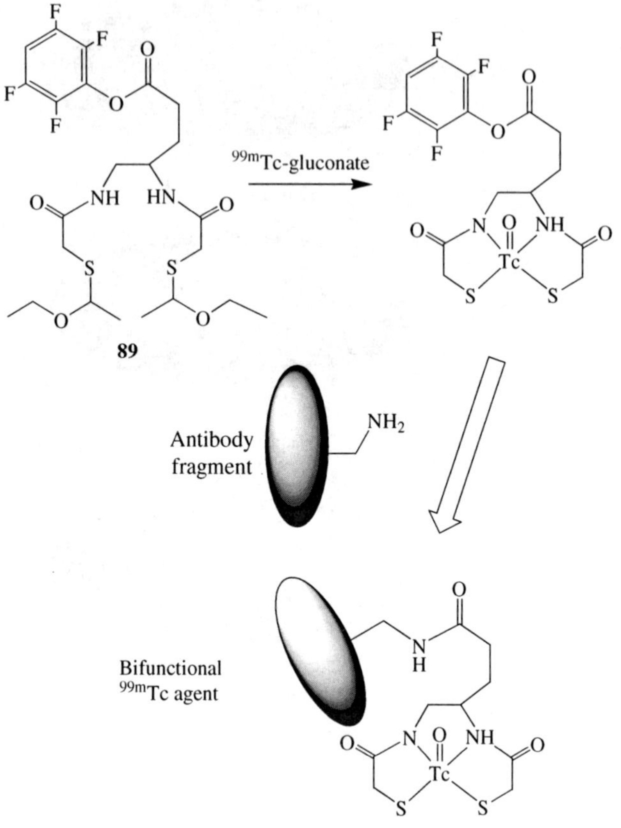

^{99m}Tc-gluconate

89

Antibody fragment

Bifunctional ^{99m}Tc agent

Scheme 11

tissue likely to be affected by stroke. The problem in developing hypoxia seeking agents is that the condition arises when blood supply is restricted. This will also limit the delivery of the agent to the target tissue so that the radiopharmaceutical may have to remain in circulation for some time in order to achieve sufficient uptake. This contrasts with the usual objective of rapid uptake and rapid clearance of the agent from the bloodstream.

Compounds containing nitroimidazole (**90**) groups have been used to carry radioactive halogens (^{18}F, ^{82}Br, ^{131}I) into hypoxic tissue where the nitroimidazole group is reduced to amine. The radiolabelled amine is then selectively retained in the hypoxic cells. Attempts have been made to extend this approach to ^{99m}Tc agents by adding nitroimidazole substituents to known ^{99m}Tc-complexes used as radiopharmaceuticals. As examples derivatives of two cerebral perfusion agents have been prepared. Replacing the boron substituent in **72** with a nitroimidazole containing group gave a compound which was reduced by xanthine oxidase under anaerobic conditions but the rate was too slow to be useful. In animal studies using a less bulky metal complex in the form of **91** (derived from **68**) gave better results in terms of relative retention in ischemic myocardial tissue. Unfortunately the compound showed high liver uptake, which partly obscures the heart in imaging procedures. A more subtle metal centred approach involved modifying the structure of **68** to change the redox potential of the complex and eliminate the need for the nitroimidazole substituent. The inclusion of an additional -CH_2- group in the central chelate ring to give **92** made the complex easier to reduce. Limited human trials showed hypoxic tumor accumulation of **92** in 7 out of 10 patients with lower liver uptake than **91**. These studies illustrate how subtle changes in ligand structure can be exploited to good effect in the development of new radiopharmaceuticals.

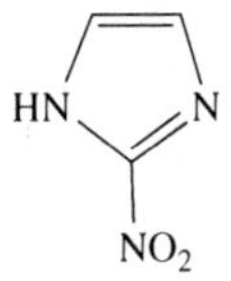

90 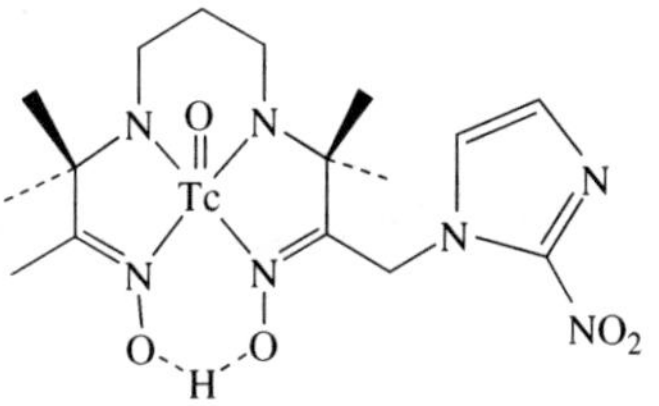 **91** 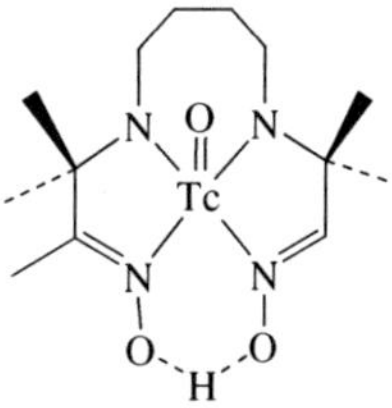 **92**

3.3.12 Synthetic Approaches to Bifunctional ^{99m}Tc Radiopharmaceuticals

The preceding Sections 3.3.11.2 and 3.3.10.2 contain some examples of what are known as bifunctional radiopharmaceuticals. That is they contain both a biologically active carrier moiety, which selectively binds to a specific receptor *in vivo*, and a radioactive metal ion. In order to form a bifunctional agent it is necessary to covalently attach, to the carrier part of the agent, a metal binding site, which can form a kinetically inert complex with the radionuclide to be used. In principle the radionuclide might be incorporated in the metal binding site before (pre-labelling) or after (post-labelling) its attachment to the carrier (Scheme 12). However, with short-lived nuclides such as ^{99m}Tc, the time required for linking the complexed metal to the carrier, and any subsequent purification step will be very limited and only rapid simple chemical procedures will be acceptable. If the procedure is at all time consuming the post labelling approach will be more appropriate. Thus the ease and efficiency of the chemistry needed to incorporate the metal into the binding site are important considerations in the design of bifunctional agents.

If a bifunctional agent is to be effective, the change in charge and structure resulting from the attachment of the metal and its binding site must not significantly impair the ability of the carrier moiety to recognise and bind to its target receptor. This may place limitations on the type of binding site used and the means by which it is attached to the carrier. Consequently, it is important to allow for flexibility in the design of the binding site. It is not only important that the binding site forms a sufficiently inert complex with the metal ion of choice, its presence must also be compatible with the carrier

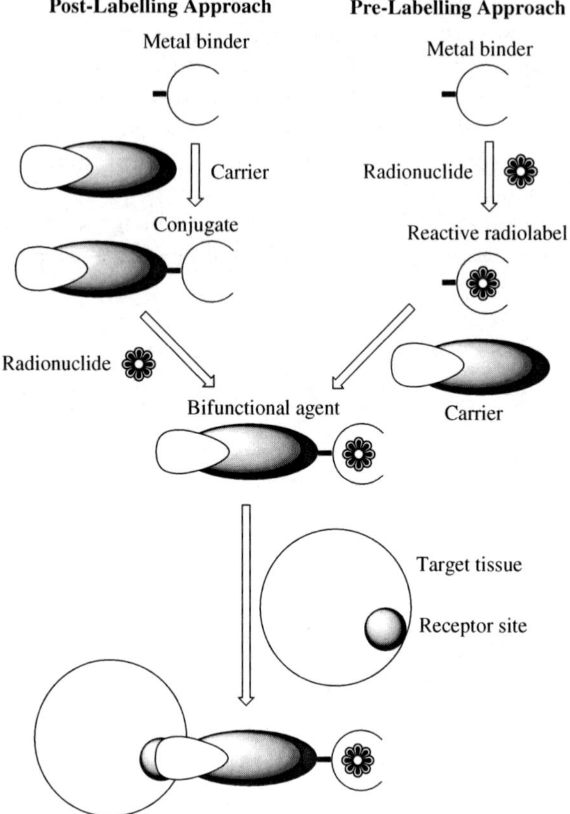

Selective radionuclide uptake at receptor site

Scheme 12

moiety. As an example dtpa anhydride can be an effective reagent for adding a metal binding site to a protein but requires a suitably located functional group, such as a free amine on the protein surface, with which to form a covalent link. In some cases the large size of the attached metal complex might affect the biological activity of the conjugate so that a less bulky binding group may be needed. It is also necessary to consider the charge and lipophilicity of the attached metal complex. The binding of a metal ion of charge 3+ to a binding site with charge 4− would add a charge of 1− on the surface of the carrier. The acceptability of such a change must be considered. Similarly, if the metal complex is more lipophilic or hydrophilic in character than the carrier the effect on the biodistribution of the bifunctional agent will need to be considered. Past experience can be a useful guide and computer modelling offers an increasingly powerful means of streamlining the development of bifunctional agents. However, a positive outcome cannot be predicted with certainty and there remains an element of trial and error in developing successful bifunctional agents. In some cases it may be necessary to change the chemical form of the metal ion or the binding group. It may, for example, be necessary to change the linking group to one reacting with a different type of functional group found only in a location more remote from the active site of the carrier.

The exceptionally good nuclear properties of ^{99m}Tc make it the usual radionuclide of choice for diagnostic imaging; furthermore the rich chemistry of Tc offers a variety of ways in which ^{99m}Tc might be attached to a biologically active carrier. Thus the development of new highly target specific bifunctional ^{99m}Tc radiopharmaceuticals is a very active area of research where coordination chemistry is being exploited in the design of imaging agents. Historically the Tc(+5) TcO^{3+} centre has been widely used in nuclear medicine and is easily accessible from $[TcO_4]^-$ Not surprisingly therefore, the design of binding groups for TcO^{3+} is an important feature of the approaches being used. However, other Tc(+5) centres are possible which offer lower charges so that $TcO_2{}^+$ and TcN^{2+} provide Tc centres with charges +1 and +2, respectively. The 'naked' Tc^+, Tc^{3+} or Tc^{4+} ions offer alternative metal centres. In particular, the Tc(+1) organometallic complexes are of interest following the development of isonitrile complexes for heart imaging and, more recently, the chemistry of the $\{Tc(CO)_3\}^+$ moiety (Section 3.3.4.4). Among the higher oxidation numbers Tc(+6) is found in the TcN^{3+} core of $[TcNCl_4]^-$ (X=Cl, Br) which can be prepared from $[TcO_4]^-$ by reaction with azide, $N_3{}^-$, in the presence of excess HX. Other Tc(+6) species tend to be unstable, readily converting to other oxidation states. The highest oxidation state, Tc(+7) is typically encountered in the form of $[TcO_4]^-$ although a few complexes containing Tc(+7) are known. The utilisation of individual Tc oxidation states in the formation of bifunctional ^{99m}Tc-radiopharmaceuticals is considered in a little more detail in subsequent sections.

3.3.12.1 Linking Groups and Labelling Methods

The choice of a suitable binding site for a metal will obviously depend on the nature of the metal ion. Hard metal ions such as Ga^{3+} or In^{3+} can be effectively bound by polydentate amine carboxylates. Closed electron shell ions like these d^{10} systems form kinetically labile complexes so to augment the thermodynamic stability it is helpful to use binding groups which fully saturate the metal ion coordination sphere and block the approach of competitor ligands. Although TcO^{3+} complexes with polyamine carboxylates are also known, the widely used TcO^{3+} centre is more usually found in complexes with 'softer' N and S donor atom systems. The partly filled d-subshell allows some π–donation from amide $>N^-$ or thiolate S^- to the Tc(+5) centre, helping to make the complex more kinetically inert (Section 2.5.2). The presence of large S-donor atoms also helps in this respect by occupying the space around the metal ion more effectively than smaller N or O donor atoms.

The efficiency and ease with which the radionuclide is incorporated into a metal binding site is an important aspect of radiopharmaceutical preparation. If high temperatures, long reaction times or high pH are necessary to incorporate the radionuclide label into the binding site this may limit the clinical application of the method. If the carrier part of a bifunctional agent contains protein with structurally important disulfide links, any reagent used to reduce $[TcO_4]^-$ must be selected so as not to denature the protein by also reducing the disulfide bonds to dithiol. Such reduction would also create competitor binding sites for Tc leading to a labelled protein which may have undesirable properties *in vivo*.

Any of a variety of methods might be used to link a metal binding group to a biologically active carrier molecule. Perhaps the simplest is to use a substituent atom, such as $-S^-$, on the carrier moiety to bind directly to the metal centre. To be effective this direct monodentate means of attachment requires the formation of a particularly kinetically inert bond to the metal but is being applied in some ^{99m}Tc agents. More often some form of polydentate chelating group will be attached to the carrier by means of a covalent bond and some examples of this have already been mentioned in Section 3.3.11.3. Specifically the use of dtpa anhydride, or an active ester of dtpa, to attach a polyamine carboxylate group and the use of an aryl thiocyanate substituent on dtpa as a means of avoiding the loss of one carboxylate group when forming a link. Some examples of the various types of chemical approach which may be used to connect a metal binding group to a biologically active carrier are shown in Scheme 13. One approach attracting particular interest is the use of short sequences of amino acids designed to provide a suitable metal binding site as in the example of Tc-Depreotide (**86**). The use of a peptide sequence in labelling a protein is attractive in that it merely involves a small extension of the protein structure. This can simplify the chemistry of adding the metal binding sites as methods of extending an amino acid sequence from the carboxylate or amine terminus are well developed. Furthermore it does not add chemically dissimilar substituents to the protein.

It is important that the link joining the carrier and metal binding site is stable under physiological conditions and that the chemistry involved in its formation is simple, efficient and suited to the particular purpose. For example linking strategies which rely on amide bond formation will require a suitable free amine to be located on the surface of the carrier. If required free thiol groups could be created by the reduction of disulfide bonds in proteins but, if the presence of any disulfide in the protein is important for the protein structure and/or function, a reductive approach would not be suitable. Thus careful selection of the linking group is important along with selection of a binding site structure appropriate for the radionuclide used.

3.3.12.2 Labelling with Tc(+1) or Tc(+2)

In the context of radiopharmaceuticals Tc(+1) and Tc(+2) are most clearly associated with heart imaging (Section 3.3.9.1) in the form of the successful Tc(+1) isonitrile complexes $[Tc(CNR)_6]^+$ (*e.g.* **55c**) and the ineffective Tc(+2) chelating diphosphine dihalide complexes $[TcCl_2\{P(R_2)CH_2CH_2P(R_2)\}_2]$ (*e.g.* **23**). The neutral monodentate isonitrile ligand itself would not appear to offer a promising binding group for incorporating ^{99m}Tc in a bifunctional agent. It does not offer additional stability through chelation and Tc(+1) is not

LINKING GROUP **CARRIER**

represents metal binding group

anhydride

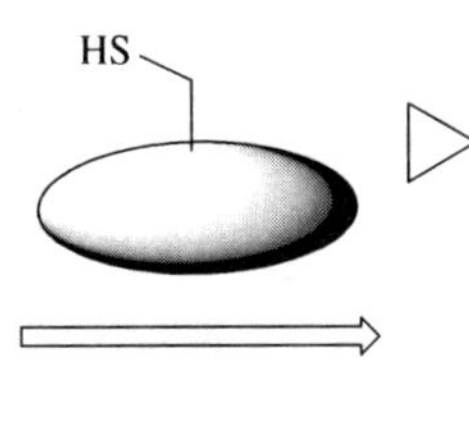

maleamide

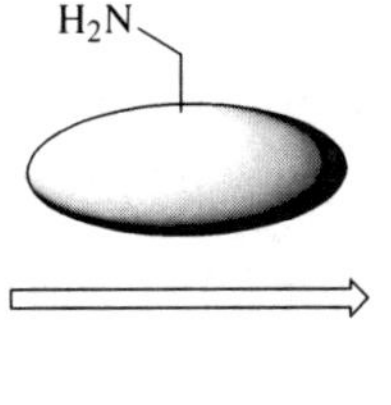

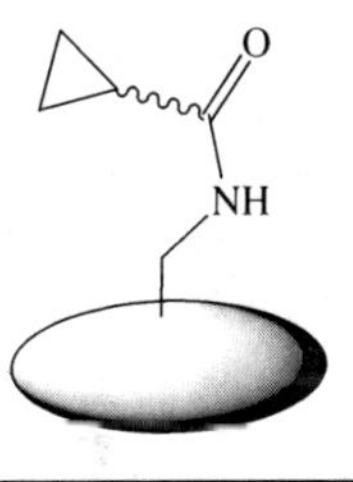

activated ester

O—Z

e.g. O—N or O—

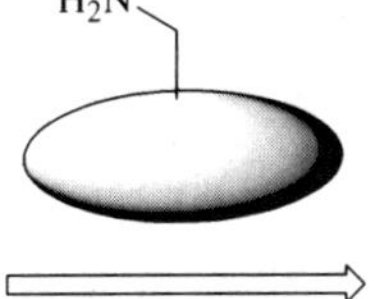

aryl thiocyanate

amine/amino acid

$-H_2O$

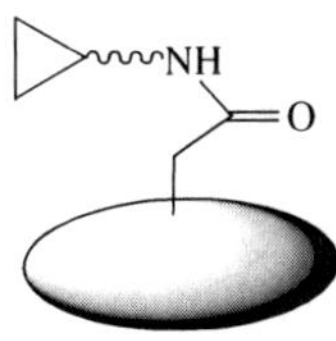

Scheme 13

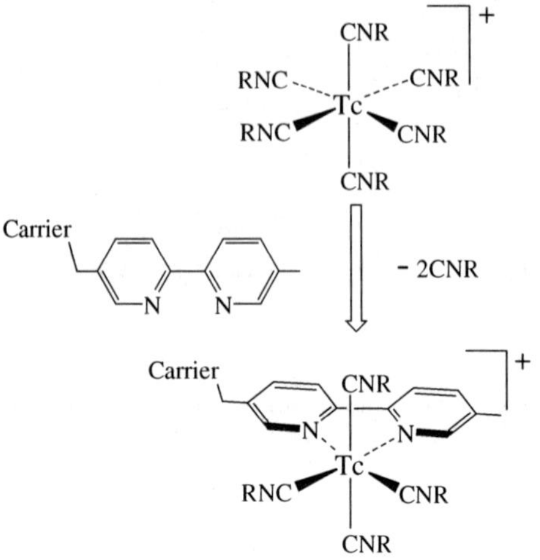

Scheme 14

93

easily accessible from $[TcO_4]^-$ using very simple chemical procedures. The use of $[Tc(CNR)_6]^+$ in which R is a group reactive towards a carrier molecule might be possible, but the risk of cross linking and other unwanted side reactions must be considered when more than one reactive centre is present. A more viable approach might be to use ligand substitution reactions of $[Tc(CNR)_6]^+$, in which R is an unreactive group, with a chelating group such as bipyridyl appended to a carrier (**93**, Scheme 14). This might offer a better potential route to bifunctional agents but an important feature of the d^6 $[Tc(CNR)_6]^+$ system is that it is sufficiently kinetically inert for radiopharmaceutical use. This suggests that the substitution reactions are likely to be slow or inefficient under the relatively mild conditions acceptable for clinical use.

A much more promising approach to labelling with Tc(+1) involves *fac*-$[Tc(CO)_3(H_2O)_3]^+$ (**42**, Section 3.3.4.4). The labile water ligands in this complex offer a potentially versatile approach to the synthesis of Tc(+1) radiopharmaceuticals (Scheme 6). Ligating groups with soft donor atoms such as S or P would seem best suited to binding the Tc(+1) centre, a particular example being provided by the preparation of a $\{Tc(CO)_3\}^+$ complex **94** with a chelating thioether substituent attached to estradiol. The estradiol carrier binds to estrogen receptors and can act as a marker for breast cancer. In another example this approach was used to prepare **95** designed to target the DAT (Section 3.3.10.2). The chelating diphosphine complexes **96** are structurally similar with respect to the mode of Tc binding and may offer another route to Tc(+1) bifunctional agents. The complexes **96** showed good *in vitro* stability and were cleared though the renal and hepatobiliary systems efficiently. These early developments with $\{Tc(CO)_3\}^+$ moiety have stimulated much interest in its use for the preparation of new radiopharmaceuticals and may be expected to bear fruit in due course.

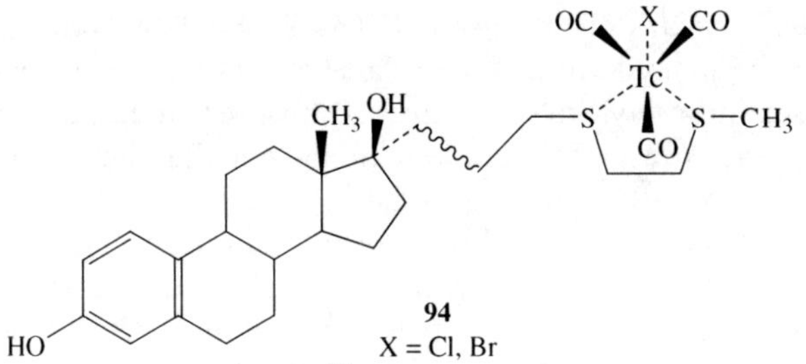

94
X = Cl, Br

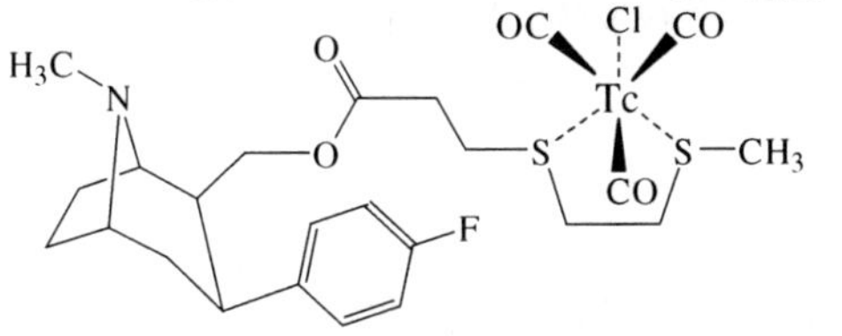

95

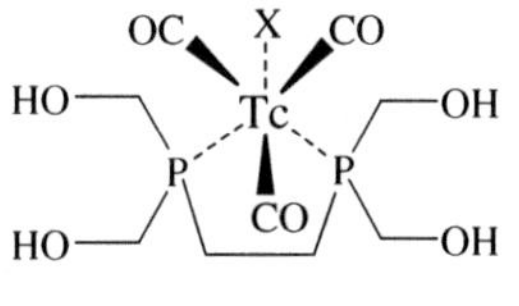

96
X = Cl, z = 0; X = H_2O, z = +1

The chemistry of Tc(+2) is not well developed, though complexes of the type $[TcCl_2\{E(R_2)\text{-}Z\text{-}Z\text{-}E(R_2)\}_2]$ (E = P, Z–Z = CH_2CH_2; E = As, Z–Z = 1,2-C_6H_4) are well known. In early studies of heart imaging agents the redox properties of the $[TcCl_2\{P(CH_3)_2CH_2CH_2P(CH_3)_2\}_2]^{+1/0}$ system were problematic in the formulation of a radiopharmaceutical (Section 3.3.9.1). Furthermore Tc(+2) compounds of this type are not simple to prepare as the $Tc(=O)_2^+$ derivatives form first and require further chemical reduction to convert them to Tc(+2) (Scheme 4). The ease of oxidation of Tc(+2) complexes and the convoluted chemistry which can be required to obtain them are probably important contributors to the absence of Tc(+2) complexes among radiopharmaceutical formulations.

3.3.12.3 Labelling with Tc(+3) or Tc(+4)

The 'naked' metal ions Tc(+3) and Tc(+4) have not featured prominently in research into bifunctional agents. Complexes of Tc(+3) which have appeared in radiopharmaceutical development include the phosphine complexes $[TcCl_2\{P(CH_3)_2CH_2CH_2P(CH_3)_2\}_2]^+$. These can be formed directly from $[TcO_4]^-$ and were first investigated as potential heart agents (Section 3.3.9.1). However, potential problems associated with using chelating phosphine ligands in the preparation of bifunctional agents are that quite forcing chemical conditions may be needed to form the Tc(+3) complex, and phosphines are prone to oxidation forming phosphine oxides [*e.g.* $(CH_3)_3P = O$] which would make them ineffective ligands. The presence of significant amounts of such oxidised ligating sites in a bifunctional agent formulation would reduce the proportion of ^{99m}Tc present in any given amount of the agent taken up by receptor sites, and hence the effectiveness of the agent.

This problem does not arise with the boronic acid tris(dioxime) ligands (R-batoH_3^-) of the type found in Cardiotec, **62**, and in **72** and which have featured in some approaches to the preparation of bifunctional agents containing Tc(+3). The use of arylthiocyanate substituted derivatives of the type illustrated by **97** provided a means of linking the R-batoH_3 binding site to a carrier. This approach has been used for antibody labelling and to prepare an agent for targeting muscarinic acetylcholine receptors. However, in the latter case the agent did not show any measurable specific binding but did show rather high non-specific binding. This was attributed to the effects of the boronic acid substituent and the lipophilicity of the Tc-RbatoH_2 moiety. In these relatively small molecules the Tc(+3) centre is 7-coordinate being bound to one Cl^- ligand and the N_3O_3 donor atom set of the R-batoH_2^{2-} ligand. This results in the addition of a large uncharged complex, which can overwhelm the binding specificity of the carrier.

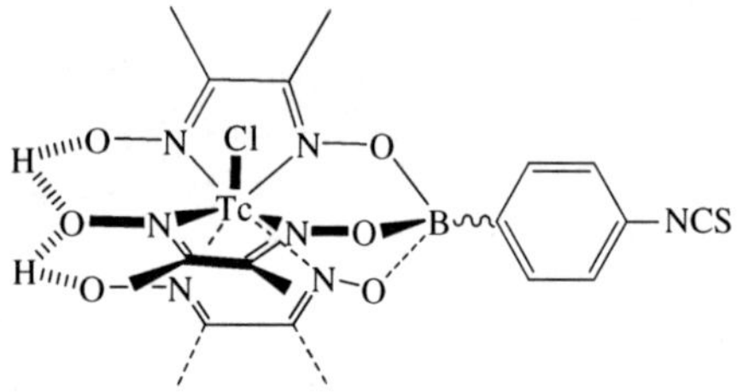

97

The Tc(+3) centre has also been reported to be present in imine diacetic acid liver imaging agents such as **53**, although it is possible that it is the TcO^{3+} centre which is actually present, as found in $[TcO(edta)]^-$ (**25**). Polyamine carboxylates would appear to offer a means of binding Tc(+3) or Tc(+4) in a bifunctional molecule but this approach does not appear to have found much application, probably because better alternatives are available involving Tc(+5) centres.

One novel approach to bifunctional agents in which Tc(+3) appears to be present involves the use of hydrazino nicotinic acid (HYNIC) derivatives, **98**. At first sight the HYNIC-carrier system would appear unattractive as it involves monodentate, or at best didentate, binding to the metal so that other coligands will be required to saturate the Tc coordination sphere. However, it turns out that the Tc–N bond is quite kinetically inert and might be thought of as involving some π-bond character as in TcO^{3+}. It also turns out that certain mixed ligand complexes can be formed easily and in high yield using the HYNIC system. Although complexes of the type [Tc(NNHC_5H_4NC(=O)NH-Carrier)$(LL)_2$] [LL=glucoheptanoate, ethylenediamine diacetate ($edda^{2-}$), or tricinate, **99**] involving two ligand types proved to be unstable in solution, and a number of different species formed in the absence of excess coligand, ternary systems involving three ligand types proved more successful. A simple example of a ternary complex is provided by [Tc(NNHC_5H_4NC(=O)NH-Carrier)(edda)Cl], **100**, but more interesting results have been obtained with the tricine system [Tc(NNHC_5H_4NC(=O)NH-Carrier)(tricinate)L] (L = substituted triphenylphosphine or substituted pyridine), **101**. In these complexes the lipophilicity/hydrophilicity of the complex can be tuned by varying the substituents on the ligand L. Groups such as -SO_3^-, CO_2H or hydroxyalkyl in ligands exemplified by **102**, **103** and **104**, afford more hydrophilic complexes, while the incorporation of alkyl substituents -$C_nH_{(2n+1)}$ would increase lipophilic character. Complexes of the type **101** proved to be very stable in solution and were formed easily in high yield and purity. The reducing properties of organohydazines offer a simple approach to labelling reactions through the reduction of preformed intermediate Tc(+5) complexes containing $^{99m}TcO^{3+}$ (*e.g.* [^{99m}TcO(tricinate)$_2$]) to Tc(+3) (Equation (8)).

$$RNH\text{-}NH_2 + TcO^{3+} \rightarrow \{Tc\text{-}N{=}NR\}^{2+} + H_2O + H^+ \quad (8)$$

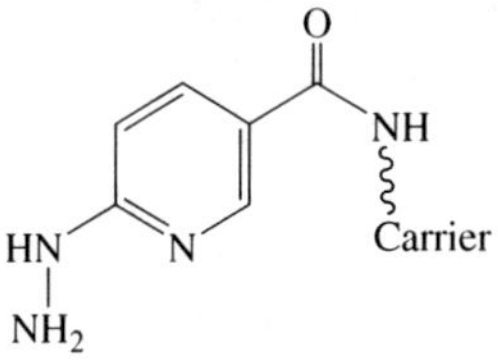

98
HYNIC-NH ∿ Carrier

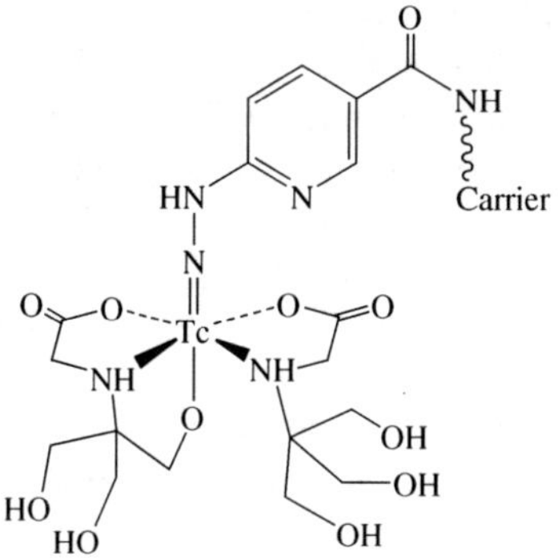

99

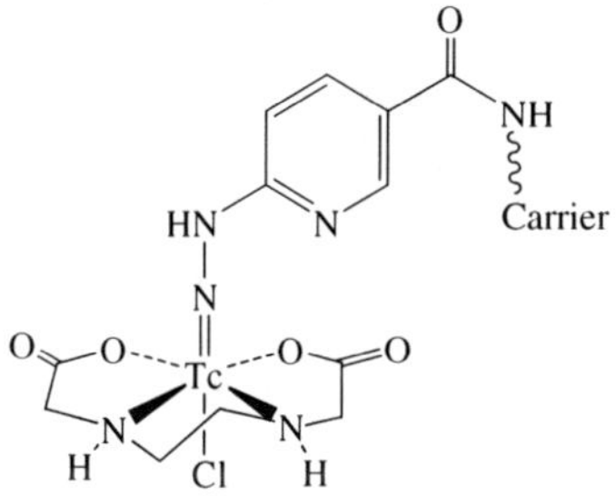

100

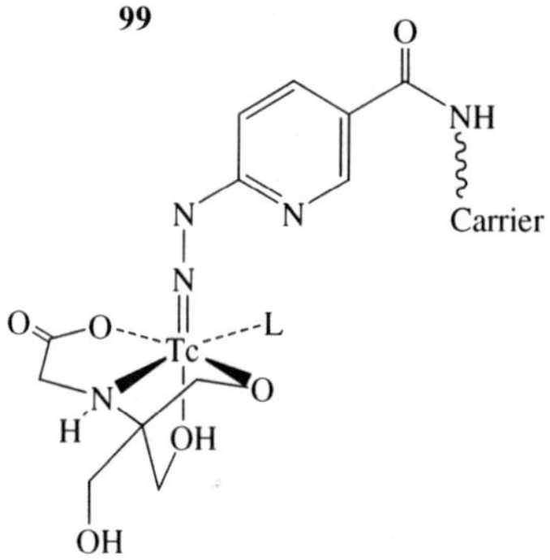

101

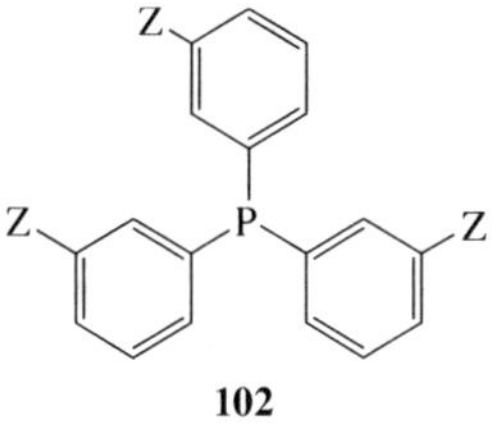

102
Z = H, SO_3Na

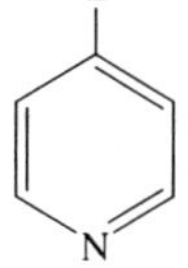

103
Z= CO_2H, CH_2CO_2H, $C_2H_4SO_3H$, $C(=O)NHC_2H_4OH$

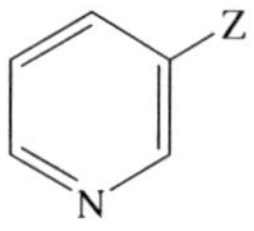

104
Z= CO_2H, SO_3H

The HYNIC system was first used to radiolabel polyclonal IgG and the bifunctional ^{99m}Tc-HYNIC-IgG agent was found to be as effective for the detection of infection and inflammation as the ^{111}In-IgG agent previously used in scintigraphy. The HYNIC approach has since been applied to the radiolabelling of various biologically active species including chemostatic peptides, antisense oligonucleotides and a GPIb/IIIa receptor antagonist. Clinical trials have been carried out on Apomate® which uses the HYNIC approach to ^{99m}Tc-labelling of recombinant human annexin. This protein shows specific binding to apoptotic and necrotic cells allowing the imaging cell death (apoptosis and necrosis). This can be used as an early indicator of the response to anticancer treatment, to determine the extent and location of myocardial injury after a heart attack and to assess the response to treatment after a heart attack. The agent was found to be biologically more stable and resistant to degradation by the liver than earlier formulations, leading to lower activity in the hepatobiliary system and intestine. Most of the ^{99m}Tc-HYNIC-annexin agent is cleared from the blood through the kidneys so that, in the absence of disease, it does not appear in the abdomen, offering the prospect of extending its application to imaging GI, ovarian and prostate cancers.

The HYNIC method has also been applied to the preparation of somostatin (Section 3.3.11.1) analogues and human trials have been carried out with a ^{99m}Tc-HYNIC-edda-octreotide radiopharmaceutical. The trial involved 40 patients with a variety of diagnosed malignant neoplasms (32 primary and 8 metastatic) including 7 with pituitary adenomas and 19 with non-small cell lung cancer. The results showed that ^{99m}Tc-HYNIC-edda-octreotide is potentially useful for imaging a range of primary and metastatic tumors.

3.3.12.4 Labelling with Tc(+5) as TcO_2^{+}, TcO^{3+} or TcN^{2+}

The most versatile oxidation state for Tc in radiopharmaceutical preparation is undoubtedly Tc(+5). The majority of ^{99m}Tc radiopharmaceuticals involve complexes containing the TcO^{3+} core but, in addition, there is interest in the development of new radiopharmaceuticals involving the TcO_2^{+}, TcN^{2+} and $\{Tc{=}N{-}N{<}\}^{3+}$ cores. The singly charged *trans*-TcO_2^{+} centre forms octahedral complexes with tetramines such as **105**, with an open chain ligand, or **106**, with a macrocyclic cyclam derivative. This type of approach has been investigated for preparing antibody labelling agents and in preparing a somostatin analogue. Unfortunately, the poor labelling efficiency of cyclam with $^{99m}TcO_2^{+}$ was a major disadvantage. Chelating phosphine ligands also form complexes with TcO_2^{+} as exemplified by Myoview, **56**, and offer an alternative approach. In a development of this approach the use of a P_2S_2 donor atom set afforded derivatives with good *in vitro* and *in vivo* stability. The inclusion of hydrophilic hydroxymethyl groups in **107** promoted water solubility and rapid renal clearance and this metal binding group has been applied to the preparation of bifunctional ^{99m}Tc-bombesin analogues. These are of interest for the detection of tumors that express bombesin/gastrin-releasing peptide receptors.

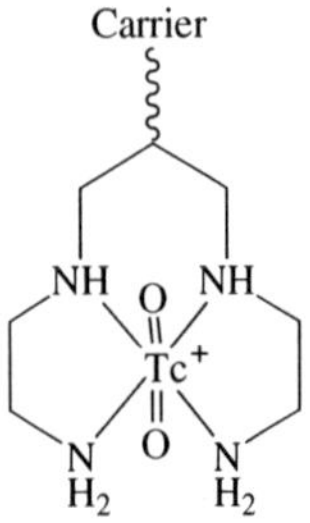

105 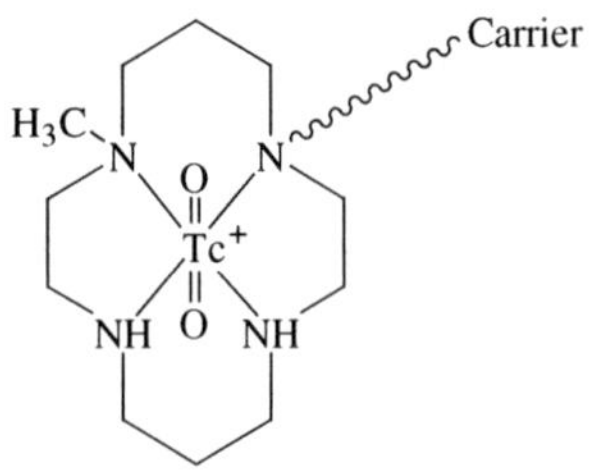 **106** 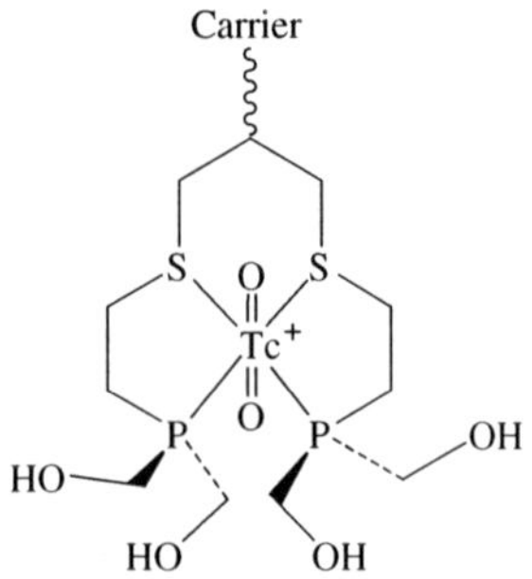 **107**

A far more extensive chemistry exists for the TcO^{3+} core than for TcO_2^+ and this is dominated by ligands containing N and S donor atoms. Examples range form N_4 to S_4 donor atom sets and, depending on the number of ionisable hydrogens present, neutral or negatively charged complexes may be obtained. Examples of the S_4 door set are provided by **108** and **78**, which were devised to target 5-HT_2-serotonin receptors associated with neurological disorders such as schizophrenia, Alzheimers disease and depression. The presence of the three thiolate groups, together with one thioether, confers neutrality on the complex formed. In these examples the so-called '3+1' strategy was used whereby a tridentate coligand occupies three of the four available coordination sites around a TcO^{3+} core, leaving one coordination site free to bind a monodentate ligand which incorporates the carrier. The use of a monodentate linking group to connect the metal to the carrier requires a very inert metal-ligand bond since no additional stabilisation is possible from the chelate effect. The '3+1' strategy could be extended to the NS_3 donor atom set by using binding sites of the type exemplified by **109** or **110**. This approach was used to prepare Tc-NS_2-Tropane, **76**, designed to target dopamine receptor sites. Tetradentate NS_3 ligand systems for TcO^{3+}, **111**, have been evaluated to establish their suitability for complex formation. These could offer another potential approach to the synthesis of bi-functional agents if a suitable linking group were attached to the ligand structure.

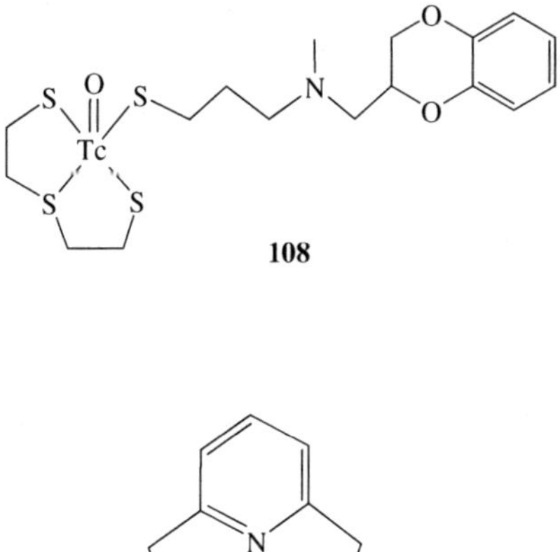

108

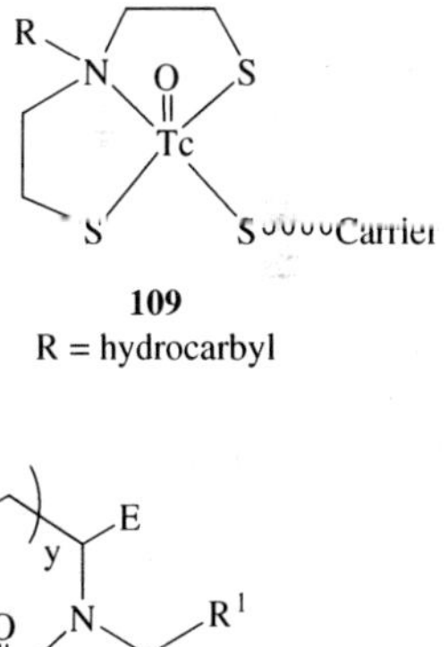

109
R = hydrocarbyl

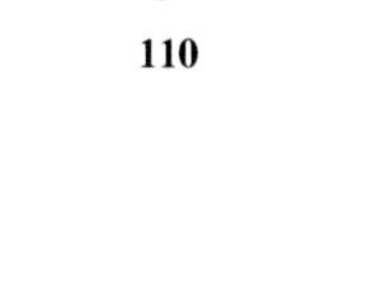

110

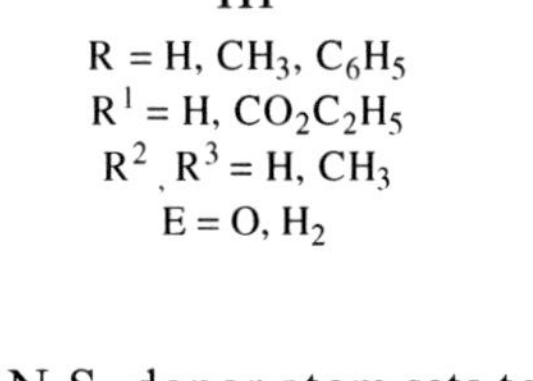

111
R = H, CH_3, C_6H_5
R^1 = H, $CO_2C_2H_5$
R^2, R^3 = H, CH_3
E = O, H_2

There are many examples of the use of N_2S_2 donor atom sets to bind TcO^{3+} and several structural variations have been demonstrated. These offer various

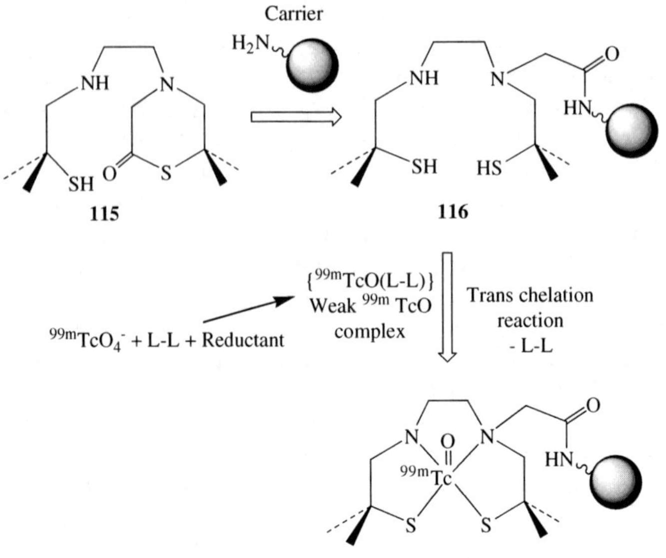

Carrier

115

116

$\{^{99m}TcO(L\text{-}L)\}$
Weak ^{99m}TcO complex

$^{99m}TcO_4^- + L\text{-}L + \text{Reductant}$

Trans chelation reaction
- L-L

117
Bifunctional agent

Scheme 15

points of attachment to the linker, differing lipophilicity for the complex and control over the charge of the complex. In **112** and **113**, for example, links are made respectively through the hydrocarbon chain in the central chelate ring or amine nitrogen. The methyl substituents confer lipophilicity on the complex and variations in the chelate ring sizes are possible. Depending on the conditions used one of the amine groups in **112** may ionise to give a trinegative $(S^-)(N^-)(NH)(S^-)$ ligand and a neutral complex or both NH groups may retain their hydrogen atoms giving a $(S^-)(NH)(NH)(S^-)$ ligand and a positively charged complex with TcO^{3+}. This type of structural motif has found application in antibody and protein labelling studies. In particular one example of the formation of a **113** type structure demonstrates an elegant combination of linking chemistry and metal binding site formation (Scheme 15). The thiomorpholinone group of **115** reacts with the amine groups of lysine residues on a protein to form an amide link to the carrier and simultaneously create a binding site for TcO^{3+}. The conjugate **116** can be labelled efficiently by ligand exchange with a ^{99m}Tc complex of a weakly bound ligand to give the bifunctional agent **117**. This methodology has been used to produce labelled peptides and muscarinic cholinergic receptor agents (Section 3.3.10.2). One disadvantage with this particular ligand structure is its high lipophilicity resulting, in part at least, from the methyl substituents. This can lead to high non-specific uptake, high liver uptake and hepatobiliary excretion. The introduction of hydrophilic substituents to the structure provides a means of moderating this problem.

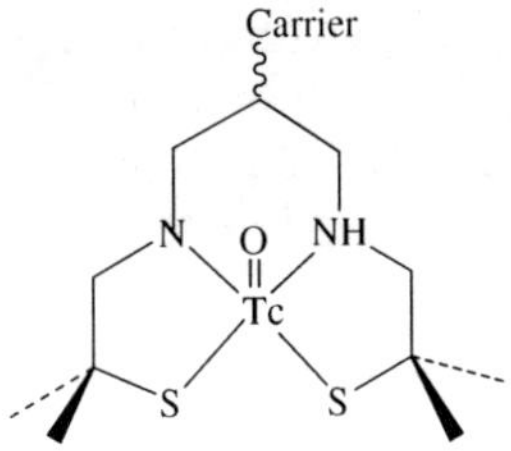

112 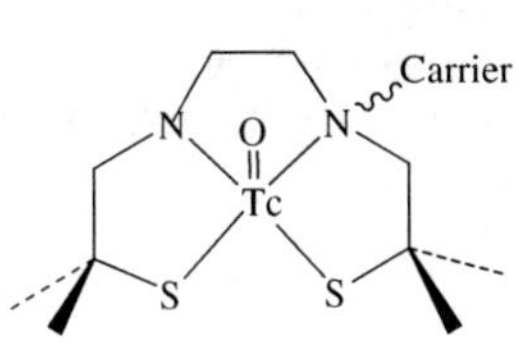 113 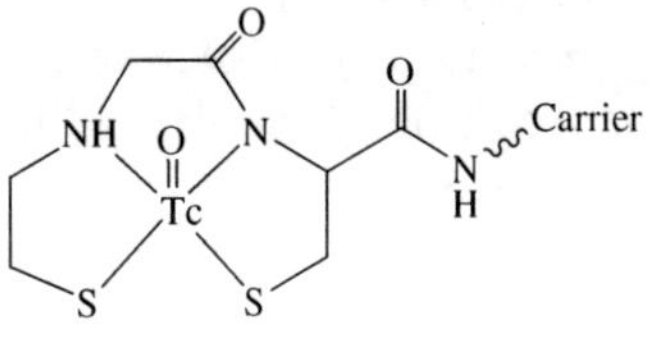 114

Amide NH groups, -CH_2NH-C(=O)-, are more easily ionised by loss of H^+ than amine NH groups, -CH_2NH-CH_2-, because the electron withdrawing >C=O group stabilises the >N^- centre. This offers a degree of control over the charge of a TcO^{3+} complex formed from an -NH- donor ligand. Thus the amine–amide N_2 donor atom set in **114** affords a neutral complex in which the amide nitrogen is ionised but the amine nitrogen is not. In contrast the diamide N_2 donor atom set in **118** is fully ionised to form a negatively charged TcO^{3+} complex. An example of a tetranegative N_3S donor ligand is provided by **119** which contains three amide N and one thiolate S donor group. The aryl thiocyanate substituted propylene imine oxide system has been used in the preparation of ^{99m}Tc-labelled biotin and somostatin derivatives, **120**, providing an example of an N_4 donor atom set used in the preparation of bifunctional agents. As with **117** high lipophilicity is an issue with these derivatives.

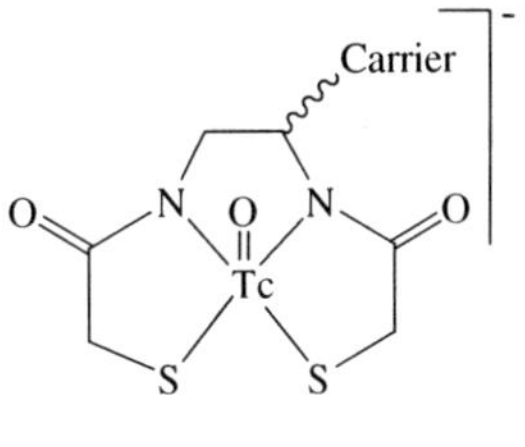

118 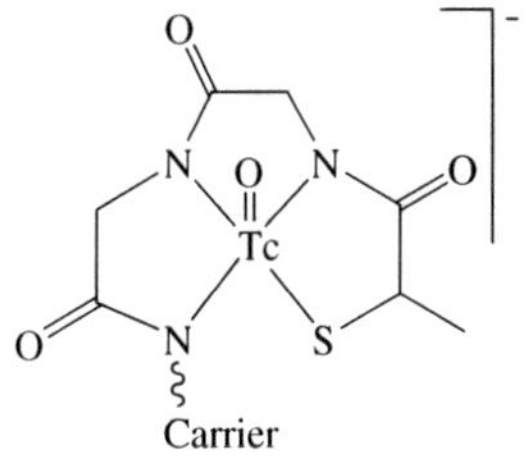 119 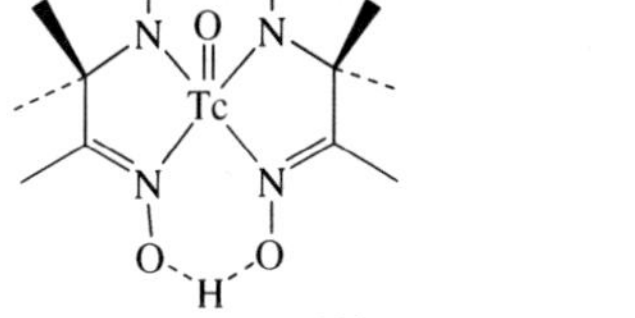 120

Small amino acid sequences present as part of protein chains, or added as extensions to proteins, offer a particularly important means of complexing metal ions. These can contain amine, amide and thiolate binding groups, all of which are well suited for forming complexes with Tc(+5). Examples include **121**, with the $N_2S_2^{4-}$ donor set, **122** and **123**, with N_3S^{3-} donor sets, and **124**, with an N_4^{3-} donor set. An advantage of this labelling approach is that routine solid-phase protein synthesis methods can be used to add the binding site to a protein. Furthermore careful choice of the amino acids allows some tuning of the hydrophilicity of the metal binding site in a manner compatible with the overall nature of a protein. It is also possible to combine amino acid residues with other groups to form chelating ligands as exemplified by the N_3S^{3-} ligand in **125** derived from picolinic acid (pyridine-2-carboxylic acid) serine and cysteine.

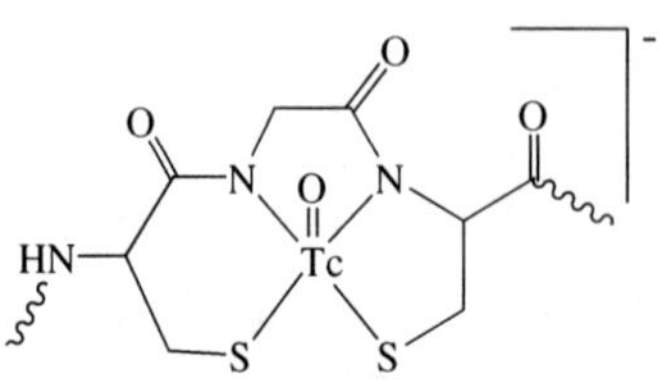

121
TcO-Cys-Gly-Cys
N_2S_2

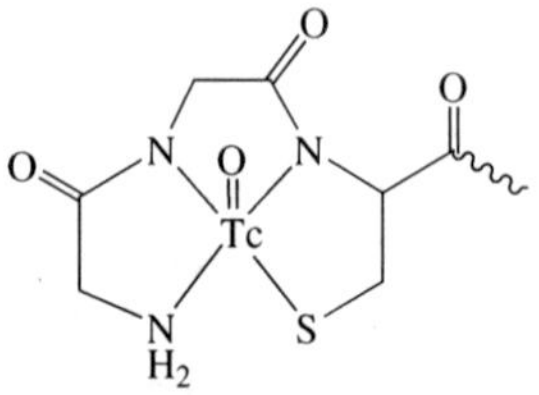

122
TcO-Gly-Gly-Cys
N_3S

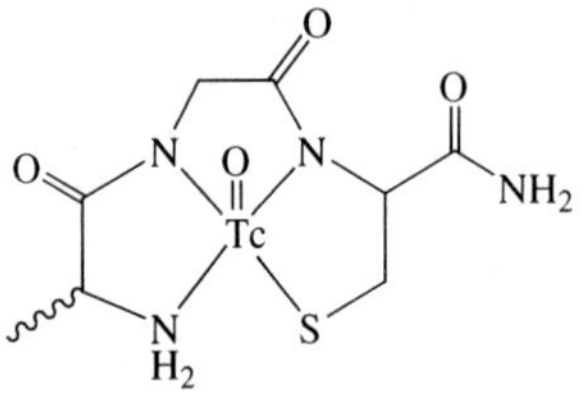

123
TcO-Lys-Gly-Cys
N_3S

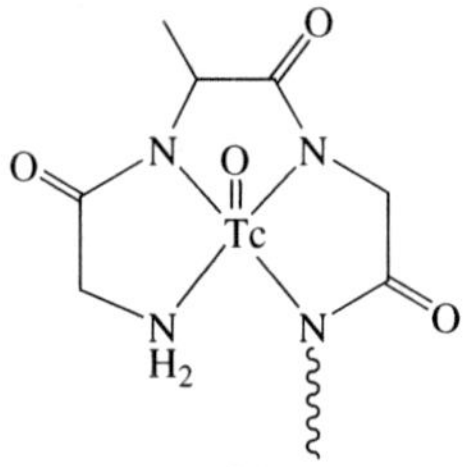

124
TcO-Gly-Ala-Gly
N_4

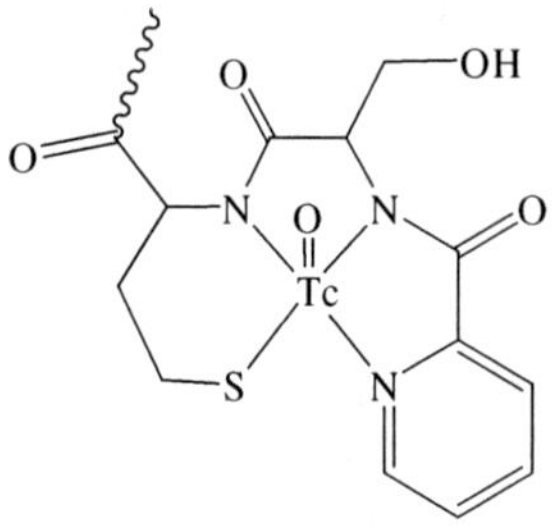

125
TcO-Pic-Ser-Cys
N_3S

The TcN^{2+} core can be prepared from $[TcO_4]^-$ by direct reaction with hydrazine, NH_2NH_2, in the presence of a suitable proligand. However, the TcN^{3+} core appears to have featured little as a means of preparing bifunctional agents. Instead pyridyl hydrazine derivatives which appear to contain Tc(+3) have been a centre of attention as described in Section 3.3.12.3.

3.3.12.5 Labelling with Tc(+6) or Tc(+7)

Despite an early report (1986) of the use of $[^{99m}TcNCl_4]^-$ to label monoclonal antibodies for imaging tumors in mice, the Tc(+6) TcN^{3+} core appears to have featured little as a means of preparing bifunctional agents. The reactivity of other species containing Tc(+6) suggests that this oxidation number would not be a natural choice for preparing bifunctional agents. Certainly controlling the reduction of $[^{99m}TcO_4]^-$ to obtain only Tc(+6) would be chemically difficult. Some complexes containing Tc(+7) are known, for example $[TcO_3Cl(L\text{-}L)]$ (L-L = 2,2′-bipyridyl or 1,10-phenanthroline), and might offer the potential advantage that no reductant is required in their formation from $[^{99m}TcO_4]^-$. However, in general, the d^0 Tc(+7) centre would seem unlikely to provide sufficiently inert complexes to offer an attractive approach to radiopharmaceutical preparation.

It is perhaps not surprising, therefore, that complexes Tc(+6) and Tc(+7) have not featured in the pursuit of bifunctional agents, although $[^{99m}TcO_4]^-$ is of course, the normal starting point for preparing such compounds.

3.4 *In vitro* Applications

One example of a diagnostic procedure involving an *in vitro* measurement of a radiopharmaceutical is the Schilling test for pernicious anaemia. This disease results from vitamin B_{12} deficiency and the efficiency vitamin B_{12} absorption *in vivo* can be investigated using Rubatrope-57 (Bristol-Meyers Squibb), The Rubatrope-57 formulation contains ^{57}Co-labelled cyanocobalamin (a form of vitamin B_{12}, Figure 21 in Chapter 2) which is administered orally. Thereafter the patient's urine is collected over 24 h and an *in vitro* measurement of the proportion of the ingested vitamin B_{12} which has been excreted is made using the γ-ray emission from the ^{57}Co labelled compound. Although the γ-ray emission from ^{57}Co could in principle be used for imaging [$T_{1/2}$ 291 days, 122 keV (85.5%), 136.5 keV (10.6%)], the amount of radioactivity required for the Schilling test is far too small for this purpose. Other radioactive metals with suitable properties can be used in a similar manner as tracers to study abnormalities in their metabolism.

Radioactive emissions are not the only type of energy which can be emitted by metal ions. Some can produce characteristic light emissions following excitation by a suitable light source. This property does not involve radioactivity and can be exploited in another type of *in vitro* measurement, fluorescence immunoassay. An immunoassay uses a biologically active probe molecule, such as an antibody or fragment antibody, to probe the concentration of another with which it has an immunological reaction. One way to monitor the concentration of the probe molecule is to attach a fluorescent dye to it. Excitation of the dye by ultraviolet light then leads to fluorescent light emission at an intensity related to the concentration of the probe molecule. Since light can be detected with high sensitivity, this is an attractive measurement method. However the sensitivity is limited by the fluorescent emission background from the test sample itself, since many organic molecules can exhibit fluorescence. The use of a label containing a fluorescent metal can offer advantages over an organic dye. Firstly, the light emitted from the metal is shifted in frequency to a greater extent making it easier to distinguish from light emitted by the organic compounds present. Secondly, and most importantly, the timescale over which the light is emitted can be much longer than for an organic dye. This allows a time resolved measurement to be made so that, following excitation by a pulse of ultraviolet light, the measurement of the emitted light can be delayed until most of the shorter lived background emission has decayed (Figure 18). This forms the basis of time resolved fluorescence immuno-assay (TRFIA) methods and related dissociation enhanced lanthanide fluorescence immuno-assay (DELFIA).

Time resolved fluorescence methods exploit the luminescence properties of lanthanide elements, particularly europium or terbium. These have especially

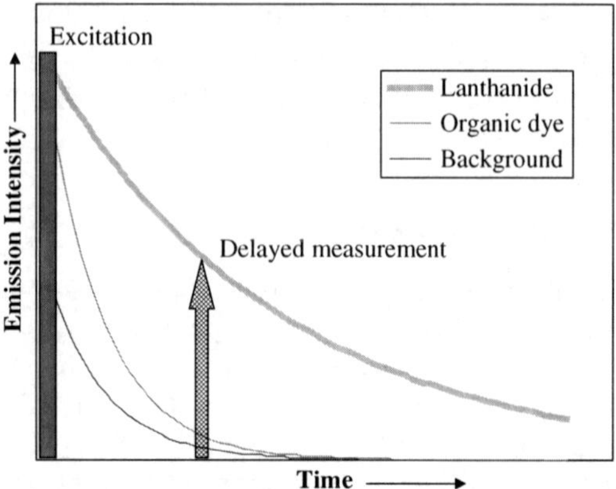

Figure 18 *Time resolved fluorescence assay in which the intensity of emission from a lanthanide ion decays much more slowly than from an organic dye giving a much higher ratio of signal to background in a delayed, or time resolved, measurement*

long luminescent lifetimes measured in milliseconds. An important requirement for long-lived lanthanide luminescence is that water is excluded from the coordination sphere of the metal ion. The vibrations of the O–H bonds in coordinated water molecules offer an efficient mechanism for extracting energy from an excited lanthanide ion in competition with the emission of light. Thus the luminescence of aquated lanthanide ions is strongly quenched. The displacement of water by a ligand system can enhance lanthanide luminescence by removing this quenching mechanism. Simple aquated Eu^{3+} or Tb^{3+} ions have been used in some applications, for example as structural probes for proteins and nucleic acids. In particular Tb^{3+} shows enhanced emission after binding to single-stranded nucleic acids but this is not seen with double-stranded nucleic acids. Replacement of Ca^{2+} in proteins by luminescent lanthanide ions can also provide a probe of protein structure since certain groups present in proteins can enhance luminescent emission when close to the bound lanthanide ion.

In order to prepare a suitable agent for use in TRFIA it is necessary to attach a suitable lanthanide ion, *e.g.* Eu^{3+} or Tb^{3+}, to an immunologically active species such as an antibody or fragment antibody. Various methods of attaching metal ions to biologically active molecules have been described in Section 3.2.1.12 and are not considered further here. In contrast to the ligand design objective for an MRI contrast agent of retaining a coordinated water molecule (Section 3.2.1), TRFIA applications require the saturation of the lanthanide ion coordination sphere to exclude water. Enzyme assays are also possible by using a substrate from which the enzyme can release a chelating proligand. The proligand is designed to react with lanthanide ions present in the aqueous medium to form luminescent complexes, which can be detected following excitation. Luminescence from uncomplexed aquated metal ions is quenched so that the emitted light provides a measure of the amount of proligand released by the enzyme. This forms the basis of the DELFIA method.

CHAPTER 4

Therapeutic Medicine

4.1 The Use of Metal Compounds in Therapy

Perhaps the most obvious use of metal compounds in therapy is to treat illness resulting from a deficiency of essential metals. As examples iron dextran complexes are used to treat anaemia due to iron deficiency and cobalt, in the form of vitamin B_{12}, is used to treat pernicious anaemia resulting from vitamin B_{12} deficiency. Many dietary supplements contain metals such as copper, manganese and zinc but coordination chemistry does not usually play a major role in the formulation of these products. The application of coordination chemistry is more relevant in pharmaceutical preparations formulated to treat specific diseases associated with problems over the management of essential metals *in vivo*. One example is provided by Menkes disease in which copper deficiency arises from a defect in intracellular copper transport. Treatment using oral supplements is ineffective as copper absorption in the gut is defective in Menkes patients. Furthermore parenteral administration of copper salts is also ineffective. The finding that exchangeable copper in human blood serum involves copper histidine complexes, which bind to serum albumin, led to the use of copper histidine coordination compounds in the treatment of Menkes patients. If administered at a sufficiently early stage, these can suppress the neurodegeneration arising from Menkes disease, although they do not eliminate connective tissue disorders associated with the disease.

In a converse sense complexing agents selective for particular metals can be used to sequester unwanted metals *in vivo* and promote their elimination from the body through excretion or decorporation. This is known as 'chelation therapy' and may be applied to cases of metal poisoning or diseases leading to excessive levels in the body of essential metals. Particular examples are the control of copper levels in patients with Wilson disease and iron levels in thalassaemia patients receiving repeated blood transfusions.

Other therapeutic applications of metals include the use of lithium to treat depression and bismuth compounds to treat gastric disorders. However, coordination chemistry does not feature prominently in the formulation of drugs containing such s-block and p-block metals. One therapeutic application in which coordination chemistry is playing a major rôle is in the development of new drugs for the treatment of cancer. These exploit the particular properties of metals such as platinum to produce a therapeutic effect. As a measure of the importance of this type of compound, the estimated global market in platinum cancer drugs in 2000 was estimated to be about US$1 billion. The particular reactivity or electron transfer properties of specific metals also find application in the treatment of rheumatoid arthritis, diabetes and problems associated with the regulation of bloodflow through the vascular system. Radioactive metals also have potential uses in therapy and, although the application of coordination chemistry in this field is not yet highly developed, there is growing interest in radiotherapy using internal sources of radiation in the form of radiopharmaceuticals.

4.2 Chelation Therapy

4.2.1 Metal Sequestration

The term 'chelation therapy' usually refers to the use of proligands as drugs to treat disorders resulting from the presence of unwanted metal ions arising from intoxication or disease. To be effective, the proligand must complex and sequester the target metal ion then promote its excretion and removal from the body in complexed form without impairing the normal biochemistry of metals *in vivo*. Consequently the proligand used must be selective for the target metal ion so as not to remove other biologically important metals. In cases where a biologically essential metal such as iron or copper is the target, the sequestering agent must not compete with any natural binding sites for the target metal to the extent that it compromises normal function and health. However, where non-essential toxic metals are involved the sequestering agent may need to compete with natural metal binding sites to prevent uptake of the target metal ion. As the name implies, the compounds used in chelation therapy exploit the chelate effect (Section 2.7.3) to sequester the target metal ion. It is also important that the electronic properties of the ligand and the metal are matched depending upon whether the metal is hard, soft or intermediate in character (Section 2.7.2).

Chelation therapy is particularly relevant to two diseases, Wilson disease which leads to excess copper levels in the body and thalassaemia which is treated with repeated blood transfusions leading to iron overload in patients. In suitable cases polydentate proligands can also be used to treat cases of toxic metal poisoning and a particular example of ligand design is provided by the development of sequestering agents for plutonium. The design of proligands for use in chelation therapy has to meet several challenges. Not only the proligand must be of low toxicity and show strong and highly selective binding to the target metal, it must also be neutral in its effect on the biological activity of essential metals. As an example iron is able to undergo electron transfer reactions involving the Fe^{3+}/Fe^{2+} couple. These can lead to the formation of reactive free radicals which can be toxic. Under natural conditions these reactions are controlled by the coordination environments of iron *in vivo*. However if the wrong type of ligand were used to sequester iron, its electron transfer properties could be promoted by complex formation and cause oxidative stress. The lipopilicity of the metal complex will also be dependent on the proligand design, affecting the ability of the metal to penetrate cell membranes and the proportions removed through the different excretion pathways. However, increased lipophilicity can also lead to increased brain uptake of the sequestered metal and, for oral treatments, increased uptake in the gut. Careful design of the proligand structure is thus important in achieving the required selectivity and pharmokinetics.

4.2.2 Macrocyclic Antibiotics

A broader view of the concept of chelation therapy might include the use of drugs to sequester metal ions in order to disrupt an unwanted biochemical process. An example of this is provided by the use of metal binding agents for the Group 1 metals Na^+ and K^+ as antibiotics. In this case the function of the drug is not to promote the excretion of the target metal ion, but rather to interfere with its use by bacteria and so produce an antibacterial effect. Antibiotics which work in this way include the macrocyclic antibiotics nonactin and valinomycin. Nonactin is a member of a family of antibiotics based on naturally occurring metal ion binding agents known as the macrotetrolides. These compounds are capable of selectively forming complexes with K^+ and transporting it through cell membranes. In this way they change the permeability of the membrane to K^+ disrupting oxidative phosphorylation and inhibiting the processing of some proteins. Valinomycin acts similarly but is structurally different from nonactin.

The Group 1 metal ions are classified as hard ions and form their most stable complexes with hard ligand donor atoms, particularly the oxygen of water so that the metal ions are strongly hydrophilic. In aqueous media conventional proligands containing nitrogen or oxygen donors bound within organic molecules are ineffective competitors for K^+ or Na^+ ions against coordinated water. However, organic compounds such as nonactin or valinomycin can exploit the macrocyclic effect (Section 2.7.3) by creating an oxygen lined cavity capable of encapsulating the hydrophilic K^+ ion. The ligand structure surrounds the hydrophilic K^+ ion with a lipophilic exterior (*e.g.* valinomycin Figure 1) solubilising it in a hydrophobic environment such as a lipid bilayer membrane. In achieving this nonactin and valinomycin exploit the principles of coordination chemistry demonstrated by a group of compounds of simpler macrocyclic structure known as crown ethers. The crown ethers are macrocyclic polyethers in which the size of the ring and number of oxygen atoms may be varied, examples being provided by 15-crown-5, **1**, dibenzo-18-crown-6, **2** and 24-crown-8, **3**, where the first number in the name refers to the number of atoms in the macrocyclic ring, and the second the number of oxygen atoms. These compounds show an unusual affinity for Group 1 metal ions and have the remarkable ability to solubilise their salts in non-polar organic solvents. As an example dibenzo-18-crown-6, **2**, renders $KMnO_4$ soluble in benzene by complexing the K^+ ion to produce the benzene soluble compound $\{K^+(\mathbf{2})\}\{MnO_4^-\}$.

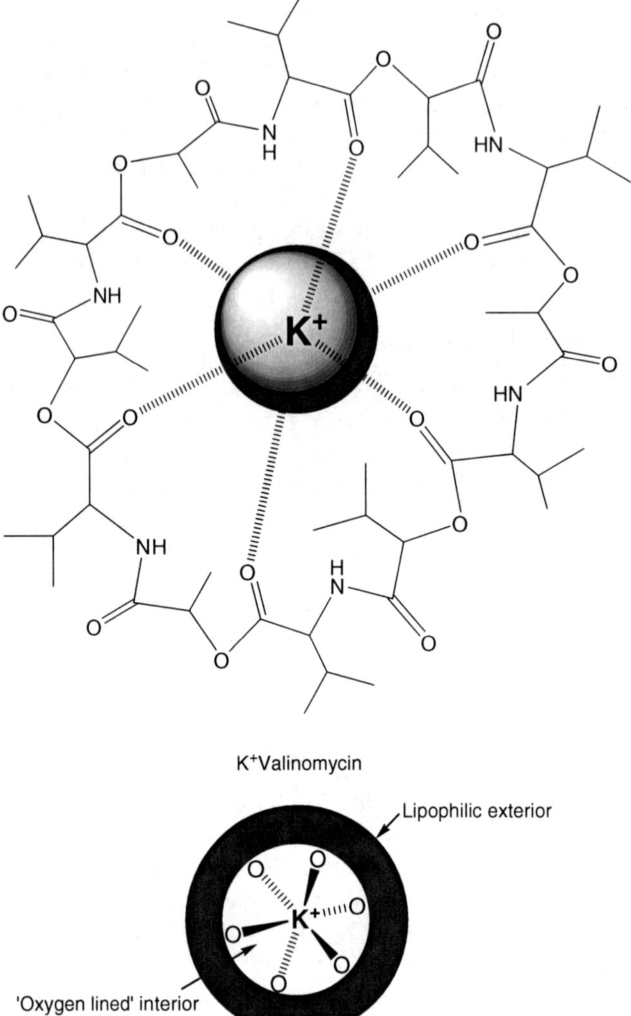

Figure 1 *The binding of valinomycin to K^+ (top) and its 'cartoon' representation (bottom)*

1 15-crown-5

2 dibenzo-18-crown-6

3 24-crown-8

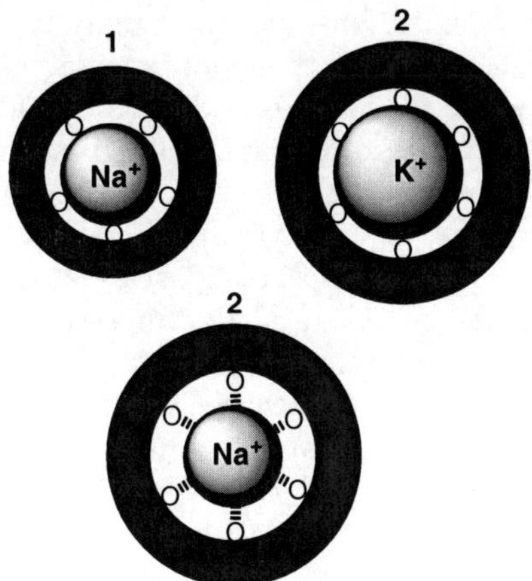

Figure 2 *Cartoons showing how the larger **2** ligand is a better fit for K^+ than Na^+ which is best fit by **1***

A further important feature of the crown ethers is their selectivity for specific metal ions. This arises from the well-defined size of the macrocyclic cavity which can discriminate between Group 1 metals according to their ionic radii. Thus in 1:1 ligand/metal complexes the size of the cavity in **1** allows an optimum interaction between the Na^+ ion and the polyether oxygens. In the case of the larger K^+ ion the larger cavity of 18-crown-6 (or the dibenzo-derivative **2** in the example shown) provides the best fit, optimising the strength of the metal-ligand interaction (Figure 2). The structure of valinomycin is optimised for K^+ binding as is that of nonactin. However, the nonactin molecule binds to K^+ through eight oxygen atoms in an approximately cubic 8-coordinate arrangement by folding around the K^+ ion, adopting the shape of a tennis ball seam. Valinomycin folds to offer approximately octahedral 6-coordination to K^+ (Figure 1).

4.2.3 Metal Intoxication

4.2.3.1 Historical Development

The use of chelation therapy to treat metal intoxication was developed during the 20th century, initially to moderate the toxicity of arsenic compounds used to treat syphilis. The use of citrate for the treatment of lead poisoning was also investigated but met with only limited success. Subsequently the polyamine carboxylic acids ethylenediaminetetra-acetic acid (see Chapter 3, **4**, $edtaH_4$)[1] and, more effectively, diethylenetriaminepenta-acetic acid (see Chapter 3, **3a**, $dtpaH_5$) were used to treat lead intoxication and for the decorporation of radionuclides. These two compounds are poorly absorbed by the gastrointestinal tract and are best administered by slow intravenous infusion. They

[1] See footnote 1 in Section 3.2.1.7 regarding the use of upper and lower case abbreviations for ligands/proligands.

distribute into extracellular regions and are rapidly excreted in urine. Toxic effects can arise from their ability to sequester essential metal ions such as Ca^{2+} and Zn^{2+} but these can be moderated by the use of $CaNa_2$(edta), Ca_2(edta), Zn_2(edta) or $ZnNa_3$(dtpa) instead of simple sodium salts. As an example, for a given dose, Ca_2(edta) is around 20 times less toxic than Na_4(edta) and, in turn, Zn_2(edta) is around 10 times less toxic than Ca_2(edta).

In the 1950s *meso*-2,3-dimercaptosuccinic acid (**4**, DMSA, $dmsaH_4$) and *d,l*-2,3-dimercaptopropane-1-sulfonic acid (**5**, $dmpsH_3$) came into use as metal sequestering agents and are registered drugs in the USA and Germany, respectively. These superseded the less biologically stable and more toxic compound British Anti Lewisite, 2,3-dimercaptopropanol (**6**, $dmpaH_2$), developed in World War II as an antidote to the chemical weapon dichlorovinylarsine (Lewisite). In addition to their low toxicity compared to $dmpaH_2$, $dmsaH_4$ and $dmpsH_3$ can be taken orally while $dmpaH_2$ must be injected and has unpleasant side effects. However, $dmpaH_2$ appears to be distributed into intracellular spaces whereas $dmsaH_4$ and $dmpsH_3$ are primarily extracellular in distribution, so to some extent the compounds address different biological compartments. All are excreted in urine and, in humans, the whole blood and urinary excretion half times of $dmsaH_3$ are less than 4 h and, for $dmpsH_3$, rather longer at 9–10 h. Sequestering agents specific to iron and copper were also developed to treat diseases which give rise to toxic effects resulting from imbalances of these metals. These include desferrioxamine (**7**, DFO, $dfoH_3$), *d*-penicillamine (**8**, $dpaH_2$) and 2,2,2-tet (**9**, $tetaH_2$ also known as TETA). These are discussed in more detail in Sections 4.2.4 and 4.2.5.

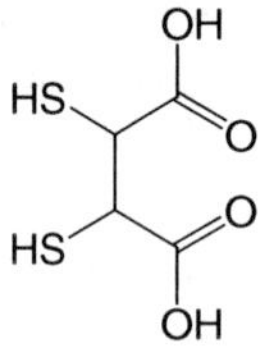

OH HS O HS O OH

4
$dmsaH_4$

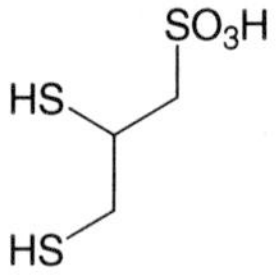

SO$_3$H HS HS

5
$dmpsH_3$

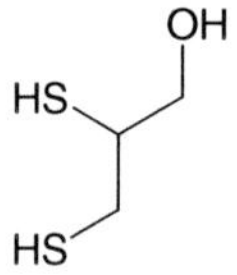

OH HS HS

6
$dmpaH_2$

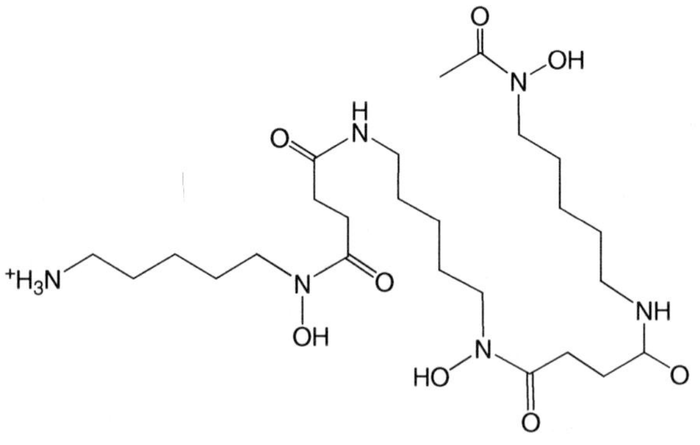

O N OH H N O ^+H_3N N O OH NH N HO O O

7
$dfoH_3$
Desferrioxamine B,
Desferal

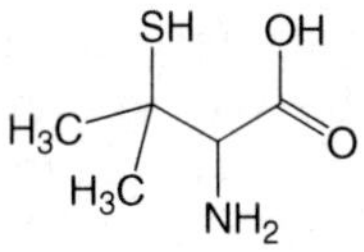

8
dpaH_2

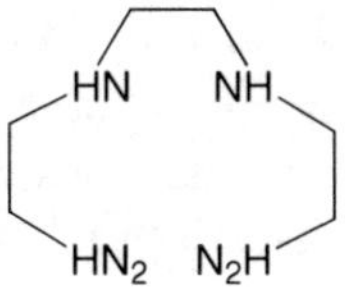

9
tetaH_2

4.2.3.2 *Choice of Ligand*

The polyamine carboxylic acids exploit the chelate effect (Section 2.7.3) in forming stable complexes through the formation of multiple chelate rings. The increase in complex stability with increasing number of chelate rings can be seen in the stability constant (β_1) variation for the series of Cd^{2+} complexes with iminodiacetic acid (**10**, idaH_2), $\beta_1 = 10^{5.7}$; nitrilotriacetic acid (**11**, ntaH_3), $\beta_1 = 10^{9.8}$; edtaH_4, $\beta_1 = 10^{16}$ and dtpaH_5, $\beta_1 = 10^{19}$. The match between the size of the chelate ring and the size of the metal ion, together with the ability of the ligand to fold around the metal to optimise metal-ligand interactions, are also important considerations. This is illustrated by the equilibrium constant (K) values for the reaction in which two tridentate *N*-methylimidodiacetate, $\{CH_3N(CH_2CO_2)_2\}^{2-}$, ligands on Cd^{2+} are displaced by a single hexadentate $\{(O_2CCH_2)_2N(CH_2)_nN(CH_2CO_2)_2\}^{4-}$ ligand in which the size of the central -$(CH_2)_n$– link is varied. As the chelate ring size increases with increasing n, the value of K decreases from the value for a 5-membered chelate ring at $n = 2$ as follows: $K = 10^{7.2} (n = 2$, edta$^{4-})$, $10^{2.4} (n = 3)$, $0.1 (n = 4)$ and 0.025 $(n = 5)$. Thus as the central chelate ring size increases the ligand structure becomes less well adapted to a 6-coordinate geometry and the hexadentate ligand becomes less competitive with two separate tridentate $\{CH_3N(CH_2CO_2)_2\}^{2-}$ ligands.

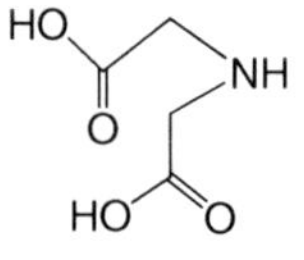

10
idaH_2

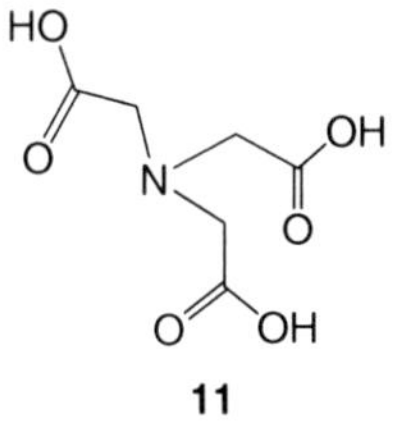

11
ntaH_3

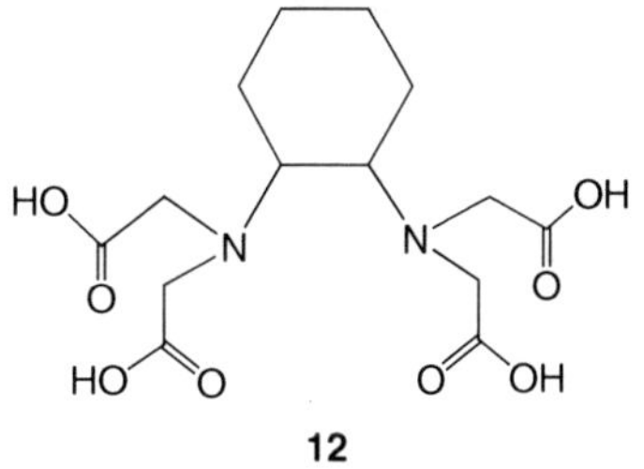

12
cdtaH_4

The rigidity of the ligand structure can also be controlled by ligand design. In edta^{4-} rotation about the C–C bonds in the central –CH_2CH_2– group allows the ligand to adopt many structural arrangements which are not suitable for complete ligand binding to a single metal ion. A more rigid structure arises if the –CH_2CH_2– group is replaced by a 1,2-cyclohexyl group as in cdtaH_4, **12**. Here the amino carboxylate groups are constrained to adopt a structure where the N atoms occupy locations more nearly suited for metal ion coordination. This can improve the stability of the complex formed through preorganisation of the ligand (Sections 2.7.3 and 3.2.1.7).

In the polyamine carboxylate ligands the amine groups are best suited to binding borderline metals while the carboxylate groups are best suited to binding hard character metal ions, although they are also effective in binding to borderline metals. The dithiol ligands dmsaH_4 and dmpsH_3 are particularly suited to binding soft metal ions (Section 2.7.2) such as Cd^{2+}, Hg^{2+} and Pb^{2+} but, again, are also able to form complexes with a variety of metals in the borderline category (Table 7, Chapter 2). Some examples of the application of different sequestering agents to various toxic metals follow. The range of metals for which information is available is somewhat limited and, although quite a lot of animal data has been collected, data obtained clinically on human subjects is mostly confined to a few industrially important metals.

4.2.3.3 Some Applications of Chelation Therapy to Metal Intoxication

4.2.3.3.1 Aluminium. Al toxicity is typically treated by dfoH_3 infusion, although the similarity in charge and radius of Al^{3+} and Fe^{3+} ions has led to the consideration of other iron chelators (Section 4.2.4) for Al detoxification.

4.2.3.3.2 Antimony. Antimony is a component of 'tartar emetic' and its compounds are used to treat Schistosomiasis by intravenous infusion. Human cases of antimony poisoning have been treated with some success using dmpaH_2, dmsaH_4 or dmpsH_3. Animal studies suggest that dmsaH_4 should be the proligand of choice for this element.

4.2.3.3.3 Arsenic. Historically dmpaH_2 or dpaH_2 have been used to treat arsenic poisoning. However animal studies indicate that dmsaH_4 or dmpsH_3 would be more effective. Human data for the latter two compounds are limited but they appear to have been used with some success.

4.2.3.3.4 Bismuth. Bismuth compounds have been used in the past for the treatment of syphilis and are used today to treat gastric disorders. In humans mixed results have been obtained using dmpaH_2 but there is evidence that dmpsH_3 offers an effective treatment.

4.2.3.3.5 Cadmium. Cadmium is widely used in a variety of applications and the treatment of Cd intoxication by chelation therapy has attracted some interest. However, toxicity issues can be a problem with novel bespoke chelating agents. It has been suggested that oral dmsa combined with parenteral $CaNa_3$dtpa might be the best approach.

4.2.3.3.6 Cobalt. Animal studies suggest that edtaH_4 or dtpaH_5 would be more effective than dmsaH_4 but information is limited.

4.2.3.3.7 Copper. Animal studies have shown that dmpaH_2, dpaH_2, 2,2,2-tet and dmsaH_4 can promote copper excretion. Human cases of copper poisoning have been treated using dmpaH_2 and dpaH_2 but it was not clear whether this, or the other supportive treatment, was responsible for the recovery. In China

$dmsaH_4$ has been used for many years to treat copper intoxication arising from Wilson disease (Section 4.2.5).

4.2.3.3.8 Gold. Historically $dmpaH_2$ and $dpaH_2$ have been used to treat gold intoxication but animal studies suggest $dmsaH_4$ and $dmpsH_3$ could also be effective and are less toxic than $dmpaH_2$.

4.2.3.3.9 Iron. The American Association of Poison Control Centers surveillance system indicates that around 20,000 cases of iron intoxication in children arise each year, mainly involving vitamin and mineral supplements. Chelation therapy with $dfoH_3$ appears to offer no benefit in mild cases. Fatalities continue to arise from acute iron poisoning due to the ingestion of iron salts, for example in concentrated iron supplements and, in these acute cases intravenous infusion of $dfoH_3$ can be an effective treatment. Iron overload in thalassaemia sufferers has traditionally been treated using $dfoH_3$ infusion but other oral chelators are coming into clinical use (Section 4.2.4).

4.2.3.3.10 Lead. Even though lead is no longer used in all of these applications, its widespread use in car batteries, buildings, paints, petrol antiknock agents and water pipes make it an important target for chelation therapy. Epidemiological evidence that environmental lead poses a risk of cognitive impairment in children has been a major driver for the application of chelation therapy and, in 1991, $dmsaH_4$ was licensed in the USA as an oral treatment for children with blood lead levels above 450 $\mu g\ l^{-1}$. Animal studies suggest that $dmsaH_4$ has advantages over $dmpaH_2$, $dpaH_2$ and $edtaH_4$ which have also been used to chelate lead. In particular $dmpaH_2$ and $edtaH_4$ form toxic complexes with iron so interfering with the simultaneous use of iron therapy to treat lead-induced iron deficiency in cases of lead poisoning.

4.2.3.3.11 Manganese. Animal studies showed increased urinary excretion of Mn when $dtpaH_5$ or, less effectively, $edtaH_4$ were administered. Parenteraly administered Mn was not mobilised by $dmsaH_4$ in rats. In chronic human intoxication by Mn neither $edtaH_4$ nor $dmsaH_4$ relieved the symptoms but $edtaH_4$ did increase Mn excretion in urine while $dmsaH_4$ was ineffective.

4.2.3.3.12 Mercury. As with lead mercury has a long history of human toxicity incidents and the methyl mercurials are particularly dangerous. A variety of chelators including $dmpaH_2$, $dpaH_2$, $dmsaH_4$ and $dmpsH_3$ have been applied to the problem and $dmpaH_2$ followed by oral $dpaH_2$ has been a traditional approach. However, $dmpaH_2$ increases brain deposition of CH_3Hg^+ and is unsuited for methylmercury poisoning. A significant amount of animal and human data has now accumulated and suggests that $dmsaH_4$ and $dmpsH_3$ should be the agents of choice for treating mercury poisoning. Chronic intoxication by inorganic mercury would seem to be best treated with $dmpsH_3$ while $dmsaH_4$ seems to be the more effective for cases of organic mercury poisoning.

4.2.3.3.13 Nickel. Dermatitis induced by exposure to soluble nickel compounds probably represents the primary incidence of toxic effects from nickel. Creams containing $edtaH_4$, diethyldithiocarbamate [edt^-, $(C_2H_5)_2NC(=S)S^-$]

or dimethylglyoxime [dmgH_2, $CH_3C(=NOH)C(=NOH)CH_3$] have been found to reduce the response of nickel allergic subjects to nickel salts in patch tests but effects on skin penetration by Ni^{2+} were variable.

4.2.3.3.14 Platinum. The introduction of platinum anticancer drugs has promoted interest in Pt^{2+} chelators to moderate the toxic side effects of these drugs. Reduced nephrotoxicity and increased biliary excretion have been found when using edt^- and, in animal studies, dmsaH_4 increased urinary excretion of Pt. The use of Amifostine, a phosphorylated aminothiol compound, for this purpose is mentioned in Section 4.3.2.3.1. Some people have a severe allergic reaction to Pt^{2+} compounds such as $K_2[PtCl_4]$ and must avoid any contact with such materials.

4.2.3.3.15 Plutonium. A particular example of the application of ligand design to toxic metal sequestration is provided by the development of agents for the decorporation of radiotoxic plutonium. As an f-block actinide element Pu(+4) is oxophilic forming its strongest complexes with hard donors. The chemistry and biological behaviour of Pu^{4+} shows some similarities to that of Fe^{3+} but the larger ionic radius of the Pu^{4+} ion can accommodate 8-coordination. The typical choice of sequestering agent for an ion of this type would be $dtpa^{5-}$ which offers the prospect of 8-coordination within a pentanegative multichelating ligand. In an attempt to improve on this agent various other polydentate ligands designed to encapsulate the Pu^{4+} ion were investigated. In mice the so-called LICAM proligands containing four 1,2-dihydroxyaryl [derived from 1,2-dihydroxybenzene or catechol,1,2-$(HO)_2C_6H_4$] chelating groups could produce similar urinary excretion of injected ^{238}Pu to $CaNa_3$dtpa Table 1, but were limited in their effectiveness by the weak acidity of the aryl OH groups. Under physiological conditions of near neutral pH the aryl OH groups remain largely protonated and, in effect, H^+ competes for the aryl-O^- groups. Better total excretion could be obtained with proligands containing more acidic hydroxamate and *N*-hydroxypyridinonecarboxylate chelating groups, exemplified by 3,4,3-lihopoH_4, **13**, and dfohopoH_4, **14**. Although these

Table 1 *Excretion of injected ^{238}Pu from mice following chelation therapy*[a]

	% injected $^{238}Pu(+4)$ citrate excreted via		
Sequestering Agent (30 μmol kg^{-1} ip)	*Urine*	*Gut (Faeces and GI tract contents)*	*Total*
$CaNa_3$dtpa	63	7	69
LICAM (Carboxylic acid derivative)	49	22	71
LICAM (Sulfonic acid derivative)	62	2.4	64.4
Lihopo	24	57	81
Fe^{3+}-lihopo complex	40	46	86
Dfohopo	19	68	87

[a] Data from D.L. White, P.W. Durbin, N. Jeung, and K.N. Raymond, *J. Med. Chem.*, *1988*, **31**, 11–18.

compounds produced lower urinary excretion than $CaNa_3$dtpa a much larger proportion of the Pu was excreted *via* the gut.

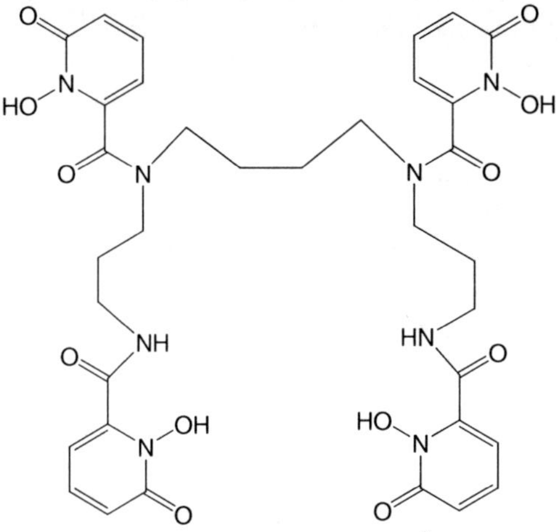

13
lihopoH_4

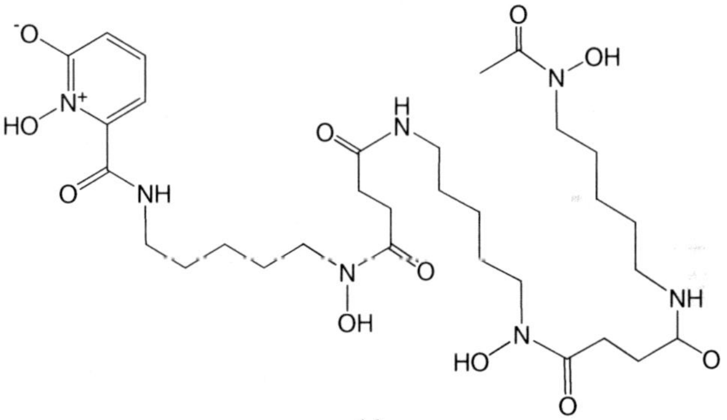

14
dfohopoH_4

4.2.3.3.16 Thallium. Although banned in many countries thallium salts were once used in rodenticides among other applications. The superficial similarity between Tl^+ and K^+ has led to the administration of K^+ salts to moderate Tl^+ toxicity and promote its urinary excretion. Prussian blue, particularly the ammonium salt $NH_4Fe\{Fe(CN)_6\}$, appears to be an effective treatment for Tl(+1) poisoning but acts as an ion exchange reagent rather than a chelator. It is of low toxicity, retained within the gut and so promotes faecal elimination of Tl^+ through its exchange with NH_4^+.

4.2.3.3.17 Zinc. Zinc is widely used, particularly as an anticorrosion coating and in skin creams and ointments. Cases of zinc poisoning have been treated with dmpaH_2 or edtaH_4 and it is known that edtaH_4 and, more efficiently, $dtpaH_5$ mobilise Zn in humans. Studies of parenteral Zn intoxication in rodents suggested that $edtaH_4$, $cdtaH_4$ and $dtpaH_5$ are the most effective treatments but $dmsaH_4$ also reduced acute mortality. A higher chelator/zinc ratio was needed to produce an effect with $dmpsH_3$.

4.2.3.3.18 Conclusions. Apart from cases resulting from disease, acute metal intoxication is comparatively rare nowadays. An effective panel of well established chelating agents is available to treat metal intoxication, including $edtaH_4$, $dtpaH_5$, $dmsaH_4$, $dmpsH_3$ and $dfoH_3$. No doubt more effective sequestering agents could be developed for specific metals but increasing controls of environmental and occupational exposure to toxic levels of metals reduces the economic incentive to pursue research of this type. However, the search for new agents to address metal intoxication resulting from diseases or their treatment offers an important avenue for research into chelation therapy agents.

4.2.4 Thalassaemia–Iron

Thalassaemia is a genetic disease, which affects a significant number of the babies born each year worldwide. In 1990 of the order of 100,000 of babies were said to be seriously affected by thalassaemia. The number of gene carriers for the disease at that time was said to be around 100 million, widely distributed across the globe but particularly in areas affected by malaria. The genetic disorder is characterised by a reduced rate of synthesis of hemoglobin protein and is particularly acute for those having two defective genes (homozygotes) for hemoglobin. Sufferers of this more severe form of beta-thalassaemia, that is thalassaemia-major or Cooley's anaemia, usually die within one year of birth if left untreated. However, those having only one defective gene (heterozygotes) for hemoglobin show increased resistance to malaria. In humans the malaria parasite spends part of its life cycle in the red blood cells and, in the heterozygotes, the lifetime of these cells is reduced and development of the malaria parasite is impaired. Unfortunately, there is a one in four chance that two heterozygote parents will have a homozygote offspring.

The life expectancy of a thalassaemia-major sufferer can be extended to around 20 years by repeated regular blood transfusions. However, each pint of blood contains around 200 mg of Fe compared to a normal daily requirement of 1–5 mg. As there is no efficient excretion pathway for this excess iron, over a period of about 10 years a transfusion dependent thalassaemia patient might accumulate over 50 g of iron. This causes damage to the heart, liver and endocrine system and produces the toxic effects of iron overload. Removal of the excess iron by chelation therapy can substantially extend the life expectancy of transfusion dependent thalassaemia sufferers.

Polyamine carboxylates such as $edta^{4-}$ might appear to offer an obvious approach to the treatment of iron overload by chelation therapy. However, these compounds are insufficiently selective for iron and remove other metals such as calcium or zinc. In fact the first chelating agent to be widely used to treat iron overload in thalassaemia patients was the naturally occurring fungal siderophore desferrioxamine B (**7**, $dfoH_3$). This compound is a potent iron scavenger in nature and is highly selective for iron having much lower affinities for copper, zinc, calcium and manganese. The three hydroxamate groups provide strong binding sites for iron and the structure of the molecule allows it wrap around Fe^{3+} offering an essentially octahedral 6-coordinate environment. Desferrioxamine B is used clinically under the name Desferal, is water soluble, not strongly bound to protein in plasma and shows an extracellular distribution. The disadvantages are that it is expensive to prepare, can have side effects and is unsuitable for oral use. The hydroxamate groups are sensitive to acid in the stomach and the compound is poorly absorbed in the gastrointestinal tract. As a consequence administration by intravenous infusion over a period of hours is necessary several times a week. The complexed iron is excreted in the urine and *via* bile through the gut, the renal excretion being bimodal with the half time for the slower excretion being around 6 h. The compound is not well tolerated by all patients and a range of side effects have been found in long-term therapy. These include allergic reactions and renal, pulmonary and neurological effects. More recently a new agent has been obtained by chemically attaching $dfoH_3$ to a modified starch polymer. Like Desferal the resulting compound shows a high affinity and specificity for Fe^{3+} but it does not cause the acute toxic effects. Promising results have been obtained from Phase I clinical trials of this preparation and it may offer a means of extending the efficacy of the $dfoH_3$ chelation approach.

Despite the undoubted success of $dfoH_3$ in treating iron overload, the expense of this chelator, the need for its intravenous administration and unacceptable side effects in some patients provide strong driving forces to develop an alternative oral treatment. The design of such new agents might be informed by consideration of the structures of other naturally occurring siderophores secreted by micro-organisms to scavenge iron. Apart from the hydroxamate moiety found in $dfoH_3$, the 1,2-dihydroxybenzene (catechol) unit found in the cyclic trimer siderophore enterobactin (**15**) offers a possible alternative approach to the sequestration of iron. Unfortunately catechol derivatives tend to be poorly absorbed in the gut and prone to oxidation so do not offer a viable oral alternative to $dfoH_3$. A structure which incorporates the structural features of both catechol and hydroxamate is provided by 1-hydroxy-pyridin-2-one (**16**) derivatives. Such compounds show a high affinity for iron ($\log\beta_3 = 27$ for **16**) and are relatively stable to acid. Higher binding constants are possible using 1-alkyl-3-hydroxy-pyridin-2-one (**17**, $\log\beta_3 = 32$) derivatives or, even better, 1,2 dialkyl-3-hydroxy-pyridin-4-one (**18**, $\log\beta_3 = 37$) derivatives. The 3-hydroxy-pyridin-4-ones show high selectivity for iron over copper and zinc compared to other chelating agents (Table 2) and are neutral ($pK_{a1} = 3.6$, $pK_{a2} = 9.9$) over the physiologically important pH range of 5.0–9.0. In solution under these conditions the 3:1 complex of **18** with iron predominates. The solid-state structure of the iron complex with **18** ($R^1{=}C_2H_5$, $R^2{=}CH_3$) reveals an distorted octahedral 6-coordinate geometry around the Fe^{3+} ion as expected. In the biological context the 1-alkyl-3-hydroxy-pyridin-2-ones do not remove

Table 2 *Selected stability constants (log β) of metal complexes*

	Proligand and log β[a]			
Metal ion	*18*[b]	*DFO-B*[c]	*Catechol*[d]	*edtaH₄*
Fe^{3+}	37 (log β_3)	31	40 (log β_3)	25
Cu^{2+}	17 (log β_2)	14	25 (log β_2)	18
Zn^{2+}	12.5 (log β_2)	11	17 (log β_2)	16
Ca^{2+}	4.5 (log β_2)	2.5		11
Mg^{2+}	7 (log β_2)	4	6 (log β_2)	9

[a] As log β_1 unless otherwise specified
[b] 3-hydroxypyridin-4-one.
[c] Desferrioxamine B.
[d] 1,2-dihydroxybenzene.

iron from hemoglobin or cytochromes but do compete effectively with the weaker albumin binding sites in blood. They show a similar affinity for iron to the iron transport protein transferrin so can reduce iron levels without greatly affecting essential iron uptake in bone marrow. In contrast to $dfoH_3$, which is too large for a molecule to enter channels in the structure of the iron storage protein ferritin, the 1,2 dialkyl-3-hydroxy-pyridin-4-ones can enter the structure and mobilise stored iron.

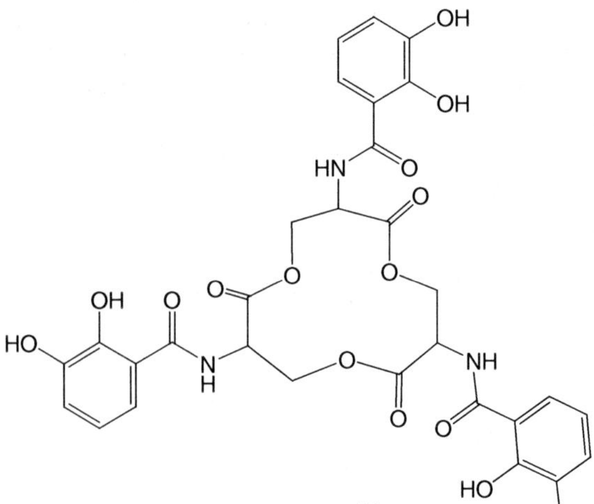

15
enterobactin

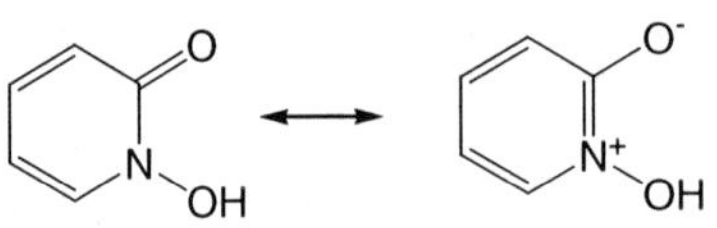

16
1-hydroxypyridin-2-one

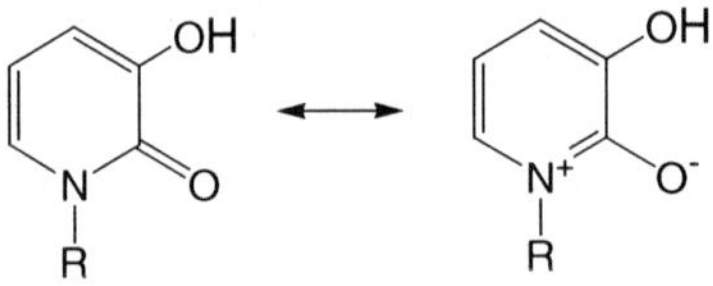

17
1-alkyl-3-hydroxypyridin-2-one
R = alkyl

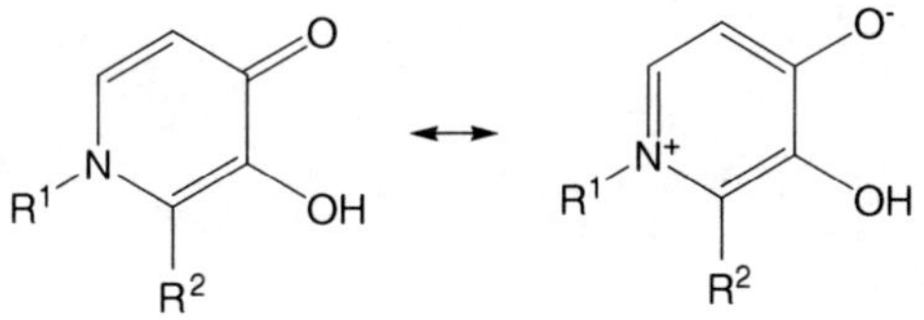

18
R^1, R^2 = alkyl
$R^1 = R^2 = CH_3$ Deferiprone

The dimethyl derivative **18** ($R^1{=}R^2{=}CH_3$) has undergone clinical trials under the name Deferiprone (L1) with promising results from several countries. However, the compound is not universally effective with discontinuation rates of 20–40% in some trials. There is also a higher risk of agranulocytosis which may restrict its use to patients for whom treatment with desferrioxamine is unsuitable. In 1999, Deferiprone was licensed in Europe for use with patients in whom treatment with $dfoH_3$ proves inadequate, and is now in use in around 50 countries. The possibility of a synergistic effect from using $dfoH_3$ and Deferiprone together has been considered and there is increasing evidence to suggest that this may be the case. Combined treatment may provide therapeutically beneficial results in that using lower doses of Deferiprone may reduce the occurrence or severity of side effects while requiring fewer $dfoH_3$ infusions each week. In addition recent studies indicate that the long-term use of Deferiprone can markedly improve cardiac function.

In order to minimise the ability of $Fe^{3+/2+}$ to enter into unwanted oxidation/reduction reactions it is important to fully saturate the iron coordination sphere to prevent the approach of reactive substrates. Increasing the size of the sequestering ligand and moving to tridentate or hexadentate, rather than didentate, chelation may offer improvements over the hydroxypyridinones. Examples of tridentate sequestering agents for iron include the siderophore Desferrithiocin (DFT, **19**, $dftH_2$) originally isolated from Streptomyces antibioticus. Animal studies showed this to be an orally available agent which can reduce acute iron overload. However there were side effects raising doubts over its suitability for long-term use in humans. A number of analogues of Desferrithiocin are now being investigated with a view to identifying less toxic alternatives. Among these, GT56-252 (**20**) has entered clinical trials with promising results. An example of the development of a hexadentate sequestering agent is provided by HBED (**21**, $hbedH_4$). This compound has a high affinity for Fe^{3+} and forms a 1:1 complex with good selectivity. Despite promising results in rodents it is insufficiently active in humans when given orally. However, in animal studies the monosodium salt $NahbedH_3$ appeared to be about twice as efficient as $dfoH_3$ in promoting iron excretion when delivered parenterally. The compound did not show significant systemic toxicity or have detrimental effects on blood pressure or heart rate. If the compound is similarly well tolerated in humans it may offer an alternative to $dfoH_3$ for treating acute iron poisoning or chronic iron overload, although parenteral administration would still be necessary. Since $hbedH_4$ is chemically different from $dfoH_3$, it may offer an important alternative for patients allergic to $dfoH_3$.

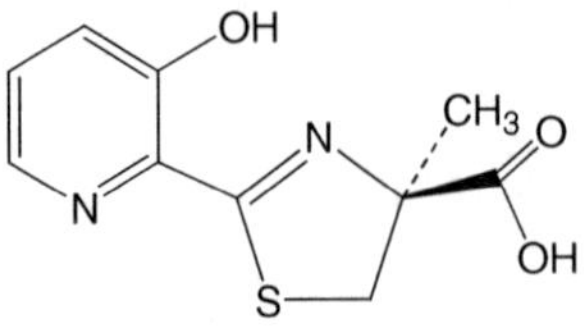

19
Desferrithiocin

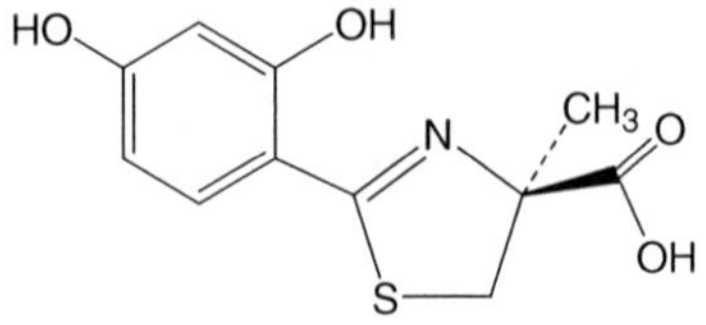

20
GT56-252

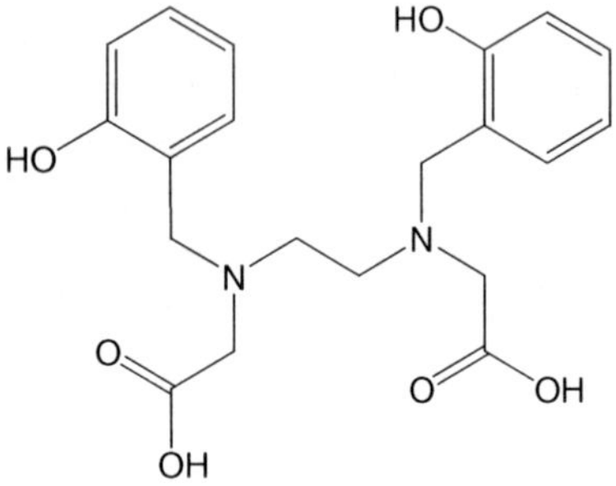

21
HBED, hbedH_4

Most notable among the variety of other iron chelators being investigated is the tridentate chelator deferasirox (**22**, ICL670). In clinical trials involving over 1000 patients this compound was well tolerated and produced a dose dependent excretion of iron almost entirely *via* faeces. In patients with thalassaemia and sickle cell disease, as well as other rare anaemias, who were receiving blood transfusions, the iron burden could be maintained or reduced using doses of 20–30 mg kg^{-1} day^{-1}. The drug has the great advantage that it can be administered orally and only once daily at that. Approval in the USA for the worldwide use of deferasirox under the trade name Exjade® was granted in November 2005.

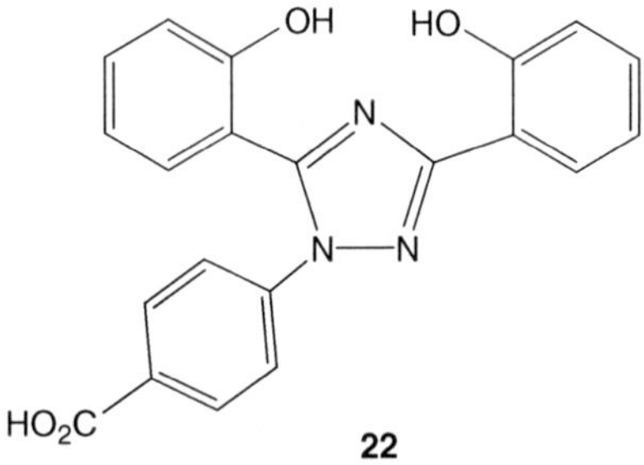

22
Deferasirox
ICL670

4.2.5 Wilson Disease–Copper

Wilson disease is a genetic disorder of gene ATP7B on chromosome 13 which affects around 1 out of every 30,000 members of the population. It is an autosomal recessive condition so that a person must inherit an altered gene from each parent if they are to develop the disease. Those with only one altered gene do not develop the disease but remain carriers who may pass the disease gene on to their offspring. The ATP7B gene is associated with the synthesis of a P-type ATP-ase used in copper transport and is expressed mainly in the liver, kidney and placenta. In Wilson disease sufferers, this genetic disorder leads to excess copper accumulating in the liver. Defective biliary excretion of copper may play an important role in the process and the build up of copper in the cytosol of hepatocytes eventually leads to necrosis. This results in the release of copper into the bloodstream where it can damage erythrocyte membranes. It also delivers excess copper to other organs, particularly the brain, leading to more widespread damage including neurological symptoms.

Copper is an essential trace metal required, for example, by enzymes such as oxidoreductases and mono-oxygenases. Most people obtain more than enough copper to meet their needs from their diet and, in order to maintain normal copper levels, the excess copper must be excreted. Ingested copper is absorbed through the wall of the intestine to enter the bloodstream as exchangeable Cu^{2+} bound to albumin and low molecular weight amino acid complexes. Reduction to Cu^{+} is thought to occur in a membrane reductase before entry to the cell *via* a membrane Cu^{+} transporter. Once inside the cell it is thought that most of the copper is bound to glutathione (GSH) as Cu^{+}. The copper is also processed by various so-called 'chaperone proteins' which mediate its transport to proteins such as cytochrome oxidase and superoxide dismutase, as well as to copper transporting ATP-ases. One of the recipient sites for copper is the Wilson ATP-ase which has been found to bind six Cu^{+} ions in the N terminal region. The defective function of this protein in Wilson disease sufferers leads to inadequate excretion of copper and its accumulation in the liver. Chelation therapy can be used to create a negative copper balance thereby preventing the development of symptoms in presymptomatic patients or relieving the symptoms where the effects of the disease are more fully developed. Since chelation therapy cannot address the underlying cause of the illness it must be continued throughout the lifetime of the subject. This places demands on the toxicity of the chelating agent used and toxicity levels acceptable for short-term treatment of acute copper poisoning would not necessarily be acceptable for long-term treatment of Wilson disease.

Under physiological conditions copper can be present either of two oxidation states, Cu(+2) or Cu(+1). The d^9 Cu^{2+} ion has a small Crystal Field Stabilisation Energy and is most usually found with coordination number 6 in distorted octahedral geometries or with coordination number 4 in square planar geometries. It is borderline in character typically forming complexes with ligands containing N or O donor atoms. In contrast d^{10} Cu^{+} has no Crystal Field Stabilisation Energy and is normally found with coordination number 4 in near tetrahedral geometries. It is a soft metal ion which is very readily oxidised and most stable when bound to soft donor atoms such as phosphorus or sulfur. One approach to the removal of extracellular copper would be to sequester it in ligands which can effect the reduction of Cu^{2+}

traces of platinum were present in the culture as a result of reactions between the platinum electrodes used and the growth medium which contained chloride and ammonium ions. Complexes such as $[PtCl_6]^{2-}$, $[Pt(NH_3)_2Cl_2]$ and $[Pt(NH_3)_2Cl_4]$ were found to be present in the solution and eventually *cis*-$[Pt(NH_3)_2Cl_2]$, known as *cis*-DDP (**23**), or subsequently the drug named cisplatin, was identified as a causative agent. This compound produced filamentation in the absence of an electric field and a similar effect was found with the platinum(+4) complex *cis*-$[Pt(NH_3)_2Cl_4]$ (**24**). *cis*-DDP had been known for over 100 years prior to Rosenberg's discovery, having been prepared by Peyrone in 1845. In 1893 Werner correctly assigned the formula *cis*-$[Pt(NH_3)_2Cl_2]$ to the complex requiring it to have a square planar geometry. [Nowadays this structure would be expected from crystal field theory for a complex of a 3rd row d^8 transition metal ion (Section 2.5.4) but Werner's work preceded any knowledge of crystal field theory].

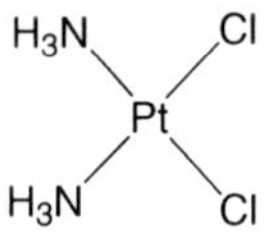

23
cis-DDP
or
cisplatin

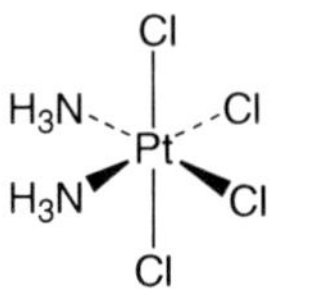

24
cis- $[Pt(NH_3)_2Cl_4]$

The suppression of cell division while not killing the cells, and so leading to filamentous growth, was recognised as a sign that platinum compounds of this type might be effective as anticancer agents. Further investigations of antitumor activity by the National Cancer Institute and the Michigan State team produced remarkable results. Mice with implanted Sarcoma 180 tumors showed tumor regression within days of being injected with *cis*-DDP and went on to have a normal lifespan. In controls not treated with *cis*-DDP death occurred after around 20 days. A typical therapeutic index (LD_{50}/ID_{50}) for *cis*-DDP was found to be about 8.

In 1971 Phase I clinical trials commenced on *cis*-DDP and in 1978 FDA approval was granted for its clinical use in the treatment of testicular and ovarian cancer. By 1983 cisplatin was the leading anticancer drug in use in the USA and is most effective against genitourinary tumors. In the case of testicular cancer, once a leading cause of death in young males, cisplatin cures nearly all patients with stage A or B carcinomas. Typically a course of treatment might involve weekly doses of around 5 mg kg^{-1} body weight of cisplatin over a period of a month. Usually cisplatin is used in combination with other anticancer drugs. The different drugs may act to disrupt different aspects of tumor development and so can have a greater combined impact than if used individually. There can also be a problem if tumors have, or acquire, a resistance to cisplatin and the use of several drugs can maintain efficacy in the treatment. Radiation therapy may also be used in combination with drug regimes involving cisplatin. In addition to testicular cancer, ovarian carcinomas, head and neck tumors, cervical cancer and non-small cell lung cancer patients may benefit from treatments which include cisplatin or related compounds.

One major limitation to the use of cisplatin is kidney toxicity. However, this can be controlled by administering large quantities of water intravenously and administering a diuretic agent to flush the compound through the kidneys and reduce nephrotoxicity. Other side effects include nausea and vomiting, tinitus and bone marrow suppression. Allergic reactions to the drug are also known but are not common. Over 3000 platinum compounds have now been tested in the search for new platinum drugs which may offer improvements in performance compared to cisplatin and, of these, about 1% have entered clinical trials.

The search for new anticancer compounds has been guided by structure activity relationships developed since the medicinal properties of platinum compounds were first recognised. These have helped guide the search for second generation platinum drugs with lower toxicity such as the now widely used carboplatin (**25**). Other examples include nedaplatin (**26**) which has been approved for use in Japan, oxaliplatin (**27**) approved for use in France, and Lobaplatin (**28**, where * denotes a chiral carbon atom) which is in Phase II clinical trials for treating cisplatin-resistant ovarian cancer, advanced head and neck cancers and small cell lung cancer. Oral drugs such as JM216 (**29**) now in Phase II clinical trials offer a further innovation. New compounds which are effective against a broader range of tumors, or against cisplatin resistant tumors, are important research targets and their further development is similarly dependent on a detailed understanding of the mechanism of action of platinum anticancer agents.

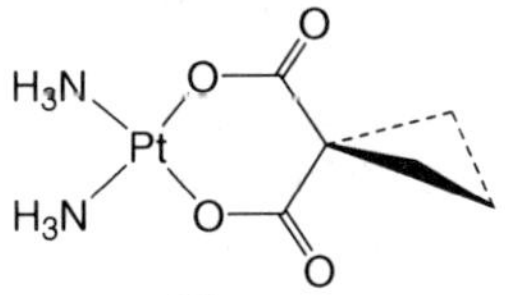

25 carboplatin

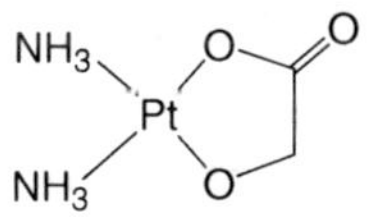

26 nedaplatin

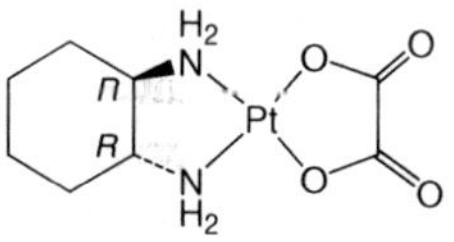

27 oxaliplatin

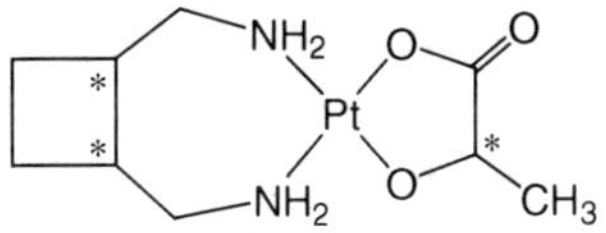

28 Lobaplatin

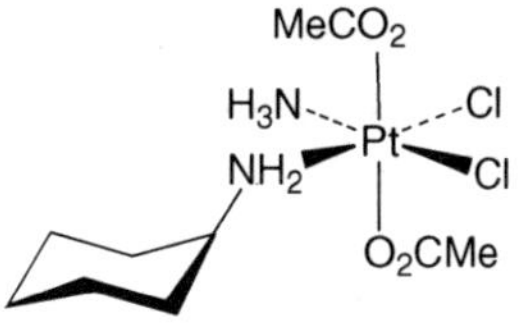

29 JM216

4.3.2.2 Structure and Activity

4.3.2.2.1 Platinum Amine Chloride Complexes. Both *cis*-DDP and its *trans*-isomer contain two labile anionic chloride ligands together with two neutral ammonia ligands, which are inert under biological conditions. Based on the results of screening a large number of compounds, some early structure activity relationships were developed for cisplatin type compounds. In the initial investigations of *cis*-DDP it was found that the platinum(+4) complex *cis*-$[Pt(NH_3)_2Cl_4]$ was also effective in suppressing cell division, probably as a result of *in vivo* reduction to form *cis*-DDP so that *cis*-$[Pt(NH_3)_2Cl_4]$ is actually acting as a prodrug. Since *cis*-$[Pt(NH_3)_2Cl_4]$ is a strong field octahedral d^6 complex it shows kinetically inert behaviour (Section 2.8.1). However, reduction to a d^8 platinum(+2) complex will increase ligand lability and the loss of two ligands to give a 4-coordinate square planar complex is to be expected following the reduction. Unlike their *cis*-isomeric counterparts, *trans*-DDP and *trans*-$[Pt(NH_3)_2Cl_4]$ were found to be inactive. Related palladium complexes were also found to be inactive, as were platinum complexes with pyridine rather than amine ligands. Similarly, when ammonia was replaced by organophosphine (PR_3; R=hydrocarbyl) or organosulfide (R_2S) as the neutral ligand, biological activity was lost. In general it was found that neutral compounds were required that contain two leaving groups in a *cis* orientation. Non-leaving groups with weak *trans* effects (Section 2.8.3) such as ammonia or organoamines worked best and the optimum lability of the leaving groups was found to be based around that of chloride. Another observation was that at least one NH group bound to platinum was needed. The nature of substituents on the amine ligands in compounds such as *cis*-$[Pt(NH_2R)_2Cl_4]$ (R=alkyl) was found to affect both the toxicity and the activity of the complex. These substituents will alter the balance between water and lipid solubility but, through variations in size, could also exert structural effects. Being octahedral, platinum(+4) complexes offer more scope for substitution and hence the enhancement of lipophilicity while retaining the ammonia ligands. This is important in oral drugs *e.g.* JM216 (**29**) as lipophilicity is considered necessary for effective intestinal uptake. However, it is thought that the platinum(+4) complexes are rapidly reduced *in vivo* to platinum(+2) compounds so that following oral uptake JM216 is presumably reduced and loses the axial carboxylate ligands to produce the pharmacologically active form of the complex.

The early structure-activity relationships developed for *cis*-DDP derivatives have provided a useful guide to focus the development of this particular class of platinum anticancer agent. However, most of the results on which these relationships are based were obtained from *in vivo* results using mice implanted with animal tumors. More recent studies involving a large panel of *in vitro* human tumor models have provided many examples of active compounds which do not conform to the structure activity relationships, established earlier for *cis*-DDP derivatives. The results of these more recent studies reveal a much more subtle relationship between structure and activity, and serve to suggest new directions for the development of platinum group metal-based anticancer agents.

4.3.2.2.2 Nature of the Metal – Palladium Complexes. The chemical similarities between palladium and platinum might lead to the expectation that related palladium complexes might also be active against some cancers. The coordination chemistry of both palladium and platinum is dominated by the metal(+2) ions, Pd^{2+} and Pt^{2+}, which have similar ionic radii. Both have the d^8 electron configuration and so a preference for a square planar coordination geometry. These ions form many complexes which share a common structure and formula, differing only in the metal ion present. However, there are some important differences between these two elements. Platinum is more readily oxidised to the oxidation number +4 so that relatively stable complexes can be formed with Pt^{4+}, although under physiological conditions reduction to Pt^{2+} is to be expected. Palladium(+4) is much more reactive and oxidation number +4 is much less important in palladium chemistry. Typically third row d-block metal ions show more kinetically inert behaviour than their second row counterparts. Thus isomerisation and ligand exchange reactions will usually be more rapid for Pd^{2+} complexes than for their Pt^{2+} counterparts. In the chemically diverse environment of within an organism, issues of kinetic stability can be very important as a complex will be exposed to a wide variety of potential ligands, some of which may compete with the ligands initially present on the metal. The palladium complex $[Pd(en)Cl_2]$ (**30**, en=$NH_2CH_2CH_2NH_2$) for example reacts spontaneously with water to form $[Pd(en)Cl(H_2O)]^+$ and $[Pd(en)(H_2O)_2]^{2+}$, hydrolysis being around 200,000 times faster than for cisplatin.

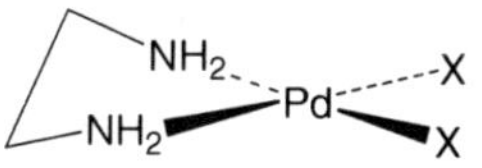

30 X = Cl
31 X = NO_3

Early investigations of palladium analogues of *cis*-DDP derivatives found them to be inactive and, as yet, no palladium complexes have entered clinical trials. However, several have now been found to show some antitumor properties (*e.g.* **31**, **32**, **33**). In the case of **31** a dose of 80 mg kg^{-1} body weight almost doubled the survival time of mice bearing the Sarcoma 180 ascitic tumor, although the drug was administered immediately after tumor implantation, contrary to the usual protocol. A common feature in each of these complexes is the presence of chelating diamine ligands. The chelating diamine ligand will show reduced lability compared to unidentate amines. Furthermore, a chelating diamine ligand cannot isomerise from the *cis*- to the *trans*-geometry and it has been suggested that the inactivity of palladium complexes containing *cis* - unidentate amines may result from isomerisation to inactive *trans*-complexes. Another example of inactivity being associated with unidentate amines is provided by **34**. This compound was found to be inactive, whereas the related chelating diamine complex **35** has been found to have similar *in vivo* activity to cisplatin. So long as it remains fully bound to the Pd^{2+} ion, the methylorotato ligand should prevent isomerism of **34**. However, if one or both donor atoms in this ligand become detached, the two ammonia ligands may then relocate. The greater lability of Pd^{2+} complexes compared to Pt^{2+} may be addressed by using more strongly bound ligands, hence chelating ligands may offer an advantage. Alternatively, changing the nature of the ligand donor atom may be beneficial. Since thiocyanate has a greater affinity for Pd^{2+} than chloride, the reactivity of **33** may be more favourable for antitumor activity than that of its counterpart **30** with X=Cl. On this basis it would seem difficult to explain the activity of **31**, which contains labile nitrate ligands. It is possible that in these cases the nitrates are rapidly replaced by hydroxide in solution, possibly leading to more inert hydroxy-bridged complexes such as **36**. Among other palladium complexes which show potential antitumor properties, **37** was found to be cytotoxic in a panel of seven human tumor cell lines. Palladium complexes of *N*-substituted 2-acetylpyridine thiosemicarbazone ligands have shown activity against leukaemia P388 and *trans*-$[PdL_2Br_2]$ (L^-=diethyl-2-quinolylmethylphosphonate) has shown cytotoxic activity with KB and L1210 cell lines.

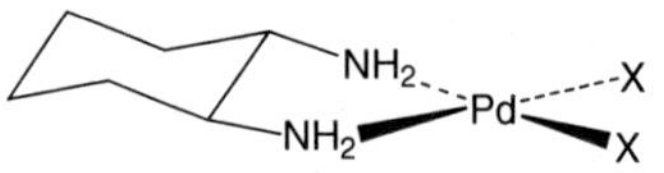

32 X = NO_3
33 X = SCN

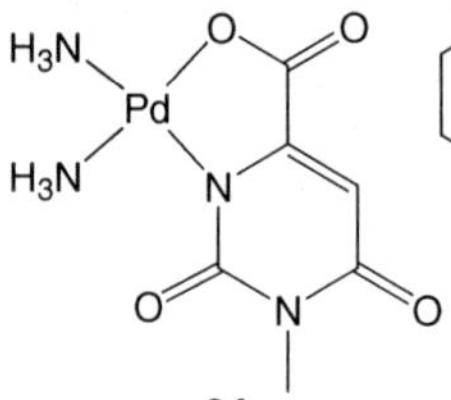

34

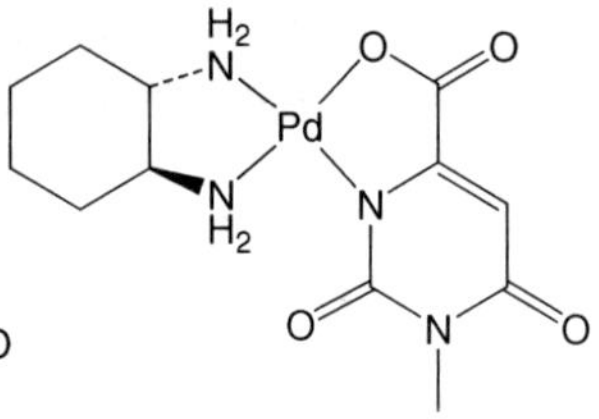

35

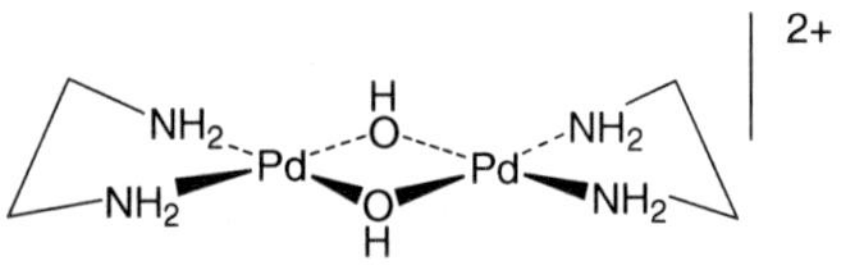

36

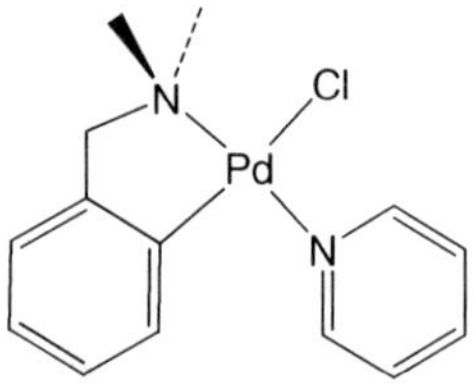

37

4.3.2.2.3 Nature of the Amine Ligands. A huge number of complexes, similar to *cis*-DDP but containing different amine ligands, have been screened for potential antitumor properties. The nature of the amine can affect the properties of the complex in three important ways. Firstly, the relative solubility of the complex in fatty, lipid, environments as compared to water can be increased by adding hydrocarbon substituents to the nitrogen. Secondly, the electron donor properties of the amine can, through the *trans*-effect (Section 2.8.3), modify the lability of the *trans*-anionic ligand. Thirdly, the size and shape of the substituents on the amine ligands can affect the reactivity of the complex by hindering the approach of solvent or other reactant molecules to the metal ion. Regarding the first of these effects, the nature of the alkyl substituents on the amine ligands in *cis*-$[Pt(NH_2R)_2Cl_4]$ (R=alkyl) was found to affect both the toxicity and activity of the complex. Larger alkyl groups R such as octyl (C_8H_{17}), adamantyl ($C_{10}H_{15}$) or branched alkyl were associated with greatly reduced activity compared to the complex with R=hexyl (C_6H_{13}). Although there was no clear correlation with solubility, complexes showing similar solubilities in water and lipids showed lower toxicity and so a higher therapeutic index, whereas higher water solubility was associated with higher activity but also higher toxicity. Careful selection of R in complexes of this type thus offers one approach to optimising the properties of a cisplatin derivative.

The second effect mentioned above is exemplified by **38**, a metabolite of the orally-active Pt(IV) complex JM216 (**29**). The chloride *trans* to the stronger donor cyclohexylamine ligand is the more labile of the two and the water ligands in the hydrolysis product **39** show *pKa* values (5.68 and 7.68) indicative of a 100-fold difference in acid dissociation constants. The effects of amine ligand structure are also seen in complexes **40** and **41**. Differences in the electron donating capacity of methyl pyridine compared to ammonia are apparent in the lower substitution rate of the chloride *trans* to the 3-methyl pyridine ligand in **40** compared to that of the chloride *trans* to the ammonia ligand, which is a little higher than for cisplatin (Scheme 1). In contrast, the 2-methylpyridine complex **41** has a substitution rate for the replacement by water of the chloride *trans* to the ammonia which is 5 times slower than for **40** and 3 times slower than for cisplatin under similar conditions. In part these results reflect differences in the size of the methylpyridine ligands compared to ammonia, in particular the increased bulk associated with the 2-methyl pyridine ligand. The crystal structure of **41** shows that, in the solid state at least, the pyridine ring is almost perpendicular to the plane of the platinum and its four donor atoms, being tilted by 103° to this plane. The 2-methyl substituent obstructs access to the vacant axial coordination site on one side of the metal centre and affects its reactivity through what is known as steric hindrance (Figure 4). In an associative reaction mechanism, typical of square planar platinum(+2) complexes (Section 2.8.3), interactions between the 2-methyl substituent and the other ligands in a 5-coordinate reaction intermediate will destabilise the intermediate and slow the reaction down. Such steric effects are not apparent in the replacement of the chloride *cis* to the sterically less demanding 3-methylpyridine in **40**. The crystal structure of **40** shows that the pyridine ring is tilted by 49° to the square plane and the methyl group is located further from the metal centre. Thus steric effects are less important in the reactivity of **40** than **41**. It is the combined effect of the electronic and steric properties of the ligands in these complexes which determine their overall reactivity.

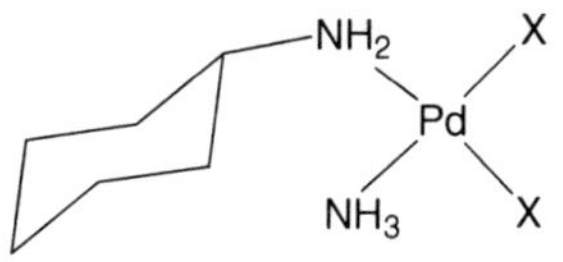

38 X = Cl
39 X = H_2O

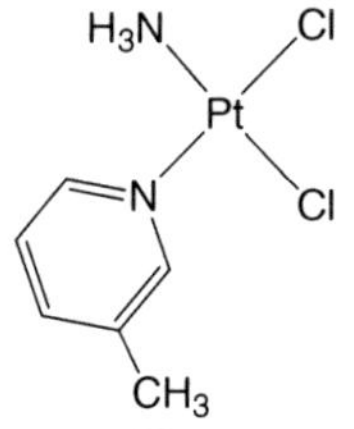

40

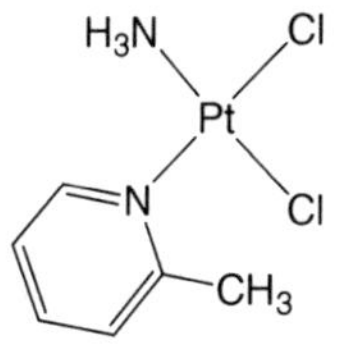

41

One further feature which may be exploited with the amine ligands is the chelate effect. The lability of the amine ligands can be reduced by using diamine ligands, such as 1,2-diaminoethane or 1,2-diaminocyclohexane, which exploit

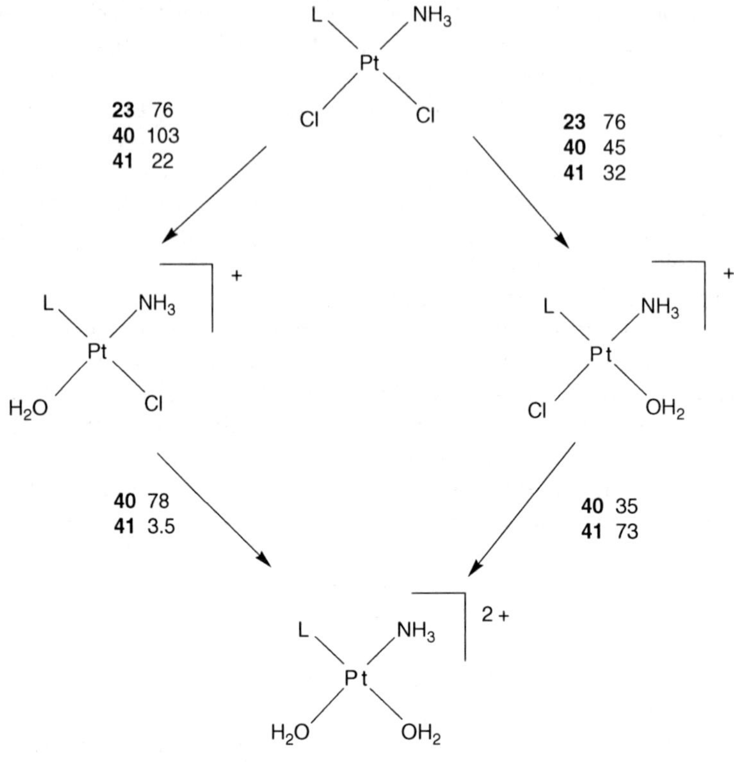

Scheme 1

the chelate effect. This approach has been used with palladium complexes, which are inherently more labile than their platinum counterparts. However, it can equally be applied to reducing the lability of platinum complexes; oxaliplatin (**27**) and Lobaplatin (**28**) providing examples of drugs with chelating diamine ligands which are in clinical use. A structural feature common to these two complexes is the presence of chiral carbon centres in the cyclic hydrocarbon part of the diamine ligand. The importance of chirality in determining activity is not well understood but this structural feature may be relevant in some cases. Data for substituted ethane-1,2-diamine complexes **42** and **43**, which contain respectively one and two chiral carbon centres, indicate substantial differences in mutagenicity between the *R,R*, *S,S*, and *R,S* isomers. In other studies the *S,S* isomers of **44** or **45** were found to be more active than the *meso* (*R,S*) isomers or racemic mixtures (*R,R* plus *S,S*) against the breast cancer cell line MDA-MB 231 *in vitro*. A suggested explanation of this is that the ligands mimic compounds capable of binding to estrogen receptors which selectively bind

40 41

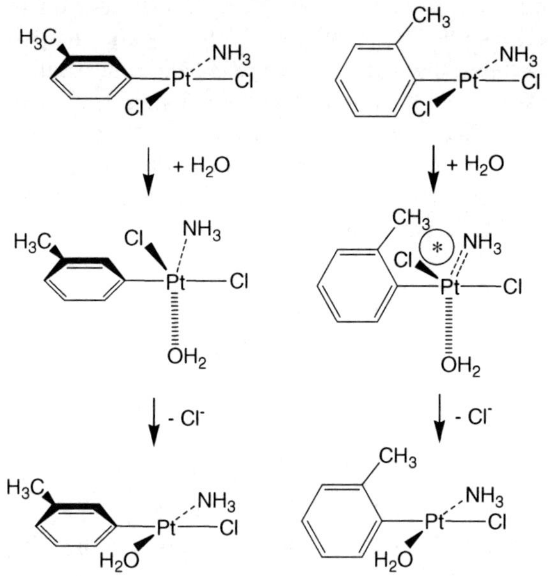

Figure 4 *An illustration of how the -Cl/CH_3- interaction in **41** can slow its aquation rate compared to **40** through a steric effect arising from the difference in ligand structure*

particular enantiomeric structures. Activity differences between isomers were also found *in vivo* against some leukaemias implanted in mice. Although the antitumor properties of the enantiomers of **46** were found to be similar, the *S* enantiomer was found to be more nephrotoxic in rats and gave 5 times higher renal platinum concentrations. Similar results have been obtained with one optical isomer of the 1,2-diaminocyclohexane (dach) complexes [Pt(dach)Cl_2] (**47**) or [Pt(dach)(SO_4)] being more mutagenic than the other.

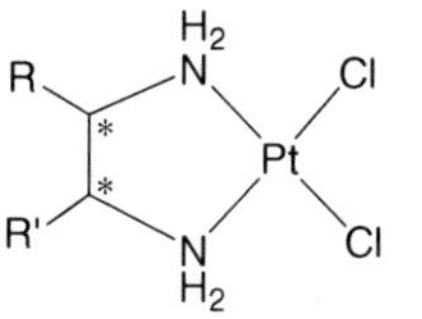

42 R = CH_3, R' = H
43 R = R' = CH_3
44 R = R' = C_6H_4OH-3
45 R = R' = C_6H_4OH-4

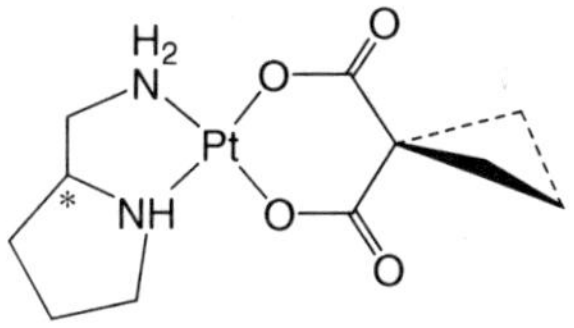

46

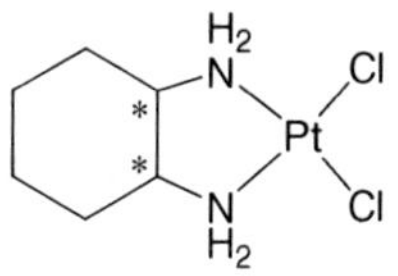

47

An extensive study has been made of the effect of substituents in the ethane-1, 2-diamine complexes **48** on their *in vitro* activity against the breast cancer cell line MDA-MB 231 and *in vivo* against leukaemia P388 implants in mice. The presence of a methyl substituent in place of H at the R^1 or R^2 position led to a decrease in activity but at the R or R′ position an increase in activity was found. The most active compounds *in vivo* were those with no substituent on the aryl ring (*i.e.* Z=H). Among aryl-substituted compounds those with *ortho* halide substitution were more active than those with *meta* or *para* halide substitution. Extending the length of the link between the aryl ring and the diaminoethane chelate ligand led to reduced activity. Clearly a number of subtle structural effects are at work in these systems and it is not yet possible to offer definitive explanations for the origins of these effects. These findings underline the importance of screening a wide variety of structural variations in any attempt to optimise drug performance.

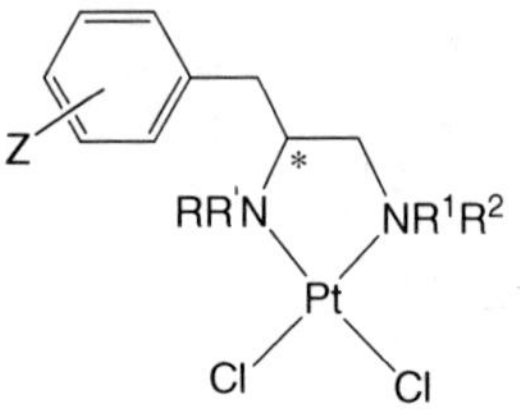

48

A number of other platinum complexes containing chelating amine ligands have been evaluated which do not conform to the structure activity relationships proposed earlier. Nonetheless they are active in *in vitro* tests. Despite the earlier finding that pyridine complexes had low activity, the pyridyl complexes **49** and **50** have shown *in vitro* activity similar to that of cisplatin. An early suggestion that at least one NH group is necessary for activity is contradicted by the activity of **51**. Further to this compounds **52** and **53** contain pyridine ligands and chelating amide/amine ligands with no NH groups, yet both are active against platinum resistant and human tumor cell lines. It is possible that the electron withdrawing tetrafluorophenyl substituents in these compounds help to labilise the chelating ligand providing a source of $\{Pt(pyridine)_2\}^{2+}$ in reactions with biomolecules. The finding that these compounds, which contain no NH groups, are active again contradicts earlier suggestions that the NH group is essential. It may be that the presence of larger substituents on the nitrogen atoms sterically hinders interactions with the biological target and it is this, rather than the absence of the NH group which leads to the loss of activity in earlier examples. In the cases of **52** and **53**, it is thought that the chelating nitrogen ligand is labile. Thus the presence of the large tetrafluorophenyl substituents will be less important as these will not remain to obstruct binding to the target site.

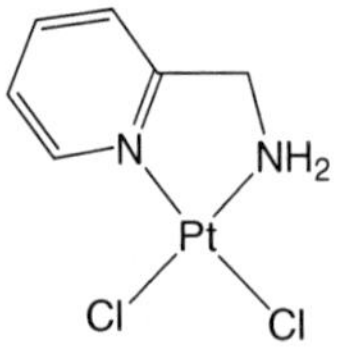

49 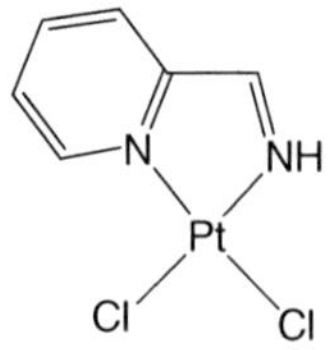 **50** 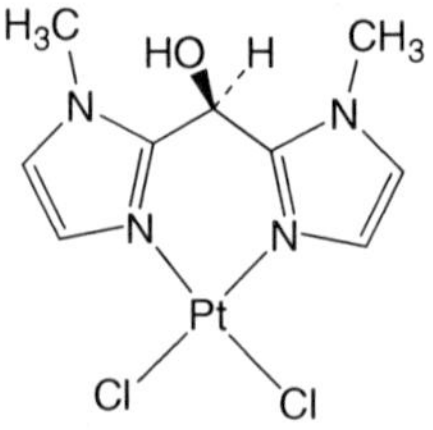 **51**

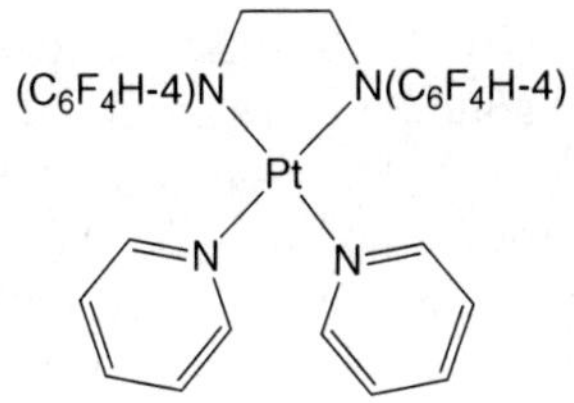

52 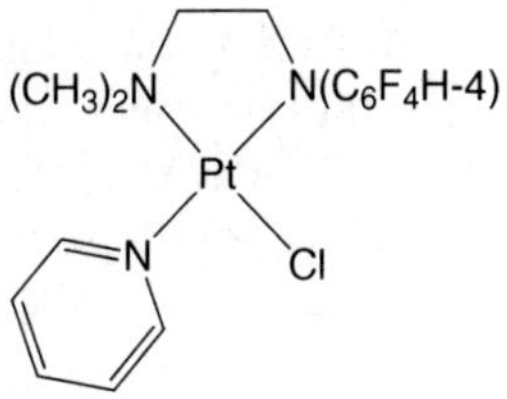 **53**

The contradictory nature of some earlier and later findings reflects, in part, differences in the ways in which antitumor activity has been evaluated. However, it also shows that structure activity relationships of this type are a rather crude tool and should be used only as a guide. The activity of a complex will vary from cell type to cell type and subtle structural changes may have a profound influence in one situation but not in another. That said such information has been, and continues to be, important in developing an understanding of the mechanism of action of the platinum anticancer drugs. Furthermore, judicious choice of the amine ligands offers one important tool for controlling the properties and reactivity of cisplatin derivatives.

4.3.2.2.4 Nature of the Anionic Ligands – Oral Drugs. The reactivity of the anionic ligands is an important feature of cisplatin type compounds. Rapid hydrolysis or substitution reactions can lead to the complex failing to reach its pharmacological target, while excessively inert groups will result in inactivity. Early structure activity relationships identified chloride as having the optimum lability when present with ammonia or organoamine ligands in Pt(+2) complexes. As indicated above, the lability of the anionic ligands can be modified to some degree by the choice of neutral coligand. However, anionic ligands such as halide or nitrate do not offer the scope for structural modification available for the amine ligands. It was realised at an early stage that complexes containing carboxylate ligands could show cytotoxicity towards tumor cells and that these do offer scope for structural variation. In particular oxalic acid (**54**), glycolic acid (**55**) and malonic acid (**56**) are proligands offering the possibility of enhanced stability associated with chelation by the dicarboxylate. Furthermore, the latter two offer opportunities for derivatisation at the CH_2 group. Hence oxaliplatin (**27**) has been produced from **54** and nedaplatin (**26**) from **55**. Lobaplatin (**28**) contains pyruvate, the methylated homologue of glycolic acid, and carboplatin (**25**) contains a malonate derivative.

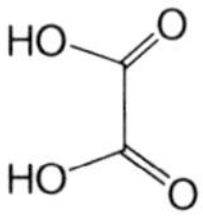

54
oxalic acid

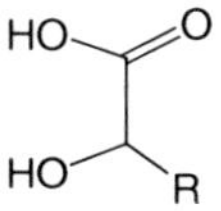

55
R = H, glycolic acid
(R = CH_3, pyruvic acid)

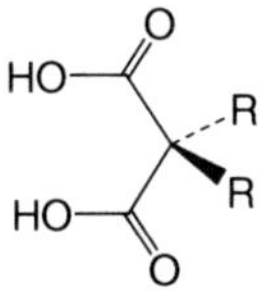

56
R = H,
malonic acid

The effect of adding long chain hydrocarbon substituents to the amine ligands of cisplatin type complexes on lipid solubility was mentioned above in Section 4.3.2.2.3. Similar modifications of carboxylate ligands are possible and have been exploited in preparing the lipophilic complex **57** which may be incorporated in liposomes. This compound is in clinical trials having shown promising activity against liver metastases and cisplatin resistant tumors. Enhanced lipophilicity through the use of carboxylate ligands has also proven important in the development of oral platinum anticancer drugs. Early comparisons of the effectiveness of intra-peritoneal and oral routes for the treatment of mice bearing the ADJ/PC6 plastocytoma showed that, although both routes were effective, the oral route required significantly higher doses and generally gave lower therapeutic indices. While lipophilic compounds showed better oral uptake, the associated loss in aqueous solubility resulted in poor absorption. Solubility studies of *cis*-$[Pt(NH_2R)_2Cl_2]$ (R=hydrocarbyl) and *cis*-$[Pt(NH_2R)_2Cl_4]$ revealed that the platinum(+4) complexes with R=*n*-butyl or cyclohexyl showed substantially higher solubilities in both water and chloroform than their platinum(+2) counterparts. A further advantage of the platinum(+4) compounds is that modification of the anionic ligands offers an opportunity to modify the lipophilicity of a kinetically inert prodrug while leaving the amine ligands unchanged, these being crucial for good antitumor activity. Following oral uptake, *in vivo* reduction to a platinum(+2) complex and loss of two of the lipophilic anionic ligands, should lead to a compound with higher aqueous solubility allowing more efficient absorption. Although helpful in stabilising the prodrug during oral uptake, the inert behaviour of platinum(+4) compounds does present difficulties regarding their synthesis. The use of carboxylate coligands offers one way round these problems. Although direct substitution of platinum(+4) precursor compounds using a carboxylic acid or its salt is ineffective, reaction of a platinum(+4) hydroxide complex with a carboxylic acid anhydride does offer a viable means of forming a carboxylate complex (Equation (1)).

$$[Pt(NH_2R)_2Cl_2(OH)_2] + 2(R'CO)_2O \rightarrow [Pt(NH_2R)_2Cl_2(O_2CR')_2] + 2R'CO_2H \qquad (1)$$

The response of the ADJ/PC6 tumor model in mice to platinum compounds is believed to be a good predictor of antitumor properties in man and was used to screen a variety of platinum(+4) complexes of general formula **58**, for activity. Among the compounds tested those with R=cyclohexyl, R′=butyl and R=propyl R′=phenyl had therapeutic indices of about 300, the value for JM 216 being 57. Ten compounds were tested by oral administration against a panel of five human ovarian carcinoma xenografts. All induced substantial antitumor growth delays of at least 60 days against the cisplatin sensitive PXN/100 model. Further studies of emetic effects in ferrets showed that the acetate compounds were tolerated best with JM216 comparable to carboplatin and substantially less emetogenic than cisplatin. Further oral antitumor tests were carried out using each of JM216 (**58**: R=cyclohexyl; R′=CH_3) and the three compounds

58 (R=cyclohexyl; R′=H; R=cyclopentyl, R=CH_3 and R=cycloheptyl, R=CH_3) against the M5076 murine sarcoma and A2780 human ovarian xenograft, JM216 being the most active against M5076. At the end of these studies JM216 was selected for further trials as it showed good antitumor activity against a variety of tumor models when administered orally. In addition, its emetic properties were mild in comparison with other similar compounds.

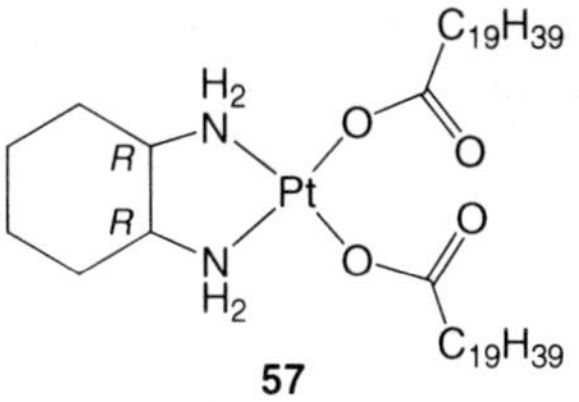

57

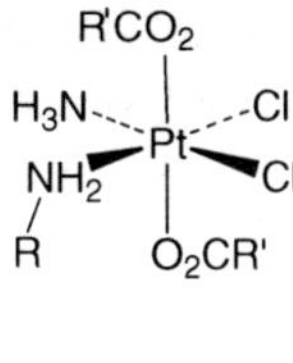

58

A further consideration in the design of an oral drug is its chemical stability. On entering the stomach the compound will be exposed to a strongly acidic environment. Here the kinetic inertness of the d^6 platinum(+4) centre is an important feature. Although acid hydrolysis through protonation of the acetate groups in JM216 does occur, the reaction is sufficiently slow that it is of little consequence in terms of oral administration of the compound. More surprisingly perhaps, the compound is prone to alkaline hydrolysis. This arises because, under alkaline conditions, the ammonia or cyclohexylamine ligands can be deprotonated to form an amide ligand. This has a very strong *trans*-effect and stimulates substitution of the *trans*-ligand through what is known as a dissociative conjugate base mechanism. In JM216 the cyclohexylamine protons are the more acidic and so the rate of substitution of the chloride *trans* to this ligand is faster than that of the chloride *trans* to the ammonia ligand. In the acid environment of the stomach such reactions are unimportant, however, they do become an issue in the subsequent metabolism of JM216 once it is removed from the acidic environment of the stomach.

4.3.2.2.5 Geometric Structure – trans-Complexes. Although the early studies of DDP and related compounds indicated that *trans* complexes such as *trans*-DDP (**59**) are not active antitumor agents, more recent work has revealed a number of *trans*-complexes which do show *in vitro* cytotoxic activity. The nature of the amine ligands is of particular importance. Thus while **60** containing one ammonia and one cyclohexylamine ligand is active, **61** with two cyclohexylamine ligands is reported to be inactive. Other examples of active platinum(+2) *trans*-complexes include the pyridine complex **62**, its analogue **63** containing thiazole, the platinum(+2) quinoline complexes **64** and **65** and the platinum(+4) quinoline complex **66**. In the case of *trans*-DDP it is likely that the higher reactivity of this compound compared to *cis*-DDP results in its deactivation through side reactions before it can exert an effect. Replacement of one ammonia ligand in *trans*-DDP by a more sterically demanding ligand such as quinoline to give **64** produces a substantial increase in biological activity and the related, more kinetically inert, platinum(+4) complex **66** has *in vivo* activity approaching that of cisplatin. It has been suggested that reducing the rate of chloride substitution in *trans*-DDP analogues may induce activity. Thus the steric effects (Section 4.3.2.2.3) of the quinoline ligand might be responsible for the activity of **64** and **65**. However, despite containing a less sterically hindered platinum centre, the isoquinoline complex **67** shows similar rates of chloride substitution to its quinoline containing counterpart **64**. This suggests that electronic factors may be as important as steric factors in modifying the activity of complexes of this type. In addition to **66**, several other *trans*-platinum(+4) complexes have found to be active *in vitro* including **68** and **69**. The cyclohexylamine complex **70** is as active as cisplatin, as is its platinum(+2) counterpart **60**.

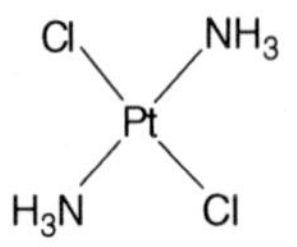

59 *trans*-DDP

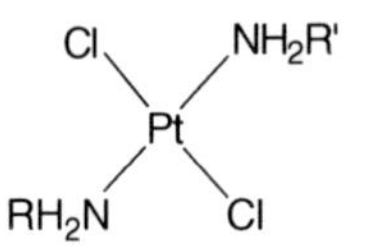

60 R = H, R' = c-C_6H_{11}
61 R = R' = c- C_6H_{11}

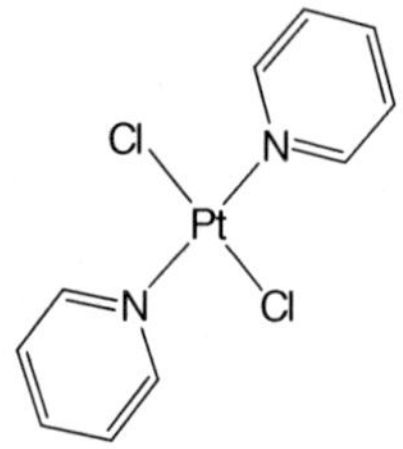

62

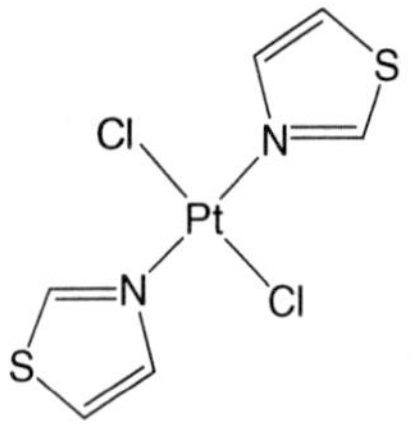

63 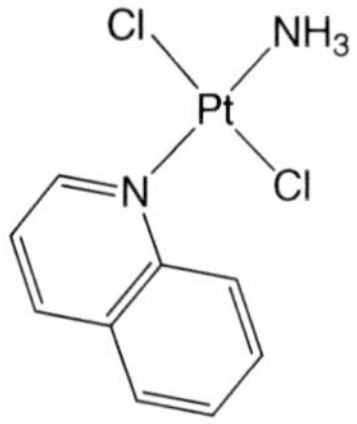 **64** 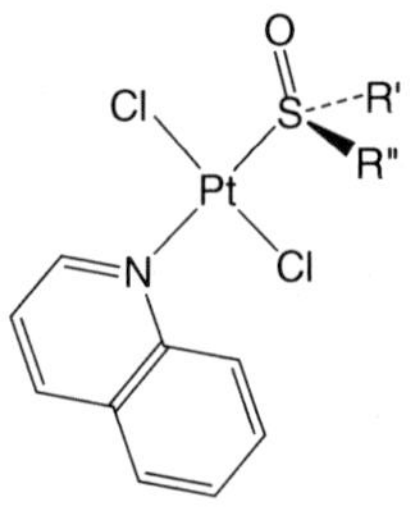 **65** R', R" = alkyl **66**

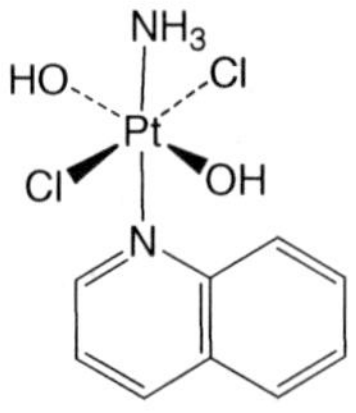

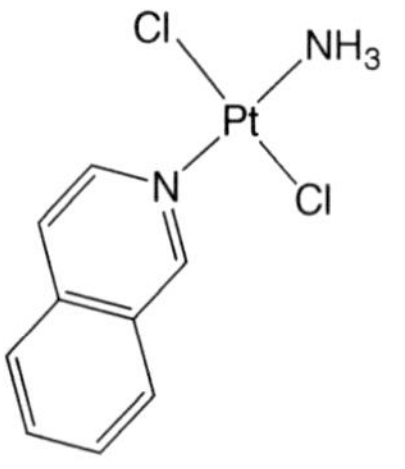

67 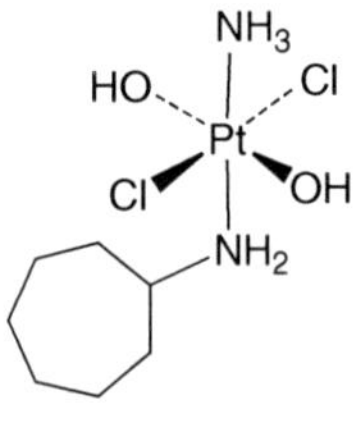 **68** 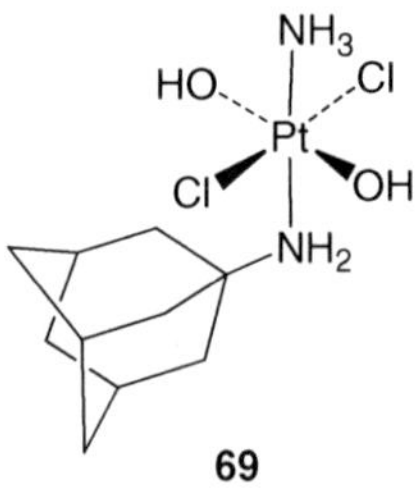 **69** 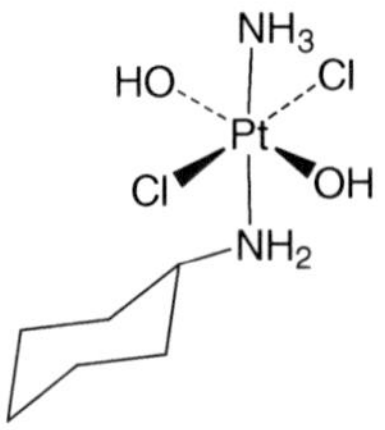 **70**

4.3.2.2.6 Charged Complexes. Early studies of *cis*-DDP derivatives indicated that charged complexes were in general less active than neutral complexes. Usually neutral complexes are more lipophilic and so better suited for passive transport into cells; also they are less rapidly excreted. However, exceptions were noted such as $[Pt(NH_3)Cl_3]^-$(**71**), although in this case it was thought that chloride substitution to give neutral $[Pt(NH_3)(H_2O)Cl_2]$ might be occurring in aqueous media. Studies of a series of cationic complexes *cis*-$[Pt(NH_3)_2(L)Cl]^+$ (L=pyridine, piperidine, pyrimidine or their derivatives; primary organoamine: **72**) revealed that several have activities comparable with cisplatin in murine tumor models. Among these the most active contained pyridine, 4-substituted pyridine, cytosine or 2′-deoxyguanosine as the ligand L. Heating the complex with L=3-methylcytosine in water led to the loss of the ammonia *trans* to the cytosine ligand to give *trans*-$[Pt(NH_3)(L)Cl(X)]^z$(where X could be Cl^- or OH^- and $z = 0$ or X might be H_2O and $z = +1$). It might be proposed that this species is the active compound by analogy with the related quinoline complex **64**. However, chloride substitution of the bromopyridine complex *cis*-$[Pt(NH_3)_2(NC_5H_4Br\text{-}4)Cl]^+$ similarly affords *trans*-$[Pt(NH_3)(NC_5H_4Br\text{-}4)Cl_2]$ but this compound is inactive against Sarcoma 180 in mice. A possible explanation for this confusing result is that the solubility of *trans*-$[Pt(NH_3)(NC_5H_4Br\text{-}4)Cl_2]$ is insufficient to convey it to the target site efficiently, but that more soluble *cis*-$[Pt(NH_3)_2(NC_5H_4Br\text{-}4)Cl]^+$ can reach the target site and be converted to *trans*-$[Pt(NH_3)(NC_5H_4Br\text{-}4)Cl_2]$ which is then active *in situ*.

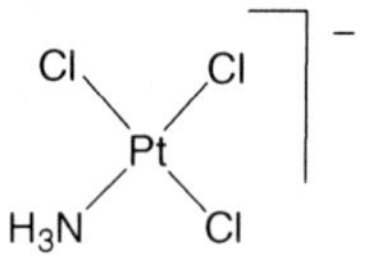

Cl Cl Pt H_3N Cl] −

71

H_3N Cl Pt H_3N L] +

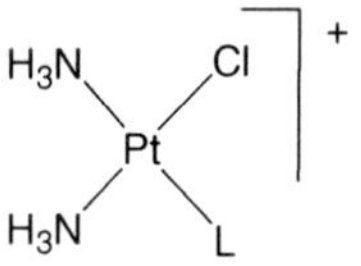

72

Other examples of active cationic complexes are provided by the binuclear complexes **73** and **74** and the trinuclear complex **75**. As these compounds contain two potentially biologically active terminal platinum centres they could be considered to be ditopic agents. As such they are considered in the next section together with neutral binuclear platinum complexes and bifunctional agents which contain a platinum centre linked to a bioactive carrier group.

4.3.2.2.7 Ditopic Complexes. The term ditopic is used here to denote a complex containing two biologically active parts. One type of ditopic complex contains two terminal platinum centres, either or both of which might be involved in binding to the biological target of the drug. In that the two active sites are the same or similar, and independently capable of acting as platinum drugs, the system might be described as homotopic. The second type of ditopic complex contains a platinum centre linked to another quite different functional group designed to interact with a different biological target, or with the same target but in a different way. These may be described as heterotopic complexes.

Charged complexes such as **73**, **74** and **75** together with neutral complexes such as **76**, **77** and **78** contain two equivalent terminal platinum centres, either or both of which may be involved in binding at the target site. Compound **76** is in effect two cisplatin molecules linked by a hydrocarbon bridge. Similarly **77** is derived from two *trans*-DDP molecules linked by a hydrocarbon bridge. The platinum coordination environments in complex **78** are similar to that **51**.

The terminal groups in complexes **76** and **77** show obvious similarities to cisplatin and *trans*-DPP, respectively, so it might be expected that such compounds might act in a similar way to cisplatin or *trans*-DDP in a biological environment. However, being ditopic and tetrafunctional through the four chloride sites, the possibility exists that they may crosslink between sites on different biomolecules, or form intramolecular crosslinks between similar sites within a single macromolecule. Compounds of this type show good antitumor activity in leukaemia L1210 and P388 models resistant to cisplatin. An obvious structural variable which could have a major influence on such reactions is the length of the hydrocarbon chain joining the donor groups bonded to the two terminal platinum centres. The optimum chain length in this type of compound was found to correspond with $n = 4$–8. The malonate derivatives **79** showed better *in vivo* activity than their chloride containing counterparts in some models (*e.g.* murine colon 26). However, potency and antitumor activity against cisplatin-resistant murine L1210 leukaemia were lower for the malonate. Tests of **79** ($n = 4, 5, 6$) against human ovarian IGROV-1 xenografts in nude mice showed no significant variation with n. The compounds were all more potent than carboplatin at a reduced dose and showed lower toxicity.

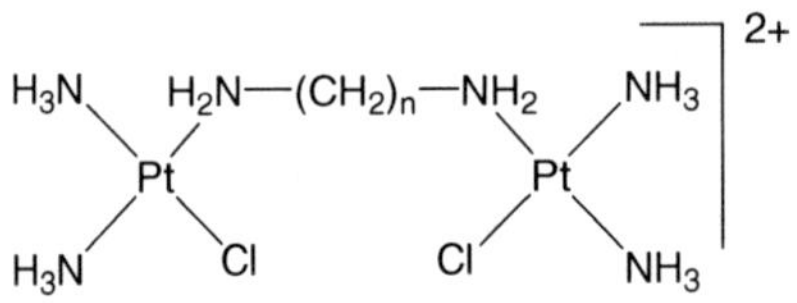

73 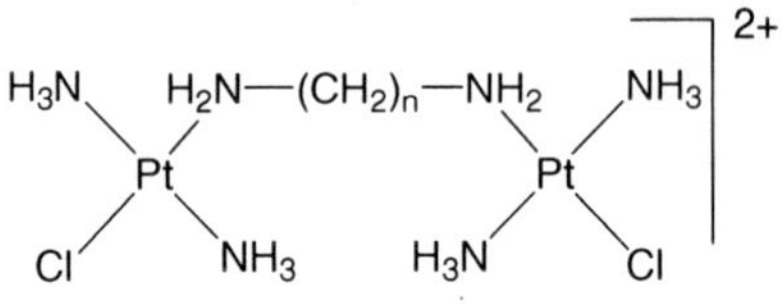 **74**

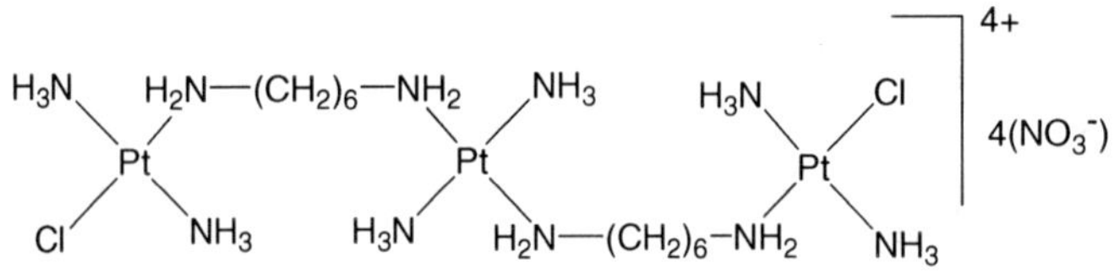

75

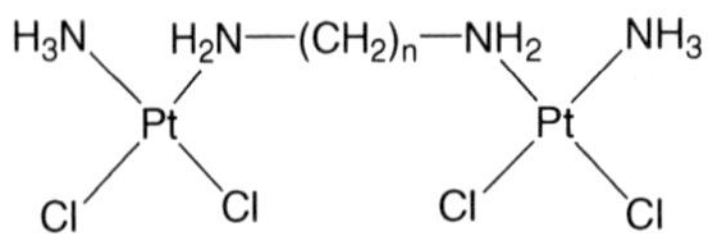

76 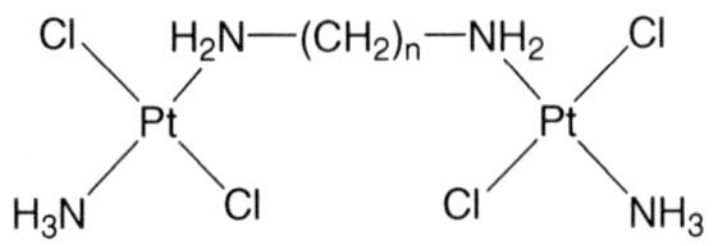 **77**

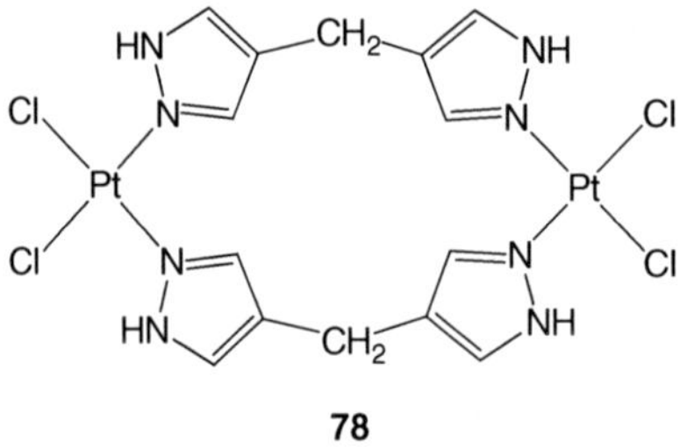

78 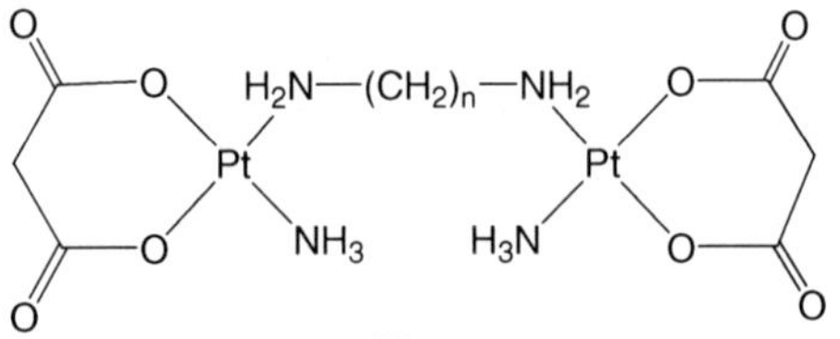 **79**

Although the terminal groups in complexes **73**, **74** and **75** also show similarities to cisplatin or *trans*-DDP, they differ importantly in being charged and having only one chloride leaving group. Thus they are ditopic and bifunctional through the two chloride sites. As with **76** and **77** the possibility exists that they may form crosslinks. A comparison of the *in vitro* cytotoxicity of **74** ($n = 4$) and **79** against a panel of human ovarian tumor cell lines revealed that the neutral complex **79** ($n = 4$) showed a similar profile of activity to cisplatin and carboplatin, **79** being more potent than carboplatin. However, the charged compound **74** ($n = 4$) showed a different activity profile, being in some cases more active than carboplatin and in others less active.

One logical approach to improving the efficacy of platinum anticancer drugs is to attach a platinum complex known to have antitumor properties to a bioactive group, which might improve its transport to the target site, or increase the likelihood of platinum exerting its effect at the target site. In pursuit of this goal hetero-bifunctional molecules containing platinum and oestrogen mimics have been prepared with a view to improving uptake by cells with appropriate hormone receptors. Such an approach has relevance to the treatment of breast or prostate cancer. Similarly compounds in which a platinum containing group is linked to amino acids, sugars or organic drugs have been prepared. The main focus of work on ditopic platinum complexes has been associated with the latter goal, the enhancement of interactions between the platinum containing drug and its supposed target site. In this context many of the systems studied so far have sought to incorporate known DNA intercalators. Polyaromatic groups derived from anthraquinone or acridine provide a well-established class of DNA intercalator and compounds such as **80–84** have been evaluated. An interesting finding of this work is that the position through which the polyaromatic moiety is connected to the platinum moiety has an effect on activity. The activity of compounds such as **80** and **82** where the link is perpendicular to the long axis of the polyaromatic group being about an order of magnitude larger than for **81** and **83**, where this is not the case.

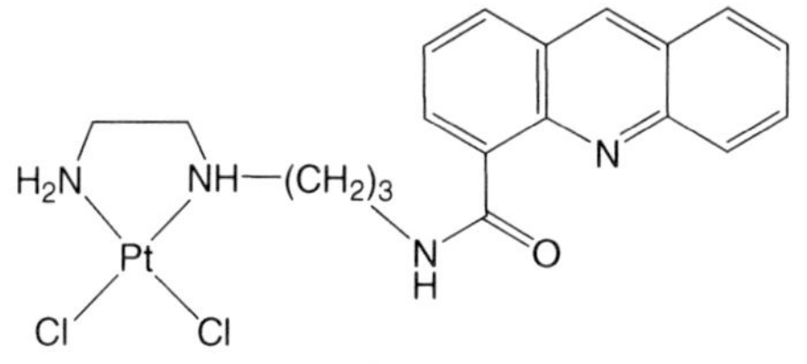

80

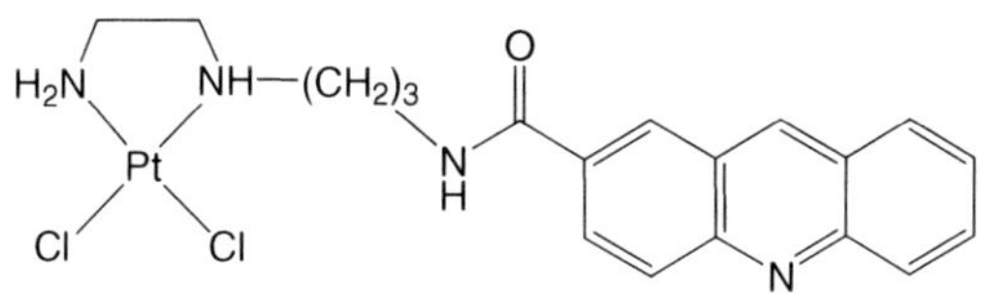

81

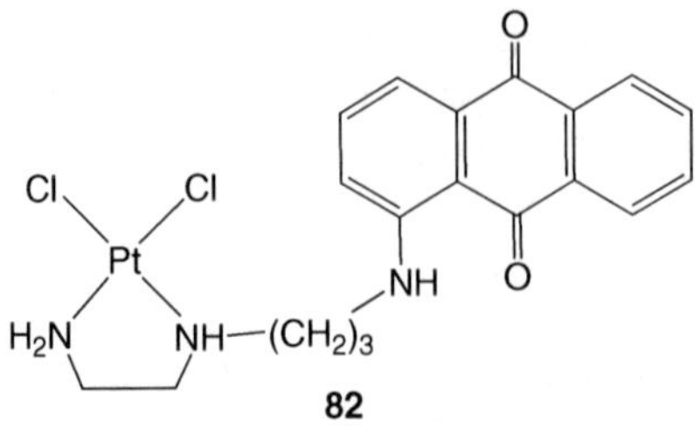

82

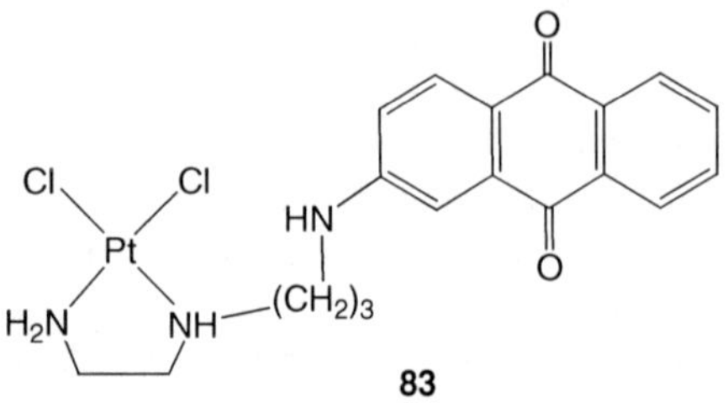

83

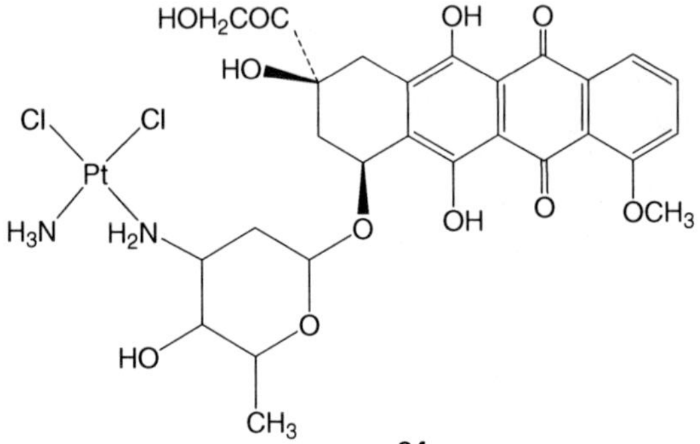

84

The use of heterotopic antitumor compounds is in its infancy and this approach has yet to produce significant clinical results. However, as the understanding of the mechanism of action of platinum drugs improves further this approach may prove a valuable tool in broadening the range of tumor types, which can be effectively treated using platinum agents.

4.3.2.3 *Mechanism of Action of Cisplatin*

4.3.2.3.1 The Reactivity of Cisplatin. In order for cisplatin to exert its effect on cellular reproduction the complex must be administered to the patient and subsequently find its way into the interior of the target cell. After administration cisplatin will enter the bloodstream and, in that medium, be dispersed around the body. When cisplatin first enters the bloodstream, it is confronted by a complex cocktail of compounds. These include sugars, proteins, lipids, salts and, of course, water, any of which might react with the complex in some way. Typical donor atoms available for metal binding will be oxygen, particularly in the form of water and carboxylate or hydroxide groups within biomolecules, nitrogen, in the form of amine groups or *N*-heterocycles such as the imidazolyl group of histidine or in purine or pyrimidine groups within nucleic acids, and finally sulfur in thioether or thiol groups such as the CH_3S-group in methionine or the thiol group in cysteine. A knowledge of platinum chemistry would suggest that sulfur would represent the most effective donor atom for a soft class b metal ion such as platinum(+2) (Section 2.7.2). In accordance with this expectation, studies have shown a strong reduction in the number of protein bound thiol groups in the kidneys following cisplatin treatment. However, there is also a mass action effect to be considered, in that water molecules and chloride ions are present in the blood at high concentrations compared to sulfur atoms. The high chloride concentration (100 mmol dm^{-3}) in the bloodstream serves to suppress the hydrolysis of the complex so that much remains available unchanged. The cellular uptake of cisplatin is found to be limited by its concentration but is not inhibited by analogues such as carboplatin or *cis*-$[Pd(NH_3)_2Cl_2]$ suggesting that uptake is not carrier-mediated. It seems likely therefore that cellular uptake occurs largely through passive diffusion.

Once in the cell the lower chloride concentration (*ca.* 20 mmol dm^{-3}) is unable to suppress hydrolysis as effectively and conversion of cisplatin to *cis*-$[Pt(NH_3)_2Cl(H_2O)]^+$ can occur. This type of solvated compound is a common intermediate in the substitution reactions of platinum(+2) complexes (Section 2.8.3) and is more reactive towards proligands, including biomolecules, than the dichloride. The aquation of cisplatin under physiological conditions appears to be half complete within about 2 h. Within the cell the platinum complex is once again exposed to an array of potential reactants. Once again sulfur compounds would seem to offer the most likely binding agents and are quite abundant in the cytosol and nucleus of the cell. However, competition studies with model compounds suggest that, although sulfur donors may be preferred kinetically, the ultimate and more thermodynamically stable binding appears to involve nitrogen. Thus the model compound $[Pt(dien)Cl]^+$ (**85**) reacts initially at the amino acid sulfur of *S*-guanosylhomocysteine (**86**) ($t_{1/2}$ *ca.* 2 h; $2<pH<6.5$) but then migrates ($t_{1/2}$ *ca.* 10 h) to form a more thermodynamically stable product through binding at the *N*7-position of the guanosyl group. Similar intramolecular isomerisations were observed with other related systems. While these intramolecular studies suggest that certain nitrogen donor sites can compete with sulfur donors *in vitro*, it is far more likely that intermolecular processes will be involved in the *in vivo* exchange of platinum between binding sites. A study of the reaction between **85** and a mixture of *L*-methionine (Met; **87**) and guanosine-5′-monophosphate (GMP; **88**) showed that during the first 40 h of reaction platination of Met occurred to give $[Pt(dien)(Met\text{-}S)]^{2+}$, there being little reaction between the platinum complex and GMP. Subsequently the platinum was transferred to the *N*7 site of GMP with a half time of *ca.* 167 h liberating Met and forming $[Pt(dien)(GMP\text{-}N7)]^{2+}$. Similar experiments involving the reaction between $[Pt(dien)(Met\text{-}S)]^{2+}$ and adenosine-5′-monophosphate, thymine-5′-monophosphate, or cytosine-5′-monophosphate showed no transfer of platinum from the sulfur to the nitrogen donor site over a period of 12.6 h. These studies suggest that sulfur compounds could, through their rapid reaction, bind to cisplatin but that the ultimate target is a nitrogen donor site, *N*7 of GMP residues being a likely binding site. It is possible that initial binding to sulfur donor sites in biomolecules forms a temporary reservoir of platinum which subsequently transfers to other sites in the cell.

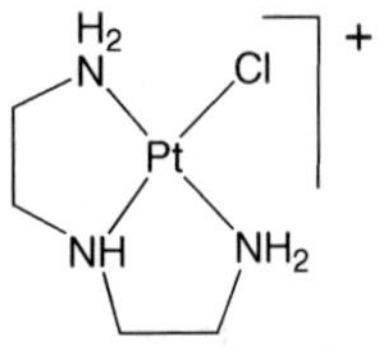

85

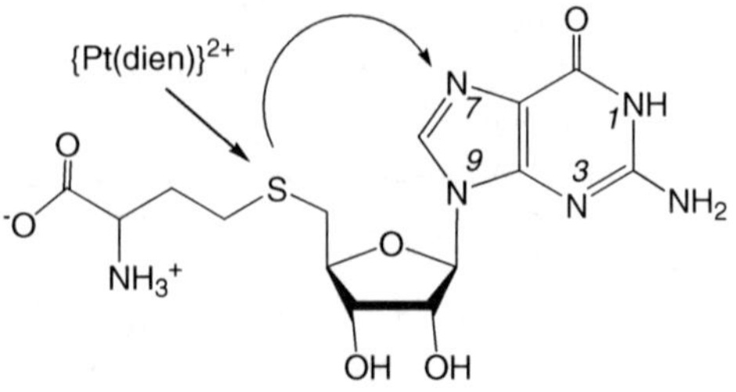

86

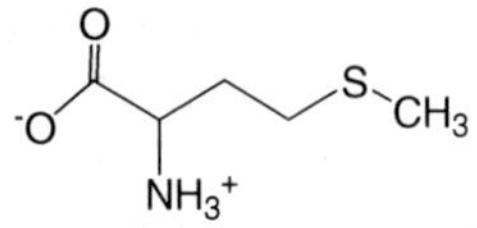

87

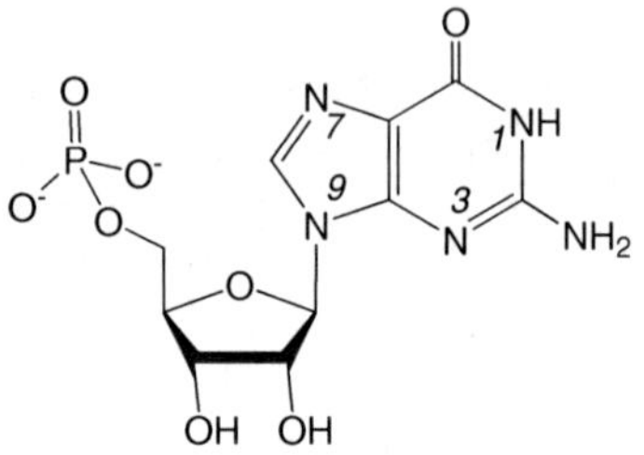

88

In view of the finding that binding to sulfur in a kinetically faster step precedes the formation of more thermodynamically stable complexes, it has been suggested that sulfur compounds might be used as 'protecting agents' for platinum complexes such as cisplatin to facilitate their transport to the active site and to reduce toxic side effects. A Phase I clinical study showed GSH (**89**) significantly reduced the renal toxicity of cisplatin without reducing its effectiveness against ovarian cancer to any significant extent. More advanced clinical trials, including Phase III studies, have also shown that the phosphorylated aminothiol 'Amifostine' (**90** also known as Ethyol and Amofostine) produces significant decreases in the renal, neurologic and hematologic toxicity of cisplatin without effects on response rates. Beneficial effects were also found when Amifostine was used in conjunction with carboplatin. Amifostine is actually a prodrug being converted to the amino thiol **91** by phosphatase enzymes and it is thought that this sulfur compound binds to the platinum.

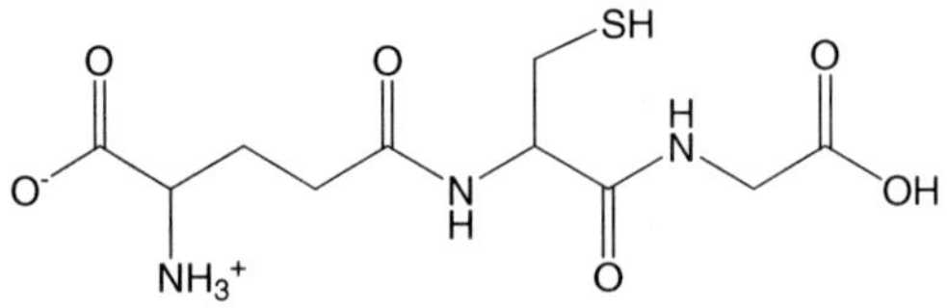

89

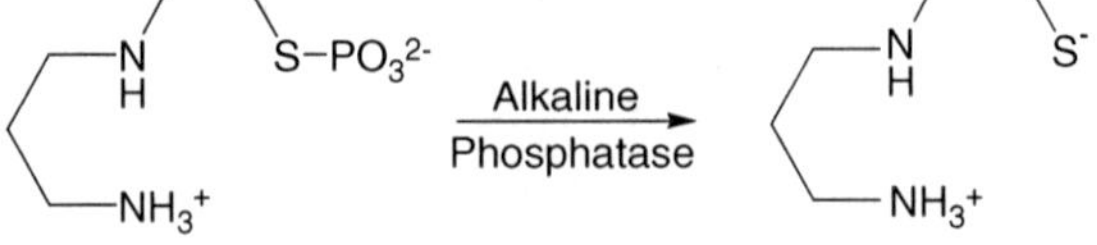

Alkaline Phosphatase

90
Amifostine

91

4.3.2.3.2 The Biological Target of Cisplatin. The fact that cisplatin causes mitotic suppression might suggest DNA as a probable target molecule through which the complex exerts its antitumor effects. In support of this view, studies of the effect of cisplatin on cells indicated that, while DNA synthesis is inhibited, RNA and protein synthesis are much less affected. Furthermore studies of ^{195m}Pt labelled cisplatin uptake in HeLa cells showed that only about 1 in 100,000 protein molecules and between 1 in 10 and 1 in 1000 RNA molecules contained a platinum atom. In contrast about nine platinum atoms were bound by each DNA molecule. These findings strongly support the suggestion that DNA is the primary intracellular target for cisplatin. This conclusion is further supported by the finding that DNA repair deficient mutants of *E. coli* or human cells are more sensitive to cisplatin than normal cells. There is some evidence that cisplatin also interacts with phospholipids and can disrupt the cytoskeleton but DNA is now generally regarded as the primary target molecule associated with the antitumor activity of cisplatin.

The double helix structure of DNA consists of two sugar phosphate backbones held together by interactions between complementary pendant base molecules attached to these backbones (Section 2.4.2.2; Figures 22 and 23). DNA is known to exist in A, B and Z structural forms of the double helix, as well as a triple helicate form. The natural form is B DNA which has base pairs stacked in an essentially parallel manner, with a separation of about 3.4 Å and with their ring planes effectively perpendicular to the helical axis. The difference in size of the complementary bases in each pair results in a right handed helical structure with a narrow minor groove and a wider major groove both of similar depth. This results in a helical repeat distance of about 3.4 nm. The walls of the minor groove contain hydrophobic sugar CH groups while the walls of the major groove contains more polar sites associated with oxygen and nitrogen atoms of the bases. In the A form of DNA the structural arrangements of the sugar units are slightly different and there are 11 bases per turn not 10 as in the B form. Groups of base pairs in adjacent helical turns become alternately tilted with respect to the helical axis and the structure is of a more open form with a central cavity in the helix. The minor groove becomes wide and shallow and the major groove deep and narrow. When alternating purine and pyrimidine bases (*e.g.* GpCpGpCpGpCpGpCpGpCpGpCp) are present in one strand of a short DNA helix the Z form adopting a left handed helical structure can be produced in solutions of high ionic strength. It is possible that sections of A and Z DNA occur naturally in cells but B DNA is the normal structure. The triple helix form of DNA is similar to the B form but with a further strand carrying pyrimidine bases wrapped around in the major groove to give a 1:1:1 purine/pyrimidine/pyrimidine triple helix. Physiological conditions are not normally suitable for the production of triply helical DNA.

The simple double helix structure represented in Figures 22 and 23 of Chapter 2 provides a basis for discussing the nature of binding between platinum and DNA. This structure offers several potential metal binding sites including oxygen atoms associated with the sugar, phosphate or base moieties and nitrogen atoms associated with the bases (Figure 5). The binding to platinum could occur through a single donor atom on DNA to give a

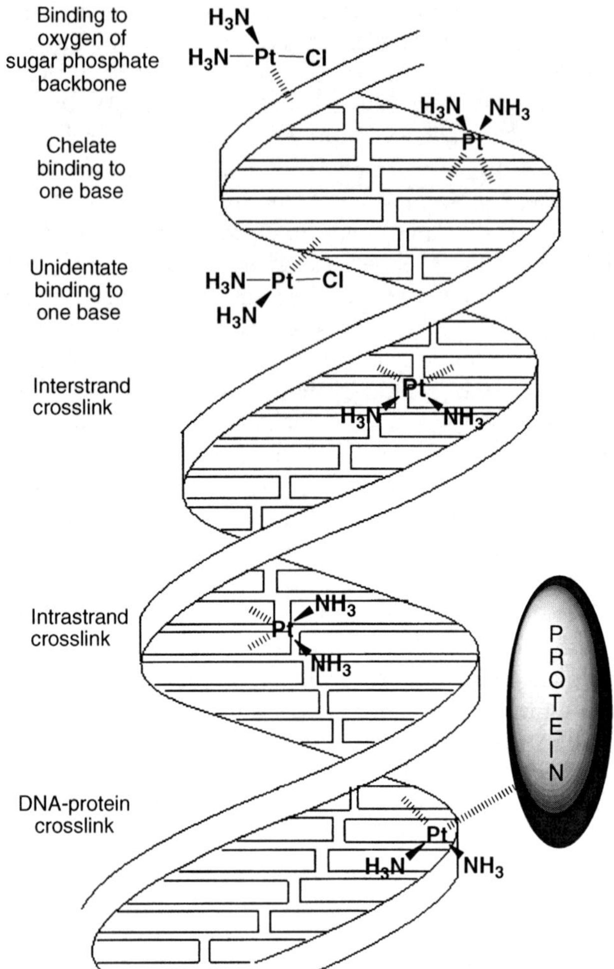

Figure 5 *Some possible cis-DDP-DNA binding modes*

monofunctional adduct of the form $[Pt(NH_3)_2Cl(DNA)]^+$ or it could involve two donor atoms to give bifunctional adduct $[Pt(NH_3)_2(DNA)]^{2+}$. As an example, binding could occur *via* one donor atom on a base or by chelation involving two donor atoms on a single base (*e.g.* guanosine as shown in Figure 6). There is also the possibility of platinum cross linking between two bases on the same sugar-phosphate strand to give an *intra-strand crosslink* or between bases on the two different sugar-phosphate strands in the duplex DNA molecule to give an *inter-strand crosslink* (Figures 5 and 6). A further possibility is that the platinum ion may crosslink DNA to second different molecule such as a protein. Given the choice between oxygen and nitrogen, the less electronegative more polarisable nitrogen centre is likely to be preferred by a soft metal ion such as platinum(+2). Complexation studies, some of which were described earlier, show a clear preference for binding at the *N7* site of guanosine bases and the order of

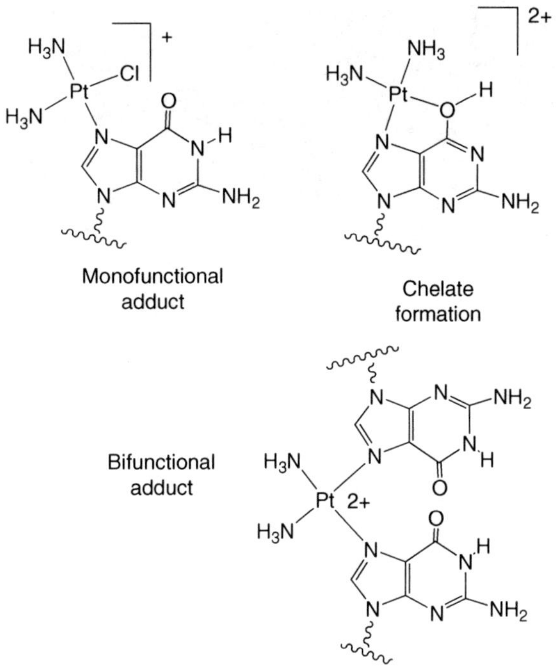

Figure 6 *Some possible binding modes of cis-[$PtCl_2(NH_3)_2$] to a guanosyl base in DNA*

preference for binding sites among nucleotides appears to be guanosine-*N*7 > adenosine-*N*7 > cytosine-*N*3 > thymine-*N*1. Both chemical and structural studies have identified the *N*7 site of guanosine as the primary target within DNA for platination by cisplatin. Furthermore 1,2-intra-strand crosslinks are found at sites within DNA at which adjacent guanosine bases are present.

Experiments were carried out in which salmon sperm DNA platinated with cisplatin was enzymatically digested, the fragments separated and their structures examined by NMR spectroscopy. About half the platinum present was found to be bound to sites involving adjacent guanosine bases on a single DNA strand, *cis*-[$Pt(NH_3)_2${d(GpG)}], indicating that 1,2-intrastrand crosslinks between adjacent guanosines constituted the primary binding mode. About one quarter of the platinum was in the form of adenosine-guanosine links, [$Pt(NH_3)_2${d(ApG)}], and about one tenth as 1,3-intrastrand crosslinks involving two non-adjacent guanosines. Similar results were obtained using [$Pt(en)Cl_2$] (**30**), instead of cisplatin. Structural studies have provided further evidence for the preferential binding of cisplatin at sites containing guanosine.

4.3.2.3.3 Structural Studies of Platinated DNA Fragments. In 1988 a single crystal X-ray diffraction study of a dinucleotide complex of platinum, [$Pt(NH_3)_2${d(pGpG)}] (Figure 7), showed how the $\{Pt(NH_3)_2\}^{2+}$ moiety might form an intra-strand crosslink between adjacent guanosine bases on a DNA strand. The presence of an O---H–N hydrogen bonding interaction between an

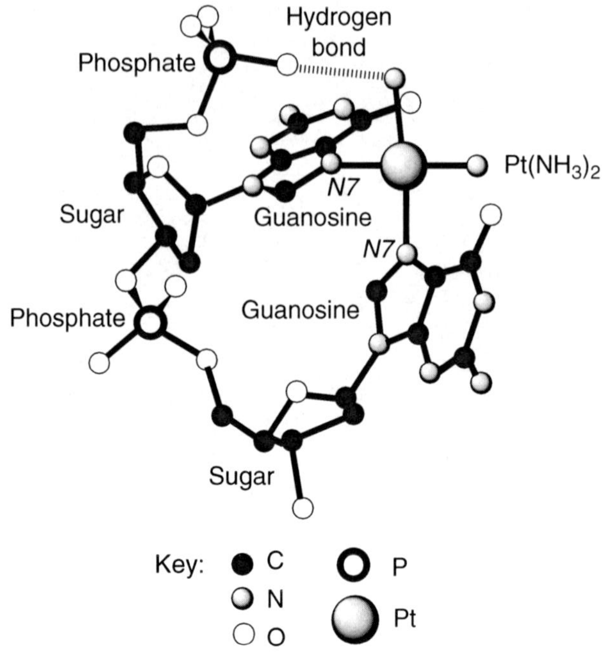

Figure 7 *The structure of a model compound in which a $\{Pt(NH_3)_2\}^{2+}$ centre is bound to two guanosyl moieties*

ammine hydrogen and a phosphate group in the molecule suggested that this might be important for effective binding of *cis*-DDP to DNA. This is consistent with earlier observations that an NH group appeared to be necessary within a platinum complex if it were to show antitumor activity. In 1995 an X-ray crystal structure of a short section of platinated DNA duplex was obtained giving further insights into the interaction between *cis*-DDP and DNA. The duplex was derived from d(CCU^{Br}CTG*G*TCTCC).-d(GGAGACCA-GAGG) where U^{Br} is a heavy atom (Br) labelled uracil used to aid the structure determination and G*G* is the site of platination. The X-ray structure revealed two independent molecules in the crystal, both of which showed substantial bending of the double helix at the site of platination (*ca.* 39° in one molecule and 55° in the other). In neither case was the base pairing disrupted to any major extent but the structures of the duplex were like A form DNA to one side of the platinum atom and like B form to the other. The general effect of this platinum induced bending towards the major groove on a section of DNA is shown schematically in Figure 8. The geometry at the metal centre was found to be substantially distorted from the ideal for a square planar complex, the platinum atom lying *ca.* 1 Å out of the guanine ring planes.

In order to take account of differences in structure between the solution and solid states, the solution structure of a similar platinated duplex was determined by a combination of Nuclear Magnetic Resonance (NMR) and molecular mechanics (rMD) studies. This showed a larger bend in the molecule

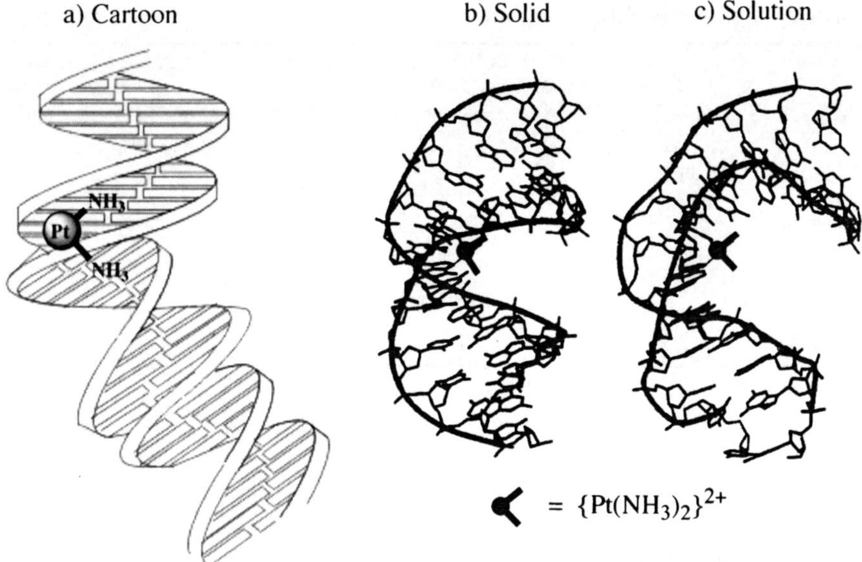

Figure 8 *Views of the effect of intra-strand cis-*$\{Pt(NH_3)_2\}^{2+}$ *crosslinking on DNA structure (a) a cartoon representation of DNA bending; (b) the solid state X-raydiffraction structure of the cis-*$\{Pt(NH_3)_2\}^{2+}$ *adduct of d(CCTCTG*G*TCCC).d(GGAGACCAGAGG) (G* denote Pt binding sites); (c) the solution NMR structure of the cis-*$\{Pt(NH_3)_2\}^{2+}$ *adduct of d(CCTCTG*G*TCCC).d(GGAGACCAGAGG)*
(Reprinted in part with permission from E.R. Jamieson and S.J. Lippard, *Chemical Reviews*, 1999, ***99***, 2467–2498. Copyright (1999) American Chemical Society).

of *ca.* 78° in solution suggesting that crystal packing places some constraints on the structure. The DNA was now all in the B form and the distortion of the platinum coordination sphere slightly less. Solution studies of two other platinated duplexes, one derived from an octamer, d(CCTG*G*TCC).d(GGACCAGG) (* denotes the platinum binding site) the other from an undecamer, d(CTCTCG*G*TCTC).d(GAGACCGAGAG) gave similar results as summarised in Table 3. In both of these examples the DNA was in the B form in the solution.

Similar studies of DNA duplexes containing 1,3-intra-strand crosslinks, a minor component of the platinated DNA, have also been carried out. In the *cis*-$\{Pt(NH_3)_2\}^{2+}$ adduct of d(CTCTAG*TG*CTCAC).d(GTGAGC-ACTAGAG) the global distortions of the helical structure were not as large as for the 1,2-intra-strand cross linked adducts, the helix being bent by 24°. However, the local base pairing near the platinum was significantly disrupted and, although the minor groove remains in the B DNA form it becomes narrowed to 5.8 Åand occupied by the intervening thymine from the G*TG* region. This may have significant effects on DNA-protein interactions. In particular the high mobility group (HMG) protein HMG1 shows specific

Table 3 *Some structural parameters from platinated duplex DNA structures*

Duplex structure	*DNA bend (o)*	*Average twist of DNA helix (o)*	*Displacements of Pt from guanine ring planes (Å)*	*Average distance between DNA repeat units*[a] *(Å)*	*Minor groove depth (Å)*	*Minor groove width (Å)*
Dodecamer	39[b]	32	1.3	5.5	3.0	9.5–11.0
X-ray	55[b]		0.8			
Dodecamer	78	25	0.8	6.9	1.4	9.4–12.5
Solution			0.8			
Octamer	58	25	1.0	6.8	3.2	4.5–7.8
Solution			0.8			
Undecamer	*ca.* 81	26	0.5	6.8	2.1	9.0–12.1
Solution			0.65			

[a] Average distance between adjacent phosphate phosphorus atoms.
[b] Two independant molecules in the crystal.

interactions with cisplatin modified DNA but does not to bind to the 1,3-intrastrand adduct of DNA.

Structural information is also available for inter-stand crosslinks involving *cis*-$\{Pt(NH_3)_2\}^{2+}$. Solution NMR and rMD studies of platinated d(CATAG*CTATG)$_2$ show that the structure of the DNA around the inter-strand platinum G-Pt-G crosslink reorganises to resemble that of Z DNA. The two cytosine bases originally paired with the guanosine residues now bound to the *cis*-$\{Pt(NH_3)_2\}^{2+}$ centre become extra-helical (Figure 9) and the DNA bends towards the minor groove by about 40°.

4.3.2.3.4 Kinetic Studies of Cisplatin Reactions. The rate at which cisplatin binds to DNA is an important consideration in that too slow a reaction might allow time for undesirable competing reactions to take place which are not associated with antitumor activity. A typical substitution reaction mechanism for square planar platinum(+2) complexes in water involves aquation followed by substitution of the coordinated water molecule (Section 2.8.3). Kinetic studies of the hydrolysis of cisplatin and its subsequent complexation by short lengths of single- or double-stranded DNA containing adjacent guanosines have been measured using ^{15}N NMR spectroscopy. The results of these measurements made in water containing 10 mol dm^{-3} sodium phosphate at pH 7.1 are summarised in Scheme 2. The labelled platinum complex $[Pt(^{15}NH_3)_2Cl_2]$ reacts with water to form the monoaquated complex $[Pt(^{15}NH_3)_2Cl(H_2O)]^+$ in a relatively slow step ($t_{1/2}$ *ca.* 2.51 h) which is followed by the comparatively rapid formation ($t_{1/2}$ *ca.* 8 min) of a monofunctional adduct with the deoxydecanucleotide 5′-d(ACATGGTACA) (***N*7G**). This then undergoes ring closure to form the bifunctional adduct at a rate some 4 times faster than competing reactions which lead to other products. The platinum complex of the corresponding duplex DNA can be obtained either by reacting $[Pt(^{15}NH_3)_2Cl($***N*7G**$)]$ with the complementary deoxydecanucleotide 3′-d(TGTACCATGT) or by reacting $[Pt(^{15}NH_3)_2Cl_2]$ with the preformed

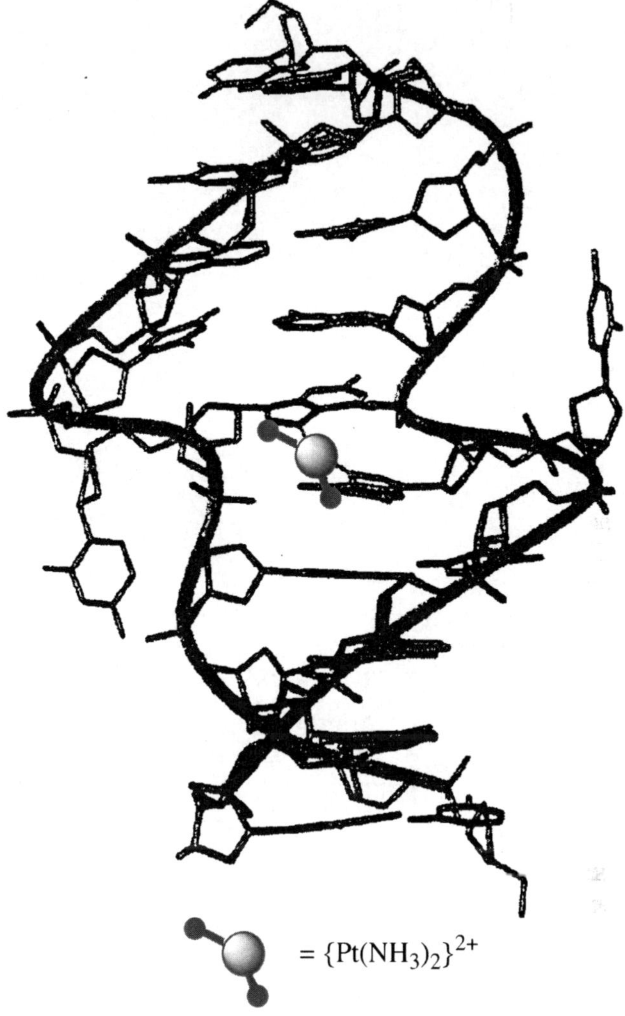

Figure 9 *An inter-strand crosslink in DNA, the NMR solution structure of the cis-*$\{Pt(NH_3)_2\}^{2+}$ *adduct of d(CATAG*CTATG).d(CATAG*CTATG) (G* denote Pt binding sites)*
(Reprinted in part with permission from E.R. Jamieson and S.J. Lippard, *Chemical Reviews*, 1999, **99**, 2467–2498. Copyright (1999) American Chemical Society).

duplex DNA d(ACATGGTACA).d(TGTACCATGT). Similar results were obtained in a ^{195}Pt NMR study of the platination of chicken erythrocyte DNA at 37 °C and pH 6.5 as summarised in Scheme 3. Again slow aquation of [$^{195}Pt(NH_3)_2Cl_2$] inside the cell is followed by rapid reaction with guanosine *N*7, a further slower aquation step removes the second chloride prior to rapid ring closure to form the bifunctional adduct. Outside the cell aquation reactions can

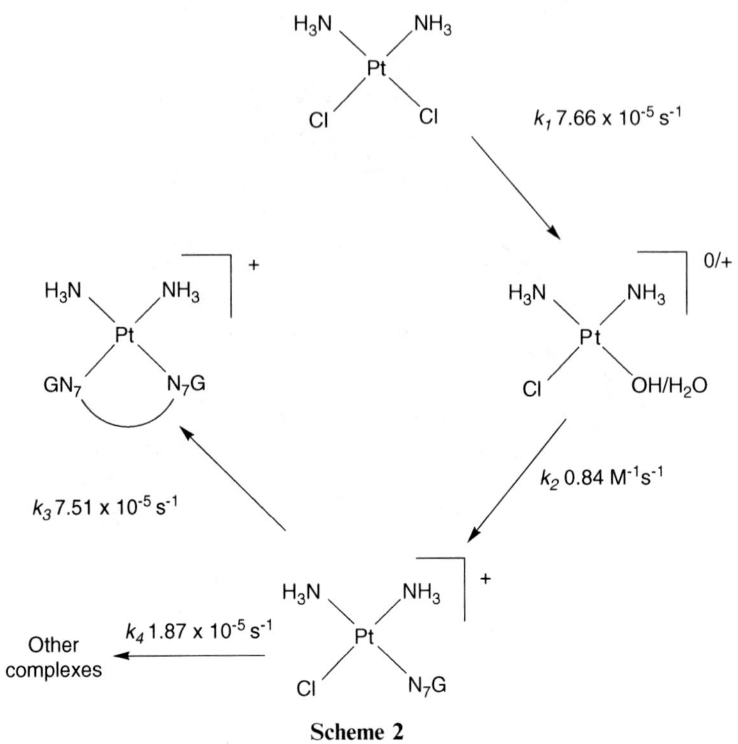

Scheme 2

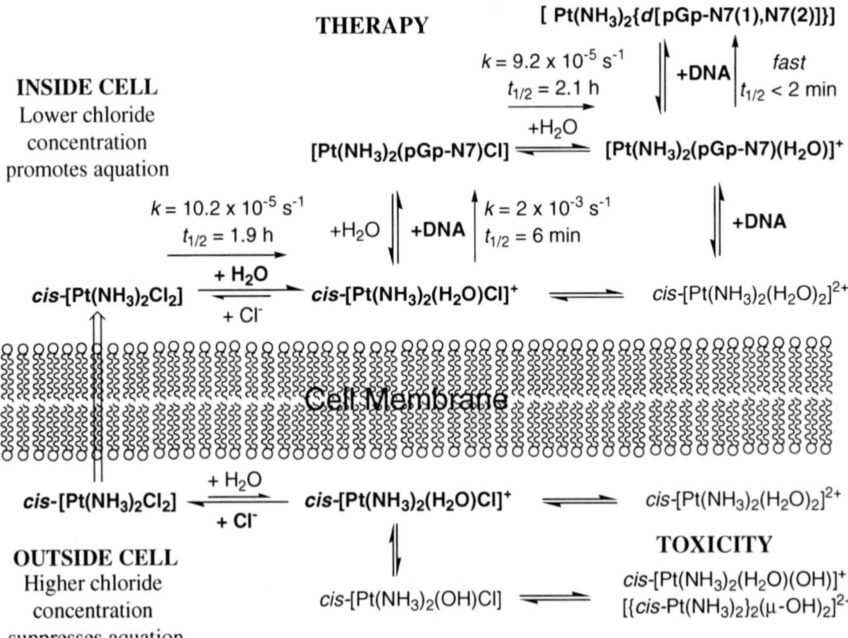

Scheme 3

also occur, leading to toxic by-products, but these reactions are suppressed by the higher extracellular chloride concentration.

4.3.2.3.5 Binding of trans-DDP to DNA. Unlike its *cis*-isomeric counterpart *trans*-DDP is not an effective anticancer agent, so it is important to understand why a fairly subtle structural change can produce such a large difference in activity. Although *trans*-DDP also binds to guanine and adenine, geometrically the *trans*-coordination sites in DDP are separated by *ca.* 5 Å compared to 3.5 Å for the *cis*-coordination sites. As a consequence *trans*-DDP is unsuitable for forming 1,2-intra-strand crosslinks between adjacent bases. Enzymatic digestion experiments on DNA, platinated with *trans*-DDP revealed the presence of interstrand crosslinks between guanine and cytosine (*ca.* 50%) or between guanine and adenine (*ca.* 10%) and 1,3-intrastrand crosslinks between non adjacent guanines (*ca.* 40%). These types of interaction are structurally quite different from the 1,2-intrastrand crosslink formed with cisplatin and apparently necessary for its antitumor activity.

In addition to structural differences there may be reactivity differences between the *cis* and *trans* isomers of DDP. Although the first hydrolysis step appears to occur at similar rates for both cisplatin and *trans*-DDP, in some experiments the formation of the second platinum-nucleotide bond is found to be slower for *trans*-DDP which cannot form 1,2-intra-strand crosslinks between adjacent bases. This provides a greater opportunity for *trans*-DDP to form crosslinks between DNA and other molecules, particularly sulfur donors such as GSH. This will result in different biological consequences compared to the formation of toxic intra- or inter-strand DNA crosslinks as found for cisplatin. A further difference is suggested by experiments showing that natural DNA repair processes are more efficient in dealing with DNA platinated with *trans*-DDP than DNA platinated with cisplatin, also it is found that cisplatin can promote the transition of DNA from the B to the Z form whereas *trans*-DDP appears to hinder this process. Thus, although the structural differences between *cis*- and *trans*-DDP may at first sight appear modest the effects of the two different isomers on the structure and the reactivity of the DNA adducts they form are substantial.

4.3.2.3.6 Consequences of DNA Platination. Assuming that DNA is the primary target molecule associated with the antitumor activity of cisplatin, and having seen structural and chemical evidence that DNA does indeed form complexes with platinum compounds, the next question must relate to how the platination of DNA might lead to antitumor activity. The platination of DNA could interfere with any of several processes important for cell replication and so inhibit cell division. The normal life cycle for a cell (Scheme 4) involves its formation from another cell by mitosis, a first gap phase (G_1) follows during which protein synthesis is occurring; next there is a DNA synthesis phase (S) in which new copies of DNA are produced, a second gap phase (G_2) with continued protein synthesis and finally a mitosis phase (M) in which another new cell is formed. This cycle continues repeatedly until the cell dies. In the DNA synthesis phase it is necessary for a duplicate copy of the nuclear DNA to

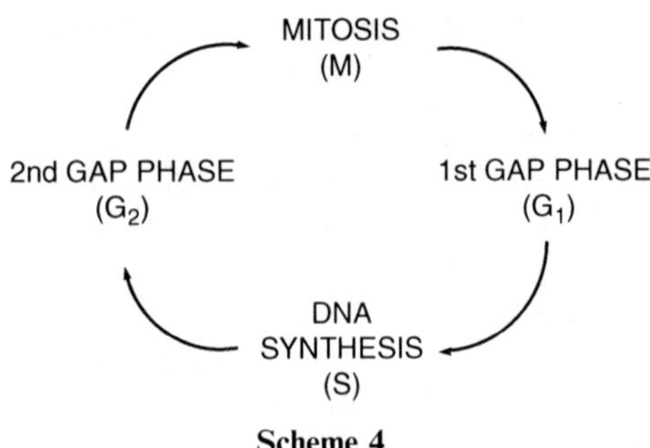

Scheme 4

be produced. This involves unravelling the DNA, separating the duplex into two strands, then using the strands as templates to synthesise new DNA for incorporation into the new cell. If synthesis of the new DNA strands by DNA polymerases were blocked by platination, cell replication might be inhibited. Another process necessary for cell division is protein synthesis. If the transcription from DNA of the genetic information necessary for the synthesis of proteins were blocked, this too could prevent cell replication. The binding of proteins to DNA is another important process occurring in the cell, and the inhibition or promotion of protein binding could have a rôle to play in the activity of cisplatin.

A further possibility is that platination of DNA leads to early cell death, apoptosis, through binding to telomeric regions of DNA at the ends of chromosomes. These telomeric regions protect the ends of chromosomes from degradation and are guanosine rich consisting of lengths of repeating 5′-TTAGGG-3′ groups in humans. Part of the telomeric region is lost with each successive cell division and eventually, when the telomeric region become sufficiently short, the cell dies. However, there is a ribonucleoprotein, telomerase, which can add sections to the telomeric regions thereby prolonging cell life. If cisplatin binding to telomeric regions results in degradation of the telomeres, or the inhibition of telomerase activity, premature cell death would result. Measurements of telomere loss in HeLa cells have shown that cisplatin can degrade and shorten telomeric regions. Cisplatin has also been shown to inhibit telomerase activity when other agents which damage DNA, including *trans*-DDP, did not show this effect. It would seem possible therefore that cisplatin can cause premature cell death through interactions affecting telomeric DNA.

Finally it is important to note that, in addition to nuclear or genomic DNA (gDNA), there is also DNA in the nucleosomes of mitochondria. Mitochondrial DNA (mDNA) is not subject to natural DNA repair processes in the same way as gDNA, so platination of mDNA may have a greater effect because its platinum adducts persist whereas in gDNA adducts may be removed by repair processes. Experiments have shown a 4- to 6-fold higher proportion of platinum adducts in mDNA than in gDNA after treatment with cisplatin. This could be attributed more extensive initial binding to mDNA or

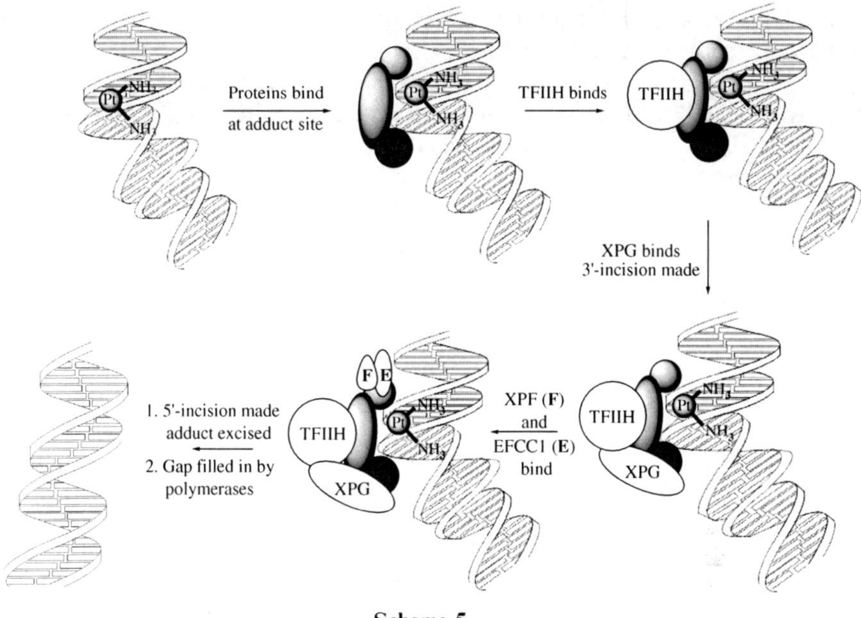

Scheme 5

to greater retention of platinum adducts as a result of deficient repair. It is not yet clear whether platination of mDNA is more or less important than platination of gDNA in the activity of cisplatin. However, cells deficient in DNA repair have been shown to be more sensitive to cisplatin than those proficient in DNA repair indicating that repair mechanisms may be important in moderating cisplatin activity.

4.3.2.3.7 DNA Repair Processes. Among the natural cellular processes for repairing damaged sequences of DNA, nucleotide excision repair (NER) appears to be important for the removal of platinum adducts. The mechanism of NER is quite well understood and involves a number of proteins which can interact with DNA and one with another. The general features of the NER process are summarised in Scheme 5. Although the exact sequence of events is uncertain several proteins (XPA, RPA, XPC, HR23B) are implicated in the initial recognition of DNA damage caused by platinum binding. After these 'damage recognition' proteins have bound to the platinated site the protein TFIIH binds to them. The presence of TFIIH allows binding of the protein XPG which makes an incision in the DNA strand on the 3′-side of the platination site. Thereafter the XPF-ERCC1 protein complex binds and makes a second incision on the 5′-side of the platination site. The process cuts out a section of the DNA strand approximately 30 nucleobases in length and the gap left as the proteins dissociate is refilled by DNA polymerases to produce the repaired DNA duplex. Where repair mechanisms of this sort are effective in removing platinum from DNA they will limit the antitumor effect of the platinum compound. Some of the proteins involved in NER appear to bind specifically to sites containing *cis*-$\{Pt(NH_3)_2\}^{2+}$ adducts and so NER may be an important factor limiting the effectiveness of cisplatin.

In an early comparison of cisplatin with *trans*-DDP it was noticed that while *cis*-$\{Pt(NH_3)_2\}^{2+}$ adducts accumulated with time, adducts with *trans*-$\{Pt(NH_3)_2\}^{2+}$ increased to a maximum then decreased. This effect reflects differences in the rates of DNA repair rather than differences in platinum uptake by DNA and some evidence suggests that repair of *trans*-$\{Pt(NH_3)_2\}^{2+}$adducts involves a different cellular mechanism. Repair of *cis*-$\{Pt(NH_3)_2\}^{2+}$-1,3-GTG adducts was found to be more efficient than that of corresponding *cis*-$\{Pt(NH_3)_2\}^{2+}$-AG or GG adducts. In other work the 1,3-*cis*-$\{Pt(NH_3)_2\}^{2+}$-GTG adduct was found to be repaired some 15–20 times more effectively than 1,2-intra-strand crosslinks. It would appear that, although subject to repair, the 1,2-*cis*-$\{Pt(NH_3)_2\}^{2+}$-GG is less efficiently removed than other intra-strand crosslinks and persists sufficiently to prevent tumor growth. However, stimulation of repair processes, perhaps resulting from repeated exposure to cisplatin, could diminish the efficacy of the drug.

4.3.2.3.8 Inhibition of DNA Polymerase. Inhibition of DNA polymerase activity will interfere with the normal cell cycle during the S phase (Scheme 4). Studies of the effects of cisplatin on DNA polymerase activity include experiments involving partially purified human DNA polymerase which showed inhibition of activity after platination of the template DNA using either cisplatin or *trans*-DDP. The extent of DNA platination required to produce 50% polymerase inhibition was found to be smaller for cisplatin than for *trans*-DDP by a factor of between 2 and 7. In other work involving the SV40 chromosome from green monkey cells 14 times as much *trans*-DDP as cisplatin was needed to inhibit replication to the same extent, although similar loadings of platinum on the SV40 DNA were found in each case. If adducts with *trans*-DDP are more prone to repair this could explain the higher concentration of *trans*-DDP required to produce the same level of inhibition as cisplatin. Experiments on human HeLa and 293 cells have shown evidence for such differential repair although this effect was not observed in experiments on Chinese hamster or African green monkey cells. *In vitro* experiments have indicated that, on average, DNA polymerases can bypass the platinated region of DNA to the extent of only about 10% with the 1,2-*cis*-$\{Pt(NH_3)_2\}^{2+}$-GG adduct being more inhibitory than its 1,2-*cis*-$\{Pt(NH_3)_2\}^{2+}$-AG or 1,3-*cis*-$\{Pt(NH_3)_2\}^{2+}$-GNG (N=nucleotide) counterparts.

Since equal loadings of cisplatin or *trans*-DDP on DNA itself can produce similar inhibition of polymerase activity it would seem that the differences in the concentrations of the isomeric compounds required in the surrounding medium to produce similar effects reflect differences in their processing by cells. Thus while polymerase inhibition occurs, and may play a part in cisplatin activity, it cannot fully account for the particular activity of cisplatin compared to its *trans* isomer.

4.3.2.3.9 Inhibition of DNA Transcription. The transcription of genetic information from a DNA template into messenger RNA (mRNA) is an essential step in protein synthesis. Experiments designed to reveal the effects of cisplatin on transcription show that, at low concentrations, cisplatin can temporarily arrest the cell cycle in the G_2 phase (Scheme 4). At higher doses G_2 arrest continued until cell death occurred. The concentration of cisplatin required to produce G_2 arrest has been found to be lower in DNA repair deficient cells than cells proficient in DNA repair. This implicates DNA transcription in the cytotoxicity of cisplatin as improved platination of the DNA leads to the cell becoming unable to transcribe genes producing proteins necessary for mitosis. More specifically *cis*-$\{Pt(NH_3)_2\}^{2+}$-GG or AG adducts, 1,3-*cis*-$\{Pt(NH_3)_2\}^{2+}$-GNG adducts and inter-strand *cis*-$\{Pt(NH_3)_2\}^{2+}$ cross-links have been found to block transcription by some RNA polymerases. In contrast, monofunctional or *trans*-$\{Pt(NH_3)_2\}^{2+}$ platinum-DNA adducts did not fully block transcription. Experiments with NER deficient cell lines showed that, compared to cisplatin, 4 times the concentration of *trans*-DDP was needed to produce similar inhibition of RNA synthesis, an effect which cannot be attributed to differential repair. Cisplatin could interfere with transcription through any of a variety of mechanisms. It seems that cisplatin can block the binding of some proteins associated with transcription and also alter the structure of chromatin so as to inhibit gene expression by interfering with the binding of promoters. A good correlation has been found between the cytotoxicity (LC_{50} the concentration producing 50% cell death) and gene expression inhibition (IC_{50} the concentration at which gene expression is reduced to 50%) of platinum compounds, some of which were active antitumor agents and some not. In other experiments cisplatin was found to be much more effective than *trans*-DDP in inhibiting β-lactamase expression while $K_2[PtCl_4]$ was inactive in the respect. There is evidence to suggest that cisplatin-DNA adducts bind certain proteins which are present as transcription factors in ribosomal RNA (rRNA) synthesis, thus preventing them from exerting their normal influence. It is clear from various studies that cisplatin does inhibit transcription and is probably involved in blocking or inhibiting more than one of the processes involved in the transcription mechanism.

4.3.2.3.10 DNA-Protein Binding. The ability of platination to modify DNA protein interactions will have important consequences for cellular processes involving DNA and is likely to be an important feature of the mechanism of action of platinum-based drugs. A number of proteins have been identified which bind to cisplatin modified DNA. These include a group of proteins associated with DNA repair, some proteins associated with the transcription of genetic information and HMG domain proteins which show a particular affinity for cisplatin modified DNA. Among the proteins associated with DNA NER the damage recognition protein xeroderma pigmentosum A, XPA and the replication protein, RPA, each bind individually and specifically to cisplatin modified DNA as well as binding cooperatively. RPA binds to DNA containing 1,3-*cis*-$\{Pt(NH_3)_2\}^{2+}$-GNG adducts almost twice as well as to DNA containing 1,2-*cis*-$\{Pt(NH_3)_2\}^{2+}$-GG adducts. XPA binding to DNA is increased by interaction with ERCC1 (Scheme 5) a protein which shows elevated expression in some cisplatin resistant tumor cell lines. Another damage recognition protein associated with NER, XPC-HR23B, has been shown to bind preferentially to cisplatin damaged DNA. The protein XPE, associated with NER but of unknown function, is induced by cisplatin and cisplatin resistant tumor cell lines over-express this protein. Other repair proteins associated with DNA mismatch repair, double-strand break repair and UV damage repair (photolase) are also known to bind to cisplatin modified DNA. The enzyme T4 endonuclease VII which cleaves branched DNA structures selectively binds and cleaves 1,2-*cis*-$\{Pt(NH_3)_2\}^{2+}$-GG, 1,2-*cis*-$\{Pt(NH_3)_2\}^{2+}$-AG and *cis*-$\{Pt(NH_3)_2\}^{2+}$-inter-strand crosslinks but does not cleave 1,3-*trans*-$\{Pt(NH_3)_2\}^{2+}$-GpTpG or *trans*-$\{Pt(NH_3)_2\}^{2+}$-inter-strand crosslinks.

The HMG domain proteins represent a large family of proteins which bind in the minor groove of DNA and cause bending, they can also recognise distorted DNA structures. Early examples were identified in extracts of HeLa cells which would bind to double-stranded DNA adducts formed from cisplatin or $[Pt(en)Cl_2]$ (**30**) but not to adducts from *trans*-DDP or [Pt(dien)Cl]Cl (**85**). Subsequently two proteins HMG1 and HMG2 were isolated, both of which bound to DNA modified with cisplatin but not that treated with *trans*-DDP. The HMG domain which binds to DNA consists of about 80 amino acids and typically more than one such domain may be present in the protein. Many other proteins which contain the HMG domain and bind to platinated DNA have since been identified including transcription factors and non-histone chromosomal proteins. HMG1 has been shown to bind specifically to 1,2-*cis*-$\{Pt(NH_3)_2\}^{2+}$- DNA adducts but not to 1,3-*cis*-$\{Pt(NH_3)_2\}^{2+}$- DNA or *trans*-$\{Pt(NH_3)_2\}^{2+}$- DNA adducts. The protein contains two HMG domains, A and B, both of which can bind preferentially to platinated DNA. There is evidence that the strength of binding to DNA varies not only with the structure of the HMG domain but also with the nucleotide sequence in which the platinated site resides. The structures of the HMG1-A and HMG1-B domains have been determined from solution NMR measurements and both contain three α-helices arranged in an L shape (Figure 10). NMR studies have shown such proteins binding to DNA in the minor groove, bending and unwinding the DNA. An X-ray crystal structure of HMG1-A bound to the 16-base pair adduct *cis*-$\{Pt(NH_3)_2\}^{2+}$-d(CCTCTCTG*G*ACCTTCC).-d(GGAAGGTCCAGAGAGG) shows the bend in the DNA to be located two base pairs away from the platinum on the 3′-side (Figure 11). An aromatic amino acid group from the protein intercalates into the DNA at a hydrophobic notch in the minor groove opposite the platinum. This feature appears to be important since mutation of this amino acid to alanine substantially reduces the binding affinity of the protein for the platinated DNA. As observed with the first model structure (Figure 7) one of the ammine ligands of the platinum is within hydrogen bonding distance of phosphate oxygen. The structural characterisation of this interaction provides highly detailed information on the interaction between one example of an HMG domain and a section of platinated DNA duplex.

Apart from structural studies, kinetic measurements of protein DNA interactions provide an important insight into protein-DNA binding. DNA is exposed

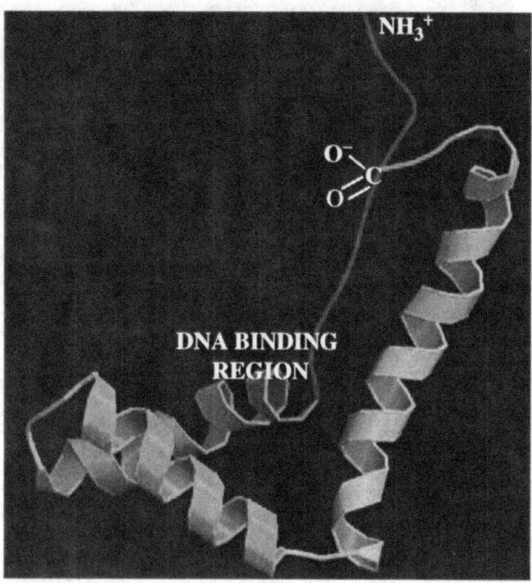

Figure 10 *A schematic view of the solution structure of the high mobility group protein HMG-1 domain B*
(Adapted from a representation of the structure in the Protein Database, H.M. Berman, K. Henrick and H. Nakamura, Announcing the worldwide Protein Data Bank. *Nature Structural Biology*, 2003, **10(12)**, 980. http://www.pdb.org; PDB ID1HME, H.M. Weir, P.J. Kraulis, C.S. Hill, A.R. Raine, E.D. Laue, J.O. Thomas, Structure of the HMG box motif in the B-domain of HMG1, *EMBO J.*,1993, **12** 1311–1319).

to a large number of different proteins and, in a competitive situation, which one binds to a particular region may depend not only on any sequence specificity but also on relative rates of binding. The rate of binding of HMG1 B to cisplatin modified DNA has been measured as $1.1 \times 10^9\ M^{-1}\ s^{-1}$, close to the diffusion controlled limit, compared to a dissociation rate of $30\ s^{-1}$. It has been suggested that selective binding of HMG1 observed in the presence of the protein RPA is due to this fast binding rate with which RPA binding cannot compete.

Two different mechanisms have been proposed to explain how HMG domain proteins might affect the sensitivity of cells to cisplatin. The first involves the 'hijacking' of a protein from its normal binding site as a result of its stronger competitive binding to platinum modified regions of DNA, hence disrupting normal cell function. The second proposal is that the binding of HMG domain proteins to the platinated regions of DNA prevents them from being recognised by damage repair systems so that the adducts persist to impair cell function. There is some experimental evidence to support both of these proposals and it seems likely that either or both processes may be important in the mechanism of cisplatin action. A more detailed understanding of the mechanism of protein-DNA interactions in the cytotoxicity of platinum drugs is needed before firm conclusions can be drawn. However, such information may pave the way to new anticancer treatments which exploit the ability of proteins to sensitise cells to platinum drugs.

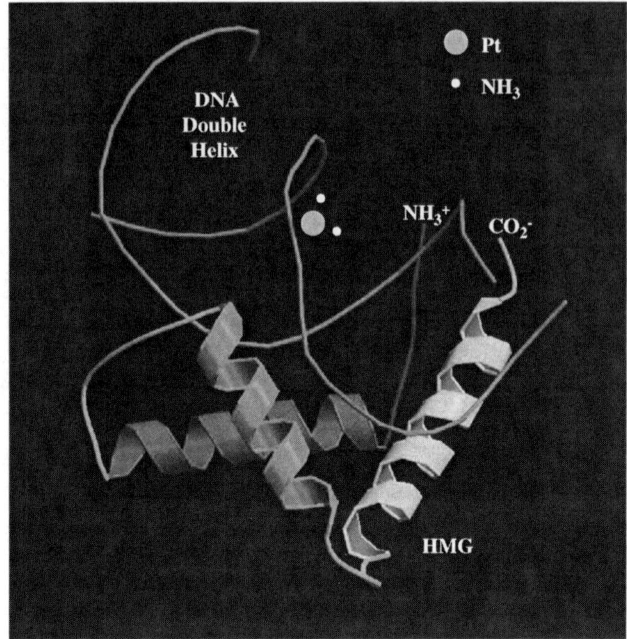

Figure 11 *A schematic view of the solid state structure of HMG-1 domain A bound to the cis-{$Pt(NH_3)_2$}$^{2+}$ adduct of d(CCTCTG*G*TCCC).d(GGAGACCAGAGG) (G* denote Pt binding sites)*
(Adapted from a representation of the structure in the Protein Database, H.M. Berman, K. Henrick and H. Nakamura, Announcing the worldwide Protein Data Bank, *Nat. Struct. Biol.*, **10(12)**, 980. http://www.pdb.org; PDB ID 1CKT, U.M. Ohndorf, M.A. Rould, Q. He, C.O. Pabo, S.J. Lippard, Basis for recognition of cisplatin-modified DNA by high-mobility-group proteins, *Nature*,1999, **399** 708–712).

4.3.2.3.11 Resistance to Cisplatin. Resistance to cisplatin is an important factor limiting the use of the drug. Some cells may be intrinsically resistant to cisplatin while others may develop resistance during treatment. Several possible causes of cisplatin resistance have been proposed and any or all may play some part in the development of resistance. Firstly, the intracellular accumulation of cisplatin may decline, possibly due to decreasing uptake of the drug coupled with release rates which do not change significantly. Some experimental results show such effects while others reveal little difference in uptake and efflux between cisplatin resistant and sensitive cells. At present it would appear that, although decreased uptake or increased efflux may have some part to play in some cases of cisplatin resistance, this does not provide a consistent explanation for resistance.

A second possibility involves the stimulation by cisplatin of the production of complexing agents such as thiols which can bind to the cisplatin and so compete with DNA binding. In particular GSH can be present in cells at concentrations of 0.5–10 mmol dm^{-3} and may bind to platinum, although it does not necessarily interfere with its normal function (Section 4.3.2.3.1). Investigations of GSH levels in cisplatin resistant cells have produced mixed results. In some cases elevated GSH levels appeared to be associated with resistance and examples were found where the reduction of GSH levels partially reversed the resistance. In other cases no difference in GSH levels was found between cisplatin sensitive and resistant cells. It would appear that, although increased GSH levels may contribute to resistance in some cases, this too does not provide a comprehensive explanation for cisplatin resistance. The thiol rich protein metallothionein is involved in removing heavy metals from cells and offers another potential thiol complexing agent for cisplatin. The experimental evidence for the involvement of metallothionein in cisplatin resistance forms a similar pattern to that found for GSH and, although the involvement of this protein in cisplatin resistance cannot be ruled out, its interaction with platinum cannot offer a complete explanation of resistance.

A third possible cause of cisplatin resistance is the stimulation of enhanced DNA repair activity resulting in the excision of platinated regions of DNA. A number of experiments on different cell lines, including human ovarian and human malignant glioma cell lines, have shown enhanced repair activity to be common among cisplatin resistant cells although, there is not a complete correlation between repair activity and cisplatin resistance. Over expression of DNA polymerase β and of proteins associated with DNA repair has been found in some resistant cell lines further implicating enhanced repair as a source of resistance. Conversely, evidence has been found that some tumor cells which are especially sensitive to cisplatin are repair deficient. Thus results from ovarian carcinoma biopsy samples and a testicular non-seminomatous germ cell line showed differential DNA repair. In addition, studies of human testis tumor cell lines has shown reduced levels of the proteins XPA and ERCC1-XPF complex which are associated with DNA repair (Scheme 5). It is possible that deficiencies in repair activity lead to cisplatin sensitivity and that this may account, in part at least, for the particular effectiveness of cisplatin against ovarian and testicular tumors. Another protein which has been implicated in cisplatin resistance is the transcription factor p53. This is involved in regulating a number of genes and its activation through DNA damage might result in cell cycle arrest and apoptosis. Disruption of p53 function is found in about half of human cancers but mutated p53 genes are not normally found in testicular tumors which are most susceptible to successful treatment with cisplatin. The effects of changes in p53 function depend upon the genetic context of the cell line. Loss of p53 function results in cisplatin resistance for some human ovarian cancer cell lines while disruption of p53 function renders breast cancer MCF-7 cells more sensitive to cisplatin. In other experiments on ovarian cancer cell lines the presence of normal or mutant p53 did not appear to influence cisplatin cytotoxicity. Because p53 is involved in regulating a variety of cellular processes, including the activation of a gene associated with the production of an NER protein, it is not possible to pinpoint a specific rôle for p53 in conferring cisplatin resistance. It seems clear that p53 has different effects on cisplatin processing in different cell types and may contribute to cisplatin sensitisation in some cases.

In summary it seems likely that a number of effects are operating in concert to produce cisplatin resistance, and no single mechanism can fully account for the observed patterns of resistance. Furthermore the relative importance of the different resistance mechanisms is likely to vary from case to case. However, enhanced DNA repair capability does emerge as an important common feature among cisplatin resistant cell lines. Conversely, deficiencies in repair activity lead to cisplatin sensitivity and this may contribute to variations in the effectiveness of cisplatin against different tumor types.

4.3.2.4 Conclusions. Although many uncertainties remain, at a general level a plausible explanation for the antitumor activity of cisplatin can be proposed. On entering the bloodstream cisplatin is exposed to a variety of potential ligands which could substitute for the more labile chlorides. However, the first step in the substitution of square planar platinum complexes in aqueous media is typically aquation. The relatively high chloride concentration in the bloodstream suppresses this reaction allowing a significant proportion of the cisplatin to circulate unchanged for long enough that some of this material may enter cells intact, probably by passive diffusion through the membrane. Once inside the cell the cisplatin is exposed to a substantially lower chloride concentration which no longer suppresses hydrolysis of the more labile chloride ligands to such an extent. In the first instance this leads to the formation of a monoaquated complex which, once formed, reacts quite rapidly with guanosine *N7* sites within DNA (Scheme 3). It is possible that the platinum may initially form more labile complexes with other ligands in the cytoplasm or migrate from base to base along the DNA strand but, where a second guanosine is present adjacent to another guanosine binding site, a second chloride substitution leads to a relatively stable 1,2-intra-strand crosslink resulting in an accumulation of platinum at GpG sites in the DNA. Other bifunctional adducts may also form but may be repaired more efficiently than the 1,2-*cis*-$\{Pt(NH_3)_2\}^{2+}$GG adducts. The presence of the 1,2-*cis*-$\{Pt(NH_3)_2\}^{2+}$GG adduct has the structural effect of inducing a substantial bend in the DNA. The consequences of this may include inhibition of DNA polymerase activity, inhibition of transcription, promotion of apoptosis, selective binding of some proteins and blocking the binding of others. Although it seems clear that DNA, and in particular GpG sites within DNA, are the biological target associated with the antitumor activity of cisplatin, the platinum adduct can interfere with many different aspects of cellular DNA processing. It is highly likely that cisplatin exerts its cytotoxic effect through more than one of these interferences so that not a single process can be defined as the whole basis for the antitumor activity of the drug. However, since tumor cells are among the most active in the body with respect to cell division and DNA processing, they are more sensitive to the effects of cisplatin, a factor leading to its antitumor activity.

Cisplatin has a fortuitous combination of properties which lead to its antitumor properties. The rates of hydrolysis of the labile ligands are such that extracellular reactions are slow enough to allow time for cellular uptake of the drug. Once in the cell, the changed environment allows more rapid reactions but the greater stability of adducts formed between cisplatin and DNA leads to accumulation at DNA binding sites in competition with other potential ligands. In particular binding to adjacent guanine sites opens the minor groove creating a more open hydrophobic site. This promotes binding of HMG proteins which interact with the minor groove through interactions with phenylalanine side chains. The bound HMG protein appears to protect the platinated sites from excision repair allowing them to persist, inhibit transcription of DNA and activate signalling pathways which lead to cell death. The high levels of HMG2 proteins in testicular tissue may be responsible for the particular sensitivity of testicular cancer to cisplatin. The recognition of the importance of HMG proteins in cisplatin action and the elevation of HMG expression by oestrogen and progesterone has led to Phase I clinical trials of oestrogen/carboplatin co-therapy for ovarian cancer. Recent detailed structural information and the improved understanding of the ways in which cisplatin can modify cellular processes will also help guide the search for new metal-based antitumor agents. It may be possible to find complexes of other metals with the right combination of properties to function as antitumor agents through a mechanism similar to that of cisplatin. However, as few of the very large number of platinum compounds tested have found use in the clinic so the requirements for activity appear to be exact and demanding. Success rates with other complexes are unlikely to be high. A more important aspect of the search for antitumor activity in complexes of metals other than platinum is that they will act by different means and so may reveal other chemical mechanisms by which cancer may be treated.

4.3.3 Non-Platinum Anticancer Agents

The importance of certain platinum compounds in the treatment of cancer begs the question of whether other metal complexes might be found which have such useful medicinal properties. Since the chemical properties of the d-block metals cover a wide range the possibility exists that compounds may be found which exert antitumor effects through mechanisms quite different from that of cisplatin. A good example is provided by redox activity, which is an important feature of d-block metal chemistry but not important in the mechanism of action of cisplatin. Redox activity does appear to be important in the action of another group of anticancer drugs, the bleomycins. These natural products are thought to be active in the form of their iron complexes and to effect DNA cleavage as a result of metal centred redox reactions.

Complexes of some early d-block metals have been found to inhibit tumors in animal models and these seem to act through a mechanism quite distinct from that of cisplatin. Among the later d-block metal complexes some ruthenium, rhodium, copper and gold complexes show antitumor activity in laboratory studies. So far complexes of rhenium, osmium and iridium have not featured significantly in studies of antitumor activity. This may reflect a lack of biological activity but these metals have probably not yet been the subject of as much research attention. The use of d-block radionuclides such as ^{99m}Tc and

^{186}Re in the diagnosis and treatment of cancer is considered in Sections 3.3.11.2 and 4.7.3, respectively.

4.3.3.1 Bleomycins

4.3.3.1.1 Bleomycins and Metals. The bleomycins belong to a group of about 200 natural products obtained from *Streptomyces verticillus.* Two of these, bleomycin A_2 and bleomycin B_2, are present in Blenoxane, a drug which may be used in the treatment of testicular carcinomas, ovarian cancer, non-Hodgkin's lymphomas or squamous cell carcinomas. Blenoxane can be used alone or in combination with other drugs such as cisplatin. The cytotoxicity of bleomycin is generally thought to be the result of DNA strand scission, a process which is found to be dependent upon the presence of certain metal ions. Bleomycin complexes of VO^{2+}, Mn^{2+}, Fe^{2+}, Ru^{2+}, Co^{3+}, Ni^{3+} and Cu^{+} have been shown to effect DNA cleavage upon suitable activation. This ability of bleomycin to effect DNA strand cleavage in the presence of redox active metal ions and dioxygen would suggest that DNA is the target molecule in the anticancer activity of Blenoxane. In the medical context it is generally thought that the Fe^{2+} complexes and dioxygen are the active agents in producing anticancer activity. The majority of the DNA strand cleavage events produced by iron-bleomycin are confined to a single DNA strand but 10–20% result in double-strand cleavage at complementary nucleotides or nucleotides staggered by only one base pair. Since such double cleavage events are less efficiently repaired than single strand breaks, double-strand cleavage may be an important feature of bleomycin anticancer activity. Although the main focus of research has been the iron-bleomycin system, bleomycins are typically isolated as their Cu^{2+} complexes, and show a higher affinity for this metal than for Fe^{2+}. Thus the possibility that complexes with metal ions other than Fe^{2+} are important in the mechanism of action of the drug cannot be ruled out.

4.3.3.1.2 The Structures of Bleomycin and its Complexes. The chemical structures of bleomycins A_2 and B_2 are known (Figure 12) and differ only in the nature of the C-terminus region. Both include a bithiazole group next to the C terminus and a metal binding region surrounding a pyrimidine core, these two regions being joined by a dipeptide link. Within the metal binding region, and attached to the pyrimidine, are β-amino alanine and β-hydroxyhistidine moieties. The latter connects to the bithiazole unit *via* the dipeptide link as well as to an appended disaccharide group. The structure of the iron complex of bleomycin has been the subject of a number of studies and several different binding modes have been proposed for the iron bleomycin interaction. Based on spectroscopic studies of bleomycin complexes containing Fe^{2+}, Fe^{3+}, {Fe(CO)} and {Fe(NO)} centres, a probable structure appears to contain 6-coordinate iron bound at five sites within the bleomycin as shown in Figure 13. Nitrogen atoms from the pyrimidine, the β-hydroxyhistidine amide and imidazole moieties and the β-amino alanine are thought to occupy equatorial binding sites with the primary amine of the β-amino alanine occupying an

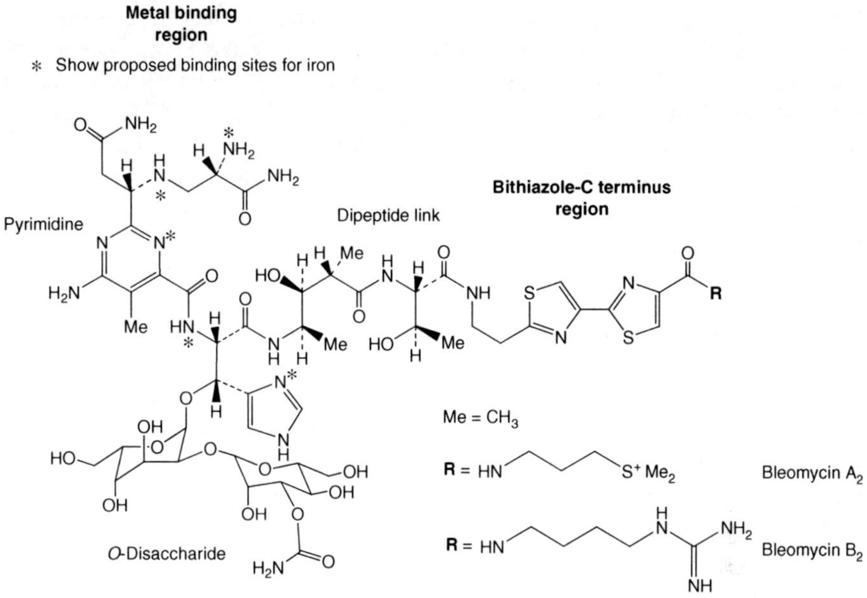

Metal binding region
* Show proposed binding sites for iron

Pyrimidine Dipeptide link Bithiazole-C terminus region

O-Disaccharide

Me = CH_3

R = HN~~~S⁺Me$_2$ Bleomycin A$_2$

R = HN~~~~NH–C(=NH)–NH$_2$ Bleomycin B$_2$

Figure 12 *The structures of bleomycins A_2 and B_2*

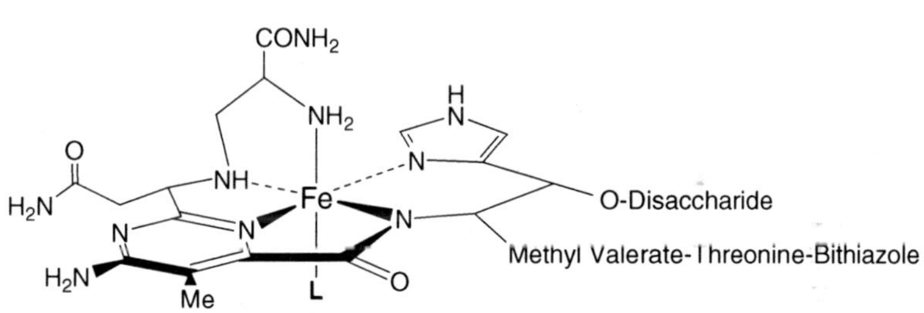

O-Disaccharide

Methyl Valerate-Threonine-Bithiazole

L = dioxygen or labile carbamoyl group from bleomycin

Figure 13 *A proposed structure for the Fe^{3+} binding region of iron-bleomycin*

axial site. The remaining axial binding site is thought to be used in reactions involving dioxygen and may be occupied by the terminal carbamoyl group of the disaccharide moiety in the precursor to the oxygenated species. The carbamoyl group is thought to be weakly bound, readily dissociating to allow binding of dioxygen during activation of the drug. Although some uncertainties still remain about the exact structures of bleomycin-metal complexes, recent evidence suggests that cobalt-bleomycin and zinc-bleomycin can adopt similar structures to that proposed for the iron complex and shown in Figure 13.

4.3.3.1.3 The Binding of Bleomycin to DNA. Metallobleomycins preferentially bind to 5′-GC and 5′-GT sequences in DNA and cause modification at the pyrimidine-C4′-H centre. This selectivity is dependent upon the nature of the metal binding region of the bleomycin but is unaffected by changes in the C-terminus. The presence of a bithiazole group in the structure of bleomycin is an important feature and sequence selectivity for 5′-GC and 5′-GT was lost when either thiazole ring was absent. The bithiazole group in bleomycin is known to be a DNA minor groove intercalator and the related compound phleomycin shows similar selectivity despite containing a structurally different bithiazole moiety. It would seem that, in bleomycin, the metal binding region is the major determinant of the sequence selectivity in DNA binding. NMR studies of the interaction between HOO-Co^{3+}-bleomycin and oligonucleotides have provided a more detailed insight into possible DNA binding modes for iron-bleomycin. Association of HOO-Co^{3+}-bleomycin with the self complimentary oligonucleotide d$(CCAGGCCTGG)_2$ gave rise to spectroscopic features consistent with partial intercalation of the bithiazole at the GC recognition site, one thiazole ring being positioned between two adjacent pyrimidines (C_6.C_7), the other between adjacent purines (G_{14}.G_{15}). A basis for the sequence selectivity of the metal binding region was found in the formation of hydrogen bonds between the 4-amino group and the N3 nitrogen of the bleomycin pyrimidine and the N3 and exocyclic 2-amino group of the DNA purine G5 to form a triplex structure (Figure 14). Importantly the hydroperoxide ligand on the cobalt ion was found to lie close to the deoxyribose C4′-H centre associated with C_6. Although it cannot be assumed with certainty that the structural

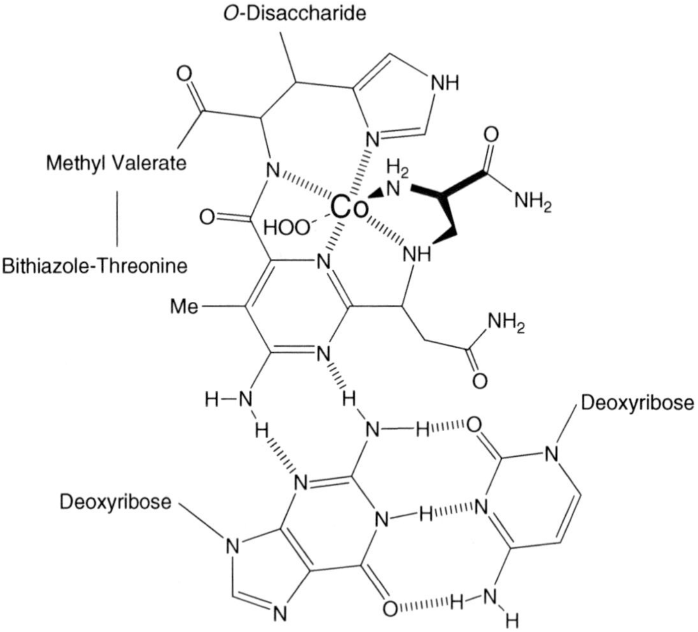

Figure 14 *A proposed interaction mode between the metal binding region of Co^{3+}-bleomycin and DNA*

features of the iron-bleomycin-DNA interaction will be the same as for HOO-Co^{3+}-bleomycin, there seems little reason to suppose that the results with the cobalt complex do not provide a good model for the iron complex.

4.3.3.1.4 The Mechanism of DNA Scission by Bleomycin. Bleomycin complexes of metal ions such as Fe^{2+}, Co^{2+} and Cu^{+}, which can be oxidised by dioxygen, are able to cause DNA scission when dioxygen is present. In contrast a bleomycin complex of Zn^{2+}, which is not redox active, did not mediate DNA strand scission. This supports the view that the mechanism of action involves reactive species derived from the reaction between a redox active metal centre and dioxygen. The reactions of dioxygen with both Fe^{2+} and Co^{2+} are well known and in simple complexes usually result in irreversible metal ion oxidation (Scheme 6). Initially dioxygen binds to an Fe^{2+} centre to form what can be formulated as an Fe^{3+}-$\{O_2\}^-$ superoxide complex. This can undergo further reduction by a second Fe^{2+} centre to form a binuclear peroxide bridged complex which undergoes irreversible decomposition to other Fe^{3+} compounds. Similar chemistry is observed for Co^{2+} but the more kinetically inert d^6 Co^{3+} centre facilitates recovery of the peroxide bridged complex. If the metal ion is structurally protected by the ligands so that a second reducing metal ion cannot approach the bound dioxygen molecule, different behaviour results. In the case of hemoglobin further reduction of the iron-dioxygen group is not possible and reversible dioxygen binding is observed. In the case of cytochrome P_{450} a second reduction by the enzyme of the iron-dioxygen group produces a peroxide ligand bound to a single metal centre. This can form a reactive

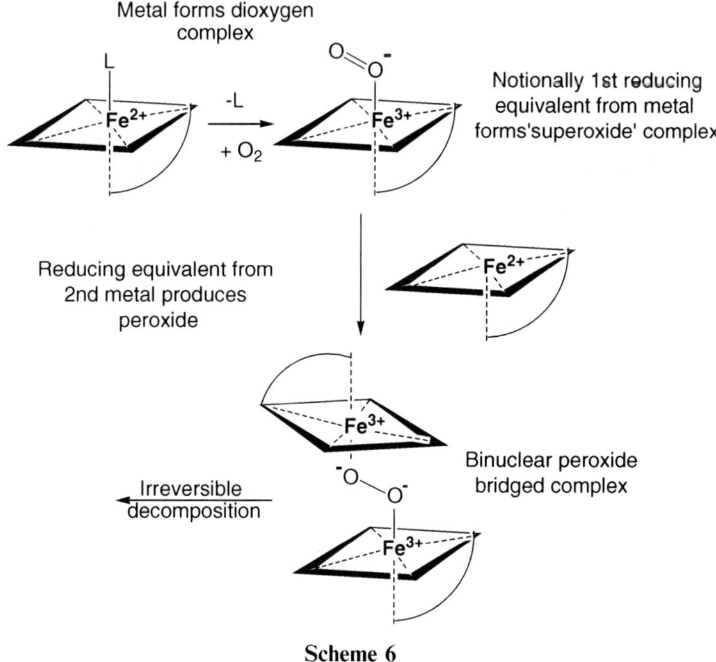

Scheme 6

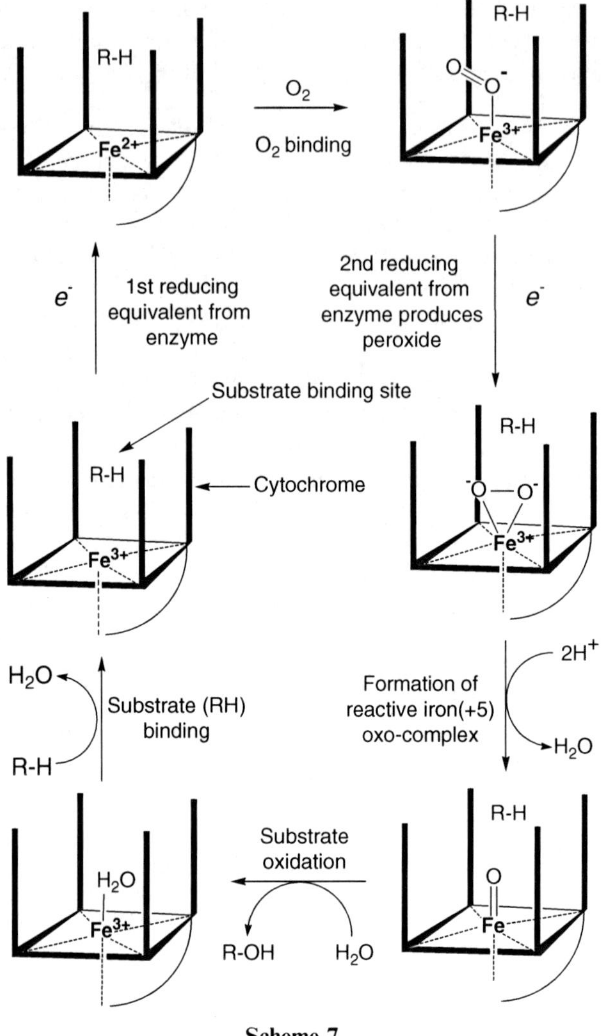

Scheme 7

iron-oxo species which can effect oxidations such as converting a C-H group to C-OH (Scheme 7). The nature of the iron coordination sites in cytochrome P_{450} is quite different from that in bleomycin, nonetheless it would seem likely that the metal binding region of bleomycin provides sufficient structural protection of the metal centre to allow a similar type of chemistry to occur. In accord with this suggestion, electrospray mass spectrometry and kinetic isotope effect studies indicate that 'activated bleomycin' is of the form HOO-Fe^{3+}-bleomycin. The proposed interaction between Fe^{2+}-bleomycin and dioxygen to produce 'activated bleomycin' is summarised in Scheme 8. After formation of O_2^--Fe^{3+}-bleomycin reduction by electron transfer from a cellular reductant would result in the formation the reactive peroxo derivative

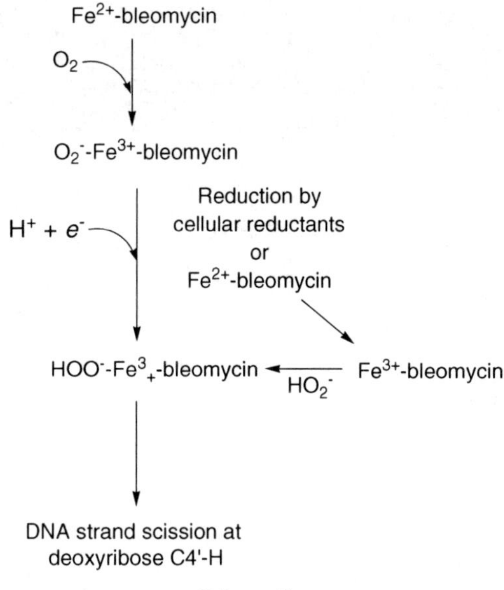

Scheme 8

O_2^{2-}-Fe^{3+}-bleomycin which, in protonated form, becomes HO_2^--Fe^{3+}-bleomycin. It is also possible that reduction to peroxide might occur through electron transfer from a second Fe^{2+}-bleomycin which is converted to inactive Fe^{3+}-bleomycin. The inactive Fe^{3+}-bleomycin could be reduced in the cell back to Fe^{2+}-bleomycin and re-enter the process through reaction with dioxygen. Fe^{3+}-bleomycin can also be converted directly to HO_2^--Fe^{3+}-bleomycin by reaction with hydrogen peroxide. The peroxy complex is chemically reactive and able to oxidatively cleave a DNA strand. The exact details of how this can occur are uncertain but NMR studies of the interaction of HOO-Co^{3+}-bleomycin with oligonucleotides revealed a structure in which the hydroperoxy group lies close to a deoxyribose C4′-H centre. Assuming a similar structural arrangement in the DNA adduct of active-iron-bleomycin, deoxyribose C4′-H appears a likely target for oxidative attack. The mechanisms of DNA scission through oxidative cleavage of deoxyribose have been studied and two reaction pathways identified. The initial rate limiting step is the formation of a C4′ radical by one electron oxidation. Reaction of the C4′ radical with dioxygen would produce a C4′-OO. group which would be converted to the hydroperoxide C4′-OOH by a one electron reduction (Scheme 9). Alternatively the C4′ radical might undergo a second oxidation to form a carbocation which adds hydroxide to form a C4′-OH group (Scheme 10). The hydroperoxide modification at C4′ is thought to result in ring rearrangements leading ultimately to strand scission (Scheme 9). In view of the relative stability of synthetic C4′-hydroperoxy nucleotides it would seem that this process might be facilitated by the presence of bleomycin. Hydroxylation at C4′ leads to quite different reactivity (Scheme 10). Ring opening leads to the release of the

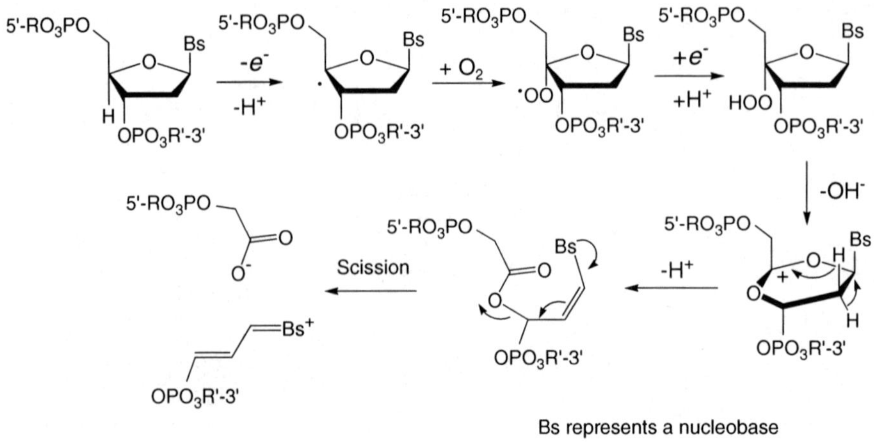

Scission

Bs represents a nucleobase

Scheme 9

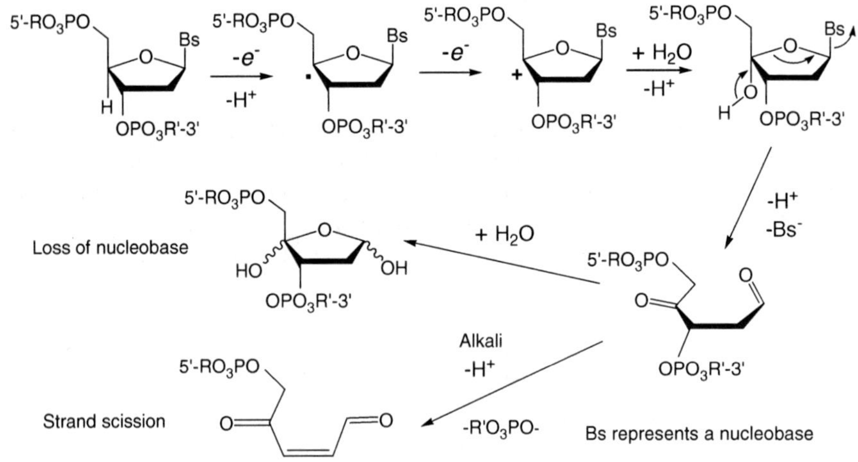

Loss of nucleobase

Strand scission

Bs represents a nucleobase

Scheme 10

nucleobase but strand scission can only be achieved though subsequent alkali attack to release the C3′-phosphate group.

The finding that some 10–20% of iron-bleomycin induced DNA cleavage events involve double-strand scission is important because these are less efficiently repaired than single-strand breaks and so potentially more cytotoxic. Double-strand cleavage does not appear to be simply an accumulation of single-strand cleavage events. Rather it may involve repositioning of the bleomycin following the first strand cleavage to effect a second cleavage nearby. It has been suggested that the bithiazole may act as a tether remaining bound to DNA after the first cleavage while the metal binding region repositions to effect the second cleavage. In order to achieve this the bleomycin must be capable of reactivation following the initial reaction. A possible explanation for this ability involves the formation of a bleomycin bound Fe(+5) centre in the form of $\{Fe(+5)=O\}^{3+}$ like that proposed for cytochrome P_{450} oxidations. The one electron oxidation of the C4′-H centre would then generate a reduced $\{Fe(+4)=O\}^{2+}$ centre. However, a pathway for the reoxidation of metal-oxo system arises if C4′-OO is formed from the C4′ radical and dioxygen. One electron reduction of C4′-OO to C4′-OO^- by $\{Fe(+4)= O\}^{2+}$ would then regenerate the $\{Fe(+5)=O\}^{3+}$ centre. In support of this proposal it was found that the primary sites of double-strand cleavage involve only C4′-OOH intermediates, products associated the alkali scission pathway were not found.

4.3.3.1.5 Conclusions Regarding Bleomycins. Iron bleomycin is a remarkable multifunctional agent. At one end the molecule contains a DNA intercalator in the form of the bithiazole group which may serve as a sort of 'tether'. At the other end of the molecule is a metal binding region which not only binds and activates a metal centre towards reaction with dioxygen, it also contributes to the selectivity of DNA binding through recognising 5′-GC or 5′GT regions in the sequence of bases. The molecule also contains a dipeptide linking region and an appended *O*-disaccharide which may contribute to molecular recognition and DNA binding. When bound to DNA the drug locates the active metal centre near to a C4′-H moiety which can be oxidised starting a process which leads to strand scission. The structure and properties of iron bleomycin offer an important paradigm for the design of new synthetic compounds which might ultimately be used in cancer therapy. Many metal complexes are known which can effect oxidative or hydrolytic cleavage of DNA *in vitro*. Photoactive complexes are also known to effect DNA cleavage. The design of metal complexes which can recognise and selectively bind to sites in DNA where they can cause strand scission constitutes an important research goal. Although this field has yet to become clinically established, remarkable advances are being made which may lead to new drug systems.

4.3.3.2 DNA Intercalators

Planar polyaromatic molecules are known to intercalate between the stacked base pairs of DNA so, unsurprisingly, some metal complexes which contain planar polyaromatic ligands have also been found to show this behaviour. Early work on phenanthroline (phen) complexes such as octahedral $[Ru(phen)_3]^{2+}$, **92**, revealed three types of non-covalent interactions which could be involved in binding: electrostatic interactions, hydrophobic minor groove binding and partial intercalation of a phen ligand into the base stack from the major groove (Figure 15). An important feature of such complexes is their chirality and this affects their DNA binding. Usually right-handed Δ-isomers preferentially bind to right-handed DNA. The preference of many metallointercalators for binding in the major groove of DNA is an important

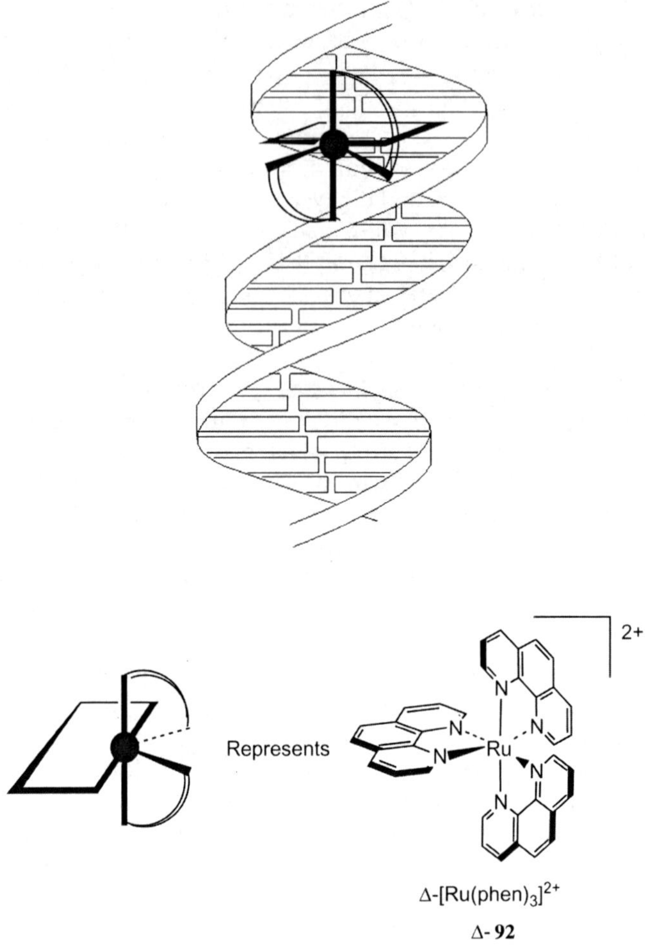

Figure 15 *A cartoon presentation of the interaction between the DNA intercalator **92** and DNA major groove*

feature, uncommon among synthetic agents. The reactivity of the metal centre is also an important feature and much of the early work in this area exploited the photo-oxidation properties of certain metal complexes. As an example photolysis of complexes $[Rh(phi)(phen)_2]^{3+}$, **93**, or $[Rh(phi)_2(phen)]^{3+}$, **94**, bound to DNA in the absence of dioxygen produces 3′- and 5′-phosphate termini as well as free bases. This is consistent with a mechanism involving radical reactions. The supposition is that light induced ligand to metal charge transfer in the DNA bound metal complex produces, in the case of **93** for example, $[Rh^{2+}(phi^{\cdot +})(phen)_2]^{3+}$ containing a ligand radical which can then extract a hydrogen from a deoxyribose unit. This initiates a sequence of events leading to DNA strand cleavage in a similar but not identical fashion to the interaction with activated bleomycin.

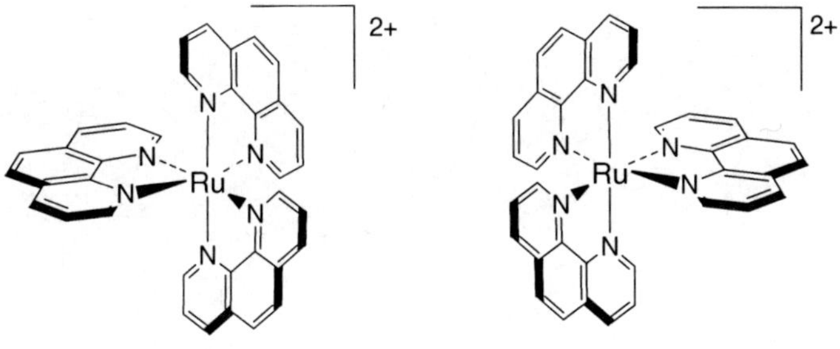

Δ-$[Ru(phen)_3]^{2+}$ Δ-$[Ru(phen)_3]^{2+}$

92

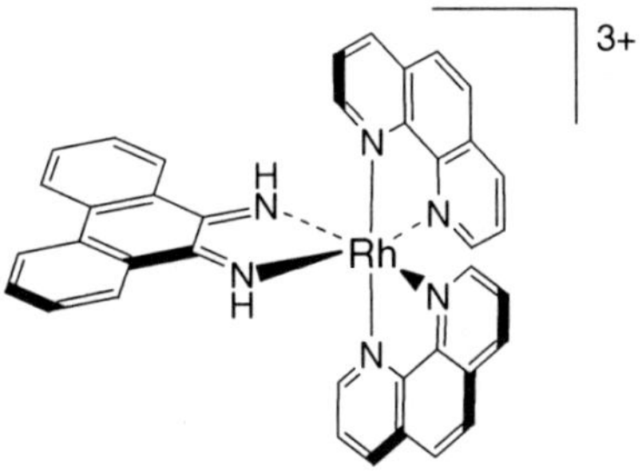

Δ-$[Rh(phi)(phen)_2]^{3+}$
93

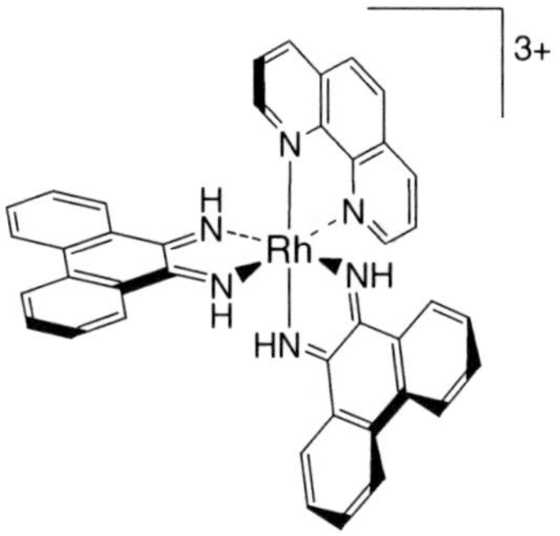

Δ-$[Rh(phi)_2(phen)]^{3+}$
94

Metalloporphyrins such as the copper complex **95** can also intercalate into DNA base stacks. In this case a structural study showed the metalloporphyrin partially intercalating at the end of a 5′-CGATCG-3′ sequence, expelling a cytosine from a G-C pair. The expelled cytosine then formed a hydrogen bonding interaction with guanine in a separate double helix so that no unpaired bases were present. Careful design of the ligand system can promote DNA binding and introduce selectivity into the DNA-metal complex interaction. In the example of complex **96** selectivity for a 5′-CATCTG-3′ sequence was found and modelling studies indicated that binding of the complex induced some unwinding of the DNA helix. The complex was also found to compete with a transcription factor protein in binding to a domain containing both metal binding and protein binding regions. DNA binding by metal complexes can occur through interactions other than intercalation of polyaromatic fragments. An example is provided by the complex **97** (Figure 16) which was designed to bind specifically to 5′-TGCA-3′ sequences through hydrogen bonding interactions between coordinated amine-NH_2 groups and guanine O6 sites. Together with hydrophobic interactions between ligand and thymine methyl groups these interactions were expected to produce selective DNA binding (Figure 16). This was confirmed by NMR and structural studies showing that it is possible to design metal complexes to produce specific DNA interactions.

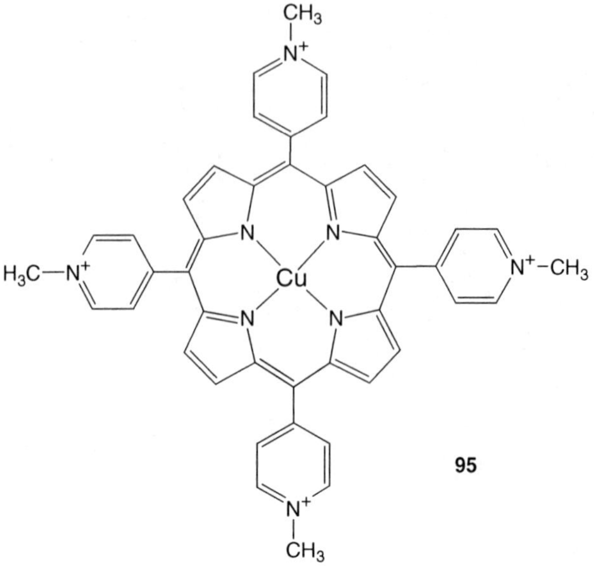

95

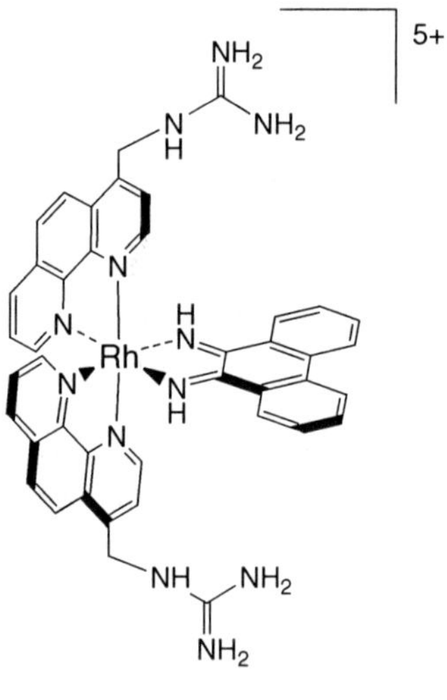

96

Another active area of research involves the so-called 'chimera' complexes which contain a metal complex capable of acting as a DNA intercalator attached to a polypeptide capable of recognising a specific base sequence in DNA. In one example, **98**, a kind of reverse DNA recognition strategy to that used by bleomycin is employed. The metal complex region acts as a non-specific DNA intercalator while the peptide sequence is a DNA binding domain from a phage P22 repressor able to recognise 5′-CCA-3′ sequences. There is particular interest in metallointercalators for DNA, which can effect oxidative or hydrolytic cleavage. However, the lack of rigidity in these 'chimeras' is a potential

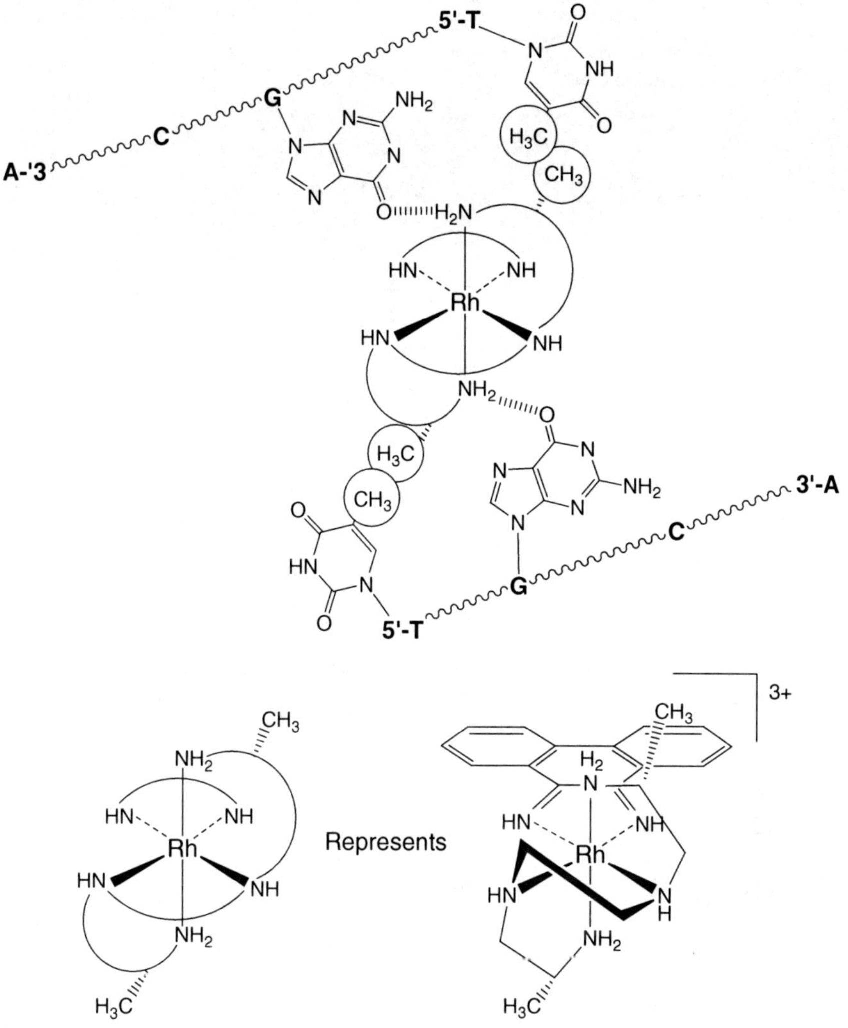

97

Figure 16 *A schematic structural representation of the interaction between* **97** *and DNA*

problem for predicting their behaviour and achieving high sequence specificity. In a further variant of this strategy a metallointercalator has been attached to a peptide chain designed to carry a Lewis acidic Zn^{2+} centre which can promote hydrolytic cleavage of DNA sugar phosphate links. Complex **99** was found to induce stereospecific DNA hydrolysis forming only 3′-hydroxyl termini. The DNA interaction was not sequence selective but the formation of 3′-hydroxyl groups was thought to be a consequence of major groove binding.

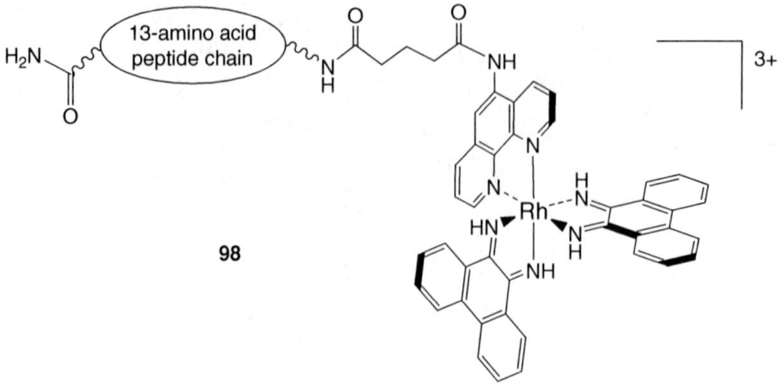

Work on molecules which combine the redox activity of metal centres, the structural selectivity of the metal complex and the biological activity of domains taken from protein or oligonucliotides is in its infancy. Many obstacles remain to be overcome before developments in this area could lead to clinically viable drugs. Nonetheless the chimera approach offers a potentially powerful and selective means of modifying DNA in a medical context. Initially *in vitro* DNA manipulation applications will be much easier to establish and will presumably precede developments of *in vivo* therapy.

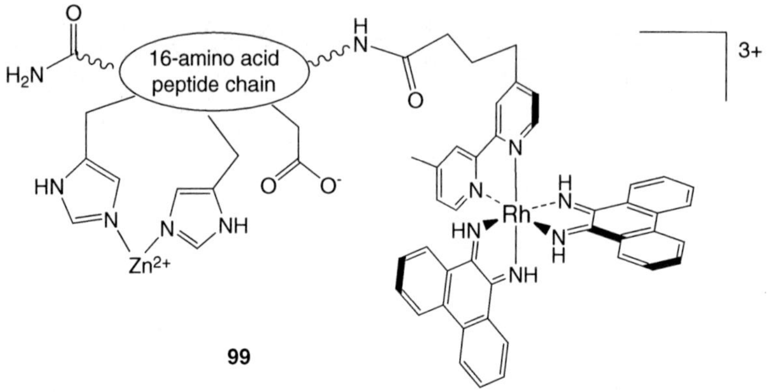

4.3.3.3 p-Block Metals

Thus far the p-block metals have, for the most part, found little use in the treatment of cancer. Gallium is perhaps the most important element of the set in this respect, although antitumor activity has been demonstrated in the laboratory for compounds of several other metals. The germanium compound **100** inhibits the synthesis of protein, RNA and DNA *in vitro,* showed cytotoxic activity in various cell cultures and some antitumor activity in rats. However, Phase II clinical trials gave disappointing results. Arsenic trioxide, As_2O_3, has proven to be effective against a rare blood disorder, acute promelocytic leukaemia, and appears to operate synergistically with all *trans*-retinoic acid in treating this disease. The arsenic derivative Melarsoprol (**101**) inhibits the growth of lymphoid leukemic cells and is being investigated for use against acute and chronic myeloid and lymphoid leukemias. Compounds of antimony and bismuth have well known antimicrobial properties but their application to cancer therapy is limited to some reports of activity against laboratory models. Bismuth(+3) complexes with derivatives of thiosemicarbazide [$NH_2NHC(=S)NH_2$] inhibit the growth of rectal (SW948) and colon (SW707) tumor cell lines, for example. There are reports that bismuth nitrate given prior to cisplatin treatment can be effective against advanced bladder tumors. Gallium and tin compounds are considered in more detail below.

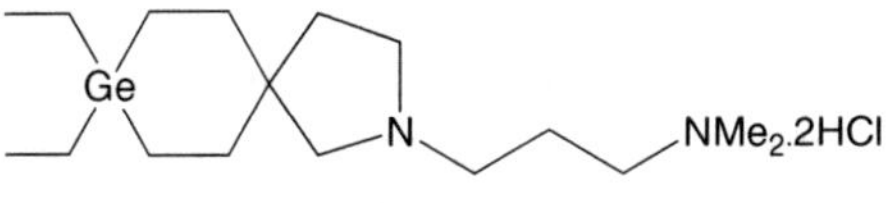

100 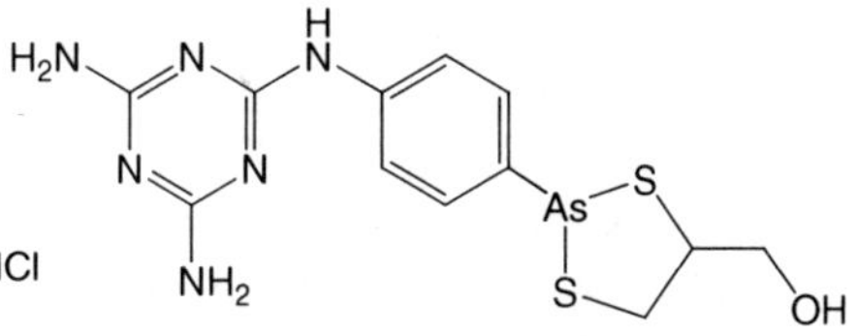 **101**

4.3.3.3.1 Gallium. Among the p-block metals, gallium has shown some clinical activity in the treatment of soft tissue tumors. Positive results have been reported from Phase II clinical trials using gallium compounds in the treatment of lymphomas and bladder carcinomas and, in combination with other agents, in the treatment of metastatic carcinoma of the urothelium and cisplatin-resistant ovarian cancer. In order to avoid nephrotoxicity problems gallium can be administered as the nitrate $Ga(NO_3)_3$ (200–300 mg m^{-2} per day) by continuous intravenous infusion. The relatively high charge radius ratio of Ga^{3+} leads to its hydrolysis at neutral pH, a process which can be suppressed by the presence of citrate. The very low solubility of gallium phosphate can lead to precipitation of Ga^{3+} as $GaPO_4$ in the kidney and uptake of Ga^{3+} by bone is also observed. The similar ionic radii of Fe^{3+} and Ga^{3+}, coupled with the absence of crystal field effects for high spin d^5 Fe^{3+}, lead to similarities in the behaviour of these two ions. The iron transport protein transferrin (Tf) binds gallium strongly, though the binding constants are some two to three orders of magnitude less than for Fe^{3+}. Thus therapeutic doses of Ga^{3+} can lead to approximately equimolar concentrations of Fe^{3+} and Ga^{3+} in serum Tf. This reduces Tf-mediated iron uptake by cells and inhibits DNA synthesis. Displacement of Fe^{3+} from the R2 subunit of ribonucleotide reductase by redox inactive Ga^{3+} would reduce the activity of the enzyme. Since this enzyme converts ribonucleotides to deoxyribonucleotides prior to incorporation into DNA, its inactivation has serious consequences for DNA synthesis.

Gallium compounds are also used to treat elevated blood Ca^{2+} levels which can result from bone cancer and to block osteolysis. Phase III clinical trials are underway to assess the value of gallium compounds for use in suppressing bone metastases from breast carcinoma.

4.3.3.3.2 Tin. The neurotoxicity of trialkyl tin derivatives is well known and various organotin compounds have been assessed in a search for anticancer activity. Studies of alkyl or aryl tin benzoate derivatives showed compounds of the type $[SnR_3(O_2C_6H_5)]$ were more active when R=phenyl than when R=*n*-butyl. Dibutyltin difluorobenzoate derivatives were more active than their tributyl tin counterparts. Several of the compounds gave good results when tested against the colon 26 tumor implanted in mice but showed higher toxicity than cisplatin. Octahedral Sn(+4) complexes of the type $[SnR_2X_2L_2]$ (R=alkyl or phenyl; X=F, Cl, Br, NCS; L=a nitrogen donor such as pyridine) showed activity against P388 lymphotic leukaemia but not other models. Dialkyl tin dipeptide complexes have also been studied. The dibutyl glycylglycinate complex **102** gave a 50% increase in life expectancy in animal models with P388 leukaemia but, again, was not effective against other tumor systems tested. Dialkyl tin pyridine 2,6-dicarboxylate derivatives have shown better activity than cisplatin against mammary tumor (MDF-7) and colon carcinoma (WiDr) cell lines with the dibutyl compound, **103**, proving to be the most active of the series tested. An unusual binuclear dibutyltinoxide carboxylate compound, **104**, has been found to be more active than cisplatin or carboplatin against human breast (MCF-7 and EVSA-T), colon (WiDr), renal (A498) and ovarian (IGROV) tumor cell lines, as well as against melanoma and non-small cell lung cancer cell lines.

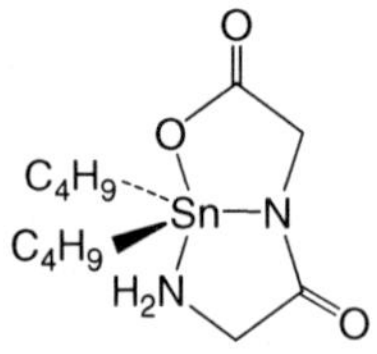

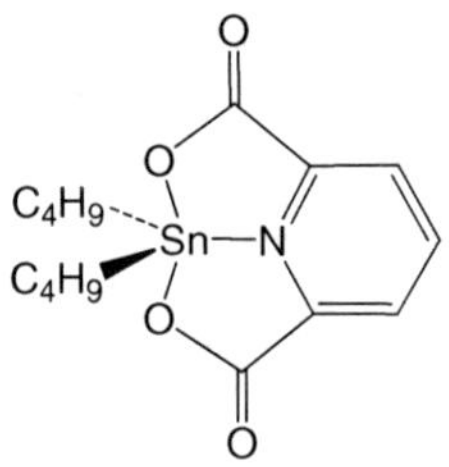

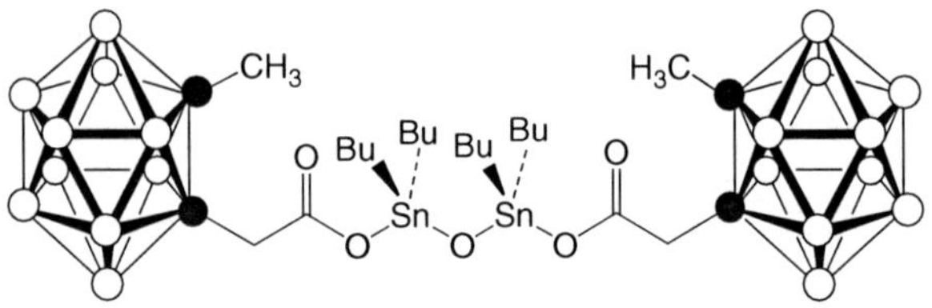

The low aqueous solubility of most organotin compounds presents a limiting factor on their therapeutic use and colloidal suspensions or addition of dimethyl sulfoxide to improve solubility may be necessary. The activity of tin compounds appears to depend on some hydrolysis occurring so that the use of hydrophilic leaving groups offers another means of improving solubility. The toxicity of organotin compounds seems to depend upon the lipophilicity of the less labile organo groups and, generally, butyl groups appear to give better results. Details of the mechanism of action of organotin compounds remain uncertain. Binding of the organotin fragments to phosphate occurred at pH values below 7 in studies with 5′-AMP, 5′-GMP, calf thymus DNA and 5′-d(CGCGCG)$_2$. Above this pH hydroxyl becomes competitive with phosphate in binding the organotin fragment. Binding to sugar hydroxyl groups is expected to require pH values above 9, although binding to appropriately oriented adjacent sugar hydroxyl groups could be assisted by the chelate effect. Phosphate would thus appear to be the most likely site for binding organotin to DNA but how this relates to their antitumor activity is unclear.

4.3.3.4 Early d-Block Metals

The early d-block metals tend to reach higher oxidation states than metals later in the block and can often attain their maximum valence oxidation states in aqueous media. In oxidation states above +3 metal oxoions can form, complexes containing oxoions such as ZrO^{2+}, VO^{2+}, $MoO_2{}^{2+}$ or ReO^{3+} providing examples. Organometallic derivatives containing direct metal to carbon bonds are also known and some of these have shown anticancer activity. Most of the metals show significant redox behaviour and this may involve the acquisition or loss of coordinated oxygen atoms. In this context vanadium is particularly interesting and is present as an essential trace element in mammalian tissues at concentration at or below about 10^{-5} mol dm^{-3}. Under biological conditions reduction reactions of $VO_2{}^+$ [vanadium(+5)] to VO^{2+} [vanadium(+4)] cores can be observed. Vanadyl (VO^{2+}) binds to proteins through both oxygen and nitrogen donor sites. Since they share the same number of valence electrons vanadium and phosphorus compounds also share certain similarities in stoichiometry. Thus vanadate esters can mimic phosphate esters and thereby inhibit kinases and nucleases. Peroxide can form side-on bound complexes with vanadium centres and this type of chemistry appears to have significant biological implications. Auto-oxidation of deoxyguanosine in air is induced by vanadyl sulfate by a mechanism which involves peroxide. Vanadyl bleomycin cleaves DNA at G-A(5′–3′) sites in the presence of H_2O_2 but the mechanism differs from that of DNA cleavage by iron bleomycin. Several peroxovanadium complexes have shown anticancer activity *in vitro* and subcutaneous injections of orthovanadate into mice carrying MDAY-D2 tumors reduced tumor growth by 85–100%. It would appear that the redox and oxygen transfer chemistry of vanadium offers a potential basis for new anticancer drugs but as yet clinical applications have not been realised. Among the other early d-block metals only compounds of titanium have come close to finding clinical acceptance, β-diketonate and metallocene derivatives having entered Phase II clinical trials.

4.3.3.4.1 Early d-Block Metals – β-Diketonate Complexes. Although Ti^{3+} can be prepared in aqueous media, the aqueous chemistry of titanium is dominated by oxidation state +4 and the tendency of free Ti^{4+} to hydrolyse and precipitate, ultimately forming insoluble TiO_2. Hydrolysis reactions can seriously limit the shelf life of preparations containing compounds of early d-block metals. However, in the presence of suitable ligands these hydrolytic reactions can be controlled and the titanium β-diketonate complex, budotitane (**105**), was the first non-platinum metal complex to enter clinical trials for the treatment of cancer. Budotitane was the most active of a series of complexes of the general formula $[M\{C_6H_5C(O)CHC(O)CH_3\}X_2]$ (X is a uninegative ligand such as Cl^- or $C_2H_5O^-$) in which activity decreased with increasingly labile binding of the ligand X and in the order by metal $Ti \approx Zr > Hf > Mo > Sn > Ge$. In animals the drug proved to be quite effective against ascites tumors and induced colorectal tumors and it was well fairly tolerated by human patients. Cardiac arrythmia was the dose limiting side effect with liver and kidney toxicity arising at high doses. The maximum tolerable dose proved to be 230 mg m^{-2} biweekly. The situation regarding the mechanism of action of budotitane is confused by the presence of three interconverting isomers differing in the relative orientation of their β-diketonate ligands. The relative proportions in solution were found to be 60% *cis, cis, cis* 21% *cis, trans, cis* and 19% *cis, cis, trans*. In water hydrolysis of the ethoxy groups occurs with a reaction half time of about 20 s whereas for the β-diketonate ligands this figure is measured in hours. Condensation of the initial hydrolysis products can produce polynuclear oxo or hydroxo bridged species. The diaqua complex binds to DNA, but evidence for DNA damage is lacking *in vitro* or *in vivo*. Thus the mechanism of action of budotitane, although unknown, appears to be quite different from that of cisplatin.

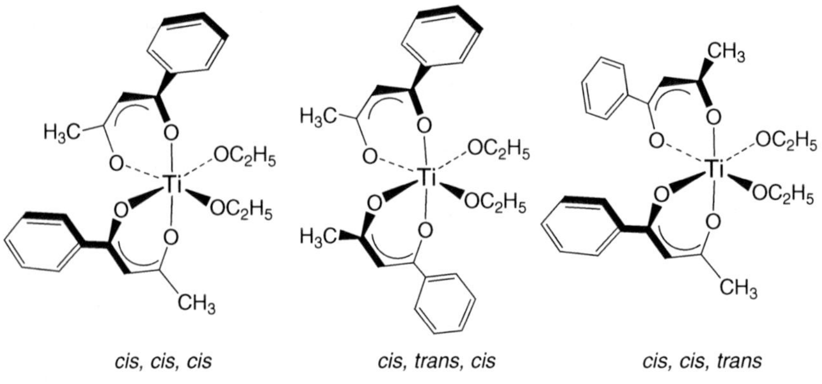

cis, cis, cis *cis, trans, cis* *cis, cis, trans*

105

4.3.3.4.2 Early d-Block Metals – Metallocene Compounds. The early d-block metals are known to form a series of organometallic complexes known as the metallocene dihalides. These have the general formula $[M(\eta^5\text{-}C_5H_5)_2X_2]$ (**106**; M is an early d-block metal and X a halogen) and contain two cyclopentadienyl ligands bound face on (η^5) *via* five carbon atoms to a formally +4 metal centre. Members of this group have shown inhibitive effects on the growth of xeno-grafted human carcinomas of breast, lung and gastrointestinal tract. Several are active against Erlich ascites tumor cells. The nature of the metal ion has a pronounced effect on the activity of $[M(\eta^5\text{-}C_5H_5)_2Cl_2]$ with the compounds containing Ti, V, Nb and Mo showing significant activity and those containing Ta, W, Zr and Hf showing little or no activity. The charged complexes $[M(\eta^5\text{-}C_5H_5)_2X_2]^+$ (M=Nb, Mo, Re) and $[Fe(\eta^5\text{-}C_5H_5)_2]^+$ also show activity. The most effective of this series of compounds was found to be titanocene dichloride, $[Ti(\eta^5\text{-}C_5H_5)_2Cl_2]$, **107**. Variations in the nature of the ligand X in $[Ti(\eta^5\text{-}C_5H_5)_2X_2]$ led to changes in toxicity but did not strongly affect activity. In contrast substitutions on the cyclopentadienyl ring led to inactivity as did linking these rings in the compounds **108**. Phase I clinical trials were carried out on **107** and nephrotoxicity was found to be dose limiting. Other side effects include increased levels of bilirubin and creatinine. In a Phase II clinical trial doses of 270 mg m^{-2} every 3 weeks for 6 weeks were ineffective against metastatic renal cell carcinoma. However, **107** shows activity against cisplatin resistant cell lines suggesting that it may find application against ovarian cancer.

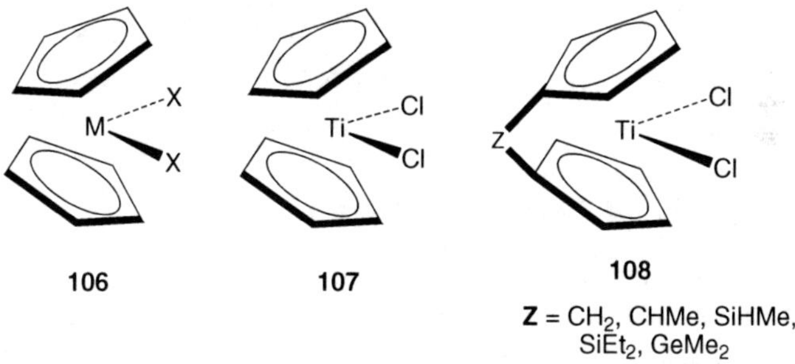

The mechanism of action of the metallocene dihalides remains uncertain but the effectiveness of **107** against cisplatin resistant cell lines suggests a different mechanism of action. Hydrolysis of **106** (X=Cl; M=Ti, V, Zr, Mo) is more rapid than for cisplatin and presumably involves successive replacement of chloride ligands by water followed by deprotonation ultimately yielding the dihydroxide **106** (X=OH; M=Ti, V, Zr, Mo). Loss of the first chloride is very rapid but the second is lost much more slowly with reaction half times ($t_{1/2}$) up to about 45 min. The rate of loss of a cyclopentadienyl ligand is slow with $t_{1/2}$ about 57 h for **107**. The molybdenum complex $[Mo(\eta^5\text{-}C_5H_5)_2(H_2O)_2]^{2+}$ is less acidic than $[Ti(\eta^5\text{-}C_5H_5)_2(H_2O)_2]^{2+}$ and exists as $[Mo(\eta^5\text{-}C_5H_5)_2(OH)(H_2O)]^+$ in water at pH 7 whereas the titanium complex becomes $[Ti(\eta^5\text{-}C_5H_5)_2(OH)_2]$. The neutral complex probably enters cells more easily than the charged complex and this may contribute to the higher activity of **107**. The hydrolysis products of **107** have a high affinity for serum proteins and bind strongly to transferrin suggesting this as a possible transport mechanism. Selective tumor uptake of **107** takes 2–3 days and the metal accumulates in cellular regions rich in nucleic acids. Collegenase type IV and protein kinase C activities are inhibited by **107**, as is the growth of new blood vessels into the tumor. Contrary to earlier findings recent studies have shown that metallocene dihalides do not bind strongly to nucleic acids at neutral pH and do not inhibit enzymes involved in DNA processing. It seems unlikely therefore that their activity involves nucleic acids directly, although a complex of $\{Mo(\eta^5\text{-}C_5H_5)_2\}^{2+}$ and 5′-AMP has been identified in which binding to both phosphate and the nitrogen base occurs, **109**. Tumor cells treated with **107** have shown a premitotic G_2 block and mitotic suppression. This could result from protein kinase inhibition affecting the regulation of cellular proliferation and the ability of metallocene dihalides to inhibit DNA topoisomerase II. However, a full mechanistic understanding of the origin of metallocene dihalide antitumor activity has yet to emerge.

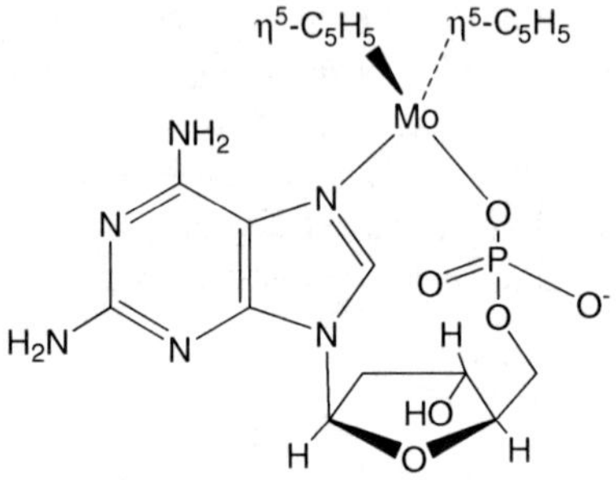

109

4.3.3.5 *Middle and Late d-Block Metals*

The mid to late d-block metals tend to show less extensive redox series than some of their earlier d-block counterparts. Under physiological conditions oxidation states +2 or +3 are the norm with higher oxidation states being chemically quite reactive. Oxoions do not usually feature in the biological chemistry of these metals with the important exception of the $FeO^{3+/2+}$ group in bleomycin and enzymes such as cytochrome P_{450}. It might also be expected on chemical grounds that ruthenium would be capable of similar chemistry and RuO^{2+} complexes with polypyridine ligands are known to cleave DNA. In the context of new anticancer compounds ruthenium and, to a lesser extent, rhodium complexes have attracted considerable recent interest but little is known of osmium and iridium in this respect. Regarding the late d-block metals, historically it has been known that some copper complexes of thiosemicarbazones show better activity than the corresponding uncomplexed proligand. Similarly some diphosphines show greater activity as their gold complexes than in the free state. Although various types of copper complex have shown anticancer activity in laboratory trials, these initial results have not yet led to clinically important developments. Because of their established clinical use in the treatment of rheumatoid arthritis, gold compounds offer an interesting prospect in the search for new chemotherapy agents and a number have shown antitumor effects in the laboratory.

4.3.3.5.1 Ruthenium Complexes. Ruthenium red $[\{Ru(NH_3)_5\}(\mu\text{-}O)\{Ru(NH_3)_4\}(\mu\text{-}O)\{Ru(NH_3)_5\}]^{6+}$ has been used as a biological stain for over a century, and contains the mixed oxidation state $\{Ru^{3+}\}(\mu\text{-}O^{2-})\{Ru^{4+}\}(\mu\text{-}O^{2-})\{Ru^{3+}\}$ core. The high positive charge on the complex promotes binding to polyanions and it has been found to concentrate in tumors. The complex can inhibit tumor growth and shows immunosuppressant activity. Following the discovery of mitotic suppression by cisplatin, chloro-ammine complexes of ruthenium also attracted interest as possible antitumor agents. Crystal Field Theory would lead us to expect that, in contrast to the square planar coordination found for d^8 Pt^2, d^6 Ru^{2+} and d^5 Ru^{3+} will normally adopt octahedral coordination geometries. Hence simple chloro-ammine complexes of Ru^{2+} and Ru^{3+} are 6-coordinate, examples being *cis*-$[Ru(NH_3)_4Cl_2]$, *cis*-$[Ru(NH_3)_4Cl_2]$ Cl and *fac*-$[Ru(NH_3)_3Cl_3]$. Although structurally different from cisplatin, all of these complexes have shown antitumor activity in laboratory models such as lymphocytic leukaemia P388 and murine L1210 tumor. Solubility is a limiting factor in some cases and this can be improved by forming more soluble anionic chloro complexes such as (imH_2)*trans*-$[Ru(imH)_2Cl_4]$ (**110**, imH=imidazole) or dialkyl sulfoxide complexes such as (imH_2)*trans*-$[Ru(imH)(dmso)Cl_4]$ (**111**, NAMI-A, dmso = Me_2SO) and *trans*-$[Ru(dmso)_4Cl_2]$. Complex **110**, and its counterpart containing indazole in place of imidazole, have shown good results against a range of tumor models including P388 lymphocytic leukaemia, Ehrlich ascites and Stockholm ascitic tumors, Walker 256 carcinosarcoma, subcutaneously transplanted B16 melanoma, intramuscularly growing sarcoma 180 and MAC15A colon tumor. These agents are particularly active against colorectal tumors and also show activity against non small cell lung, renal and breast cancers. It is thought that some Ru^{3+} complexes are actually prodrugs being reduced to active Ru^{2+} complexes under physiological conditions.

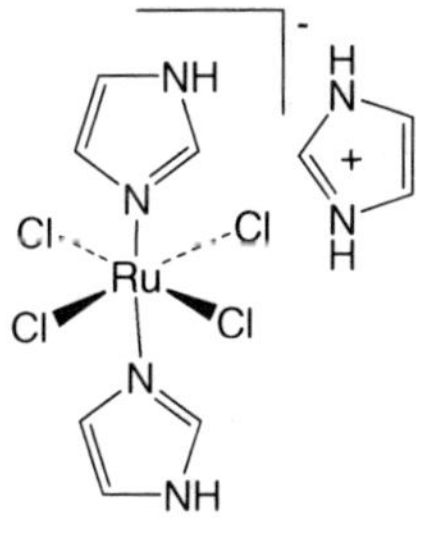

110 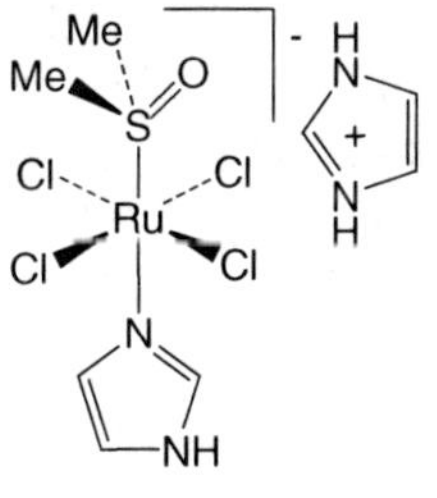 **111** 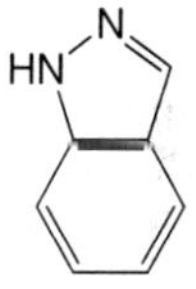 Indazole

The dimethyl sulfoxide (dmso) complexes behave differently from the ammine complexes. Both *cis*- and *trans*-$[Ru(dmso)_4Cl_2]$ are only marginally active against the primary tumor of the Lewis lung carcinoma and some of the dmso complexes are inactive against P388 lymphocytic leukaemia and L1210 murine leukaemia. However, they do show activity against tumor metastases, a particularly important property which has stimulated interest in compounds such as **111**. The dmso complexes generally show relatively low toxicity with LD_{50} values up to *ca.* 1 g kg^{-1} body weight but this advantage is offset by the relatively high doses needed for therapeutic effects. Another group of ruthenium complexes with improved water solubility involve polyaminecarboxylate ligands exemplified by 1,2-propylene diamine tetra-acetic acid (1,2-pdtaH_4). The labile complex *cis*-$[RuCl_2(1,2\text{-pdta})]^{3-}$ shows good antitumor activity and appears to lose chloride and bind to transferrin and albumin. A guanine (GH) complex $[Ru(GH)_2(1,2\text{-pdta})]^-$ has been isolated suggesting that such compounds may be capable of crosslinking DNA *via* guanine interactions. However, $[RuCl_2(1,2\text{-pdta})]^{3-}$ also stimulates NADPH oxidase and a respiratory burst in phagocytic neutrophils so antitumor activity might be the result of the release of toxic oxygen metabolites from phagocytic cells infiltrating the tumor.

Di, tri and terachloro complexes of Ru^{3+} appear to be transported in the blood by human serum albumin (HSA) to the larger extent (*ca.* 80%) and by transferrin to a lesser extent. However, it is transferrin which is important in transporting ruthenium to tumors. Ruthenium is bound to transferrin histidines and, as there are 17 of these, each transferrin molecule is capable of transporting several ruthenium ions. The rate of uptake of a complex by transferrin depends on its exact structure and $(indH_2)$*trans*-$[Ru(indH)_2Cl_4]$ (indH=indazole), the indazole containing counterpart of **110**, binds to transferrin in minutes rather than hours as found for **110**. Release of ruthenium from transferrin is an important step and it has been found that DNA binding and the cytotoxicity of *trans*-$[Ru(imH)_2Cl_4]^-$ or *cis*-$[Ru(NH_3)_4Cl_2]^+$ is increased in reducing hypoxic regions, implying that reduction of Ru^{3+} to Ru^{2+} is important for this process. Interactions of ruthenium complexes with GSH are also possible and, under aerobic conditions, $[Ru(NH_3)_5Cl]^{2+}$ reacts initially to produce GSSG and $[Ru(NH_3)_5(OH)]^{2+}$ while under anaerobic conditions $[Ru(NH_3)_5(OH)]^{2+}$ is converted to $[Ru(NH_3)_5(GS)]^{2+}$. Reduction of $[Ru(NH_3)_5Cl]^{2+}$ by GSH may stimulate DNA binding through the formation of the labile reduced complex $[Ru(NH_3)_5(H_2O)]^{2+}$. However, excess GSH can compete by coordinating to the Ru^{2+} centre prior to oxidation to form $[Ru(NH_3)_5(GS)]^{2+}$.

In aqueous media the imidazole complex **110** undergoes stepwise loss of two chlorides by aquation (initial rates: $9.6 \times 10^{-6}\ s^{-1}$ at 25 °C and 5.26×10^{-5} at 37 °C) leading to precipitation in serum, possibly through acid dissociation of bound water. At serum chloride concentrations chloride dissociation from *cis*-$[Ru(dmso)_4Cl_2]$ is suppressed and it is the oxygen bonded dmso ligand which is labile. In the case of *trans*-$[Ru(dmso)_4Cl_2]$ dissociation of sulfur bonded dmso precedes chloride dissociation leading to *fac*-$[Ru(dmso)_2(H_2O)_3Cl]^+$. Complexes of both Ru^{2+} and Ru^{3+} with dmso are likely to be accessible *in vivo* and rapid dmso dissociation is also observed for the Ru^{3+} complexes *trans*-$[Ru(dmso)_2Cl_4]^-$ and *mer*-$[Ru(dmso)_3Cl_3]$ in aqueous solution. Similarities in the rates of aquation of cisplatin and of some of these complexes, together with evidence for guanine binding reactions with *trans*-$[Ru(dmso)_4Cl_2]$, might suggest a similar mechanism of action. However, this would not explain the activity against cisplatin resistant strains observed in some cases. Nonetheless there is evidence for DNA inter-strand crosslinking by *mer*-$[Ru(dmso)_3Cl_3]$, *mer, cis*-$[Ru(dmso)(H_2O)_2Cl_3]$, *trans*-$[Ru(dmso)_2Cl_4]^-$ and other ruthenium complexes.

Polypyridine complexes of ruthenium have long been used to probe DNA structures and to effect strand cleavage. The 2,2′:6′2″-terpyridine (terpy) complex *mer*-$[Ru(terpy)Cl_3]$ (**112**) forms inter-strand crosslinks with DNA and is active against human cervix carcinoma HeLa and murine L1210 tumor cell lines. Purine N7 binding to *trans* sites vacated by chloride is possible in this *mer* complex whereas the binding of two purines to *cis* sites would be blocked by the bulky polypyridine ligand. Binding to polypyridine complexes is generally thought to be specific for guanine sites. The cytotoxicity of the chloro complexes *cis*-$[Ru(NH_3)_4Cl_2]Cl$ and **110** correlates with DNA binding and it was found that $[Ru(NH_3)_4(py)(H_2O)]^{2+}$ (py=pyridine) binds specifically to guanine(N7), whereas $[Ru(NH_3)_5(H_2O)]^{2+}$ binds only selectively to guanine(N7) with binding to adenine and cytosine also observed. Electrostatic interactions favour cation binding to DNA and variations in binding rates to RNA with ionic strength suggest that ion pairing occurs prior to binding. Ruthenium complexes may cause damage by one of several mechanisms. The metal ion could simply act as an acid stimulating hydrolysis of the *N*-glycosidic bond. Under aerobic conditions the metal centre might enter into redox reactions resulting in oxidation of the purine or sugar moieties. This might involve Ru^{4+} formation through disproportionation of Ru^{3+} or the reactions of oxo-ruthenium species. The hetero-bimetallic complex *cis, fac*-$[\{Ru(dmso)_2Cl_3\}\{\mu\text{-}NH_2(CH_2)_4NH_2\}$-*cis*-$\{Pt(NH_3)Cl_2\}]$ forms crosslinks to DNA repair proteins.

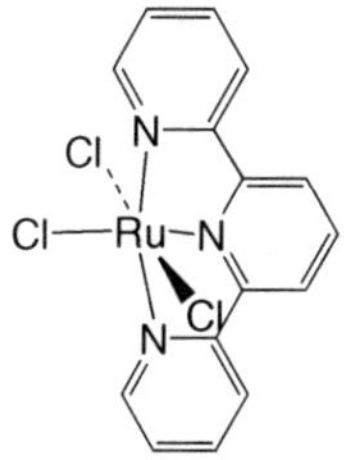

112

A particularly important development arising from studies ruthenium complexes is the finding of antimetastatic activity particularly by Na{*trans*-$[Ru(imH)(dmso)Cl_4]$} (NAMI, **111** with Na^+ in place of imH_2^+). Such complexes could find application in controlling the growth of hard to detect micrometastases following surgery or radiotherapy. NAMI is effective against a range of tumors including Lewis lung carcinoma, B16 melanoma and MCa mammary carcinoma. The salt has good water solubility, can be given orally and has significant antimetastatic effects in animals at doses of 22–66 mg kg^{-1} body weight per day. The complex does not appear to affect DNA unlike its more lipophilic counterpart **113** which causes DNA fragmentation. In aqueous media NAMI undergoes a number of aquation and redox processes (Scheme 11) which greatly complicate its chemistry *in vivo*. In fact only a very low proportion of NAMI reaches the tumor and its effect appears to be independent of its concentration in the tumor. When present at levels which dramatically reduce lung metastases, NAMI changes the relative proportion of the mRNAs of a metalloproteinase which degrades the extracellular matrix (MMP-2) and the specific tissue inhibitor of this enzyme, TIMP-2. This leads to an increase in extracellular matrix components around the tumor parenchyma and the tumor blood vessels. This presumably hinders blood flow to the tumor and interferes with metastasis formation. The imidazolium salt NAMI-A (**111**) offers better

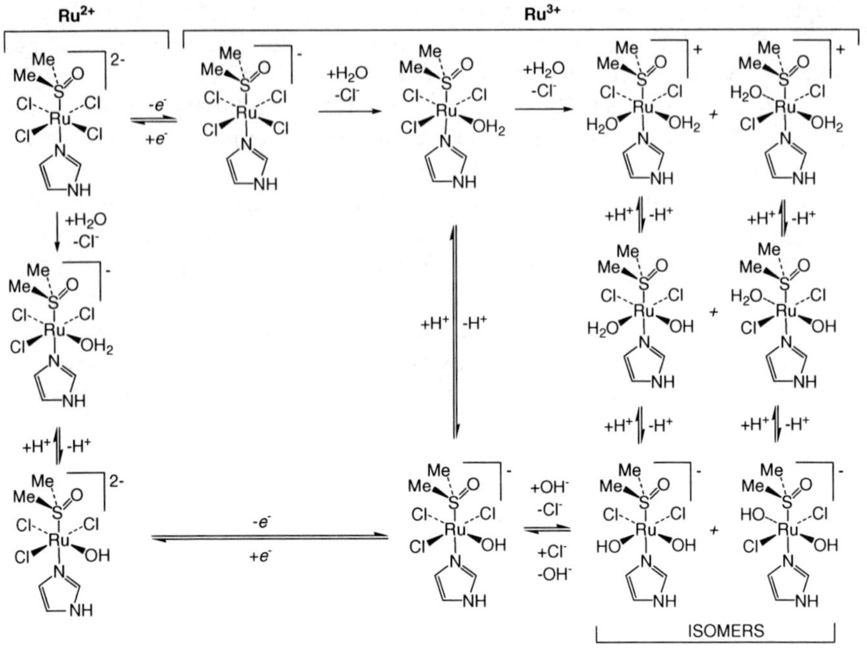

Scheme 11

pharmacological properties than NAMI, being more stable and reproducible in its preparation. NAMI-A, appears to be less toxic than cisplatin and acts differently causing a transient cell cycle arrest of tumor cells in the premitotic G2/M phase. Its effectiveness is similar to that of NAMI causing a reduction in mammary carcinoma MCa metastases to lung and interfering with the growth of Lewis lung carcinoma, MCa mammary carcinoma and TS/a adenocarcinoma metastases present in the lung. NAMI-A, like NAMI, interferes with MMP-2 activity and increases the thickness of connective tissue in the tumor and around tumor blood vessels.

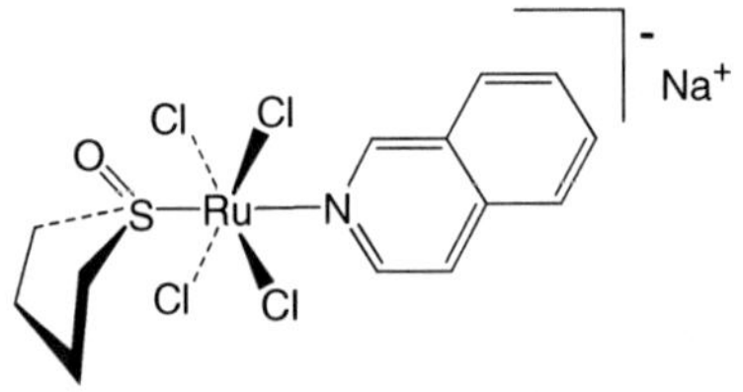

113

In summary ruthenium chloro-ammine, and related complexes, do show antitumor properties. Transport of ruthenium to the tumor by transferrin seems likely and reduction of Ru^{3+} to Ru^{2+} complexes may be important in the release and antitumor action of the ruthenium. Redox chemistry involving GSH is possible and GSH may be a competitor for ruthenium binding. Unlike cisplatin where only Pt^{2+} is important in the mechanism of action, both Ru^{2+} and Ru^{3+} are accessible under physiological conditions making the biological chemistry of the Ru systems more complicated. Furthermore, there is a possibility that ruthenium-oxo species might be formed in highly oxidising regions with the prospect of chemistry like that seen for iron in P_{450} or bleomycin occurring. As with cisplatin, binding to G sites in DNA seems favoured where DNA interactions occur but octahedral ruthenium complexes are not well suited to forming intrastrand crosslinks. Rather there is evidence of inter-strand crosslinks through purine binding to *trans* sites in the ruthenium complex. The observation of antimetastatic activity by **111** is potentially a very important lead, showing that enzymes can be an effective target for antitumor complexes.

4.3.3.5.2 Rhodium Complexes. Mononuclear rhodium complexes having structures similar to those of active ruthenium compounds have been studied and show some antineoplastic activity in laboratory studies. However, this is generally less significant than for the ruthenium compounds. Low solubility can be a problem with rhodium complexes, as with ruthenium complexes, and the rhodium dmso complexes appear to be more toxic than their ruthenium counterparts. The different redox properties of rhodium complexes may also contribute, Rh^{3+} being less readily reduced in aqueous media than Ru^{3+}. The metals also differ in that d^6 Rh^{3+} complexes are likely to be more kinetically inert than their d^5 Ru^{3+} counterparts. Binuclear structures of the general formula $[M_2(\mu\text{-}RCO_2)_4LL']$ (**114**) containing four bridging carboxylate ligands are found for complexes of various metals spanning the middle to late d-block. Ruthenium and rhodium complexes of this type have shown antitumor activity in laboratory tests; $[Ru_2(\mu\text{-}RCO_2)_4(\mu\text{-}Cl)]_n$($R{=}CH_3$, C_2H_5) are active against P388 lymphocytic leukaemia and $[Rh_2(\mu\text{-}RCO_2)_4(H_2O)_2]$ ($R{=}CH_3$, C_3H_7, C_4H_9) are active against Ehrlich ascites tumor, Sarcoma 180 ascitic tumor, and intraperitoneally transplanted P388 with little activity against L1210 leukaemia and B16 melanoma. The activity of the rhodium compounds was found to increase with increasing complex lipophilicity from $R{=}CH_3$ to $R{=}C_4H_9$. However, the utility of these compounds is limited by their toxicity. The binuclear rhenium complex $[Re_2(\mu\text{-}C_2H_5CO_2)_2Br_4(H_2O)_2]$ showed significant activity against an intraperitoneally transplanted P388 leukaemia and against subcutaneously transplanted Sarcoma 180 and B16 melanoma.

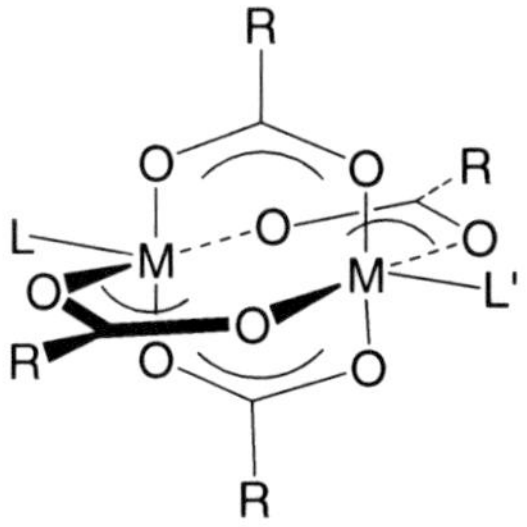

114

Examples:-
M = Re, R = C_2H_5, L = L' = Cl
M = Mo, R = CH_3, L = L' = nothing
M = Ru, R = CF_3, L = Cl, L' = nothing
M = Cu, R = CH_3, L = L' = H_2O

The binuclear rhodium complexes bind to proteins through cysteine thiolate or histidine imidazole groups inhibiting enzymes with cysteine at the active site. The binuclear $(Rh)_2^{4+}$ core appears to remain intact on binding to thiolate in the form of GSH and, in water, binding to adenosine N7 is favoured over guanosine, cytidine or uridine. Structural studies of model compounds show axial binding of two adenosines *via* N7 to the $\{Rh_2(\mu\text{-}CH_3CO_2)_4\}$ moiety but with 9-ethyl guanine (etgua) two carboxylate bridges are replaced by two bridging N7, O6 bound guanine units to give $[\{Rh_2(\mu\text{-}CH_3CO_2)_2(\mu\text{-etgua})_2\}(CH_3CO_2)(H_2O)]$. Evidence for DNA binding of the dirhodium complexes is limited; early work suggested binding to single-stranded DNA and poly(dA) but not to double-stranded DNA or poly(dG). NMR and molecular modelling studies of binding to dGpG and single-stranded d(5′-CCTCTGGTCTCC-3′) suggest that an intra-strand crosslink is possible through adjacent G(N7,O6) binding which could have similar effects to cisplatin binding. However, such possibilities remain speculative.

One interesting aspect of the behaviour of rhodium complexes is the very specific nature of DNA binding by some rhodium metallointercalators. In particular a complex of chrysene quinone diimine (chrysi), Δ-$[Rh(chrysi)(bpy)_2]^{3+}$, **115**, shows remarkable selectivity for base pair mismatches in DNA. The complex selectively binds to a single mismatch in a 2725 base pair plasmid. The bulk of the chrysi ligand is such that it cannot intercalate between normal base pairs but only where the structure is deformed by a mismatch. Complexes of this type offer the possibility of developing chemotherapeutic agents which can detect mutations and possibly even act as agents which can target specific transcription sites.

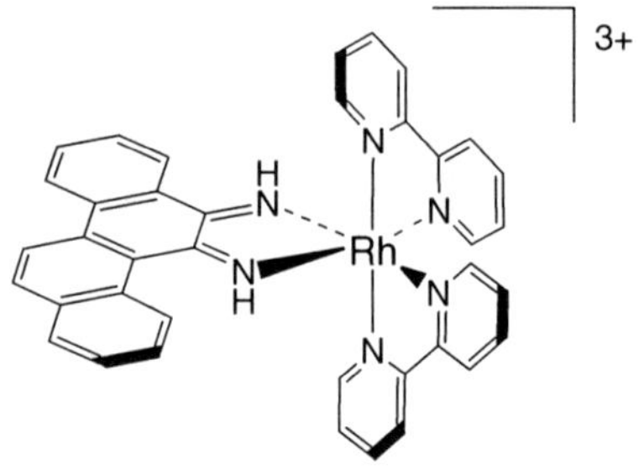

Δ-$[Rh(chrysi)(bpy)_2]^{3+}$

115

4.3.3.5.3 Gold Complexes. The established clinical use of gold(+1) drugs such as Auranofin in the treatment of chronic polyarthritis, coupled with their observed cytotoxicity, led to their being investigated as potential antitumor agents. Unfortunately, promising *in vitro* results did not translate into significant *in vivo* results. Gold(+1) complexes with the ligand bisdiphenylphosphinoethane (dppe) $[(AuCl)_2(\mu\text{-dppe})]$, **116**, showed significant activity against a variety of tumor cell lines, sometimes greater than that of dppe, itself a known antitumor agent. The complex [Au(dppe)$_2$]Cl, **117**, was similarly active against P388 leukaemia but only if applied directly to the tumor, intravenous, subcutaneous or intraperitoneal administration were ineffective. The complex inhibits protein synthesis more than DNA or RNA synthesis, forms DNA-protein crosslinks and causes DNA strand breaks. Unfortunately the cardiovascular toxicity of this class of compound has precluded clinical trials.

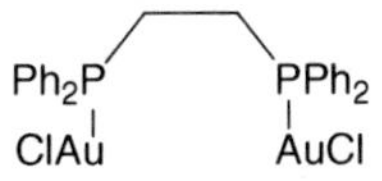

Ph$_2$P PPh$_2$ ClAu AuCl

116

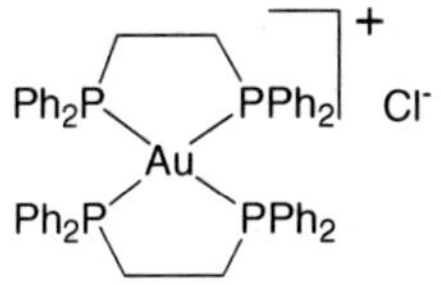

[Ph$_2$P PPh$_2$ Au Ph$_2$P PPh$_2$]$^+$ Cl$^-$

117

The electronic similarity between d^8 Au^{3+} and d^8 Pt^{2+} leads to them both adopting square planar coordination geometries and suggests the use of Au^{3+} compounds as possible anticancer agents by analogy with cisplatin. Complexes of the form [AuLCl$_3$] (L=*N*-methylimidazole, methylbenzoxazole or dimethylbenzoxazole) show reasonable cytotoxicities and antitumor activity *in vivo*. However, *in vivo* reduction of Au^{3+} to d^{10} Au^+ presents a significant problem in the application of Au^{3+} complexes and the choice of suitable ligands is important in controlling the chemistry of the system. The presence of the *ortho*-metallated benzylamine ligand (damp) in [Au(damp)X$_2$] (**118**; X=Cl, O_2CCH_3, SCN; X$_2$=C_2O_4, $CO_2CH_2CO_2$) prevents reduction of the Au^{3+} centre by thiols and these compounds are active against a panel of tumor cell lines. The acetate and malonate complexes show activity similar to cisplatin against the HT1376 xenograft *in vivo*. Their mechanism of action is different from that of cisplatin raising the possibility that such compounds might be of potential use against cisplatin resistant tumors.

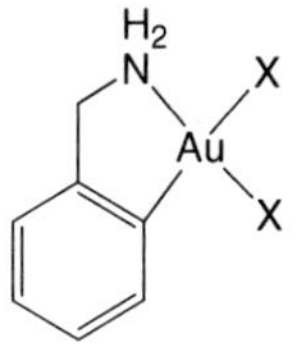

H$_2$ N X Au X

118

4.4 Rheumatoid Arthritis

4.4.1 Introduction

Rheumatoid arthritis is a debilitating inflammatory condition which affects about 1% of the overall population but is more common in women than men. Treatments may range from non-drug regimes such as changes in diet and taking relevant exercise, through the administration of non-steroidal anti-inflammatory compounds such as ibuprofen up to therapy using corticosteroids or specific disease modifying anti-rheumatism drugs. Included in this latter category are gold compounds, the use of which is sometimes referred to as chrysotherapy, a name derived from Chryses, a golden haired heroine in Greek mythology, It has long been know that certain gold compounds have anti-inflammatory properties and, in the 1930s, gold compounds began to be used as a treatment for rheumatoid arthritis. Chrysotherapy is not universally effective and around one third of patients treated this way may not benefit significantly. Side effects are, in the main, mild but in a few cases can be severe and even life threatening for about 1 in 10,000 patients. Despite this chrysotherapy can be highly beneficial in suitable subjects, leading to retardation and even remission of the disease. Careful monitoring of blood Au levels provides a means of controlling side effects. The therapeutic effects of chrysotherapy cannot be attributed to anti-inflammatory activity alone, so the gold drugs are considered to be disease modifying agents, although the exact mechanism by which this is achieved remains unclear.

4.4.2 Gold Chemistry

In the Periodic Table gold lies in Group 11 below copper which is an essential element for humans. However, gold is a much heavier element than its 1st row d-block counterpart copper and much closer in atomic mass to uranium. The electron configuration of elemental gold, contains a core with the xenon electron configuration in addition to filled $4f^{14}$ and $5d^{10}$ subshells, a single 6s electron and an empty 6p subshell. Ionisation of the 6s electron to give Au(+1) requires more energy than the 1st ionisation of Cu but the 2nd ionisation energies are similar so that the $5d^{10}$ subshell of Au can also considered part of its valence shell. In elemental form Au is rather unreactive but, in the presence of suitable complexing agents, the element dissolves in water under oxidising conditions. As an example, in the presence of H_2O_2, cyanide solutions dissolve Au(0) to form $[Au(CN)_2]^-$. Similarly thiolates such as cysteine react with Au(0) in air to give Au(+1) thiolate complexes. In fact examples of compounds with Au in oxidation numbers ranging from −1 to +5 are known but, in aqueous media and the context of biology, the most important arc +1 and +3. These oxidation numbers correspond respectively with $5d^{10}6s^06p^0$ and $d^86s^06p^0$ electron configurations for the gold ion. The (+2) oxidation number is not important in the chemistry of Au and compounds which at first sight may appear to contain Au(+2) generally contain both Au(+1) and Au(+3).

A consideration of crystal field theory leads to the expectation that, in the presence of strong field ligands, d^8 Au(+3) should typically show coordination number 4 with a square planar geometry (Section 2.5.4). The added stabilisation arising from complex formation allows an aqueous Au(+3) chemistry which is not seen with Cu(+3). Many square planar Au(+3) complexes are known, examples being provided by anionic $[Au(SCN)_4]^-$ and cationic $[AuCl_2(NC_5H_5)_2]^+$. In the case of d^{10} Au(+1) the filled 5d-orbitals are unable to contribute to bonding and only the 6s and 6p orbitals need be considered so that linear 2-coordination, trigonal 3-coordination or tetrahedral 4-coordination might be expected. In fact, because gold is such a heavy atom, its electrons are subject to additional effects predicted by the theory of relativity. These lead to a contraction of the Au(+1) ion so that the radius of Au in Au(+1) compounds is less than that of Ag in Ag(+1) compounds despite Ag having fewer electron shells. There is also an increased difference in energy between the 6s and 6p subshells in Au compared to expectations from a non-relativistic model. This leads to less involvement of the 6p subshell in bonding so favouring linear 2-coordination for Au(+1) over higher coordination numbers.

In water the Au(+1) ion is a strong reducing agent unless incorporated in a complex with an appropriate ligand such as CN^-, thiolate, or phosphine. Since Au(+1) is a large polarisable, 'soft' (Section 2.7.2), metal ion it tends to form its most stable complexes with 'soft' donor atoms such as P or S. In addition Au(+1) has a strong affinity for cyanide because of its negative charge and π-acceptor properties. Both CN^- and thiolate are important ligands in the medicinal chemistry of Au(+1). Disproportionation of Au(+1) to Au(0) (elemental gold) and Au(+3) is possible, particularly where a suitable ligand for stabilising Au(+1) is absent. This provides a pathway for converting Au(+1) to elemental gold even where direct reduction of Au(+1) to Au(0) should not occur. The Au(+3) ion is isoelectronic with Pt(+2) and forms many similar complexes but is more prone to reduction. The compound potassium tetrachloroaurate, $K[AuCl_4]$, dissolves in water and undergoes hydrolysis to form $[AuCl_3(OH)]^-$ in a similar manner to Pt(+2) chloro complexes. However, unlike $[PtCl_4]^{2-}$, $[AuCl_4]^-$ can also oxidise water to produce dioxygen. Thiolates readily reduce Au(+3) to Au(+1) forming disulfides (Equation (2)) and, depending on the nature of the ligands present, disulfides or thioethers may also reduce Au(+3) (Equation (3), for example).

$$[Au(L)_4]^{(4z-3)-} + 2RSH \leftrightharpoons [Au(L)_2]^{(2z-1)-} + RSSR + 2L^{z-} + 2\,H^+ \quad (2)$$

(L^{z-} represents a unidentate ligand and R a hydrocarbyl group)

$$[Au(L)_4]^{(4z-3)-} + R_2S + H_2O \leftrightharpoons \quad (2)$$

$$[Au(L)_2]^{(2z-1)-} + R_2S{=}O + 2L^{z-} + 2\,H^+ \quad (3)$$

4.4.3 Gold Compounds for the Treatment of Rheumatoid Arthritis

Various gold compounds have been used in the treatment of rheumatoid arthritis. All are Au(+1) complexes containing thiolate ligands, although the exact structures of all these compounds are not known. Sanochrysin, $Na_3[Au(S\text{-}SO_3)_2]$ **119**, is a purely inorganic compound and contains the bisthiosulfatogold(+1) anion in which Au^+ is bound to the terminal S atoms of two thiosulfate ions. Other clinically important gold drugs are the thiomalate complex Myochrysin, **120**, the thioglucose complex Solganol, **121**, and the thiopropane sulfonate complex Allochrysine®, **122**, all of which are oligomeric

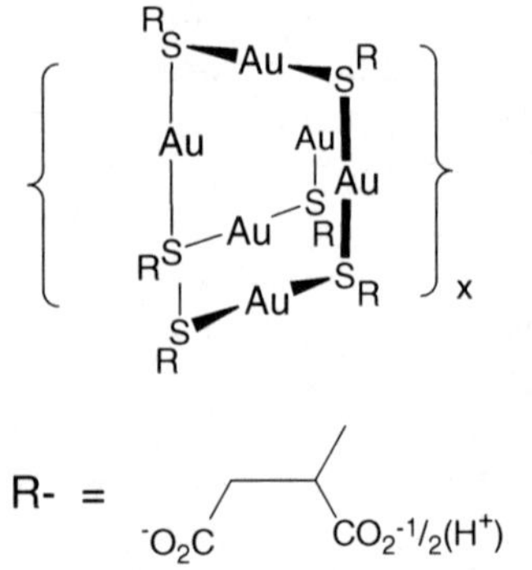

Figure 17 *A part of one of the two interpenetrating chains in the solid state structure of* $Na_{2n}Cs_n[Au_2\{SCH(CO_2^-)CH_2CO_2^-\}\{SCH(CO_2^-)CH_2CO_2H\}]_n$

and administered by intramuscular injection. The complexes form chains of notional formula $\{Au(SR)\}_n$ containing alternating Au^+ ions and RS^- ligands with the sulfur atoms acting as bridges between the Au^+ centres. Cyclic and open chain structures are, in principle, possible for these oligomers and the exact compositions of Myochrysin, and Solganol are more complicated than the notional formula $\{Au(SR)\}_n$would suggest, the Au:RS ratios not being exactly 1:1. The solid state structure of $Na_{2n}Cs_n[Au_2\{SCH(CO_2^-)CH_2CO_2^-\}\{SCH(CO_2^-)CH_2CO_2H\}]_n$, a Myochrysin analogue, has been determined and this contains interlocking spiral like chains with near linear S-Au-S angles of 170° and 179°and Au-S-Au angles of 99° (Figure 17). This structural arrangement is

Injected Drugs $(AuSR)_n$ Oligomers

119 Sanochrysin

120 Myochrysin RS =

121 Solganol RS =

122 Allochrisine RS =

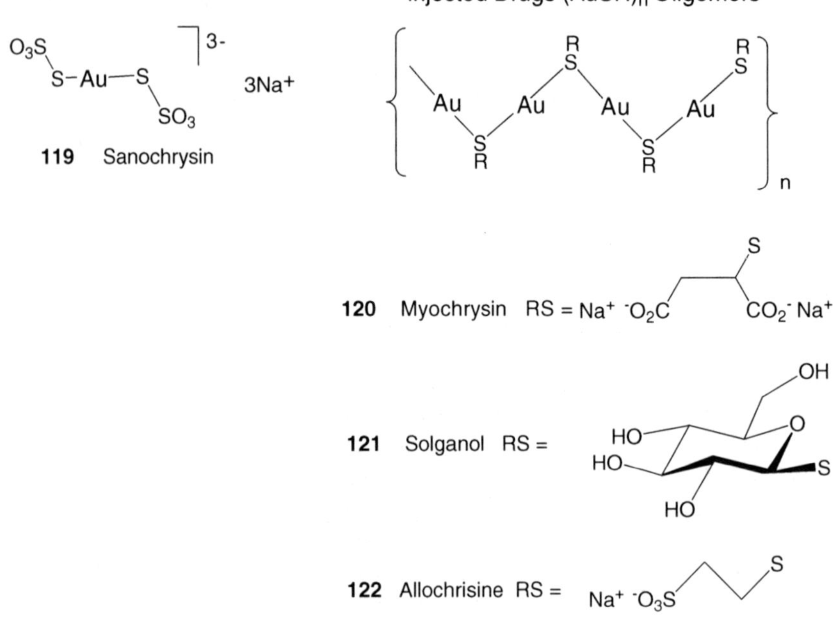

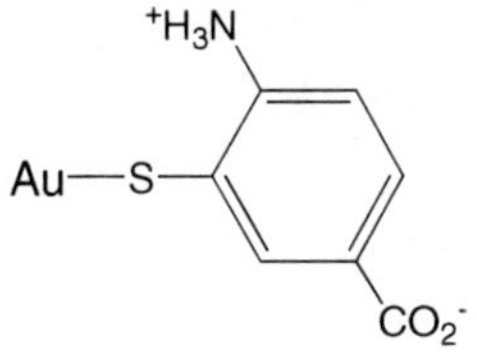

Oral Drug

123 **Krysolgan**

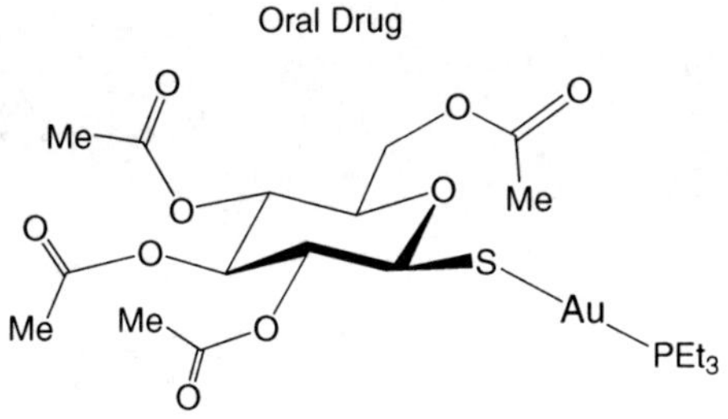

124 **Auranofin**

typical of Au(+1) thiolates and a number of model compounds similarly have near linear S-Au-S cores. Of course, once the compound is dissolved the structures of the complexes present in solution may be quite unlike those in the solid state, although oligomeric species are still thought to be present. Another example of a gold compound administered by injection is Krysolgan, **123**, which was initially used to treat tuberculosis early in the 20th century and is of historic interest. The most recently introduced drug Auranofin (also known as Ridaura), **124**, may be taken orally and has a monomeric structure which contains an essentially linear (174°) P-Au-S core in the solid state.

It turns out that the gold compounds should really be thought of as pro-drugs in that they do not represent biologically active species but rather they provide a source of biologically available gold. Thus the primary function of the ligands used in the gold drugs is to provide a soluble gold complex suitable for administration to the patient and sufficiently stable for transportation and storage prior to clinical use. Improved aqueous solubility is conferred by the carboxylate groups in Myochrysin, the hydroxyl groups in Solganol and the sulfonate group in Allochrysine. The triethyl phosphine ligand in the oral drug Auranofin is lipophilic and confers membrane solubility on the complex. After administration to the patient the ligands are soon lost from the complex *in vivo* and the Au^+ ion transferred to other binding agents. The affinity of Au(+1) for thiolate ligands suggests that and free thiol groups associated with proteins or GSH in the blood would be the first to compete with the thiolate ligands in the pro-drug for the Au^+ ions. The serum albumin is found to carry some 80–95% of gold circulating in the blood and, in the case of Auranofin added to whole blood, the transfer of the Au^+ to protein occurs within about 20 min.

In the structure of the principle form of albumin, mercaptoalbumin ($albS^-$), a cysteine residue is present at position 34 (Cys34) which carries an acidic thiol group. This is normally deprotonated at physiological pH and offers a potential binding site for Au^+. However, kinetic and spectroscopic studies of the interaction between mercaptoalbumin and Auranofin indicate that a rearrangement of the albumin structure to an active form, alb^*S^-, is required before the Au^+ ion is bound (Figure 18, Equation (4)). This involves the relocation of the Cys34 thiolate group within the protein to the protein surface where it is accessible to Au^+ species in solution.

$$\mathrm{albS^-} \leftrightharpoons \mathrm{alb^*S^-} \tag{4}$$

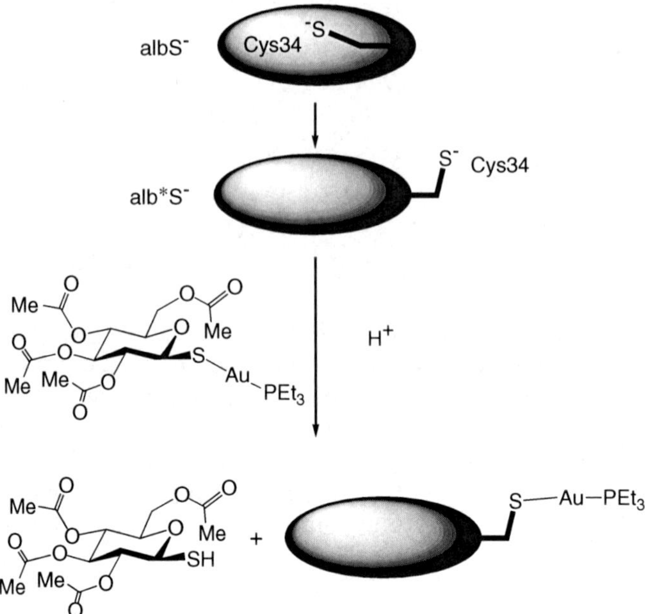

Figure 18 *A schematic view of the binding of Au^+ to an externalised thiolate group on albumin*

The exposed thiolate group of the activated protein is then able to compete with the tetra-acetoglucosethiolate ($tagS^-$) for the Au^+ (Equation (5))

$$alb^*S^- + tagS\text{-}Au\text{-}P(C_2H_5)_3 + H^+ \leftrightharpoons alb^*S\text{-}Au\text{-}P(C_2H_5)_3 + tagSH \quad (5)$$

The complex of the activated protein then relaxes to its final form (Equation (6)).

$$alb^*S\text{-}Au\text{-}P(C_2H_5)_3 \leftrightharpoons albS\text{-}Au\text{-}P(C_2H_5)_3 \quad (6)$$

In patients undergoing chrysotherapy with Auranofin the gold concentrations *in vivo* may be 5–15 μM compared to *ca.* 400 μM for mercaptoalbumin so that mercaptoalbumin can compete effectively with $tagS^-$ for the Au^+. The reactions are sufficiently rapid that the process appears to be first order with a rate constant of 2 s^{-1}. Similar reactions are thought to occur with Myochrysin in that a thiolate group of mercaptoalbumin binds to an Au^+ centre in a gold thiomalate oligomer. Redistribution of the remaining Au^+ centres to other mercaptoalbumin molecules follows ultimately transferring all the Au^+ from the oligomeric pro-drug to protein molecules.

The mechanism by which the Au^+ subsequently produces its anti-inflammatory and antiarthritic effects remains uncertain and various explanations have been proposed. These lie beyond the scope of this discussion but three aspects of gold chemistry deserving mention can be found among the proposals. One involves the binding of Au^+ to thiolate groups present naturally *in vivo*, a second to the formation of Au(+1) cyanide complexes and the third to the formation of Au(+3) complexes.

4.4.3.1 Gold Thiolate Complexes

The importance of Au^+ thiolate interactions is illustrated by two proteins known as Jun and Fos. These can combine to form the Jun–Fos conjugate and the homodimer Jun–Jun both of which can bind to DNA and stimulate the expression of genes which promote an inflammatory response. Jun and Fos both contain cysteine residues which appear at the DNA binding sites of Jun–Fos or Jun–Jun. Both gold(+1) thioglucose and gold(+1)thiomalate have been shown to react with Jun–Fos and Jun–Jun inhibiting their ability to bind to DNA. This effect can be reversed by adding excess thiolate such as GSH which presumably competes with the protein for the Au^+. Furthermore, replacement of the cysteines at the DNA binding sites of Jun–Fos and Jun–Jun with serine affords proteins which bind to DNA more strongly than the native proteins. However, the binding of these mutant proteins to DNA is no longer inhibited by gold(+1)thiomalate. This suggests that Au^+ binding to cysteine thiolate groups at the DNA binding sites of Jun–Fos or Jun–Jun inhibits their binding and so suppresses their ability to trigger inflammatory effects.

4.4.3.2 Gold Cyanide Complexes

Because of its high toxicity, cyanide is perhaps a surprising ligand to find bound to gold *in vivo*. However, gold cyanide complexes have been found in the urine of patients treated by chrysotherapy and appear to be formed at sites of inflammation. There is evidence that cyanide can be produced naturally *in vivo* from glycine and from thiocyanate present in extracellular fluids. Smoking offers a less natural source of CN^- and inhaled tobacco smoke may contain up to *ca.* 1700 ppm HCN which can enter the bloodstream *via* the lungs. Both Au(+1) and Au(+3) form complexes with cyanide and the linear Au^+ complex, $[Au(CN)_2]^-$ and the square planar Au^{3+} complex $[Au(CN)_4]^-$ are well known. The binding of CN^- to Au^+ is particularly strong and overall stability constant values of $\log \beta_2 = 36.6$ and 39 have been reported for the formation of $[Au(CN)_2]^-$ from Au^+ and CN^-. In a biological context CN^- can compete with the ligands in the gold pro-drug for the Au^+ centre but an equilibrium will arise between the CN^- and thiolate groups present. This leads to the formation of different complexes depending on the relative concentrations of the species present. The apparent equilibrium constant for the reaction of gold(+1) thiomalate, $[Au(Stm)]_n$, with HCN according to Equation (7) is reported to be 6×10^2 at pH 7.4.

$$1/n[Au(Stm)]_n + 2HCN \leftrightharpoons [Au(CN)_2]^- + HStm + H^+ \quad (7)$$

In the case of Auranofin the thiolate ligand is replaced by CN^- in preference to the phosphine ligand but if CN^- is present in excess both ligands can be substituted to form $[Au(CN)_2]^-$ as shown in Equations (8) and (9).

$$[Au\{P(C_2H_5)_3\}(tagS)] + HCN \leftrightharpoons [Au\{P(C_2H_5)_3\}(CN)] + tagSH \quad (8)$$

$$[Au\{P(C_2H_5)_3\}(CN)] + HCN \leftrightharpoons [Au(CN)_2]^- + P(C_2H_5)_3 + H^+ \quad (9)$$

Mixed ligand Au(+1) complexes containing a CN^- ligand as well as a thiolate (RS^-) or phosphine [*e.g.* $P(C_2H_5)_3$] are also known and undergo ligand redistribution reactions as shown in Equations (10) and (11) leading to an equilibrium mixture of the three complexes in each case.

$$2[Au(SR)(CN)]^- \leftrightharpoons [Au(CN)_2]^- + [Au(SR)_2]^- \tag{10}$$

$$2[Au\{P(C_2H_5)_3\}(CN)] \leftrightharpoons [Au(CN)_2]^- + [Au\{P(C_2H_5)_3\}_2]^+ \tag{11}$$

The formation of $[Au(CN)_2]^-$ appears to be important in the metabolism of gold drugs and may contribute to the cellular uptake of gold. Notably smoking has been found to increase the uptake of gold drug metabolites in red blood cells. This is attributed to the higher cyanide concentrations in the blood of smokers and red cells exposed to 4.5 μM $[Au(CN)_2]^-$ can absorb up to 95% of the gold present. Albumin also binds $[Au(CN)_2]^-$ and *in vitro* experiments suggest that as many as 10 ions may bind to an albumin molecule. Since albumin has multiple binding sites for anions such as Cl^-, Br^- and SCN^- this is not surprising, although $[Au(CN)_2]^-$ is a much larger ion. Thus albumin might transport gold, either through direct binding of Au^+ to thiolate residues, particularly at Cys34, or through the binding of $[Au(CN)_2]^-$.

The ease of cellular uptake of gold is dependent on the form of gold present in the extracellular medium. Oligomeric or mononuclear gold thiolates, $[Au(SR)]_n$ or $[Au(SR)_2]^-$, are not readily taken up by cells but Auranofin and $[Au(CN)_2]^-$ are taken up. It is thought that a shuttle operates in which Au^+ from extracellular gold complexes is exchanged onto thiolate groups associated with membrane transport proteins at the cell surface. These proteins transfer the Au^+ across the membrane to the inner surface where the Au^+ exchanges onto thiol groups associated with molecules within the cytoplasm. In the case of Auranofin the thioglucose ligand is lost before the Au^+ enters the cell but the $P(C_2H_5)_3$ ligand can accompany the Au^+ into the cell. Once in the cell the $\{Au[P(C_2H_5)_3]\}^+$ moiety may return to the extracellular medium by a similar process to its entry, or the $P(C_2H_5)_3$ ligand may be displaced and oxidised to $O{=}P(C_2H_5)_3$. The Au^+ then remains bound to thiolate groups present inside the cell but may undergo exchange with thiolate groups of the membrane transport protein once again and be transported back to the extracellular medium. The transport of gold across the cell membrane is not an active-transport process so that the concentrations of gold inside the cell and in the extra-cellular fluid ultimately reflect the thermodynamic equilibrium for complexation at intra- and extra-cellular gold binding sites.

4.4.3.3 Gold(+3) Complexes

It might be expected that gold(+3) would be unimportant in a medicinal context since many compounds present *in vivo* have the capacity to reduce Au(+3) to Au(+1) and Au(+1) is the form normally observed in biological media. However, it has been found that Au(+3) compounds can produce a popliteal lymph node assay response in animals treated with gold thiomalate for several weeks. Similar effects are not produced by gold thiomalate itself. This suggests that an Au(+3) metabolite is involved in the immunological response. It has also been found that human T-cells from chrysotherapy patients are sensitive to Au(+3) but not Au(+1). A mechanism by which Au(+3) might be produced *in vivo* has been proposed involving hypochlorite, OCl^-, generated by the enzyme myeloperoxidase during oxidative bursts. It has been shown that OCl^- can oxidise gold thioglucose and Auranofin to $[AuCl_4]^-$ or $[Au(CN)_2]^-$ to $[Au(CN)_2X_2]^-$ (X=Cl or OH) depending on the pH of the medium. Another relevant observation is that changes in tissue concentrations of thiols, proteins and metals during chrysotherapy are much larger in responding patients than the tissue concentration of gold. This implies that a redox cycle is operating in which Au(+1) is oxidised as a result of enzyme activity then reduced by thiols or other reducing species in the tissue so that repeated cycles of gold oxidation and reduction consume reductants in super-stoichiometric amounts compared to gold.

The exact role of Au(+3) produced in this way is uncertain but Au^{3+} has been shown to form peptide complexes such as **125** and **126**. This implies that Au^{3+} can bind to proteins and any such binding will change the structure and possibly the *in vivo* behaviour of the protein. It is also possible that Au(+3) could oxidise proteins, for example thiol groups might be oxidised to form disulfide links or the thioether group of methionine might be oxidised to sulfoxide, either process inducing a structural change. If a gold induced change in the structure of a protein prevents its being recognised as foreign by the immune system, this could result in immunosuppressive or anti-inflammatory effects. Binding of Au(+1) to proteins might also cause structural changes with similar consequences. In model systems, devised to test the ability of gold compounds to modify proteins or protein fragments which induce an immune response, a variety of other metal ions were found to exert an immunosuppressive effect. The utility of the gold drugs may arise from their having more suitable chemistry *in vivo* for allowing them to be transported to their site of action rather than that they have unique properties for suppressing the immune response.

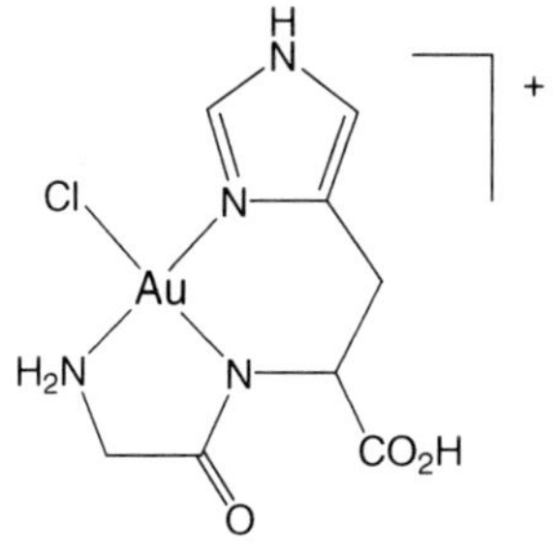

125 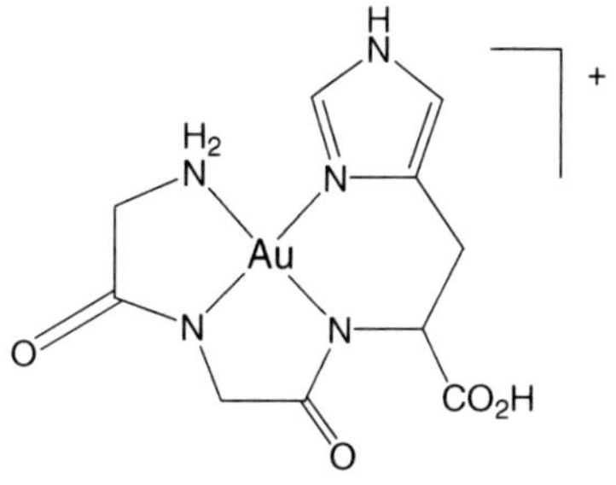 **126**

4.5 Diabetes

4.5.1 Vanadium and Diabetes

The signalling hormone insulin is essential for the metabolism of carbohydrate and fat. It is secreted by the pancreas in response to elevated blood glucose levels and promotes glucose uptake by the liver, gut or muscle tissue leading to either its storage or use in energy production as required. A deficiency of insulin, or cellular resistance to its function, results in diabetes in humans. In Type I diabetes, known as insulin dependent diabetes, the pancreas secretes insufficient insulin and regular injections of insulin that are necessary to compensate. In Type II diabetes, which accounts for some 90% of cases, sufficient insulin is secreted but the cellular response is impaired leading to hyperglycemia. Type II diabetes is therefore described as insulin independent and patients show 'insulin resistance'. Administering doses of exogenous insulin to the patient offers a means of treating both types of diabetes and is the primary treatment for almost all Type I and many Type II cases. Insulin is a protein and, as a consequence, orally administered insulin does not deliver a biologically active hormone so that unpleasant and less convenient subcutaneous injections are necessary. Many other types of compound have been investigated for the treatment of diabetes and an oral drug which can mimic the effect of insulin is particularly desirable. Unfortunately, the required combination of good absorption, low toxicity, stability *in vivo* and insulin mimetic behaviour has proven elusive. In many examples promising *in vitro* results have not been translated into effective treatments *in vivo*.

The discovery by Lyonnet and Martin in 1899 that diabetic patients excreted less glucose in their urine after treatment with vanadate, containing V(+5) in VO_4^{3-}, indicated that transition metal compounds may have an important rôle to play in the treatment of diabetes. However the subsequent discovery of insulin directed the focus of research away from inorganic compounds and only more recently has interest in metal compounds revived following the finding that micromolar vanadate could inhibit phosphohydrolases. In fact several transition metal compounds have been found to promote glucose uptake and these are sometimes referred to as insulin mimics. However, such compounds do not exhibit the full functionality of insulin and, for example, they do not counteract catabolic hormones such as glucagon or suppress glucose production in the liver. Nonetheless there is continued interest in metal complexes which might have therapeutic applications for Type II diabetes where insulin is secreted but ineffective. In particular metal compounds offer the possibility of an oral drug which could overcome insulin resistance and, in this respect, vanadium complexes have become a focus of research.

4.5.2 Vanadium Chemistry

As an early 1st row d-block metal vanadium can exhibit a wide range of oxidation states among which +3, +4 and +5 are accessible under physiological conditions. The oxidation state (+3) species typically contain simple V^{3+} ions in an approximately octahedral coordination geometry, although 7-coordinate complexes are also known. The oxidation state +4 species typically contain the vanadyl ion, VO^{2+}, in a square pyramidal coordination environment. Vanadium in oxidation state +5 is present in the tetrahedral vanadate ion, VO_4^{3-} which is usually present as a mixture of the protonated species HVO_4^{2-} and $H_2VO_4^-$ rather like the hydrogen phosphates HPO_4^{2-} and $H_2PO_4^-$. Complexes containing the vanadium(+5) dioxo ion, VO_2^{3+}, are also known and may exhibit coordination numbers ranging from 5–7 depending on the nature of the ligands present. Vanadium exhibits a rich redox chemistry but in the medical context the oxidation states V(+4) and V(+5) appear to be of primary importance, both being found to participate in extra- and intra-cellular equilibria. Vanadium has been shown to be important for development and growth in animal studies and the human body pool is typically in the range 100–200 micrograms. The toxic effects of vanadium depend on the route of administration and the chemical form in which the metal is initially present. The principle adverse effect is gastrointestinal distress. Vanadium accumulation in tissue, particularly in bone, does not appear to have significant harmful effects but could be a potential problem with its use in therapy if acceptable clearance rates cannot be demonstrated.

4.5.3 Vanadium Salts as Insulin Mimics

A variety of vanadium compounds have shown promising insulin mimetic activity *in vitro* but most have not proven to have suitable *in vivo* properties for therapeutic applications. Tests for *in vivo* insulin mimetic activity often involve the use of diabetic rodents as models. The diabetes may be spontaneously present in rats bred for the purpose or may be chemically induced. One widely accepted test for *in vivo* insulin mimetic activity involves rodents in which diabetes has been induced by treatment with streptozotocin (STZ). STZ is an antibiotic, which attacks the insulin secreting cells in the pancreas leading to reduced insulin secretion and the development of diabetic characteristics, including high glucose levels in blood and urine. Although not a complete model for Type I diabetes in humans, the STZ-diabetic rat does offer a reproducible and reliable test for *in vivo* insulin mimetic behaviour. Data from human studies is rather limited but studies of some vanadium compounds have been undertaken. Several inorganic vanadium compounds have been evaluated in limited clinical trials with otherwise healthy Type I and Type II diabetic subjects. Doses of 125 mg day^{-1} of sodium orthovanadate (Na_3VO_4) over 2 weeks produced increased mean glucose metabolism rates in two out of five Type I subjects and all five Type II subjects. Treatment of Type II subjects with 100 mg day^{-1} of vanadyl sulfate for 3 weeks produced improved insulin sensitivity, increased glucose disposal and reduced hepatic glucose production. These effects persisted for 2 weeks following the cessation of treatment but in both studies there was mild gastrointestinal intolerance. In another trial using

100 mg day^{-1} of $VO(SO_4).3H_2O$ over 6 weeks, improved insulin sensitivity was observed in three out of five Type II subjects who were receiving oral hypoglycemics. Lower doses of vanadyl sulfate have been tried and at 25 mg V day^{-1} there was no change in glucose and lipid metabolic parameters. However at 50 mg V day^{-1} increased insulin sensitivity was found but with no change in plasma glucose levels.

Compared to the limited range of inorganic vanadium compounds which might be used in therapy, incorporating vanadium into a complex with organic ligands offers a far more versatile means of delivering the metal. The absorption, biodistribution, tissue uptake, retention and insulin mimetic activity of vanadium might be improved by the correct choice of ligand and useful therapeutic effects might be achieved at lower doses. In recent years a variety of complexes have been synthesised and tested for insulin mimetic properties. Related molybdate and tungstate compounds generally show lower insulin mimetic activity than vanadium compounds but are less toxic and may warrant further investigation.

4.5.4 Vanadium Coordination Compounds as Insulin Mimics

One of the most studied insulin mimetic vanadium complexes is the neutral water soluble vanadyl complex of maltol (maH) BMOV, **127a** $[VO(ma)_2]$. This complex has a square pyramidal coordination geometry and the stability constants for its formation are given by $\log K_1 = 8.80$, $\log K_2 = 7.51$, $\log \beta_2 = 16.31$. The complex is oxidised by dioxygen to the V(+5) dioxo complex $[VO_2(ma)_2]^-$ which does not show insulin mimetic properties. BMOV was found to be two to three times more effective as an insulin mimic than vanadyl sulfate. In STZ-diabetic rats 50% blood glucose lowering was achieved at an initial oral dose in drinking water of 0.4 mmol kg^{-1} day^{-1} decreased to 0.2 mmol kg^{-1} day^{-1} for maintenance. No evidence of toxicity was found over a 6-month period. BMOV has also been found to prevent some pathological consequences of diabetes, such as cardiomyopathy, and to attenuate hyperinsulinemia and hyperlipidemia in genetically diabetic rats. Vanadium residence times of 31 days, 7 h and 4 h were found respectively in bone, liver and kidney from studies with radiolabelled $[^{48}VO(ma)_2]$.

Compared to BMOV its ethyl substituted counterpart BEOV, **127b** $[VO(ema)_2]$, shows longer turnover times, is a little less soluble, has slightly higher lipophilicity and is more stable towards hydrolysis. This illustrates the way in which small structural changes in the organic ligands can be used to modify the properties of a metal complex and improve its suitability for therapeutic applications. BEOV successfully completed Phase 1 clinical trials in 2000 and doses to healthy human subjects approached therapeutic levels at 90 mg day^{-1}. The compound was well tolerated by the gastrointestinal tract, liver and kidneys and no serious side effects were found at three times the expected therapeutic levels in blood. Administration prior to meals improved oral availability of the drug.

127 a BMOV, R = CH_3, $[VO(ma)_2]$
127 b BEOV, R = C_2H_5, $[VO(ema)_2]$

A variety of other vanadium complexes have been investigated as potential insulin mimics. These include picolinic acid derivatives which afford square planar V(+4) complexes such as VOPA, **128a**, and its more lipophilic analogue VOMPA, **128b**, as well as the 5-coordinate dipicolinate V(+5) complexes $[VO_2(dipic)]^-$, **129a**, and $[VO_2(dipic\text{-}OH)]^-$, **129b**. In STZ-diabetic rats VOPA, given at an oral dose of 10 mg V kg^{-1} day^{1} for 2 days followed by 5 mg V kg^{-1} day^{-1} for 11 days produced normal plasma glucose levels after 7 days. The rats remained almost normoglycemic for a further 30 days without further treatment. A higher oral dose of 25 mg V kg^{-1} day^{-1} gave faster glucose lowering but with diarrhoea as one side effect and other tests at 60 mg V kg^{-1} day^{-1} produced evidence of gastrointestinal irritation. In comparison VOMPA administered orally at a dose of 10 mg V kg^{-1} day^{-1} produced a more sustained response persisting for 80 days after treatment ceased. Compared to BMOV, VOPA has lower aqueous solubility and, at an equivalent dose, appears to produce more gastrointestinal irritation.

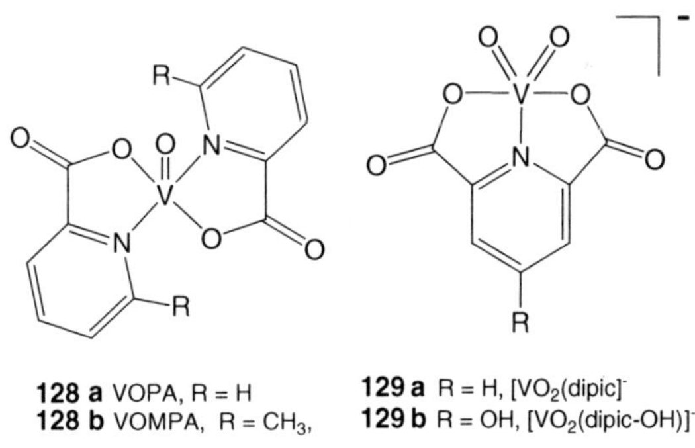

128 a VOPA, R = H
128 b VOMPA, R = CH_3,

129 a R = H, $[VO_2(dipic]^-$
129 b R = OH, $[VO_2(dipic\text{-}OH)]^-$

Other V(+4) complexes which have been investigated include the cysteine methyl ester complex VCME, **130a** and the related cysteine octylamide complex Naglivan, **130b**, a dithiocarbamate complex V-P, **131**, the tetradentate ligand derivative [VO(salen)], **132**, and the β-diketonate derivatives **133a** and **b**. Some dipeptide derivatives and [VO(edda)], **134**, have also been studied. At a dose of 10 mg V kg^{-1} day^{-1} VCME was more effective in normalising glucose levels in STZ-diabetic rats than related complexes with malonate, oxalate or tartarate ligands but, although toxic effects were not apparent at the effective dose, the compound was lethal at 10 times this dose. At this dose V–P produced normoglycemia in 2 days and this was maintained using a dose of 5 mg V kg^{-1} day^{-1}. However, oral administration was less effective than intraperitoneal injection which had the side effects of weight loss and increased bilirubin levels. [VO(salen)] was tested in alloxan-diabetic rats for 30 days at an oral dose of 7.6 mg V kg^{-1} day^{-1} and reduced blood glucose levels from hyper- to hypoglycemic levels. The complex [VO(edda)], **134**, has been tested *in vitro* for the inhibition of lipolysis and showed significant dose dependent activity. Similar chiral Δ-configuration complexes inhibited free fatty acid release at concentrations which decreased with increasing complex lipophilicity, an effect not found with similar achiral complexes. Several peroxo-complexes of vanadium have also been tested and shown promise in *in vitro* trials but have limited hydrolytic stability. They can also participate in reactions which may lead to radical formation making them unsuited for clinical use.

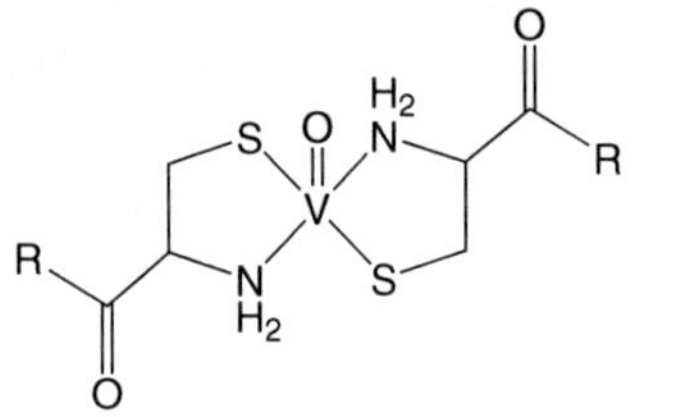

130 a VCME, R = OCH_3,
130 b Naglivan, R = NHC_8H_{17},

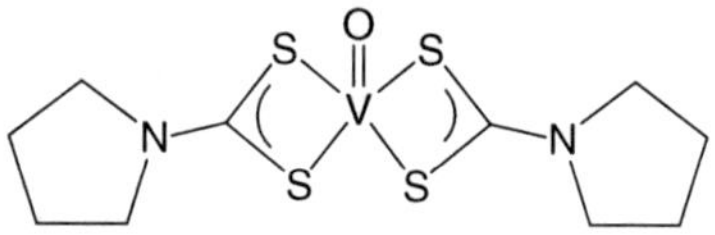

131
V-P

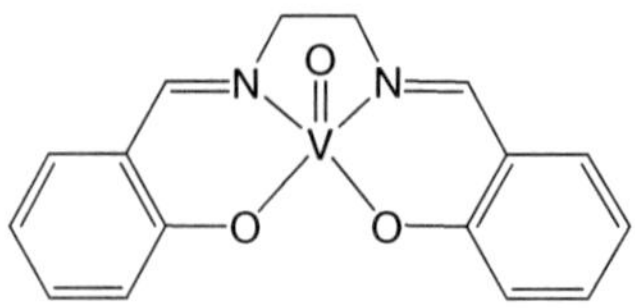

132
[VO(salen)]

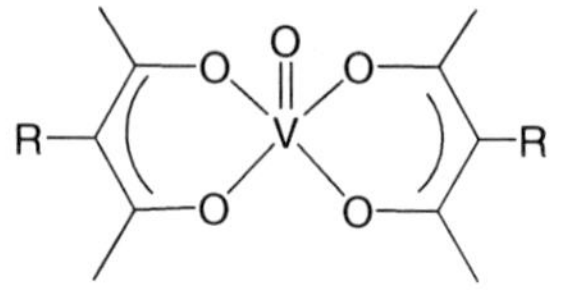

133 a R = H, [VO(acac)$_2$]
133 b R = C_2H_5, [VO(Etacac)$_2$]

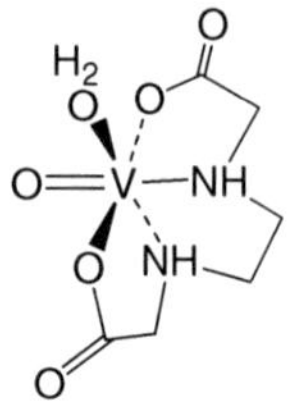

134
[VO(edda)]

An attempt has been made to demonstrate a synergistic effect by coordinating vanadium to Metformin (metfH), a ligand with known insulin enhancing properties in its own right. At an oral dose of 30 mg V kg^{-1} day ^{1} the complex [VO(metf)$_2$], **135a**, normalised plasma glucose levels in STZ-diabetic rats within 24 h but this effect was not maintained. At this dose level metfH had no effect. The limited data available showed the glucose lowering capacity of [VO(metf)$_2$] to be similar to that of BMOV.

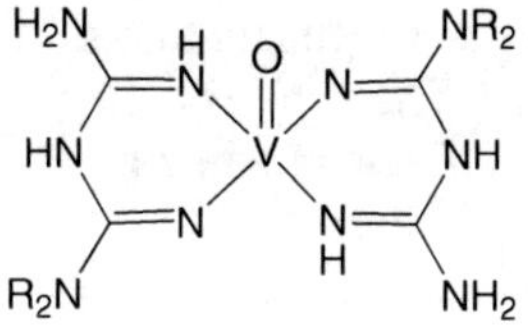

135 a R = CH_3, $[VO(metf)_2]$
135 b R = H, $[VO(big)_2]$

Compared to V(+4), few V(+5) compounds have been shown to have insulin mimetic properties. The V(+5) complex $[VO_2(dipic)]^-$ (**129a**) as been shown to produce glucose-lowering in cats and STZ-diabetic rats but is labile at pH 7 and is less effective than $VOSO_4$. Its hydroxyl substituted analogue $[VO_2(dipic\text{-}OH)]^-$ (**129b**) is more stable at neutral pH and this is associated with blood glucose lowering effects comparable with those of $VOSO_4$ but at a lower dose. A dose of 1.6 mmol V kg^{-1} day^{-1} of $VOSO_4$ resulted a reduction in blood glucose from *ca.* 4.7 mg ml^{-1} to *ca.* 3.0 mg ml^{-1} after 6 days. Similar results were obtained using $[VO_2(dipic\text{-}OH)]^-$ but at a dose of 1.0 mmol V kg^{-1} day^{-1}.

4.5.5 Mechanism of Action of Vanadium Compounds

Some vanadium compounds can mimic most of the metabolic effects of insulin *in vitro* and the exact mechanism by which vanadium produces this behaviour has been the subject of intensive study in recent years. However, the exact mechanism by which vanadium produces insulin mimetic effects *in vivo* is not yet fully understood, although some important biological effects of vanadium have been explained. A common feature of the organic complexes of vanadium, compared to inorganic salts, appears to be improved tissue permeation by passive diffusion without causing significant changes in toxicity indices.

At a superficial level, since both vanadium and phosphorus have five valence electrons, vanadate, VO_4^{3-}, exhibits an obvious similarity to phosphate, PO_4^{3-}. To some extent vanadate, which is mostly in the form $H_2VO_4^-$ under physiological conditions, can mimic $H_2PO_4^-$. However, the elements V and P are otherwise chemically rather dissimilar. In particular the redox properties of vanadium are quite different from those of phosphorus and, while the V(+4) $\{V{=}O\}^{2+}$ moiety can exist under physiological conditions, there is no counterpart for this in phosphorus chemistry. The vanadyl ion, $\{V{=}O\}^{2+}$ has also been said to resemble to Mg^{2+} and vanadium may have important effects on intracellular Ca^{2+} metabolism.

As with many other d-block metals, *in vivo* vanadium readily associates with proteins such as transferrin, albumin and hemoglobin in addition to GSH. Two proteins in the blood may be important in vanadium transport. One is serum albumin which has many potential metal binding sites, at least 5 of which can enter into weak non-specific binding ($\log K = 4.38$) of VO^{2+}. There is also a strong Cu^{2+} binding site which binds VO^{2+} more strongly ($\log K = 6.41$) than the non-specific sites. BMOV also appears to bind to serum albumin but only at the strong binding Cu^{2+} site. The other blood protein to consider is the iron transport protein transferrin. Apo-transferrin has two Fe binding sites and is known to transport two VO^{2+} units.

Both V(+5) vanadate and V(+4) vanadyl mimic the effects of insulin on hexose uptake and glucose metabolism in rat adipocytes. Initially inhibition of Na^+, K^+-ATPase by vanadyl(+4) was thought to be important but this has been shown not to be the case. More recent *in vitro* studies have shown that vanadate inhibits several enzymes in the liver, muscle and adipose tissue. These act collectively in the storage or utilisation of glucose and their function may be modified by the presence of vanadium.

Vanadate also blocks the actions of hormones which oppose the action of insulin. It has been shown that insulin binding to the cell surface site of the insulin activated protein tyrosine kinase leads to the autophosphorylation of three tyrosine moieties located in the β subunit of the insulin IGF-1 receptor. This triggers tyrosine kinase activity leading to the phosphorylation of tyrosine in several cytosolic docking proteins which are then recognised by effector molecules including P13-kinase. Woortmannin, a strong inhibitor of P13-kinase, blocks the activation of glucose uptake, glucose oxidation, lipogenesis and glycogenesis by vanadate but does not block its antilipolytic action. A cytosolic tyrosine kinase (cyt-PTK) that is activated by vanadate has been obtained from rat adipocytes and partly purified. Cyt-PTK is activated by V(+5) vanadate but not by V(+4). It has been suggested that cyt-PTK is involved in the activation of glucose metabolism by vanadate but not the activation of hexose transport or its antilipolytic action. It also appears that vanadium complexes can inhibit phosphotyrosine phosphatases so that enzymes remain in their phosphorylated active state.

An important feature of insulin mimetic vanadium compounds is the multiplicity of their effects in treating diabetes. In experimental animals not only were the primary symptoms alleviated but secondary effects such as cardiomyopathy, cataract formation, changes in kidney morphology and thyroid imbalance were also moderated. One important application of coordination chemistry in this context will be to use ligand design to improve vanadium bioavailability and *in vivo* distribution.

4.6 Cardiovascular System

4.6.1 *In vivo* Management of Gaseous Compounds

Two important applications of metal containing pharmaceuticals relating to the cardiovascular system are the management of nitric oxide, NO and dioxygen, O_2. Both exist as gases under ambient atmospheric conditions but must be stored or transported in an aqueous liquid medium. The other gaseous compound transported by the cardiovascular system is carbon dioxide, CO_2. This compound dissolves in aqueous physiological fluids in the form of carbonic acid H_2CO_3 which exists in equilibrium with bicarbonate, HCO_3^-. The enzyme carbonic anhydrase catalyses the hydration of CO_2 to H_2CO_3 facilitating its dissolution and allowing its efficient transport in dissolved form from the point of production to the lungs where it is expelled. A similar dissolution mechanism is not available for NO and O_2 and these gases are far less soluble in water than CO_2. It would be difficult to properly regulate the distribution and availability of NO and O_2 by managing them in the form of dissolved gases. Instead NO is produced where needed through the enzymatic oxidation of arginine and O_2 is transported or stored through binding to the iron centres in the metalloproteins hemoglobin and myoglobin.

Nitric oxide has a variety of functions *in vivo* and is typically produced where needed by members of a group of enzymes known as Nitric Oxide Synthase (NOS) enzymes. One function of NO is as a neuromodulator, for example in the hippocampus of the brain, where it is implicated in the development of short-term memory. Another important function of NO is to regulate the dilation of the blood vessels in the cardiovascular system. Excessive production of NO in response to infection or injury can result in arterial expansion and may ultimately lead to cardiovascular collapse. In order to maintain acceptable cardiovascular function under such conditions, it is important for the availability of NO to be controlled efficiently. This can be achieved using inhibitors for the enzymes producing NO *in vivo* but NO scavenging by suitable metal complexes offers another possible treatment.

Dioxygen supports cellular respiration in mammals and is transported throughout the body by the cardiovascular system. In the bloodstream O_2 is carried on the iron containing protein hemoglobin to its point of use. The ultimate product of dioxygen reduction is water but a proportion of the O_2 is converted to the reactive superoxide, O_2^-, ion. Normally O_2^- is rapidly removed by a SuperOxide Dismutase (SOD) enzyme to prevent damage arising from unwanted reactions of O_2^-. However, there can be occasions where O_2^- regulation is compromised, for example during reperfusion following acute myocardial infarct or stroke. Superoxide has also been implicated in the development of arthritis and neurological disorders such as Alzheimer's and Parkinson's diseases. If occasions arise when a patient's normal capacity for managing levels of O_2^- or NO is compromised, pharmaceutical intervention may be beneficial. The ability of certain d-block metal ions to react with small molecules such as O_2^- or NO, and to undergo electron transfer reactions, offers an important opportunity to develop metal containing pharmaceutical agents which address these issues.

4.6.2 Superoxide Dismutase Mimics

During normal respiration O_2 is consumed in living mammalian tissues but the biochemical processes involved may also produce superoxide, O_2^- or, in more acidic media, HO_2. Both of these are very reactive species capable of damaging DNA and initiating the auto-oxidation of membrane lipids and so are potentially damaging to tissues. Superoxide can also react with NO to produce highly reactive peroxynitrite, $O{=}NO{-}O^-$ which is also very damaging to tissues.

Protection from the destructive effects of superoxide is achieved *in vivo* by the rapid conversion (dismutation) of O_2^- into O_2 and H_2O_2 according to Equation (12).

$$2O_2^- + 2H^+ = O_2 + H_2O_2 \tag{12}$$

This reaction is catalysed by the SOD group of enzymes which exploit the electron transfer properties of certain metal ions. Examples are known with Mn, Fe, Ni or both Cu and Zn at the active site. The dioxygen produced in the reaction is, in effect, recycled while the peroxide is rapidly consumed *in vivo* by peroxidase enzymes carrying out biochemically controlled oxidations. Many forms of SOD enzymes are found in animal and plant species. In humans there are two main types. The enzyme found in extracellular spaces contains both Cu and Zn at the active site, while that found in mitochondria contains Mn. SOD enzymes containing Fe or Ni are also known to occur but in other organisms.

The human SOD in plasma and extracellular spaces effects the dismutation of O_2^- to O_2 and O_2^{2-} through successive electron transfers involving the copper centre according to Equations (13) and (14) (Scheme 12).

$$O_2^- + Cu^{2+}(SOD) \rightarrow O_2 + Cu^+(SOD) \tag{13}$$

$$HO_2 + Cu^+(SOD) + H^+ \rightarrow Cu^{2+}(SOD) + H_2O_2 \tag{14}$$

The d^{10} Zn^{2+} ion cannot undergo electron transfer reactions but has structural and electronic effects on the Cu^{2+} site. In the first reaction the Cu^{2+} centre is reduced to Cu^+ by O_2^- which is converted to O_2. In the second reaction the Cu^+ formed reduces a second O_2^- to O_2^{2-} reforming Cu^{2+}. These reactions are very fast *in vivo* so that the rate of removal of superoxide is essentially determined by the rate at which O_2^- can diffuse to the enzyme. Mitochondrial SOD employs a similar catalytic cycle involving Mn^{3+} and Mn^{2+} while iron containing SOD similarly utilises a cycle involving Fe^{3+} and Fe^{2+}. SOD enzymes containing Fe or Mn as the reactive metal react with O_2^- a little more slowly than the Cu/Zn enzyme.

In some cases of disease or trauma the production of O_2^- exceeds the capacity of the available SOD to carry out dismutations. This can result in tissue damage

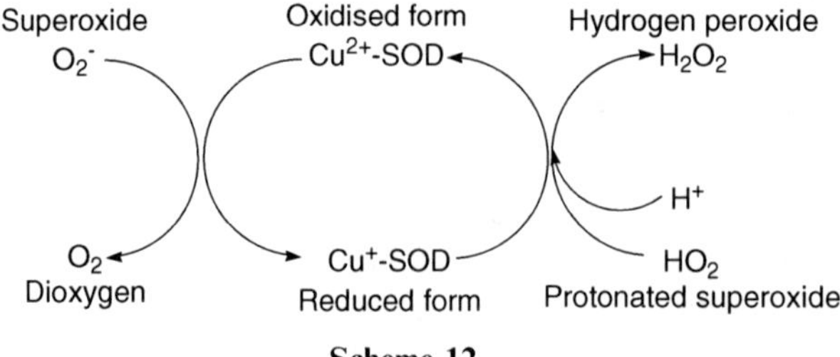

Scheme 12

through processes such as the oxidation of lipid membranes and the site selective oxidation of DNA. In such cases clinical administration of natural SOD enzyme might be used to alleviate the problem. However, the use of synthetic low molecular weight SOD mimics as pharmaceutical agents to dismute excess O_2^- may offer some advantages over the use of the SOD enzyme itself. The human enzyme preparations may induce immunological responses, are stable in blood for only short periods, are unsuitable for oral administration and are expensive to produce. Synthetic SOD mimics should produce no immunological responses, be more stable in blood and they might be expected to have better access to intercellular spaces and show better permeability into cells. It is also possible that compounds could be developed which would be suitable for oral administration. These possibilities have driven a large body of research into synthetic SOD mimics containing metal ions for potential use in pharmaceutical products.

The selection of a suitable metal for use in a SOD mimic has focused on those found in the naturally occurring SOD enzymes, particularly Cu, Fe and Mn. Free aquated Cu^{2+} ions are a very effective catalysts for O_2^- dismutation but would rapidly bind to serum proteins and other species *in vivo*. Similar problems arise with free aquated Fe^{2+}, Fe^{3+} or Mn^{2+} ions. There also particular toxicity issues with Cu^{2+}, Fe^{2+} and Fe^{3+} in that they can react with peroxide, a product of O_2^- dismutation, leading to the formation of reactive hydroxyl radicals which contribute to toxicity. In order to develop a SOD mimic for clinical use it is necessary to incorporate the metal in a suitable complex which has the required catalytic effect while showing low toxicity combined with acceptable biodistribution and pharmokinetics. The known structure of the Cu/Zn SOD enzyme might be thought a good starting point for the design of a SOD mimic containing Cu^{2+}. However, the structure of the Cu-Zn enzyme is elaborate and any attempt to reproduce the active site structure introduces an unwelcome level of complexity to the design of a simple synthetic SOD mimic. A variety of simpler copper complexes have been investigated as potential SOD mimics but evaluation of their catalytic performance has been complicated by the fact that aquated Cu^{2+} ions alone are very effective catalysts for O_2^- dismutation. Consequently it is often difficult to be sure that the activity arises from the complex not traces of free aquated Cu^{2+}. Iron complexes have also been investigated but slower dismutation kinetics and toxicity issues have limited their suitability. The metal emerging as the most promising for clinical use is Mn. The $Mn^{3+/2+}$ system is less prone to form hydroxyl radicals and has lower toxicity in the free aquated form.

In order to be of pharmaceutical value the metal complex should be relatively easy to prepare, very stable so as not to release the metal ion and produce unwanted toxicity and, importantly, the ligand system chosen needs to be resistant to attack by O_2^-, O_2 or HO_2^-. As with all metallodrugs, the biodistribution and pharmokinetic properties of the complex also need to be acceptable. Since high spin Mn^{2+} is a d^5 ion, weak field Mn^{2+} complexes would have no CFSE and be more prone to dissociation than strong field complexes.

The additional stability conferred by polydentate macrocyclic ligands is thus an important consideration for catalytic cycles involving Mn^{2+}.

A variety of ligand types have been investigated in the search for a suitable Mn based SOD mimic. Complexes containing macrocyclic porphyrin ligands **136** have been found to show catalytic properties. The complex Mn(tmpyp) catalysed O_2^- dismutation with an apparent rate constant of 10^7 M^{-1} s^{-1} at pH 7.6 ± 0.2 and 21 °C while Mn(Br_4tmpyp) is said to have a SOD activity of about 12% that of the Mn enzyme itself. Patents have been taken out for the therapeutic use of some of these compounds but there are limitations on the suitability of this class of compound. Under oxidising conditions porphyrin complexes of this type are prone to the formation of dimers through the formation of O^{2-} or OH^- bridges. These dimers are not active SOD catalysts. However, dimer formation could be suppressed by incorporating bulky substituents which prevent the close approach of two molecules necessary for dimer formation. Although the Mn(+3) complexes appear quite stable in competition experiments with $edta^{4-}$ this is not always true of the Mn^{2+} complexes which appear in the catalytic cycle. Furthermore the porphyrin ligands can be prone to oxidative degradation in some cases. In addition to these chemical issues there are problems with toxicity. Porphyrin complexes can act as DNA intercalators, and can show phototoxicity or hepatotoxicity. Thus, although Mn(+3) porphyrin complexes do represent potential SOD mimics in that they have the necessary catalytic activity, to obtain a suitable therapeutic agent it is necessary to overcome the toxicity and reactivity issues associated with porphyrin complexes. This presents a substantial challenge in ligand design.

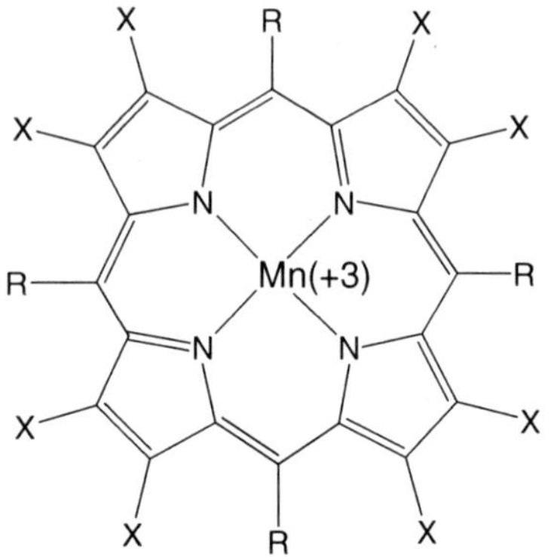

136

Mn(tmpyp) X = H, R = 4-pyridyl-N^+-CH_3

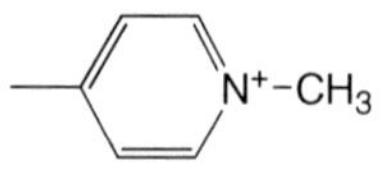

Mn(Br_4tmpyp) X = Br, R = 4-pyridyl-N^+-CH_3

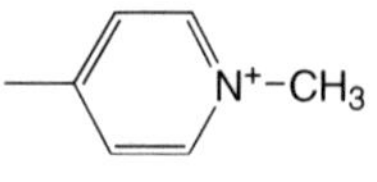

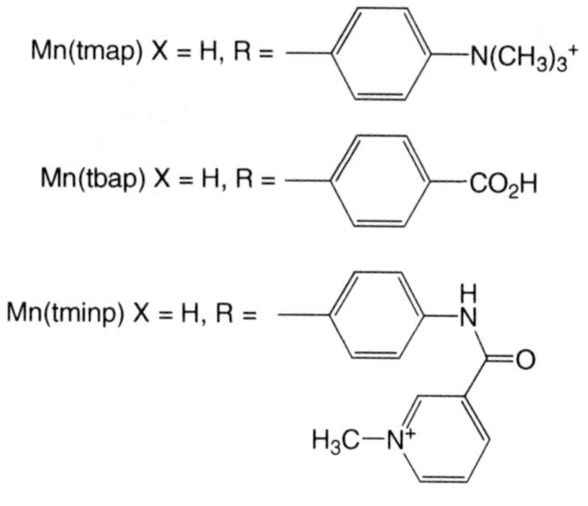

Mn(tmap) X = H, R = —C_6H_4—$N(CH_3)_3^+$

Mn(tbap) X = H, R = —C_6H_4—CO_2H

Mn(tminp) X = H, R = —C_6H_4—NH—C(=O)—(pyridinium, H_3C—N^+)

Another group of Mn complexes which have been investigated as SOD mimics contain what are known as 'salen' type ligands. These ligands are obtained from the condensation reactions of a hydroxybenzaldehyde and 1,2-diamino ethane to produce a tetradentate proligand,which forms a complex containing Mn^{3+} (Scheme 13). The structures and properties of the complexes may be varied by adding substituents to the ligand framework and through modification of the hydroxybenzaldehyde or diamine precursors (**137**, **138**, Scheme 13). Patents covering the therapeutic use of such compounds have been taken out by Eukarion Inc. and their development undertaken in collaboration with Glaxo-Wellcome. Promising results were obtained in animal models for oxidative stress including reperfusion injury and neurological disorders. However, kinetic studies reveal relatively low catalytic activity for these compounds, the rate for **137** (R=H) being less than 8×10^5 M^{-1} s^{-1} at pH 8. Furthermore the compounds do not appear to be genuine SOD mimics but rather their effect appears to arise from a more complicated mechanism which may not be catalytic in nature but may involve reactions with peroxide. Similarly, studies of $\{Fe(salen)\}^+$ reduction by O_2^- show that the reaction is too slow for the iron complex to be a competent SOD mimic. In addition to the uncertainties over the mechanism of action salen complexes also show low water solubility, which is problematic for clinical applications.

The most promising types of Mn^{2+} complex for use in SOD mimetic applications are the polyazamacrocyclic Mn(+2) complexes exemplified by **139**. In the solid state this complex contains 7-coordinate Mn^{2+} with a *trans* arrangement of the Cl^- ligands. It is quite thermodynamically stable with $\log K = 10.7$ at pH 7.4 and is also kinetically stable. Stopped flow kinetic analysis showed that the complex catalysed O_2^- dismutation with a rate of 4×10^7 M^{-1} s^{-1} at pH 7.4 in the presence of $>$100:1 excess $[O_2^-]/[$**139**$]$. The complex is an effective anti-inflammatory agent *in vivo* and was found to block cardiac reperfusion injury in a canine model. Investigations of the effects of ligand structure on the catalysis of O_2^- dismutation have shown that adding

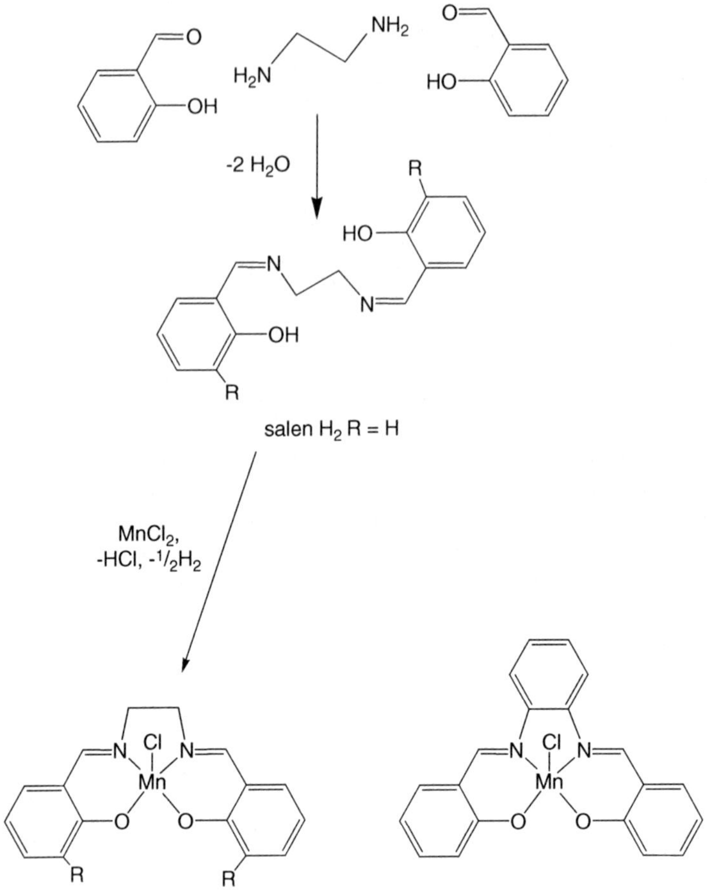

Scheme 13

substituents to the nitrogen donor atoms destroys the catalytic activity. However, adding substituents on the carbon atoms of the macrocycle can increase chemical stability. Adding one fused cyclohexane ring to give **140** results in a 2-fold improvement in kinetic stability, increases the thermodynamic stability to $\log K = 11.6$ and increases the catalytic rate to 9.1×10^7 M^{-1} s^{-1} at pH 7.4. Further improvements were achieved by adding a second fused cyclohexanyl ring to give **141**, the all *R*-isomer having $\log K = 13.3$ and a catalytic rate of 1.2×10^8 M^{-1} s^{-1} at pH 7.4. However, the structural effects of substituents can be quite subtle and the *R,R,S,S*-isomer of **141**, despite having similar stability, has almost no catalytic activity for O_2^- dismutation. Addition of two further methyl substituents to **141** further improves the thermodynamic and kinetic stability of the complex but reduces slightly the catalytic rate for O_2^- dismutation.

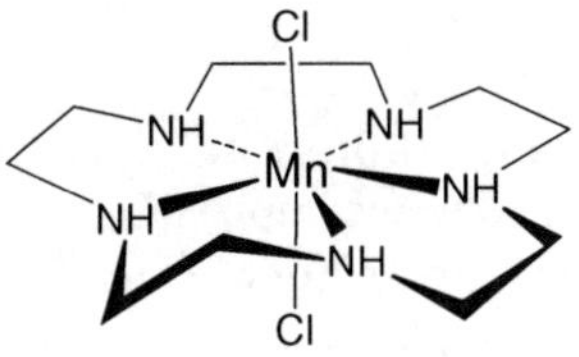

139 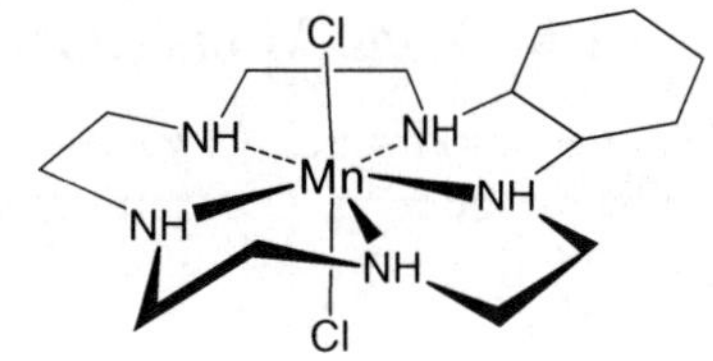 **140**

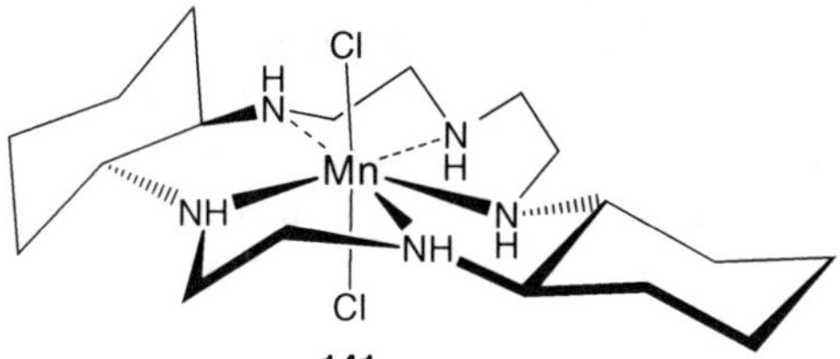

141

Mechanistic studies indicate that the O_2^- dismutation reaction with these Mn(+2) complexes involves two pathways, a minor inner sphere pathway and a faster outer sphere pathway. Both require oxidation of Mn(+2) to Mn(+3) and appear to involve a 6-coordinate intermediate the formation of which involves folding of the macrocyclic ligand. Molecular mechanics calculation show that the all *R*-isomer of **141** favours the folded geometry while the *R,R,S,S*-isomer is constrained to a planar arrangement favouring 7-coordination. In this way the substantial difference in reactivity between the two isomers can be explained.

The all *R*-complex **141** inhibits ischemic and reperfusion injury in anaesthetised cats and further elaboration of the ligand through the incorporation of a pyridyl ring into the structure has provided the SOD mimic M40403 (**142**). This compound has been developed by Metaphore Pharmaceuticals Inc. and clinical trials undertaken. Favourable results were obtained in a Phase II clinical trial of the enhanced analgesic effects of combining M40403 with morphine for post-operative pain relief from dental operations. A similar trial of this combination approach was conducted on a bunionectomy post-operative pain model and a further trial is in progress to study the relief of moderate to severe pain in cancer patients. The development of these SOD mimics demonstrates how it can be possible to design small molecules to mimic the function of much larger enzyme systems with the goal of obtaining metal complexes with therapeutic applications. It also illustrates the importance of ligand design in the formulation of such therapeutic agents.

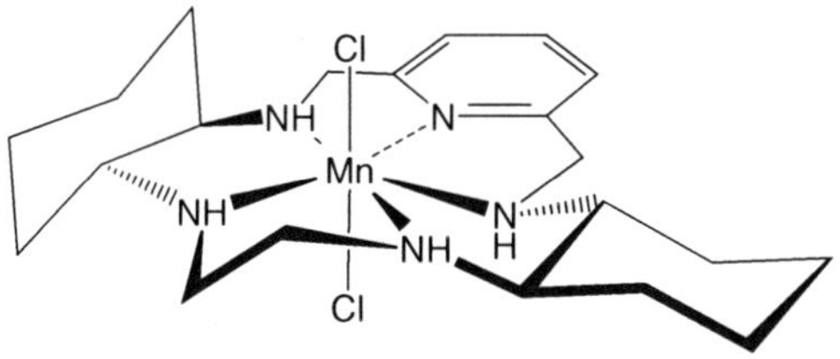

142
M40403

4.6.3 Nitric Oxide Management

Nitric oxide is important in a variety of physiological processes. Examples are provided by its action as a neuromodulator in the hippocampus of the brain, its production by macrophages to attack bacteria, fungi or tumor cells as part of the immunological response and its action as a vasodilator in the cardiovascular system. *In vivo* NO is produced by NOS enzymes through the oxidation of arginine to citrulline and NO. Endothelial, neuronal and immunological classes of NOS are known. These enzymes act to incorporate oxygen from O_2 into the NO and citrulline formed from the oxidation of argenine. *In vivo* NO has toxic effects, probably through its binding to intracellular iron or iron in non heme proteins and, although this may be beneficial in the immunological context, excessive NO production also has deleterious effects. One of these is to cause vasodilation and, in toxic-shock syndrome, excess NO production in macrophages produces arterial expansion which, in the extreme, can lead to cardiovascular collapse. Metal complexes which can absorb NO may thus play a rôle in treating toxic shock. Conversely it is sometimes necessary to stimulate vasodilation, to reduce blood pressure for example, and in such cases a metal complex capable of releasing NO under appropriate conditions would be of pharmacological value.

The complex used clinically to treat hypertension is sodium nitroprusside, $Na_2[Fe(CN)_5(NO)].2H_2O$ available as NITROPRESS® or NITROPRUSSIN. This compound is most often used because its effect is almost instantaneous and careful adjustment of the dose can allow a smooth reduction in blood pressure. However, because of its very rapid action, it is important that patients are carefully monitored and doses regulated properly to avoid irreversible ischemic injuries. Another potential problem with the use of $[Fe(CN)_5(NO)]^{2-}$ is the release of cyanide from the complex in conjunction with NO. If excessive doses are used it is possible for toxic levels of cyanide to be reached. The usual range of dose rates used is 0.5–10 pg kg^{-1} min^{-1} and at dose rates of <2 pg kg^{-1} min^{-1} cyanide toxicity is not usually an issue. However, infusion at the maximum dose rate for more than a few minutes might lead to unacceptable cyanide levels.

The release of NO from $[Fe(CN)_5(NO)]^{2-}$ (**143**) follows reduction of the complex *in vivo*. The bonding in $[Fe(CN)_5(NO)]^{2-}$ is complicated by the 'non-innocent' behaviour of the NO ligand. The complex can be thought of as containing a d^6 Fe(+2) ion bound to NO^+ acting as a 2-electron ligand in similar manner to CO. Calculations suggest that reduction of the complex adds an electron to an orbital which has both NO antibonding π^* and metal t_{2g} character. In the synergic bonding model (Section 2.5.2) addition of an electron to the NO antibonding π^* orbitals opposes back bonding from the metal to the NO weakening the Fe–N(nitric oxide) bond through reducing the π-interaction while leaving the σ-interaction intact. The resulting increase in electron density at the metal labilises the CN^- ligand *trans* to the NO. When this *trans*-CN^- dissociates to form a 5-coordinate complex the metal d_{z^2} orbital decreases in energy and becomes partially populated at the expense of the NO π^* orbitals. Since d_{z^2} contributes to the e_g^* orbital which is σ^*-antibonding with respect to the metal-ligand bonding, this has the effect of labilising all of the ligands leading to the release of NO and CN^-. The action of nitroprusside depends upon the subtle interplay between the π-acceptor properties of the NO^+ ligand, the π-donor properties of the metal ion and the lability of NO, which is, in effect, formed from coordinated NO^+ by reduction of the complex.

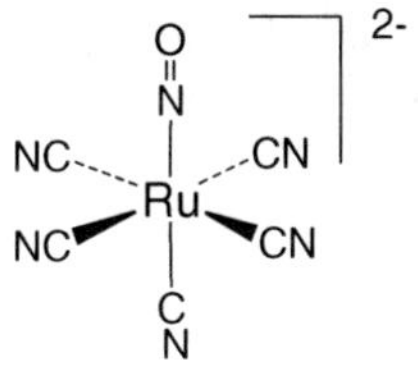

143

It would clearly be advantageous to develop a metal nitrosyl complex which could act as a source of NO *in vivo* without the associated toxicity effects due to cyanide. Attempts to identify such compounds have focused on Ru complexes. This metal lies immediately below Fe in the Periodic Table, it has a well established chemistry involving the NO ligand, its complexes tend to be more kinetically inert than those of Fe and it its compounds tend to have low toxicity. A simple cyanide free counterpart to $[Fe(CN)_5(NO)]^{2-}$ is provided by the complex $[\textit{trans}\text{-}Ru(NH_3)_4(L)(NO)]^{3+}$ (**144**) in which L is a ligand chosen to optimise the lability of the *trans*-NO. Examples of L are provided by $P(OC_2H_5)_3$ or aromatic nitrogen heterocycles such as pyridine or *N*-methyl imidazole. Such complexes do show vasodilator properties so, for example, $[Ru(NH_3)_4\{P(OC_2H_5)_3\}(NO)]^{3+}$ is easily reduced to $[Ru(NH_3)_4\{P(OC_2H_5)_3\}(NO)]^{2+}$ which undergoes a hydration reaction to form $[Ru(NH_3)_4\{P(OC_2H_5)_3\}(H_2O)]^{2+}$ releasing NO. In a similar manner to $[Fe(CN)_5(NO)]^{2-}$, in an ionic bonding model, $[\textit{trans}\text{-}Ru(NH_3)_4(L)(NO)]^{3+}$ could be thought to contain NO^+ coordinated to d^6 Ru(+2). Unfortunately the π-acceptor properties of $P(OC_2H_5)_3$ compete with *trans*-NO^+ ligand for electron density on the metal ion making it more susceptible to attack by OH^-. This leads to the formation of coordinated nitrite, NO_2^-, which readily dissociates to leave the Ru(+2) complex $[Ru(NH_3)_4\{P(OC_2H_5)_3\}(H_2O)]^{2+}$ but without NO evolution (Equation (15)).

$$[Ru(NH_3)_4\{P(OC_2H_5)_3\}(NO)]^{2+} + 2OH^- \rightarrow [Ru(NH_3)_4\{P(OC_2H_5)_3\}(H_2O)]^{2+} + NO_2^- \quad (15)$$

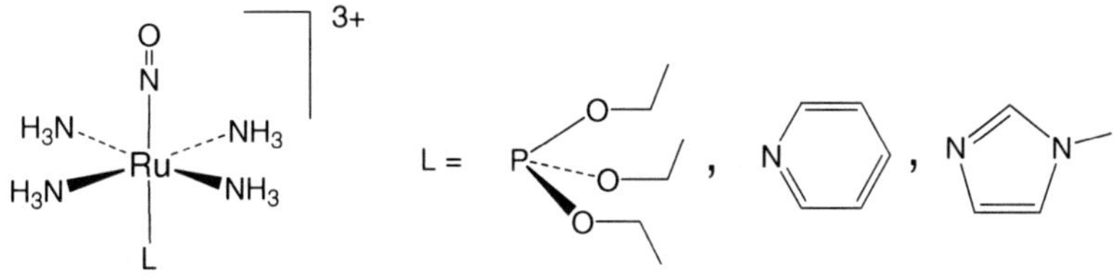

144

This problem can be alleviated by replacing the $P(OC_2H_5)_3$ligand with Cl^- which does not have π-acceptor properties. In addition the ammonia ligands can be replaced by the macrocyclic ligand cyclam to give a more robust platform for carrying NO in the form of $[Ru(cyclam)Cl(NO)]^{2+}$(**145**). Following reduction, this complex dissociates NO more slowly than $[Fe(CN)_5(NO)]^{2-}$ and produces a hypotensive effect in Wistar rats lasting some 20 times longer. After reduction the complex releases NO with a rate of $2.2 \times 10^{-3}\ s^{-1}$ at 35 °C with no associated cyanide release.

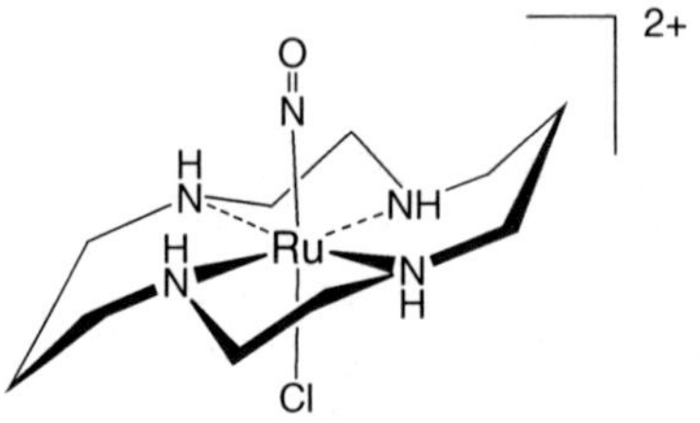

145 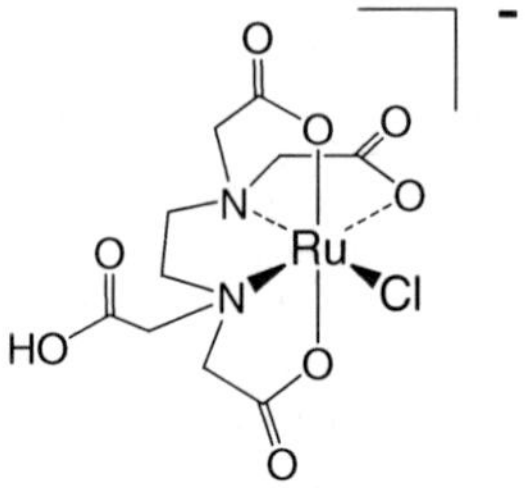 **146**

The presence of the neutral cyclam ligand in $[Ru(cyclam)Cl(NO)]^{2+}$ leads to the formation of a positively charged complex in which π-donation from the metal centre to the NO ligand is inhibited. By increasing the negative charge on the ligand set around the metal centre to promote back donation to coordinated NO, it is possible to strengthen the binding of NO in the complex. This provides a basis for the design of an NO scavenger to treat toxic shock arising from high NO levels in the bloodstream. Polyamine carboxylate ligands offer one approach in that they possess multiple negative charges and confer high thermodynamic stability on their metal complexes through a multiple chelate effect. The negatively charged carboxylate groups help polarise the molecule to encourage π-back donation to π-acceptor ligands such as NO^+. The water soluble Ru(+3) ethylenediaminetetra-acetic acid ($edtaH_4$) complex $[Ru(edtaH)Cl]^-$ (**146**) has been found to act as an NO scavenger and to inhibit the vasodilation effects of NO-releasing agents. In this chloro-complex one carboxylic acid group of the $edtaH^{3-}$ remains uncoordinated so that the Ru(+3) ion is 6-coordinate. In aqueous media the complex is easily hydrated and an equilibrium is established between $[Ru(edtaH)Cl]^-$ and $[Ru(edtaH)(H_2O)]$. The latter complex reacts with NO in a second order reaction (rate = $2 \times 10^{-7}\ M^{-1}\ s^{-1}$ at pH 7.4, 7 °C) to give [Ru(edtaH)(NO)] which can be thought of as containing Ru(+2), NO^+ and $edtaH^{3-}$. Acid dissociation of the pendant carboxylate affords the negatively charged ion $[Ru(edta)(NO)]^-$. Tests in rats used to model toxic shock syndrome showed that treatment with K[Ru(edtaH)Cl] reduced the time taken to return to normal blood pressure from *ca.* 26 to 9 h. These results demonstrate the potential for using Ru complexes to regulate NO levels *in vivo*. Careful selection of the ligands used allows the properties of the complexes to be modified to suit NO scavenging or NO releasing behaviour. These properties have potential applications in the treatment of toxic shock, the formulation of vasodilators and in treatments for neurological disorders arising from imbalances in NO levels.

4.7 Therapeutic Radiopharmaceuticals

4.7.1 Radiation Therapy

Conventional radiotherapy procedures involving external radiation sources play a vital rôle in the treatment of some forms of cancer. Typically the subject will be moved around in a narrow beam of radiation so that the cancerous region always lies within in the beam but other tissue is only transiently irradiated. In this way the whole body radiation is minimised while giving a therapeutic cytotoxic dose to the diseased tissues. Unfortunately this approach cannot be applied when the cancer is dispersed or metastatic. Radiopharmaceuticals provide a potential means of selectively targeting diseased tissue for irradiation through internal processes. These accumulate the radioactive agent in the diseased tissue where it can deliver a cytotoxic radiation dose. This general approach to managing the *in vivo* distribution of radionuclides is well established in non-invasive diagnostic medicine (Section 3.3). However, the use of cytotoxic radiopharmaceuticals in therapy is much more challenging because of the necessarily high radiotoxicity of the doses used.

In order to be acceptable for clinical use a therapeutic radiopharmaceutical must accumulate in the target tissue with high selectivity and at a rate which does not lead to unacceptable non-target tissue radiation doses. The agent must also be rapidly cleared from non-target tissues and excreted. It is also important that the products of chemical degradation and radioactive decay of the agent are of low toxicity and/or rapidly cleared form the body. These demands place restrictions on the nature of the radionuclide used in terms of its half life and decay products as well as on the biodistribution and pharmokinetic behaviour of the radiopharmaceutical agent itself. The energy and type of radiation emitted by the radionuclide, together with its cost, availability and chemistry, will also have a bearing on its acceptability for clinical use. The types of radiation emitted by radionuclides are summarised in Section 3.3.2.1 and radionuclides which emit either α or β-particles are of potential use in therapeutic applications. However, most clinical studies have focused on β-emitters and regulatory approval has not yet encompassed any α-emitting radionuclides.

4.7.2 α-Emitters

Radionuclides which emit α-particles are of potential interest in situations where very short range (<0.1 mm) cytotoxic effects are sought. As an example irradiation of cancerous bone surfaces to control pain while not irradiating the blood forming bone marrow could be of interest. However, few α-emitters have suitable half lives for therapy or can be obtained in sufficient quantity and radiochemical purity. The two which have attracted most interest are ^{211}At

($T_{1/2} = 7.2$ h) and ^{212}Bi ($T_{1/2} = 1$ h). The chemistry of astatine is similar to that of iodine and antibody labelling studies with this radionuclide have exploited this similarity. Although accumulation of ^{211}At-labelled antibodies in tumors has been found, ^{211}At levels in normal tissue were a cause for concern suggesting that intravenous administration of ^{211}At agents may be problematic. Direct injection into tumors may offer a more viable approach.

Antibody labelling with $^{212}Bi^{3+}$ has focused on complexation with bifunctional dtpa conjugates (Section 3.3.11.3) following the chemical precedent of $^{111}In^{3+}$ radiopharmaceuticals. The short 1 h half life of ^{212}Bi is a limitation on its use as complex formation can be slow and the time taken for efficient antibody uptake at the target site is a further problem with short half life radionuclides. An interesting solution to this problem might be offered by the use of ^{212}Pb ($T_{1/2} = 10.6$ h,) which decays to ^{212}Bi by β^--emission (0.17, 0.35, 0.589 MeV). In using this approach the chemical effects of the change from Pb^{2+} to Bi^{3+} will need to be considered in case they lead to unacceptable radionuclide redistribution on the timescale of the treatment.

Although there has been some research interest in the therapeutic use of α-emitters, this is a very challenging technology and obtaining regulatory approval for clinical use may not be easy. An alternative method of using α-particles in therapy is possible which does not involve administering radioactive materials to the patient. Rather it exploits the ability of non-radioactive ^{10}B, which constitutes 20% of natural boron, to capture a thermal neutron and convert to ^{7}Li with the release of an α-particle. Coupled with fact that ^{10}B presents a much larger cross section for neutron capture than the other light elements present in the body this offers a different approach to radiotherapy by an *in vivo* source. To a thermal neutron the ^{10}B nucleus appears enormous (3838 barns) compared to those of the other elements which are the main components of human tissue; *e.g.* ^{1}H (0.33 barns), ^{12}C (0.0034 barns), ^{14}N (1.8 barns), ^{16}O (0.0002 barns), and of bone ^{31}P (0.18 barn), ^{40}Ca (0.43 barns). If ^{10}B is incorporated in a compound, such as a tumor specific monoclonal antibody, which is selectively taken up in tumor tissue it can act as a target for thermal neutrons allowing selective α-irradiation of the tumor.

In practice the advantage of the high neutron capture cross section of ^{10}B in a pharmaceutical must be offset against its much lower concentration *in vivo* compared to other elements, even when selectively taken up by tumor tissue. The interaction of neutrons with other elements, cannot be ignored and, in particular neutron capture by ^{1}H and ^{14}N respectively results in the emission of γ-rays and protons. A ^{10}B concentration of around 30 mg kg^{-1} is needed for 85% of the neutron induced radiation dose to arise from ^{10}B α-emissions. This corresponds with a concentration of around 10^9 boron atoms per cell so that a large number of ^{10}B containing molecules need to be delivered to each target cell, selectively. Even though it is relatively easy to prepare compounds containing 10 or 12 boron atoms, and attach one or more of these units to a biologically active carrier, this presents a major challenge. Despite this substantial progress is being made in the development of bifunctional agents for what is called Boron Neutron Capture Therapy (BNCT). Boron compounds are well known to act as ligands towards transition metals and some complexes are quite stable. However, coordination chemistry is not essential to the synthesis of boron labelled pharmaceuticals and does not have a central role in producing agents for BNCT investigations. A nucleus with an even higher neutron capture cross section (225,000 barns) than ^{10}B is provided by ^{157}Gd present to the extent of 16% in natural gadolinium. Neutron capture in this nuclide produces γ-rays and low-energy electrons with very short range in tissue. There has been some research interest in complexes of ^{157}Gd for neutron-capture therapy but these typically contain only one metal ion compared to the multiple atoms easily possible for boron compounds. This reduces the benefit of the high neutron capture cross section.

Apart from the difficulty of loading enough of a suitable neutron absorbing nuclide into the target tissue, other major limitations on this potentially elegant approach to radiotherapy are the limited penetrating power of thermal neutrons, and the lack of suitable thermal neutron sources. The thermal neutron flux in human tissue falls to about 50% of that incident at the surface after about 3 cm, so it is difficult to treat deep-seated tumors by this means. At present nuclear reactors offer the principle source of sufficiently intense neutron beams but are very inconvenient for clinical applications. An efficient and compact particle accelerator driven neutron source, *i.e.* a spallation source would be better suited for medical use. Until these issues are resolved neutron capture therapy cannot enter general clinical practice although it continues to attract significant research interest.

4.7.3 β-Emitters

4.7.3.1 Choice of Radionuclide

The longer range (up to more than 1 cm) of β-particles in living tissue offers the advantage that a degree of 'crossfire' can enhance the irradiation of regions not well permeated by the radiopharmaceutical. Of course this also has the obvious disadvantage that non-target tissues close to the site of radionuclide uptake will also be irradiated to some extent. Examples of some β-emitting metallic radionuclides with nuclear properties relevant to therapeutic applications are shown in Table 4. These offer a range of half lives and β^--particle energies which affect the range of the radiation in tissue. A plot of the average range of β^--particles against their average energy taken from the data in Table 4 is shown in Figure 19. This shows the strong dependence of range on energy and so the importance of matching β^--particle energy to the dimensions of the target. The clinical utility of a radionuclide is also determined by its chemical properties in that there must be a suitable means by which they can be delivered to the target tissue efficiently. Currently the main approved therapeutic use of β^--radiopharmaceuticals is in the palliation of pain from bone metastases. However, there is much research interest in using bifunctional agents containing β-emitters for cancer treatment by what is known as radioimmunotherapy. The first FDA approval for the clinical use of a radiopharmaceutical of this type was granted in 2002.

Table 4 *Examples of β-emitting radionuclides of potential interest for therapy applications*

Radionuclide	Source	$T_{1/2}$ (days)	γ energy (keV)	γ yield (%)	β energy (MeV)	β yield (%)	Average β energy (MeV)	Average range (mm)	Maximum range (mm)
^{47}Sc	Cyclotron	3.4	159	68	0.6	40			
^{64}Cu	Cyclotron	0.5	511	38	0.57	40			
^{67}Cu	Cyclotron	2.6	184	48	0.57	20			
			92	23					
^{89}Sr	Reactor	50.5			1.46	99	0.58	2.4	6.7
^{90}Y	^{90}Sr decay	2.7			2.27	100			
^{105}Rh	Reactor	1.5	319	19	0.25	20			
			306	5	0.57	75			
^{111}Ag	Cyclotron	7.5	342	6	1.05	93			
^{117m}Sn	Reactor	13.6	159	86	0.13	Conversion		0.22	0.29
					0.15	electrons		0.29	
^{149}Pm	Reactor	2.2	286	3	1.07	89			
^{153}Sm	Reactor	1.9	103	29	0.68	32	0.22	0.55	3.4
					0.7	48			
					0.81	20			
^{166}Ho	Reactor	1.1	810	6	1.76	47	0.67	3.3	8.6
					1.84	52			
^{177}Lu	Reactor	6.7	113	6	0.5	86	0.14	0.35	
			208	11					
^{186}Re	Reactor	3.8	137	9	1.08	71	0.33	1.05	4.7
^{188}Re	$^{188}W/^{188}Re$ generator	0.71	155	10	2.12	100	0.64	3.8	11

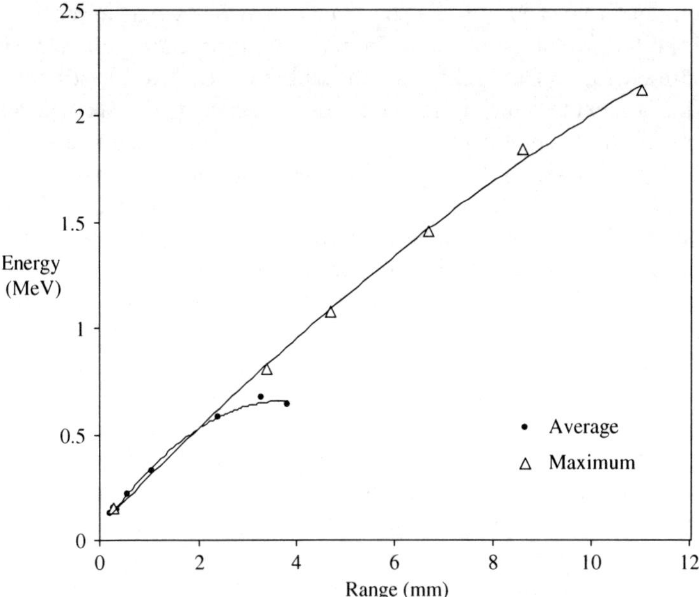

Figure 19 *A plot of range against energy for β-particles from some radionuclides used in therapy*
(Data taken from Table 4).

4.7.3.2 Palliation of Pain from Bone Metastases

Each year in the USA alone the combined total number of cases of breast and prostate cancer exceeds 350,000 and, of these, up to about 80% will develop bone metastases. These bone lesions are difficult to manage and a major cause of pain associated with cancer. Since bone contains phosphate as a major component, the non-metallic β^- emitter ^{32}P, in the form of ^{32}P-phosphate, has been evaluated as a means of alleviating this pain. However, the 1.71 MeV energy of the β^- particles emitted by ^{32}P makes them sufficiently penetrating to give unacceptably high doses to the blood forming bone marrow. The other major component of bone is calcium and the chemically related element strontium has a radionuclide, ^{89}Sr, with a β^- emission energy of 1.46 MeV which is less penetrating. Like its smaller ionic radius counterpart, the Sr^{2+} ion has a high affinity for newly forming bone, as found in metastases, and is deposited at or near the bone surface. Since Sr^{2+} is a hard, closed electron shell ion showing highly labile kinetics it is not well suited for incorporation in a kinetically stable complex to control its *in vivo* distribution. The Sr^{2+} ion is quite mobile *in vivo* and, having a lower charge, it is less prone to hydrolysis reactions than ions such as Fe^{3+} or Ga^{3+}, so that the suppression of hydrolysis by complex formation is unnecessary. It is over 50 years since $^{89}Sr^{2+}$ was first shown to relieve pain from skeletal metastases but FDA approval for its general use in the USA was only granted in 1995. The form of ^{89}Sr available for clinical use is an aqueous solution of $^{89}SrCl_2$ (Metastron, Amersham then GE Healthcare).

It is thought that $^{89}Sr^{2+}$ is taken up in the hydroxyapatite matrix of bone through similar mechanisms to Ca^{2+} uptake, and through exchange with Ca^{2+} already present. Like Ca^{2+}, Sr^{2+} is redistributed back into blood from normal bone. This appears to be a slow process as experiments incorporating the γ-emitter ^{85}Sr have shown that $^{85/89}Sr$ levels in bone remain essentially unchanged over a period of 100 days after the initial injection. The experiments also showed that the $^{85/89}Sr$ was 5 times more concentrated in bone lesions than in normal bone. A typical dose of 2.2 MBq kg^{-1} ^{89}Sr, or 150 MBq for a 68 kg subject, gives radiation doses of about 2.5 Gy to the bone surface and 1.6 Gy to the bone marrow. Patient response to the treatment takes about 2–4 weeks and may last for up to 6 months with myelosuppression as the main side effect.

In 1997 the ^{153}Sm complex of EDTMP (**147**, edtmpH_8) was the second radiopharmaceutical to receive FDA approval for bone pain palliation in the USA (lexidronam *a.k.a.* Quadramet, DuPont Pharma then Bristol-Myers Squibb). The average energy of the β^--particle from ^{153}Sm is less than for ^{89}Sr giving it a shorter range and resulting in a smaller radiation dose for a given amount of radioactivity. The edtmpH_8 proligand is similar to edtaH_4 except that the carboxylic acid groups, $-CO_2H$, have been replaced with phosphonic acid groups, $-P(=O)(OH)_2$. This change serves two purposes. Firstly, the phosphonate groups are compatible with phosphate in the hydroxyapatite matrix of bone and are thought to allow a transmetallation reaction which transfers Sm^{3+} to the bone where insoluble Sm-phosphates form. Secondly, the phosphonate group, $-P(=O)O_2^{2-}$, can offer two donor atoms to coordinate to a metal ion, whereas one is more usual for $-CO_2^-$. Thus while edta^{4-} can offer 6-coordination to a metal ion, it cannot satisfy the coordination number of 8 typically found for the Sm^{3+} ion. The edtmp^{8-} ligand is capable of satisfying the coordination requirements of Sm^{3+}. The Rb^+ salt of the Sm^{3+} complex with edtmp^{8-} has been formulated as $Rb_5[Sm(edtmp)].3H_2O$ on the basis of spectroscopic and analytical measurements and was found to be mononuclear in solution.

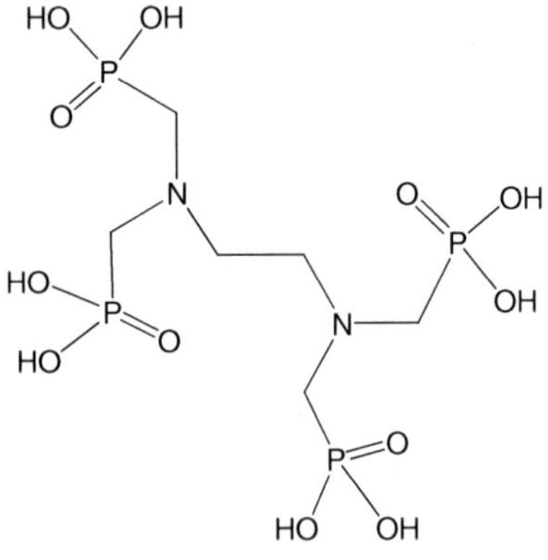

147
edtmpH_8

Formulations of ^{153}Sm-EDTMP usually contain a large excess of edtmpH_8 over $^{153}Sm^{3+}$ (up to 300:1) in order to suppress hydrolysis and liver uptake of the radionuclide. Under these conditions the agent shows high uptake in bone and non-localised agent is rapidly excreted *via* the kidneys. The agent appears to be excreted intact in urine and most of the non-bone radioactivity appears in the urine within 30 min of injection. The standard dose of ^{153}Sm-EDTMP is 37 MBq kg^{-1} and for a 68 kg subject this represents radiation doses of about 17 Gy to the bone surface and 3.8 Gy to bone marrow. The time for patient response to the treatment is shorter than for ^{89}Sr, about 2 weeks, but also shorter lived, up to 3 months. Again myelosuppression is the main side effect.

A third radiopharmaceutical for the palliation of bone pain has been approved in Europe and is under investigation in the USA. This is ^{186}Re-hydroxyethylenediphosphonate or ^{186}Re-HEDP (^{186}Re-Etidronate, Mallinckrodt/Tyco Healthcare; the proligand is hedpH_4, **148**). The preparation shows a similar biodistribution to that of ^{99m}Tc-MDP (Section 3.3.6) and, like ^{99m}Tc-MDP, ^{186}Re-HEDP is not a discrete complex but a mixture of oligomers. The standard dose of ^{186}Re-HEDP is 1.3 GBq and represents radiation doses of about 1.8 Gy to the bone surface and 1.7 Gy to bone marrow. The shorter half life of ^{186}Re compared to ^{89}Sr delivers the radiation dose more quickly resulting in shorter patient response times, some responding within 2 days. However, this also means that the treatment is effective for a shorter period, up to 2 months compared to as much as 6 months with ^{89}Sr. Once again myelosuppression is the main side effect. There has been some interest in the use of ^{188}Re-HEDP since ^{188}Re offers the convenience of being available from a generator. However, the higher energy more penetrating nature of the ^{188}Re β^--particle suggests that myelosuppression could pose a much bigger problem.

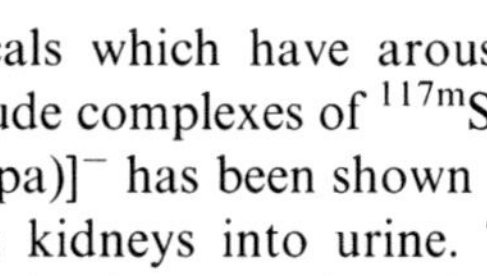

148
hedpH_4

Other radiopharmaceuticals which have aroused research interest in the context of bone uptake include complexes of ^{117m}Sn and ^{166}Ho. The negatively charged complex [^{117m}Sn(dtpa)]$^-$ has been shown to have high *in vivo* stability and is cleared through the kidneys into urine. The complex is thought to undergo transchelation at the bone surface depositing $^{117m}Sn^{4+}$ with the phosphate of the hydroxyapatite matrix. The advantage of using ^{117m}Sn is that it emits conversion electrons with specific low energies. These have a well-defined short range of about 0.3 mm offering reduced bone marrow radiotoxicity. Preliminary results from clinical trials with ^{117m}Sn suggest that this is indeed the case with a smaller percentage of patients experiencing moderate to severe myelotoxicity than with ^{89}Sr, ^{153}Sm or ^{186}Re.

The high-energy penetrating β^--emission from ^{166}Ho is unsuitable for the palliation of pain from bone metastases but has been shown to have potential for the treatment of patients with multiple myeloma. The complex ^{166}Ho-DOTMP (proligand dotmpH$_8$, **149**) localises in the skeleton and rapidly undergoes exclusively renal clearance to be excreted in urine. The long-range β^--emission have a strong myelosuppressive effect and can be used to eradicate multiple myeloma cells in the bone marrow. Like Sm^{3+}, Ho^{3+} is a lanthanide ion and can be complexed by edtmp^{8-} to suppress hydrolysis reactions and effect transport to the bone surface. However, both Sm^{3+} and Ho^{3+} form kinetically labile complexes so a ligand/metal ratio of 300:1 is required in the edtmp^{8-} formulation to prevent hydrolysis *in vivo*. The amount of ^{166}Ho needed to deliver a therapeutic radiation dose to the bone marrow would require the use of unacceptably large doses of edtmpH$_8$ so a more kinetically inert complex was required. The preorganised macrocyclic structure of dotmpH$_8$ can effectively saturate the Ho^{3+} coordination sphere and offers the required degree of kinetic stability, reducing the required proligand/metal ratio. The clinical utility, or otherwise, of this approach will no doubt be established in due course.

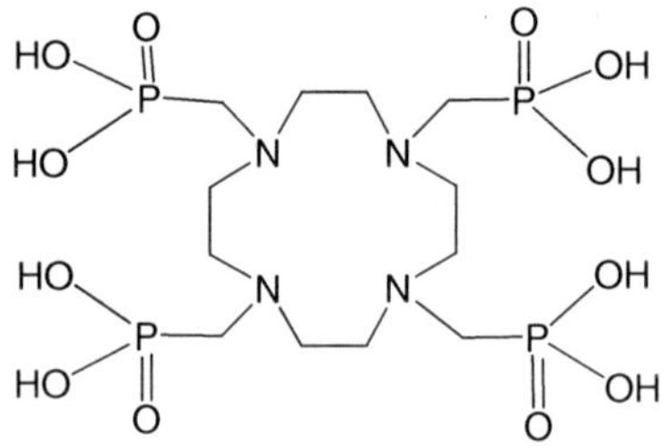

149
dotmpH$_8$

4.7.3.3 *Radioimmunotherapy Agents*

The concept of a 'nuclear magic bullet' carrying a radioactive element specifically to a diseased cell to deliver a cytotoxic radiation dose originated many years ago. Use of a radionuclide rather than a toxic chemical to kill the cell has the advantage that the 'bullet' does not have to enter the cell to be cytotoxic. Unfortunately developing a means to selectively deliver sufficient radionuclide to target tissue has proven to be a major challenge. The development of monoclonal antibodies and antibody fragments specific to particular tumor cell types provided a means of selectively carrying a radionuclide to its target for what is known as radioimmunotherapy. The requirements for a clinically acceptable radioimmunotherapy agent are very demanding. It is necessary for the antibody, or its fragment, to transport sufficient radionuclide to tumor cells to deliver a therapeutically effective radiation dose. This must be achieved with a specificity and on a timescale which does not deliver unacceptable radiation doses to non-target tissue. In order to minimise radiation doses to non-target tissues the radiolabel also needs to be securely attached to the antibody. However, the radiolabel should be readily released and excreted when the antibody reaches the liver, otherwise unacceptable radiation doses to the liver may result. In the past ^{131}I has been the radionuclide most widely used by researchers to prepare radiolabelled antibodies or their fragments for therapeutic purposes. Covalent bond formation between antibody and an ^{131}I compound offers a sufficiently chemically stable linkage. More recently interest has grown in the use of metallic radionuclides in radioimmunotherapy. The radionuclide most frequently chosen for this purpose is ^{90}Y, which is obtained by the decay of ^{90}Sr, itself a fission product of ^{235}U used in nuclear power generation. Thus ^{90}Y is readily available and can be obtained as a sterile high specific activity no carrier added preparation. The half life (2.7 days) of ^{90}Y is suitable for radioimmunotherapy purposes, the high β^--energy (2.27 MeV) gives it a range of over 1 cm and the decay product is the non-radioactive nuclide ^{90}Zr. Furthermore well tried methods for attaching $^{90}Y^{3+}$ to antibodies exist following the precedent of approved bifunctional imaging agents incorporating $^{111}In^{3+}$ (Section 3.3.11.3).

The polyaminecarboxylate ligand $dtpa^{5-}$ has been used as a basis for preparing bifunctional agents in which $^{111}In^{3+}$ ions are complexed by the dtpa part of a dtpa-protein conjugate formed using **6** of Chapter 3 (see Scheme 2, Chapter 3). A similar approach can be applied to the preparation of $^{90}Y^{3+}$ labelled antibodies. Both Y^{3+} and In^{3+} form thermodynamically stable complexes with $dtpa^{5-}$ but the complex with the larger Y^{3+} ion is kinetically somewhat labile. This can lead to losses of $^{90}Y^{3+}$ from the agent *in vivo* if **88** of Chapter 3 (R=H) (see Scheme 9, Chapter 3) is used to form the antibody-dtpa conjugate. The freed $^{90}Y^{3+}$ tends to accumulate in bone and can deliver unacceptable radiation doses to bone marrow. The addition of a methyl group to give **88** of Chapter 3 (R=CH_3) (see Scheme 9, Chapter 3) makes the backbone of the ligand less flexible and hinders the approach of competitor ligands rendering the complex more kinetically inert. This reagent has been used in the preparation of Zevalin® (Ibritumomab) which received FDA approval in 2002 for the 'treatment of relapsed or refractory low grade, follicular, or transformed B-cell non-Hodgkin's lymphoma'. The radiopharmaceutical targets the CD20 antigen on the surface of mature B cells and B-cell tumors and causes cellular damage both in the target and in neighbouring cells. The approval of this compound for clinical use in the USA heralds the regulatory acceptance of bifunctional radiopharmaceuticals in cancer radioimmunotherapy and the arrival of the 'nuclear magic bullet'.

Needless to say the story does not end with Zevalin®. Continued improvements are sought in the chemical methodology available for preparing immunoconjugates consisting of an antibody, or its fragment, and an attached metal binding site. Further kinetic stabilisation of metal complexes with dtpa-type binding groups is possible through increasing the rigidity of the ligand structure and introducing a degree of pre-organisation (Section 2.7.3). Thus in **150** a cyclohexyl group forces two nitrogen donors to adopt a structure pre-organised for chelation. This increases the kinetic stability of its complexes with Y^{3+} but the ligand system has two isomeric forms which show different stabilities *in vivo* when present in Y-labelled immunoconjugates. Although further optimisation of the dtpa-conjugate approach to radioimmunotherapy agents is possible an even greater degree of stabilisation through pre-organisation is possible by using dotaH_4 (see **5a**, Chapter 3) derivatives to incorporate a binding site for Y^{3+} or tripositive lanthanide ions such as Sm^{3+}. The dota^{4-} ligand forms very stable complexes with these metal ions and derivatives such as **151** offer a means of attaching this binding group to an antibody. The distance between the carrier and the metal binding site can be increased to reduce the likelihood that the complexed metal will interfere with the specificity or binding of the antibody to its target and examples include **152** and **153** although the latter has the disadvantage of using one of the-CO_2H binding sites to form the link to the antibody.

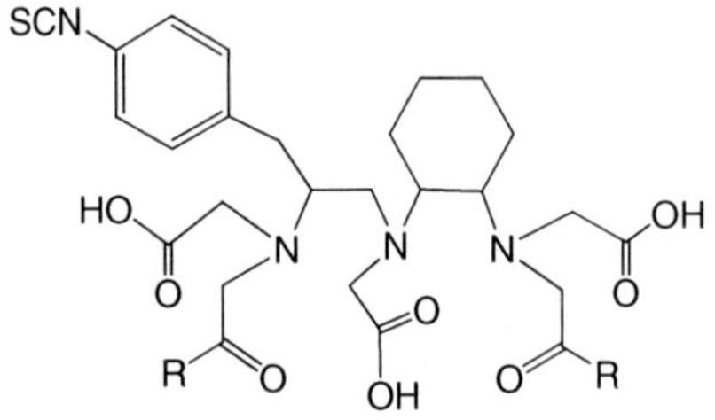

150

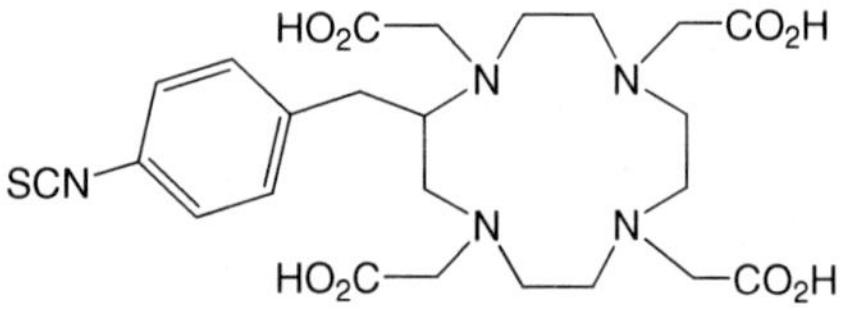

151

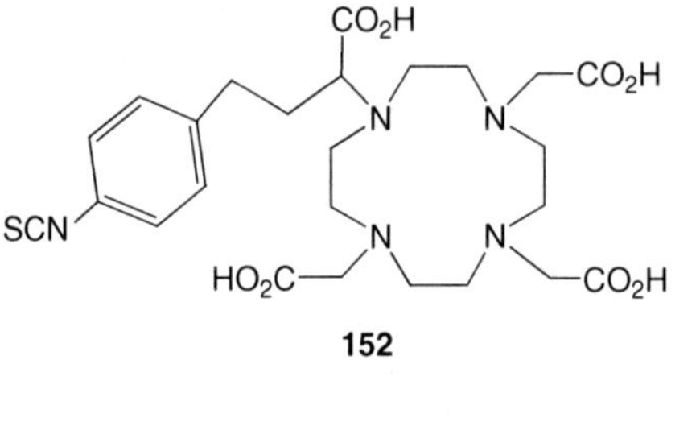

152

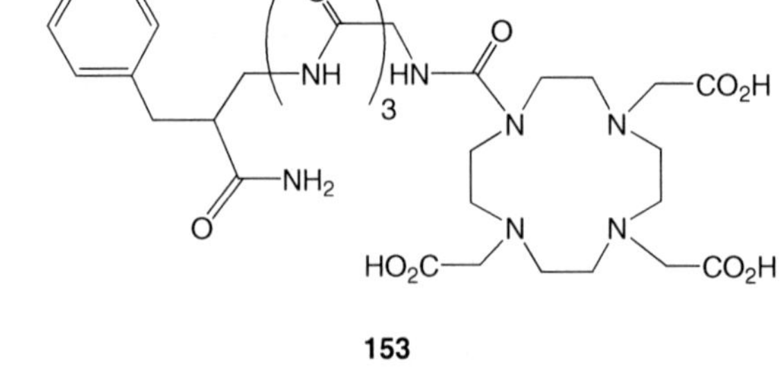

153

One problem found in experiments with radioactive metal-labelled antibody systems is the uptake of radiolabelled metabolic products in the cells of normal tissues. This can lead to unacceptable radiation doses to non-target tissues such as the liver and kidneys. Developments in the nature of the linking group between the ligand and the antibody may help alleviate this problem. For example, the chemical structure of the link might be designed to be easily metabolised in the liver to release the complexed radionuclide in a form which is rapidly excreted. This would help reduce the radiation dose to non-target tissue. The nature of the metal binding site can also be important in this respect. An example is provided by studies of ^{64}Cu and ^{67}Cu radiopharmaceuticals. The macrocyclic ligand cyclam, **154**, forms stable complexes with Cu^{2+} and has provided the basis for preparing bifunctional agents containing antibodies linked to $^{64}Cu^{2+}$ or $^{67}Cu^{2+}$. Antibodies coupled with the cyclam derivative **155** and labelled with $^{64}Cu^{2+}$ or $^{67}Cu^{2+}$ generally showed higher accumulations in kidney and liver than their counterparts in which **156** was used to incorporate the metal binding site. This was attributed to intracellular trapping of $^{64/67}Cu^{2+}$-cyclam containing metabolites, and it was though that the overall negative charge arising from the availability of four carboxylate groups in conjugates derived from **156** allowed more rapid *in vivo* clearance. Clinical trials have been carried out on the treatment of non-Hodgkin's lymphoma with antibodies labelled using the $^{67}Cu^{2+}$–**155** approach, and on the treatment of colorectal and bladder cancer with antibodies labelled using the $^{67}Cu^{2+}$–**156** approach.

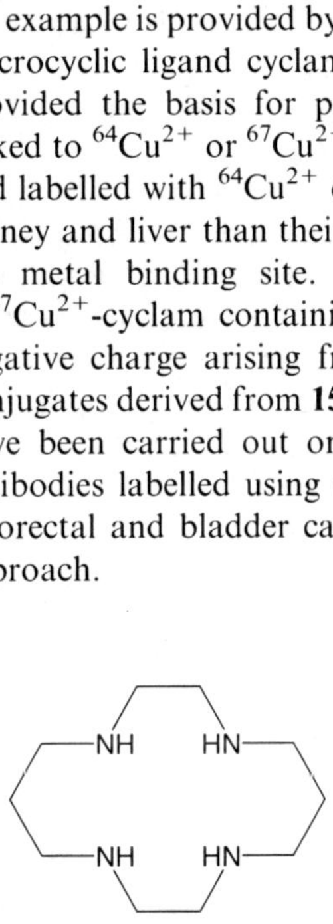

154

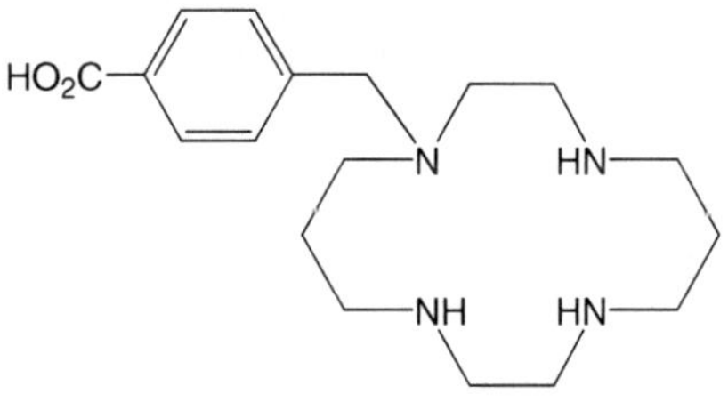

155

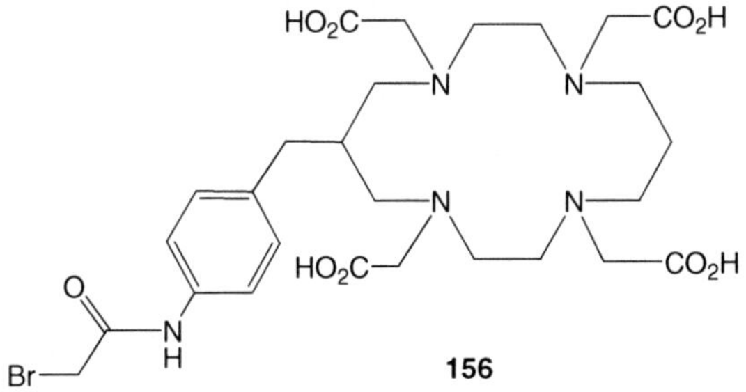

156

Although ^{90}Y has been the primary metallic radionuclide used in radioimmunotherapy, there is also much research interest in ^{186}Re and ^{188}Re.

In particular $[^{188}ReO_4]^-$ is available in high specific activity from $^{188}W/^{188}Re$ generators. These two radionuclides offer the possibility of using the same chemistry but with the option of two different radiation sources. The higher energy radiation from ^{188}Re is better suited to the treatment of larger tumors but the shorter half life requires faster clearance from blood and non-target tissues. Biomolecules which do not show such rapid clearance from the blood may be better labelled with longer half life ^{186}Re. An important driver for the interest in ^{186}Re and ^{188}Re is their chemical similarity to ^{99m}Tc which has many applications in diagnostic imaging. Both elements share the $(n+1)s^2nd^5(n+1)p^0(n = 4$ for Tc and 5 for Re) electron configuration, and the contraction in ionic radii across the lanthanide series of elements results in Tc and Re ions having very similar sizes. This gives rise to many similarities in their chemistry but there are some important differences. Complexes of Re tend to be more kinetically inert than their counterparts containing Tc, so complexation reactions may be less rapid. It might be expected that this property would improve the stability of Re containing radiopharmaceuticals but this is offset by another difference in behaviour. Higher oxidation states are generally more stable for Re than Tc and the formation of $[ReO_4]^-$ offers a mechanism for releasing Re from its complexes. This can be both advantageous and problematic. The formation of $[ReO_4]^-$ can promote excretion of Re *via* the kidneys as the radiopharmaceutical is metabolised and eliminated from the body. However, it can also lead to early release of the radiolabel so that Re-based radiopharmaceutical can show a greater tendency for Re redistribution *in vivo* than would be expected for Tc complexes. Ideally these factors need to be considered in the design of ligands systems for ^{186}Re and ^{188}Re radiopharmaceuticals.

Methods for producing bifunctional agents containing $^{186/188}Re$ mostly parallel the precedent set with ^{99m}Tc labelled bifunctional agents (Section 3.3.12). Thus direct labelling is possible through reduction of disulfide groups in protein followed by labelling with $^{186/188}[ReO_4]^-$ in the presence of Sn^{2+} as a reducing agent. Alternatively a biomolecule-proligand conjugate can be prepared, and labelled, or a radionuclide complex can be prepared in advance then attached to the protein. One of the most common approaches to the incorporation of $^{186/188}Re$ in bifunctional agents is to bind Re(+5) as $\{ReO\}^{3+}$ to the N_3S donor atom set in a mercaptoacetylglycylglycylglycine (MAG_3, **157**) based reagent. The activated ester **158** offers an example of a reagent used for attaching this binding group to amine groups in a protein. Another metal binding approach which is attracting interest is the use of $[fac\text{-}Re(CO)_3(H_2O)_3]^+$, the Re analogue of **42**, Chapter 3, as a precursor to very stable octahedral d^6 Re(+1) complexes. Hydrophilic bombesin-linked Re analogues of the hydroxymethylphosphine derivative **107** of Chapter 3 are also under investigation. These have been shown to have high affinity for their target cancer cells together with efficient clearance from the blood, mostly through the kidneys. Although it is still very much early days in terms of clinical applications, the $^{186/188}Re$ radionuclides do appear to offer considerable promise for developing radioimmunotherapy agents.

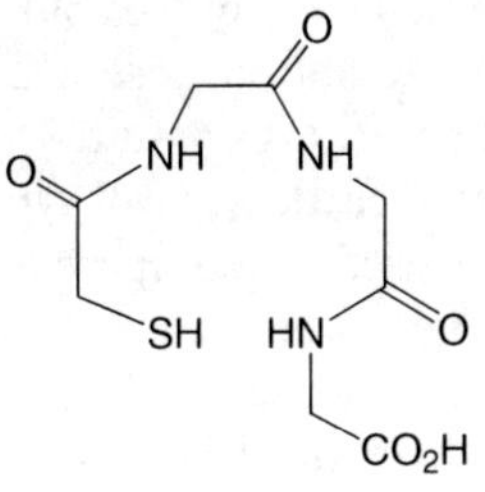

157 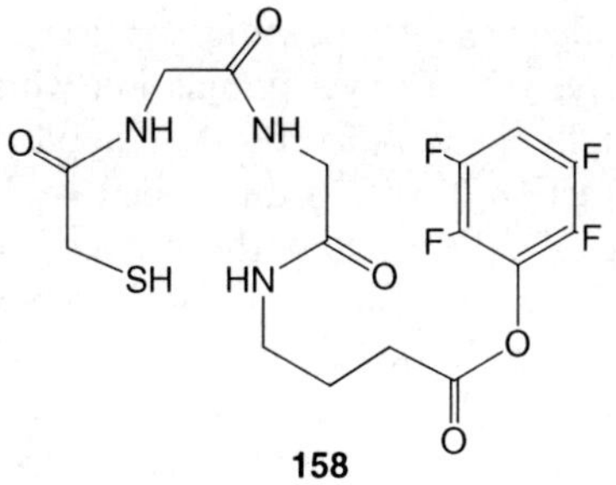 **158**

4.7.4 Radiosensitisers

It has been estimated that more than half of the damage caused to biological molecules by ionising radiation results from highly reactive species formed through the radiolysis of water, rather than from the direct interaction of the radiation with the biological molecule itself. Initially the interaction of ionising radiation with H_2O produces hydrated electrons, highly reactive H atoms and highly reactive $HO^{\bullet}$ radicals. The neutral $HO^{\bullet}$ radicals are extremely reactive and damaging to living tissue. In addition, the hydrogen atoms and hydrated electrons react very rapidly with O_2 to give, respectively, HO_2 and O_2^{-}, both of which are also highly reactive and damaging. Thus O_2 also plays an important rôle in promoting radiation damage through water radiolysis. The presence of O_2 is also important in promoting damage to species formed by the direct interaction of biological molecules with radiation. In the absence of O_2 these may revert to their original form but in the presence of O_2 non-reversible reactions occur to produce peroxide or hydroperoxide derivatives in which the damage becomes fixed. Thus the presence of O_2 is important in obtaining the maximum cytotoxic effect from a given radiation dose. The radiation dose required to kill 90% of cells in a population is around 3 times higher under an N_2 atmosphere than under normal aerobic conditions where O_2 is present.

Many tumors contain hypoxic regions where there is limited perfusion of oxygen. This can arise where tumor growth is too rapid for the development of capillaries delivering blood to keep up with the increasing tissue mass, or where irregularities in structure lead to impaired blood supply. The reduced O_2 levels in hypoxic regions make them radioresistant compared to aerated tissues. Since hypoxic regions can constitute up to a third of the tumor mass they can significantly reduce the effectiveness of radiotherapy treatments. In order to address this problem radiosensitisers have been sought which can induce improved radiotoxicity in hypoxic tissues. The basic requirements for an effective radiosensitiser are that it should have low toxicity to normal tissue, show preferential uptake in hypoxic and malignant tissue and undergo reduction so as to act as O_2 mimics. Because nitro groups in organic compounds and quinones are readily reduced these have often featured in the development of radiosensitisers. As an example the nitroimidazole derivative **159** is an effective radiosensitiser and has entered Phase III clinical trials.

The effectiveness of a radiosensitiser can be evaluated from curves plotting the percentage of cells in a population which survive a radiation dose against the magnitude of the radiation dose applied. The radiation dose required to kill a given fraction (for 90% this would be the LD_{90}) of a particular cell type in the absence of oxygen, and the absence of radiosensitiser, is compared to that required when the radiosensitiser is present. This ratio is known as the Sensitiser Enhancement Ratio (SER) and allows different radiosensitisers to be compared. The ratio of the radiation doses required to produce the LD_{90} value in the absence and presence of O_2 is called the Oxygen Enhancement Ratio (OER).

A wide variety of compounds have been investigated for use as radiosensitisers, including a number of metal complexes. In particular metal complexes with tumoricidal properties have featured prominently in research studies and their electron transfer properties have been a feature of interest. Some selected examples follow to illustrate the kind of results which have emerged. A large number of Pt complexes have been studied, some containing the nitroimidazole group as in **159**. The incorporation of this type of ligand in complexes containing the $\{PtCl_2\}$ unit, found in the antitumor drug cisplatin, **23**, has revealed that the monocomplex **160** is a better radiosensitiser than its counterpart in which both NH_3 ligands are replaced by the nitroimidazole ligand. The isomeric form of the complex can also have a bearing on its activity and the amine-coordinated isomer **161** is a poorer radiosensitiser than its counterpart **162**. Similarly the *trans*-isomer of the 5-nitroquinoline containing complex **163** is a more effective radiosensitiser under hypoxic conditions than the *cis*-isomer **164** which has SER =1.7. The free 5-nitroquinoline ligands were ineffective as radiosensitisers. Nitroimidazole complexes of other metals can also show activity and the Ru(+2) complex **165** was found to have an SER of 1.6 compared to 1.4 for 2-nitroimidazole alone.

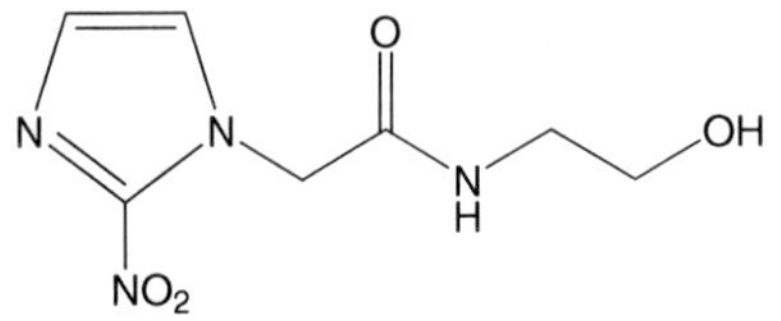

159 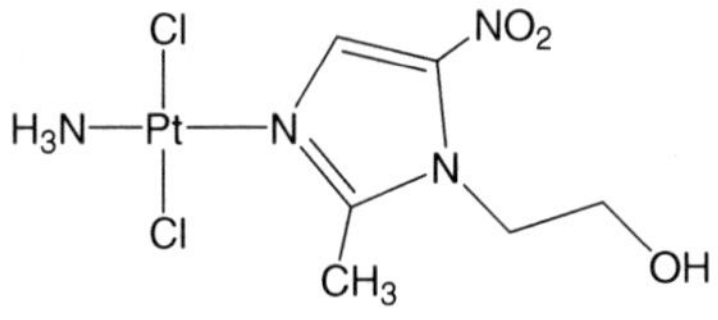 **160**

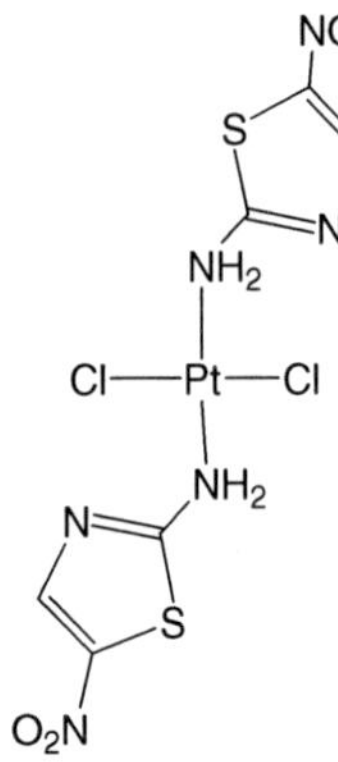

161 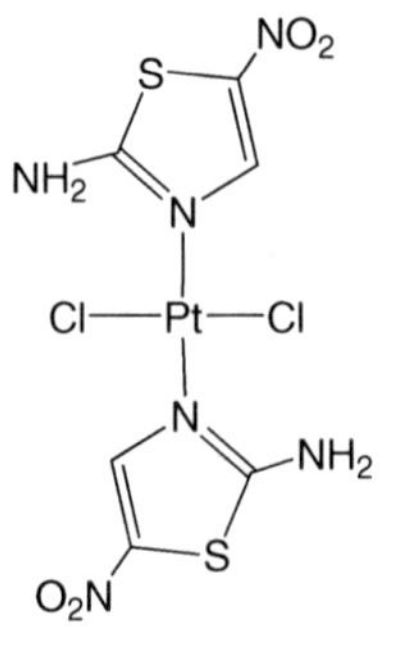 **162** 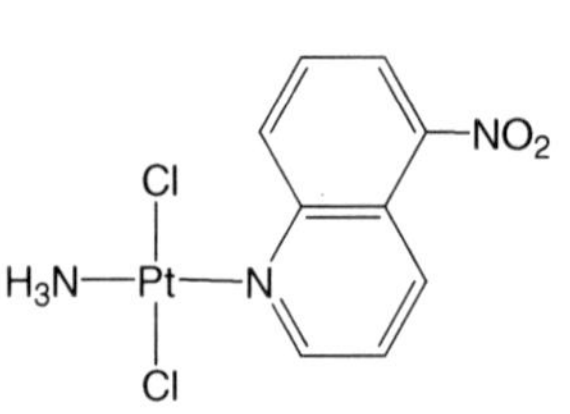 **163**

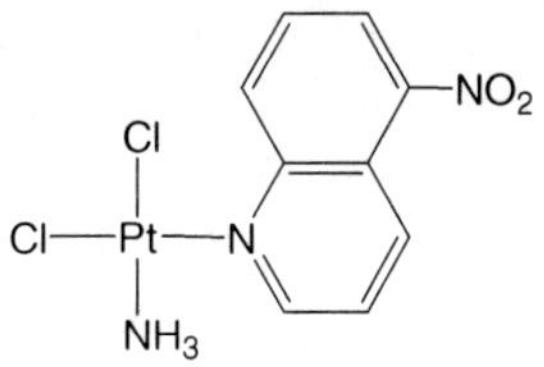

164 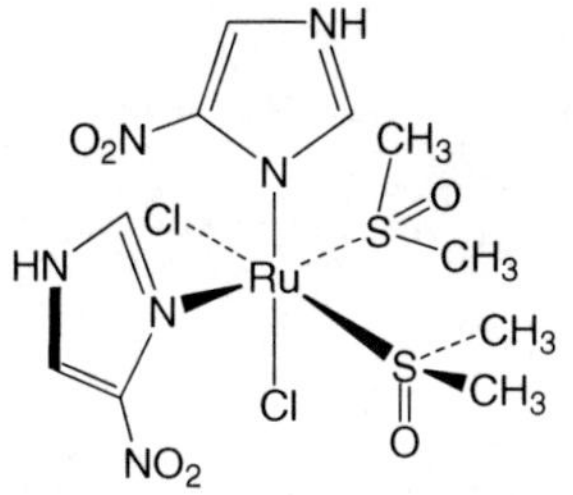 **165**

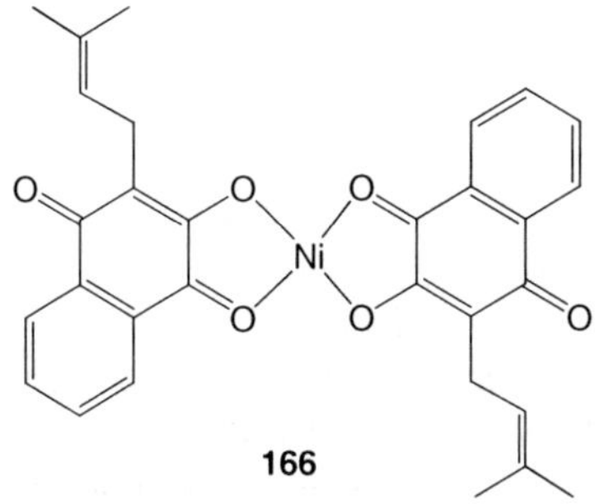

166 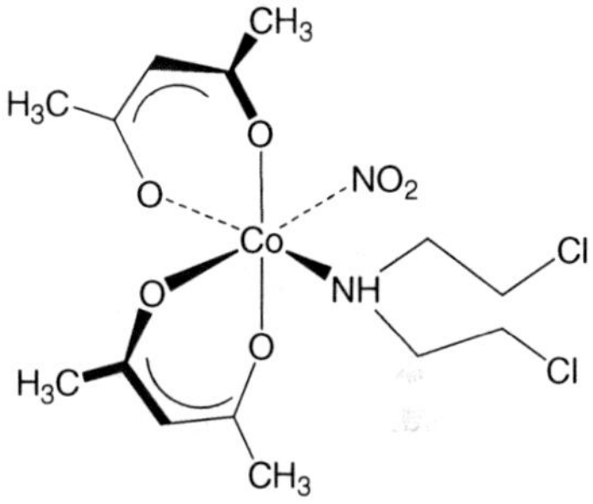 **167**

The importance of the metal as a contributor to radiosensitising is revealed in a series of quinone complexes. The Ni^{2+} complex, **166**, has an SER of 3 and the activity of complexes with this ligand decreases in the order Ni > Cu > Co > Zn. In addition to nitroimidazoles and quinones, another group of ligands which has aroused interest in radiosensitiser applications is the so-called 'nitrogen mustards'. These compounds are of the general form $RN(CH_2CH_2Cl)_2$ and are highly cytotoxic acting as potent cross-linkers of DNA. It has been found that Co(+3) complexes of nitrogen mustards can act as hypoxia selective cytotoxic agents and, under hypoxic conditions, **167** is an efficient radiosensitiser (SER = 2.4). The extent to which metal complexes can contribute to the effectiveness of clinical radiotherapy procedures has yet to be established but the targeting of hypoxic tissue is of interest not only for radiotherapy, but also in imaging applications (Section 3.3.11.4).

CHAPTER 5

Metallopharmaceuticals Design

5.1 Introduction

Since the middle of the last century a variety of powerful techniques have become available for investigating the chemistry of metals within biological systems. As a result an increasingly detailed understanding of the rôles of metals in living organisms is now emerging. In a medicinal context one particular example is provided by work on the mechanism of action of the anticancer drug cisplatin (Section 4.3.2). Knowledge about the chemistry associated with the transport of this drug to its site of action and the nature of its interaction with DNA, has stimulated new avenues of research into platinum anticancer drugs. It has also influenced the design of other potential anticancer agents. Similarly many other discoveries in bioinorganic and medicinal chemistry are informing the design of new metallopharmaceuticals. Targeted design has now become an important aspect of the development of new metallopharmaceuticals and the structure activity relationships used to guide this work need to take account of the particular properties of the metal involved.

5.2 Structure Activity Relationships

An understanding of the relationship between the structure of a compound and its biological activity is an important tool in the search for new pharmaceutical agents and the improvement of existing drugs. Unfortunately, the diverse character of metal compounds used in medicine does not lend itself well to the development of any generally applicable structure activity relationships. Nonetheless decisions on how to develop the structures of lead compounds in order to optimise their pharmacological properties require an appreciation of the ways in which structure may usefully be varied. To some extent this appreciation must arise from knowledge of the different ways in which the structures of metal complexes, and the ligands they contain, have been adapted in pursuit of enhanced clinical performance.

Structure activity relationships can be developed at two levels. In the first instance it may be possible to identify certain essential structural or compositional features through screening a large number of compounds. Within the group of compounds studied more or less systematic variations in structure or composition may reveal some general features necessary to attain the biological response of interest. As an example a series of metal complexes of the same generic formula, and differing only in the identity of the metal present, may show that only certain metals give active compounds. At a more quantitative level, in a series of systematically related compounds it may be possible to correlate a specific structural parameter, or physical property, to the biological response produced by the compound. This approach is well established for organic compounds but less well developed for metallopharmaceuticals, although the subsequent sections provide some examples of the structure activity relationships found for metal complexes.

5.2.1 Organic Molecules

Usually an organic drug molecule will need to interact with a specific receptor site *in vivo* so that the shape and size of the drug molecule, together with the positions of important functional substituents, will need to meet specific design criteria. The structures of organic molecules used as drugs are often amenable to systematic structural variation, facilitating the development of structure activity relationships. A simple example is provided by the antibacterial activity of the alkyl resorcinols, 1,3-$(HO)_2$-4-R-C_6H_3 (R = -$(CH_2)_nH$, **1** in Figure 1). In these compounds the 1,3-dihydroxy-benzene moiety is conserved but the length of the hydrocarbon chain in the group R can be systematically increased by adding -CH_2- groups. A plot of bactericidal activity against n, the number of -CH_2- groups in R, can be made and an optimum value of n defined (Figure 1). Depending on the compound and its application, any of a variety of other structural features or biological responses might be used to correlate structure with activity. For example the size of a ring structure within a molecule might be correlated with its ability to inhibit a particular enzyme.

It is also possible to develop quantitative relationships between a parameter describing a physical property of a molecule and its behaviour *in vivo*. As an example the relative preference of a compound for oily organic regions, such as membranes, compared to aqueous media, such as blood serum, will be reflected in a partition coefficient describing the distribution of the compound between water and a suitable oily liquid which does not mix with water. The partition coefficient can be measured in the laboratory by determining the ratio of the concentrations of the compound in a selected oily phase and water when a mixture of the two phases containing the compound is allowed to reach an equilibrium. The partition coefficient obtained in this way may then be correlated with some pharmacological parameter. This could be the minimum concentration required to induce a particular physiological response or, as another example, it could be the ratio of the proportion of compound excreted *via* the liver to that excreted *via* the kidney. In the latter case increasing

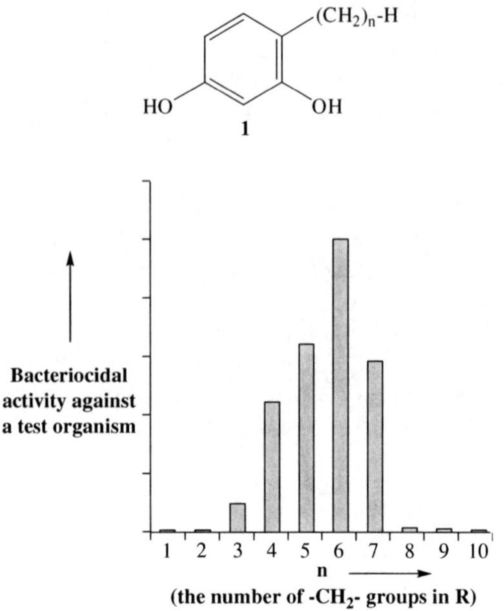

Figure 1 *A plot of bacteriocidal activity against n, the number of -CH_2- groups in R of **1**, with an optimum value of n = 6 in the example shown*

lipophilicity in a class of compound, evidenced by an increasing octanol/water partition coefficient, might correlate with increasing excretion *via* the liver as compared to the kidneys. Where such correlations are possible experimental data can be used to define an optimum range for the property of interest, in this example an octanol/water partition coefficient. Once the form of the relationship has been established suitable new compounds could be selected for further investigations on the basis of their partition coefficients.

It is known that hydrocarbon groups tend to confer fat solubility on a compound and charged groups such as sulfonate, phosphonate or carboxylate confer water solubility. Thus by adding appropriate substituents the chemical structure of the compound may be modified to optimise the partition coefficient. It is possible to define a substituent lipophilicity parameter (π) as a measure of the relative contribution of different substituents (X) to the partition coefficients of molecules of a particular type. Some examples are presented in Table 1. Thus replacing -H with a -CH_3 group with a positive π-value will generally increase the partition coefficient and lipophilicity of a molecule, but using an -NH_2 group with a negative π-value will typically have the reverse effect. Various other parameters associated with substituent effects, or properties of the molecule as a whole, may be quantified to aid in structure activity relationships. These might relate to the polarity of the molecule, its polarisability, its shape or its acidity. However, the magnitude of a substituent effect, and the order of numerical values within a given series of X, may not be not

Table 1 *Examples of values for the lipophilic substituent constant π*

Chemical species	*Lipophilic substituent constant, π, for substituents X*[a]							
Substituent X =	*Cl*	*CH_3*	*F*	*H*	*OCH_3*	*NO_2*	*OH*	*NH_2*
XC_6H_5	0.71	0.56	0.14	0.00	−0.02	−0.28	−0.67	−1.23
1,4-XC_6H_4OH	0.93	0.49	0.31	0.00	−0.12	0.50	−0.87	−1.63
1,4-$XC_6H_4NO_2$	0.54	0.52	–	0.00	0.18	−0.39	0.11	−0.46

[a] $\pi = \log P_X - \log P_H$ where P_X is the octanol/water partition coefficient for the chemical species containing X and P_H is the related value for the case in which X = H.

conserved between different types of compound so that the application of these values can prove to be quite complicated. Nonetheless, information of this type can be used to guide the design process so that the search for new active compounds can be focused on those with structures most likely to prove effective.

5.2.2 Metallopharmaceuticals

5.2.2.1 Examples from Diagnostic Imaging Agents

The design principles applied to organic pharmaceuticals might also be applied with effect to some types of metallopharmaceuticals. The development of the heart-imaging agent Cardiolite (**55c**, Chapter 3) provides one example in which a substituent was varied to optimise performance. The original agent, $[Tc\{CNC(CH_3)_3\}_6]^+$ (see Chapter 3, **55a**), contained the highly lipophilic -$C(CH_3)_3$ substituent. This gave good heart uptake but poor pharmokinetic behaviour. Prolonged retention in the lungs required delayed imaging and the subsequent washout back into the blood interfered with studies of transient ischemic flow under stress due to exercise. High liver accumulation was a further problem as this obscured the apex of the heart during imaging. Replacement of the -$C(CH_3)_3$ substituent of the isonitrile ligand by an ester group -$C(CH_3)_2(CH_2OCOCH_3)$ reduced the lipophilicity of the radiopharmaceutical. The ester group also provided a mechanism for more efficient clearance of the agent from the liver through enzymatic hydrolysis to form hydrophilic carboxylic acid substituents. Further optimisation through testing a large number of structural variants led to the -$C(CH_3)_2(OCH_3)$ substituent present in Cardiolite. This compound is highly lipophilic but shows decreased lung and liver uptake giving better target to non-target uptake ratios. The presence of the -OCH_3 group also provides a reactive site allowing more facile metabolism and hepatobiliary clearance compared to **55a**, Chapter 3.

Variations in the ligand structure were also used to develop effective cerebral perfusion agents based on the diamine dioxime ligands. The original complex [Tc(O)(PnAO)] (see Chapter 3, **68**) was found to diffuse into the brain but was not retained long enough to obtain useful SPECT images. Simply changing the distribution of the six methyl substituents on the ligand periphery to give [TcO(HMPAO)] (**69**, Section 3.3.10) led to a viable imaging agent but revealed a further structural subtlety. The *meso*- and *dl*-forms of **69** of Chapter 3 diffuse into the brain equally well but the *meso*-form is more stable and diffuses back out. In contrast the *dl*-form is converted to a hydrophilic ^{99m}Tc complex which does not diffuse out and so becomes trapped within the brain. This subtle relationship between structure and activity is of a type often encountered with organic drug molecules but, in this case, the presence of the metal is a further essential structural feature. The free ligand itself does not accumulate in the brain and both the metal and the ligand must be present to produce the required biological effect. The agent is a true example of a technetium essential radiopharmaceutical in which the clinical utility requires both a specific ligand structure and the presence of the metal.

More quantitative structure activity relationships have also been developed for some radiopharmaceuticals, an example being provided by the development of the iminodiacetic acid liver imaging agents (see Chapter 3, **53**). The ligand $CH_3N(CH_2CO_2H)_2$ produces a technetium complex which is excreted primarily through the kidneys. However, replacement of the methyl group by a phenyl or substituted phenyl rings increases the lipophilicity of the technetium complexes and so the proportion excreted through the hepatobiliary system. It was found that adding substituents adjacent to the $-N(CH_2CO_2H)_2$ group of the benzene ring (R^1 in **53**, Chapter 3) promoted ligand dissociation from at least one coordination site and increased protein uptake of ^{99m}Tc compared to the unsubstituted compound. Increasing alkyl substitution at other sites increased hepatobiliary over renal excretion and a correlation was found between octanol/water partition coefficients and the ratio of hepatobiliary to renal clearance. In the case of *para*-substituted derivatives (R only varied, $R^1 = R^2 = R^3 = H$ in **53**, Chapter 3) a linear correlation was found between the logarithm of the mass to charge ratio of the complex and the proportion of hepatobiliary excretion.

The development of gadolinium complexes as contrast agents for MRI offers another example of metallopharmaceutical design (Section 3.2.1). In this application the relatively high doses of metal involved require very stable complexes of low charge, which do not liberate Gd^{3+} in a form which might accumulate in the body. The 'hard' nature of Gd^{3+}, its labile character and its preference for coordination numbers of 8 or 9 place some restrictions on the ligand design. The 'hard' nature of Gd^{3+} can be addressed by selecting ligands containing oxygen and nitrogen donor atoms and stability can be improved by using negatively charged polydentate ligands. The use of a macrocyclic ligand structure exemplified by $dotaH_4$ (see Chapter 3, **5a**) can improve stability compared to acyclic systems such as $dtpaH_5$ (see Chapter 3, **3a**). The lipophilicity of $dotaH_4$ derivatives can be increased by either adding lipophilic substituents to the macrocyclic ring, as in **2**, or to $-CH_2-$ in the $-CH_2CO_2H$ groups. The inclusion of four negative binding groups in $dota^{4-}$, as opposed to five in $dtpa^{5-}$, has the further advantage of reducing the charge on the complex. This improves the osmolality of the complex, an important consideration where high doses are involved. Further improvements can be made in this respect by constructing a neutral complex such as [Gd(hp-do3a)] (see Chapter 3, **5b**) which also has higher lipophilicity than its $[Gd(dota)]^-$ counterpart. In 1992

ProHance® ([Gd(hp-do3a)]) was the first nonionic MRI contrast agent to be marketed and has since been used safely in over a million patients.

The acid dissociation rate of Gd^{3+} from a complex like [Gd(hp-do3a)] can be reduced still further by using more acidic phosphonate binding groups in place of carboxylate. The respective acid dissociation rate constants of Gd^{3+} from [Gd(dtpa-bma)] (see Chapter 3, **3d**), [Gd(hp-do3a)] (see Chapter 3, **5b**) and [Gd(dotmp-mbua)] (dotmp-mbuaH_3 = **3**) are > 20, 0.064 and 0.0013 in units of $s^{-1} \times 10^3$ at $[H^+] = 0.1$ mol l^{-1} revealing the effects of converting, firstly, from an acyclic to a macrocyclic ligand structure and, secondly, from carboxylate to phosphonate binding groups. The lipophilicity of the phosphonate system can be altered by changing the substituent R in the amide group and/or by replacing the methyl groups on the phosphorus atoms with other substituents.

2

3

H_3dotmp-mbua, R = $(CH_2)_3CH_3$

Serendipity and experimentally led design have always been important features in the development of new pharmaceuticals for imaging but computational modelling methods are becoming increasingly important. This is particularly so in the design of bifunctional agents where the part of the molecule recognised by a biological receptor may be designed in a similar manner to that used for organic drugs. However, for metallopharmaceuticals the modelling will also have to take into account the need to bind the metal securely and the effect of the metal and its binding site on the biological activity of the molecule as a whole.

5.2.2.2 Examples from Therapeutic Agents

Selectivity for a particular metal ion is extremely important in the design of metal sequestering agents for use in chelation therapy. It is also necessary to form stable rapidly excreted complexes of the target metal in competition with other binding agents *in vivo*. Naturally occurring compounds such as desferrioxamine (see Chapter 4, **7**) and enterobactin (see Chapter 4, **15**) provided leads for the development of simpler iron sequestering agents to treat iron overload (Section 4.2.4). The bidentate catechol (**4**) chelating group found in enterobactin is capable of forming an octahedral 3:1 complex with Fe^{3+} but is a weak acid with limited binding power at physiological pH. This basic structure was adapted to improve its properties through structural variations based on 1-hydroxypyridin-2-one (**5**), 1-alkyl-3-hydroxypyridin-2-one (**6**) and 1-alkyl-3-hydroxypyridin-4-one (**7**) chelating groups. These preserved the core structure of the catechol unit, but were more acidic and the 1-alkyl-3-hydroxypyridin-4-one unit was found to show high selectivity for Fe^{3+}. In this way the chemical structure was optimised to leading to the pharmaceutical Deferiprone (see Chapter 4, **18**).

4 **5** **6** **7**

R = alkyl

The development of sequestering agents for Pu^{4+} (Section 4.2.3.3.15) involved the construction of a larger ligand structure capable of 8-coordinate binding and offering further possibilities for structural variation (Figure 2). Early work using catechol as the chelating group C_6H_4XY explored the effects of variations in the structures of the linking groups including the preparation of H-shaped molecules. Modification of the benzene rings with sulfonate substituents improved water solubility and offered an effective low toxicity sequestering agent. The replacement of the catechol binding groups with 1-hydroxypyridin-2-ones further improved Pu^{4+} binding at physiological pH and eventually the compound lipoH_4 (see Chapter 4, **13**) emerged as an effective agent for Pu^{4+} decorporation. The development process followed here exemplifies a fairly typical approach to metallopharmaceutical development. In that the outcome of any particular structural change cannot be predicted with certainty,

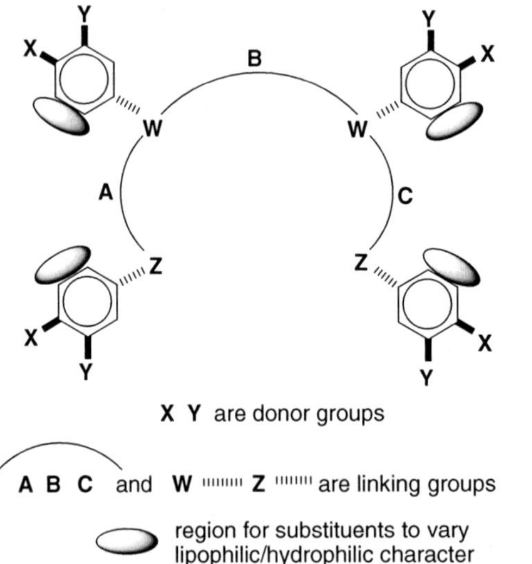

Figure 2 *Some options for structural variations in the design of an octadentate actinide metal sequestering agent based on an acyclic linear core carying four bidentate binding groups*

intelligent design is used to guide what is ultimately a trial and error approach to optimising clinical utility.

A more metal focused approach to metallopharmaceutical optimisation is found in the development of some anticancer agents (Section 4.3.2.2). Rather than needing to simply encapsulate the metal to transport it to particular locations or excretion pathways, anticancer agents generally need to exploit the particular reactivity of the metal. Thus the form of complex used needs not only to facilitate transport to the active site but, once there, also to promote the appropriate reactions involving the metal. Early structure activity relationships for compounds of the cisplatin type emerged from general observations and some of these have since proven to be somewhat misleading. In particular it was found that *trans*-$[PtCl_2(NH_3)_2]$ was ineffective as an antitumor agent *in vivo* so it was thought that *cis*-leaving groups (*i.e.* Cl^- in cisplatin) were important. However, now that the mechanism of action of cisplatin and related compounds is much better understood, there is interest in the development of *trans*-complexes for cancer therapy. It was also thought that an NH group attached to Pt^{2+} was needed and that tertiary amines lacking the NH moiety would be ineffective. There was also some structural evidence to support this view. Despite this active compounds of the cisplatin family are now known which do not contain an NH group adjacent to the platinum ion. Thus, although general structure activity relationships of this type can serve as a useful guide to further development, they should not be considered absolute criteria unless sound mechanistic evidence exists to support that view. It is always possible that two apparently structurally similar metal complexes may act by quite different mechanisms.

Although an element of trial and error coupled with rational design has been applied to the development of anticancer drugs in the cisplatin family, a more detailed understanding of the mechanism of action is now an important driver for new developments. The recognition that cisplatin (**23**, Chapter 4) hydrolyses to form toxic byproducts *in vivo* led to the development of the second generation drug carboplatin (see Chapter 4, **25**) in which a chelating ligand replaces the two chlorides. This helps to control the extracellular hydrolysis of the complex, so reducing side effects, but does not prevent its binding to DNA through the formation of the $\{Pt(NH_3)\}^{2+}$ moiety. Further advances driven by improved mechanistic understanding include the development of *trans*-complexes and binuclear compounds designed to circumvent intrinsic and acquired cisplatin resistance in some tumor types.

Mechanistic understanding and structure activity relationships are much less well developed for non-platinum antitumor agents, although some general relationships have emerged from studies of the metallocene derivatives $[M(\eta^5\text{-}C_5H_5)_2Cl_2]$ (Section 4.3.3.3.2). The nature of the metal, M, is important, some being inactive, but it is not clear why certain metals are more effective than others. Both neutral and positively charged complexes of this formula can produce antitumor effects, so a simple correlation with charge or metal oxidation state is not present. The nature of the anionic ligands X in $[M(\eta^5\text{-}C_5H_5)_2X_2]$ does not appear crucial for activity but substitution on the

C_5H_5 ligand eliminates the anticancer effect. Since, in the active compound $[Ti(\eta^5\text{-}C_5H_5)_2Cl_2]$ (see Chapter 4, **107**), hydrolysis with loss of Cl^- is rapid compared to the displacement of $C_5H_5^-$ it would appear that a $\{Ti(C_5H_5)_2\}^{2+}$ or $\{Ti(C_5H_5)\}^{3+}$ moiety must be conserved to maintain activity. It is notable that the complex budotitane (see Chapter 4, **105**) is constitutionally similar to $[Ti(\eta^5\text{-}C_5H_5)_2Cl_2]$ except that the $C_5H_5^-$ ligand is replaced by a β-diketonate and Cl^- by ethoxide, OC_2H_5. Again the nature of the anionic ligand in budotitane does not seem crucial as Cl^- F^-, Br^- can be substituted for ethoxide without eliminating activity.

One shared structural feature of cisplatin, budotitane and $[Ti(\eta^5\text{-}C_5H_5)_2Cl_2]$ is the presence of labile anionic ligands occupying two *cis*-coordination sites. It would seem that the active unit in all these complexes is of the type $\{ML_2\}^{2+}$ [M = Pt, L = NH_3; M = Ti, L=C_5H_5 or $C_6H_5C(O)CHC(O)CH_3$]. However, here the similarity ends. The retained, less labile, *cis*-ammonia ligands of cisplatin are structurally and chemically quite different from the *cis*-cyclopentadienides in $[Ti(\eta^5\text{-}C_5H_5)_2Cl_2]$ and the *cis*-β-diketonates in budotitane. This is consistent with the view that the mechanism of action of cisplatin is unlike that of $[Ti(\eta^5\text{-}C_5H_5)_2Cl_2]$ or budotitane. Furthermore $C_5H_5^-$ and $C_6H_5C(O)CHC(O)CH_3^-$ are also quite different in size and shape. The relative positions and orientations of the respective C_5H_5 and C_6H_5 rings in the $\{Ti(C_5H_5)_2\}^{2+}$ and $\{Ti[C_6H_5C(O)CHC(O)CH_3]_2\}^{2+}$ fragments are quite different. There is some evidence that $[Ti(\eta^5\text{-}C_5H_5)_2Cl_2]$ does not bind to DNA at physiological pH whereas $[Ti\{C_6H_5C(O)CHC(O)CH_3\}_2\}(H_2O)_2]^{2+}$ does, although without causing damage to the DNA. Thus the superficial compositional similarities between these three $[MX_2L_2]$ complexes, and their shared tendency to lose X^-, appear to conceal important differences in their ultimate biochemical effects. This illustrates the need for care when interpreting general structure activity relationships in the absence of a detailed understanding of the mechanism of action of different types of metallopharmaceutical.

5.2.2.3 *Design Issues for Metallopharmaceuticals*

A wide variety of issues relating to the complexation of the metal need to be considered when designing the structure of a metallopharmaceutical. These will differ from application to application but some general points to be considered include:-

The stability of the complex

This will depend upon the nature of the metal ion, the ligand structure and the compatibility of the donor atoms with the metal. Hard metals such as Gd^{3+} form more stable complexes with hard donors such as water or $\text{-}O^-$ in carboxylate or hydroxide. Soft metals such as gold form more stable complexes with thiolates or phosphines than with oxygen donors. Metals such as technetium and rhenium, which can exist in different oxidation states, exhibit more complicated behaviour. Typically low oxidation states of +3 and below are best suited to soft donor atom ligands, although Tc(+5) as TcO^{3+} forms stable thiolate complexes and Re(+5) behaves similarly. Strong π-acceptor ligands are needed to stabilise very low oxidation states such as Tc(+1). Where high complex stability is needed the chelate and macrocyclic effects can be exploited through using polydentate and macrocyclic ligands. This will be essential to stabilise metals which are kinetically labile since high thermodynamic stability is needed to resist competition from potential metal binding agents *in vivo*. An example of this approach is provided by $[Gd(dota)]^-$ ($dotaH_4$ = **5a**, Chapter 3) containing a polydentate ligand with a macrocyclic core. The brain imaging agent [TcO(HMPAO)] (**69**, Chapter 3) contains a polydentate ligand which attains macrocyclic character through the formation of a hydrogen bond to close the macrocycle.

The use of bulky ligands to saturate the metal coordination sphere provides a means of slowing down exchange reactions with other potential complexing agents and further stabilising the metal complex. In the case of kinetically inert metal centres the use of polydentate ligands to prevent ligand exchange may be unnecessary. The bulky monodentate isonitrile ligands in Cardiolite and large crystal field stabilisation energy of the d^6 Tc(+1) centre make the complex kinetically inert and thermodynamically stable. However, in cases such as this it is important to remember that oxidation of the metal will result in reduced π-backbonding which will reduce the stability of the complex. If this occurs at an inappropriate stage *in vivo* the low oxidation state approach may prove ineffective.

In some applications it may be necessary for a metal to be transferred to a biological binding agent *in vivo*. In such cases the complex stability needs to be kept in a range which prevents unwanted reactions, such as hydrolysis or general binding to serum proteins, but allows the target binding agent to compete with the ligands present and gain access to the metal. An example of this is provided by the use of ^{68}Ga-citrate where the citrate ligand suppresses hydrolysis of the $^{68}Ga^{3+}$ but allows facilc uptake of the metal ion *in vivo* by transferrin.

The charge on the complex

Broadly speaking complexes with high negative charges will be hydrophilic and less amenable to passive diffusion through membranes. Neutral complexes are likely to be more lipid soluble and require that the charges of the metal ion and ligand set cancel. Positively charged complexes are typically more prone to aquation and hydrolysis than neutral or anionic species. However, the Cardiolite (see Chapter 3, **55c**) system presents an exception since the metal ion is encapsulated within a set of six bulky lipophilic ligands which inhibit access to the central metal ion. The use of a low oxidation state organometallic complex changes the chemistry of the system to give behaviour atypical of positively charged metal complexes.

Unless very low doses are involved, high charges on complexes will lead to unfavourable osmolalities which may present problems in administering the compound. The charge on the complex can often be controlled by variations in the ligand structure, although these must be made judiciously so as not to compromise the performance of the metallopharmaceutical. An example is provided by the series of MRI contrast agents $[Gd(dtpa)]^{2-}$, $[Gd(dota)]^-$ and

[Gd(hp-do3a)] (proligands **3a**, **5a** and **5b** of Chapter 3). These show improved stability in going from $[Gd(dtpa)]^{2-}$ to $[Gd(dota)]^{-}$ and improved osmolality in going from $[Gd(dota)]^{-}$ to [Gd(hp-do3a)].

Ligand dynamics and reaction kinetics of the complex

The kinetics of complex formation can sometimes be an issue in the formulation of radiopharmaceuticals involving short half-life nuclides. Time for preparing the complex from the radioactive precursor metal can be quite limited so that long reaction times or complicated synthetic procedures may be unacceptable. This has been an issue in the preparation of some low oxidation state technetium complexes.

The kinetics of ligand dissociation are an important issue for some metallopharmaceuticals particularly the platinum anticancer drugs. In these compounds the substitution of the anionic ligands is crucial in the formation of DNA adducts and the complex must be sufficiently labile for this to occur. However, if the leaving ligands are too labile hydrolysis reactions leading to toxic byproducts may result in unacceptable side effects. The lability of the anionic ligand can be modified by changing the nature of the anionic ligand itself. This is exemplified by the replacement of chlorides in cisplatin (see Chapter 4, **23**) by a dicarboxylate in carboplatin (see Chapter 4, **25**). Alternatively the nature of the less labile ligands may be changed so that the *trans*-effect can be exploited (Section 2.8.3).

Where complexes are required to remain intact until they are excreted, or metabolised after fulfilling their function, ligand design can contribute to achieving kinetically inert behaviour. The use of preorganised, more rigid ligands can lead to lower dissociation rates and, where the application allows, the use of bulky substituents can increase the activation barrier for ligand substitution. The subtle effects that substitution patterns can have on reactivity are demonstrated in the differing behaviour of isomers of ^{99m}Tc-HMPAO brain imaging agents (Section 3.3.10).

Electron transfer properties

In some applications the electron transfer properties of a metal may be important. An obvious example is provided by superoxide dismutase mimics (Section 4.6.2). The reduction potential of a metal ion can be modified by changes in ligand structure. In particular the donor atoms can have a significant effect. Atoms such as S or P tend to stabilise lower oxidation states and facilitate reduction compared to an equivalent complex containing N or O donor atoms. Reduction potentials are also sensitive to other structural features of ligands such as chelate ring sizes and substituents.

Another important feature of changes in metal oxidation state is its effect on the lability of ligands. A well known example is provided by Cr(+3) which forms kinetically inert complexes. Reduction of Cr(+3) in a complex to Cr(+2) makes the ligand set more labile. If the new Cr(+2) complex formed can reduce the original Cr(+3) complex a tiny amount of Cr(+2) may catalyse the decomposition of all the Cr(+3) complex present. This reductive labilisation of ligands is seen in the action of $[Fe(CN)_5(NO)]^{2-}$ (**143**, Section 4.6.2) where *in vivo* reduction leads to the release of NO to produce a vasodilation effect.

5.3 Future Directions

In the past most pharmaceutical companies have shown relatively little interest in metal compounds. Possible reasons for this are concerns over perceived toxicity or metal accumulation during long-term therapy, concerns over metallodrug specificity and unfamiliarity with the chemistry of metals. In addition there has been a significant degree of serendipity in the creation of new metallopharmaceuticals rather than a rational design approach to attack specific targets. This picture is starting to change as knowledge of bioinorganic chemistry and the mechanisms of action of metallodrugs increases.

In 2000 a meeting entitled 'Metals in Medicine: Targets, Diagnostics and Therapeutics' was sponsored by a group of US agencies supporting research into health and medical issues. This demonstrated a recognition by these agencies of the potential importance of metalloenzymes, metal metabolism and metallopharmaceuticals in health care applications. The meeting set out to consider three topics

(i) The current utilisation, by the pharmaceuticals industry, of research discoveries concerning the biological and pharmaceutical behaviour of metal compounds.
(ii) The obstacles to the development of metallopharmaceuticals and the opportunities for research and for identifying new metallopharmaceuticals.
(iii) The opportunities for developing new agents to target metal metabolism and metal-regulated cellular processes.

In outline, the meeting concluded that applications of metal compounds in MRI, radiology and radiation therapy represent expanding areas of activity that will benefit from basic research on metal coordination chemistry and biological inorganic chemistry. Therapeutic applications of metal complexes represent an under-developed area of research as metallopharmaceuticals may offer unique opportunities for therapy. However, problems with metal toxicity and accumulation can present significant obstacles and these need to be addressed through further research. Studies of metalloenzyme structure, function, inhibition and mechanism of action were thought to be well supported and are already producing information which is exploited in metallopharmaceutical development. One area identified as ripe for research growth was metal metabolism, including studies of the mechanisms of metal homeostasis and the roles of metals in the regulation of cell function and cell–cell interaction. The lack of a set of basic principles to guide the development of new metallopharmaceuticals was noted. However, in view of the diverse nature of current metallopharmaceutical products and the variety of unique properties they incorporate, this seems hardly surprising. At the chemical level at least, coordination chemistry offers one set of basic principles to guide metallopharmaceutical development.

Metal complexes in general could be viewed as a more sophisticated version of organic drugs. They offer access to magnetic, optical and nuclear properties with important applications in diagnostic medicine but unavailable from organic compounds. Chemically they offer reactivities of a different character from those of organic compounds and provide many more stereochemical options to allow binding moieties to be oriented to match receptor sites. The clearance of breakdown products containing the metal may be potentially problematic in some cases but careful complex design, or the use of post treatment sequestering agents, offer means of dealing with this issue. The growth in application of metal complexes in medicine since the mid 20th century demonstrates the added value which can accrue from the incorporation of metal complexes into the portfolio of pharmaceuticals. This trend will no doubt continue.

It would be foolhardy in the extreme to try to predict which future developments will prove the most important for clinical applications of metallopharmaceuticals. Undoubtedly there will continue to be incremental improvements to some existing classes of metallopharmaceutical formulation. Beyond that it is possible to identify some areas of active research where coordination chemistry is making a continuing contribution. These illustrate the diversity of the research in progress and provide examples of the approaches from which new metallopharmaceutical agents may emerge.

In diagnostic imaging an important future development for MRI contrast agents would seem to lie in tissue targeted applications and agents suitable for functional studies. A major hurdle to the development of new contrast agents with specificity for a particular target tissue is the relatively high concentration required in the target tissue. At the relaxivities attainable with current soluble contrast agents, micromolar concentrations or higher are needed to achieve useful contrast effects. This poses problems of toxicity and of designing agents which can reach target receptor sites in sufficient quantity. Major advances might be possible if the relaxivity of soluble contrast agents could be increased substantially. Current theory suggests that optimising the rates of water exchange, molecular rotation and electronic relaxation in contrast agents could lead to a near 100-fold increase in relaxivity. Thus, fundamental coordination chemistry studies to show how complex structure can be manipulated to optimise relaxivity could make a useful contribution. It is known, for example, that more rigid ligands and symmetric complexes can lead to longer relaxation times so that $[Gd(dota)]^{-}$ derivatives might be favoured over those of $[Gd(dtpa)]^{2-}$. Increasing water exchange rates, or the number of bound water molecules would also be useful provided the stability of the complex is not compromised. Another approach to improving relaxivity is to reduce the rotation rate of the molecule. The attachment of a Gd^{3+} complexes to a macromolecule such as a protein, polysaccharide or nucleic acid would lead to a paramagnetic species with a greatly reduced rate of rotation. If, in addition, the macromolecular part has useful properties in terms of its biological distribution, new specifically targeted, bifunctional, high relaxivity MRI contrast agents might be developed.

In radiopharmaceutical imaging bifunctional agents, including ^{99m}Tc labelled monoclonal antibodies, or their fragments, offer an important means of targeting specific tissues. Accordingly, the development of improved means of attaching ^{99m}Tc to functional biomolecules is an active area of research. In particular a simple procedure for obtaining the $\{Tc(CO)_3\}^+$ fragment offers a potentially versatile means of preparing bifunctional agents as well as molecular technetium essential radiopharmaceuticals. Technetium complexes of small peptides are also attracting considerable interest and finding application in the diagnostic imaging of thrombosis, tumors and inflammation. Although specific technetium binding sites may be attached to small peptides to give bifunctional agents, some amino acid sequences are potential ligands in their own right. Where present these might obviate the need to attach a separate technetium binding site to the peptide. Coordination of a peptide to a technetium centre will change the structure of the section of the peptide involved but, by choosing the appropriate amino acid sequences, it might be possible to produce a structure recognised by specific receptors. Computer modelling offers an efficient means of identifying promising peptide sequences for further investigation and allowing rational design of new candidates for evaluation.

The use of complexes formed from β^+ emitters, particularly ^{68}Ga ($T_{1/2}$ = 68 min) and ^{62}Cu ($T_{1/2}$ = 9.8 min) offers opportunities in PET imaging applications. One advantage of PET imaging is that it is quantitative rather than qualitative allowing the dose required for each patient to be estimated accurately – the personalised medicine paradigm. Both ^{68}Ga and ^{62}Cu are available from generator systems and offer the prospect of rapid radiopharmaceutical preparation using 'shake and shoot' kits. In applications requiring long timescales the longer half-life radionuclide ^{64} Cu ($T_{1/2}$ = 12.9 h) might be employed using the same chemistry as for ^{62}Cu.

Among therapeutic applications of metallopharmaceuticals, the development of better anticancer drugs and the search for new active compounds is an area of intense research. Much attention is focused on platinum compounds for which quite detailed information is available regarding their mechanism of action. Compounds of this type can show sequence selectivity in DNA adduct formation and the extent to which this may be controlled by ligand design, to modify hydrogen bonding patterns for example, is of considerable interest. There are also issues with intrinsic or acquired cisplatin resistance which might be addressed by using other structural motifs, such as *trans*-complexes or binuclear systems, which invoke other mechanisms of action. In a different approach, the observation that hormones such as oestrogen can increase sensitivity to cisplatin through their effect on HMG protein regulation has prompted a clinical study of the use of cisplatin in combination with oestrogen or progesterone to treat ovarian cancer. Work on elucidating the mechanism of action of non-platinum anticancer agents is also an important area of study. Even though these compounds may not prove to be clinically useful therapeutic agents in their own right, the mechanistic information gained may contribute to the development of new antitumor agents through informing the rational design of other systems. The study of complexes which intercalate DNA (Section 4.3.3.1.6) is also an important developing area of study and could lead to new sequence selective DNA binding agents with pharmaceutical applications.

In the field of radiotherapy two β-emitting rhenium radionuclides, ^{186}Re and ^{188}Re, are of particular interest because of the close similarities in chemical behaviour between Tc and Re. The exploitation of coordination chemistry in the development of ^{99m}Tc-radiopharmaceuticals for diagnostic imaging applications is very well established. In principle very similar chemical approaches might be applied to the development of targeted $^{186/188}Re$ containing radiopharmaceuticals, both bifunctional and rhenium essential systems being possible. Many Re compounds show more inert chemical behaviour than their Tc counterparts reducing the risk of metal being released for non-specific binding. However, Re compounds are more easily oxidised and the formation of hydrophilic ReO_4^-, following metabolic degradation of a radiopharmaceutical, can be advantageous since ReO_4^- is readily excreted *via* the kidneys offering an effective elimination route. The nuclear properties of ^{186}Re and ^{188}Re offer a further advantage in that the β-particles emitted by the former have a longer range in tissue than those from the latter. This potentially allows the radionuclide to be chosen to better match the size of tumor or volume of tissue to be irradiated, while using the same proligand in the radiopharmaceutical preparation. The γ-rays emitted by these radionuclides are of energy suitable for imaging suggesting the possibility of simultaneous imaging and radiotherapy so that the uptake of the therapeutic agent could be monitored. Some lanthanide radionuclides are also showing promise for radiotherapy applications. In particular ^{166}Ho, ^{153}Sm and ^{177}Lu offer essentially the same chemical properties but half lives ranging from 1.1 to 6.7 days and β^- maximum energies ranging from 0.5 to 1.6 MeV. Again this potentially offers access to a variety of radiation effects while applying a common pharmaceutical chemistry approach.

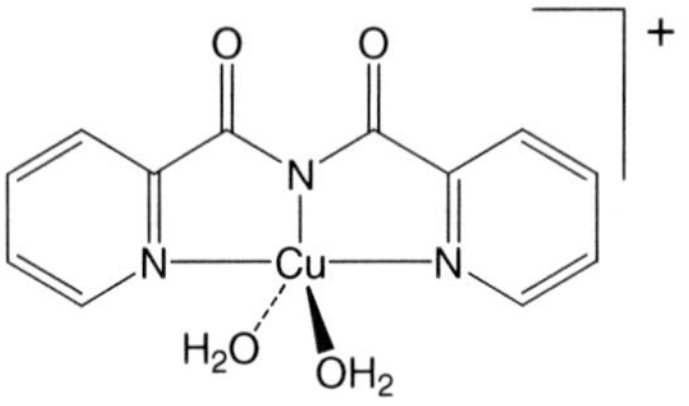

8

The information emerging from mechanistic studies of bioinorganic systems and metallodrug action should create new opportunities for metallopharmaceutical development. The ability of certain metal complexes to act as enzyme inhibitors can have medicinally useful effects. Some of these have potential applications in microbiocidal and antiviral agents, although space has not allowed a consideration of these applications. As an example some hard metal ions, including Fe^{3+}, UO_2^{2+} and VO_3^-/VO_4^{3-}, have been found to act as inhibitors of the protease enzyme from human immunodeficiency virus, HIV-1. They appear to act as non-competitive inhibitors and may bind to the enzyme together with the substrate. The Cu(+2) complex **8** acts as a competitive inhibitor of HIV-1 protease and, in this case, modelling suggests that the structure of the complex allows it to fit into the active site of the enzyme. Many other examples of the biological activity of metal ions or their complexes could be cited, each of which contributes something to our knowledge of the behaviour of metal compounds in biological systems. This in turn informs research into the pharmaceutical applications of metal complexes. At least some of these avenues of research, may be expected to bear fruit in the coming years. This being so coordination chemistry will remain an important discipline underpinning the development of new metallopharmaceuticals.

Further Reading

I Books

Inorganic chemistry

C.E. Housecroft and A.G. Sharpe, *Inorganic Chemistry*, Pearson Education Ltd., Harlow, 2001.

D.F. Shriver and P.W. Atkins, *Inorganic Chemistry*, 3rd edn, Oxford University Press, Oxford, 1999.

B.E. Douglas, D.H. McDaniel and J.J. Alexander, Chapter 18, *Concepts and Models of Inorganic Chemistry*, 3rd edn, Wiley, New York, 1994.

J.E. Huheey, E.A. Keiter and R.L. Keiter, Chapter 19, *Inorganic Chemistry*, 4th edn, Harper Collins, New York, 1993.

F.A. Cotton and G. Wilkinson, *Advanced Inorganic Chemistry*, 5th edn, Wiley, New York, 1988.

N.N. Greenwood and A. Earnshaw, *Chemistry of the Elements*, Pergamon Press, Oxford, 1984.

Coordination chemistry

S.F.A. Kettle, *Physical Inorganic Chemistry, A Coordination Chemistry Approach*, Spektrum Academic Publishers, Oxford, 1996.

J.A. McCleverty and T.J. Meyer (eds), *Comprehensive Coordination Chemistry II*, Elsevier, Oxford, 2004.

R.D. Gillard, J.A. McCleverty and G. Wilkinson (eds), *Comprehensive Coordination Chemistry*, Pergamon Press, Oxford, 1987.

Bioinorganic chemistry

D.E. Fenton, *Biocoordination Chemistry*, Oxford Chemistry Primers 25, Oxford Science Publications, Oxford, 1995.

I. Bertini, H.B. Gray, S.J. Lippard and J.S. Valentine, *Bioinorganic Chemistry*, University Science Books, Mill Valley, CA, 1994.

W. Kaim and B. Schwederski, *Bioinorganic Chemistry: Inorganic Elements in the Chemistry of Life*, Wiley, Chichester, 1994.

S.J. Lippard and J.M. Berg, *Principles of Bioinorganic Chemistry*, University Science Books, Mill Valley, CA, 1994.

M.N. Hughes, *The Inorganic Chemistry of Biological Processes*, 2nd edn, 1981, Wiley, Chichester, 1981.

H. Sigel and A. Sigel (eds), *Metal Ions in Biological Systems: The Lanthanides and Their Interactions with Biosystems*, vol 40, Marcel Dekker, Basel, 2003.

Medicinal chemistry

G. Thomas, *Medicinal Chemistry*, Wiley, Chicester, 2000.

Metallopharmaceuticals

N Farrell (ed), *Uses of Inorganic Chemistry in Medicine*, Royal Society of Chemistry, Cambridge, 1999.

M.J. Clarke and P.J. Sadler (eds), *Metallopharmaceuticals I: DNA Interactions*, Springer, Berlin, 1999 (Topics in biological inorganic chemistry).

M.J. Clarke and P.J. Sadler (eds), *Metallopharmaceuticals II: Diagnosis and Therapy*, Springer, Berlin, 1999 (Topics in biological inorganic chemistry).

Radiopharmaceuticals and nuclear medicine

G.B. Saha, *Fundamentals of Nuclear Pharmacy*, 5th edn, Springer, New York, 2005.

R.J. Kowalsky and S. Falen, *Radiopharmaceuticals in Nuclear Pharmacy and Nuclear Medicine*, American Pharmacists Association, Washington DC, 2004.

Diagnostic imaging

R.R. Edelman, J.R. Hesselink, M.B. Zlatkin, *Clinical Magnetic Resonance Imaging Online*, 3rd edn, Saunders/Elsevier, Amsterdam, 2005.

A.E. Merbach and E. Toth (eds), *The Chemistry of Contrast Agents for Medical Magnetic Resonance Imaging*, Wiley, London, 2001.

E.L. Kramer and J.J. Sanger (eds), *Clinical SPECT Imaging*, Raven Press Ltd., New York, 1995.

Cancer therapy

B.K. Keppler (ed), *Metal Complexes in Cancer Chemotherapy*, VCH, Weinheim, 1993.

II Review Articles on Specific Topics

Metallopharmaceuticals

Z. Guo and P.J. Sadler, Metals in medicine, *Angew. Chem. Int Ed.*, 1999, **38**, 1513–1531.

C.S. Allardyce and P.J. Dyson, Ruthenium in medicine: current clinical uses and future prospects, *Platinum Metals Rev.*, 2001, **45**, 62–69.

A.Y. Louie and T.J. Meade, Metal complexes as enzyme inhibitors, *Chem. Rev.*, 1999, **99**, 2711–2734.

Anatomical imaging

Magnetic resonance imaging

D. Parker, Excitement in f-block: structure dynamics and function in nine-coordinate chiral lanthanide complexes in aqueous media, *Chem. Soc. Rev.*, 2004, **33**, 156–165.

P. Caravan, J.J. Ellison, T.J. McMurry, R.B. Lauffer, Gadolinium(IIII) chelates as MRI contrast agents: structure, dynamics and applications, *Chem. Rev.*, 1999, **99**, 2293–2352.

L. Thunus and R. Lejeune, Overview of transition metal and lanthanide complexes as diagnostic tools, *Coord. Chem. Rev.*, 1999, **184**, 125.

S. Aime, M. Botta, M. Fasano and E. Terreno, Lanthanide(III) chelates for NMR biomedical applications, *Chem. Soc. Rev.*, 1998, **27**, 19–30.

X-ray imaging

S.-B. Yu and A.D. Watson, Metal based X-ray contrast media, *Chem. Rev.*, 1999, **99**, 2353–2377.

Functional imaging

C.J. Anderson and M.J. Welch, Radio-labelled agents (non-technetium) for diagnostic imaging, *Chem. Rev.*, 1999, **99**, 2219–2234.

S.S. Jurisson and J.D. Lyndon, Potential technetium small molecule radio-pharmaceuticals, *Chem. Rev.*, 1999, **99**, 2205–2218.

S.Liu and D.S. Edwards, ^{99m}Tc-labelled small peptides as diagnostic radio-pharmaceuticals, *Chem. Rev.*, 1999, **99**, 2235–2268.

J.R. Dilworth and S.J. Parrott, The biomedical chemistry of technetium and rhenium, *Chem. Soc. Rev.*, 1998, **27**, 43–55.

S. Jurisson, D. Berning, W. Jia and D. Ma, Coordination compounds in nuclear medicine, *Chem. Rev.*, 1993, **93**, 1137–1156.

Chelation therapy

Z.D. Liu and R.C. Hider, Design of iron chelators with therapeutic application, *Coord. Chem. Rev.*, 2002, **232**, 151–171.

O. Andersen, Principles and recent developments in chelation treatment of metal intoxication, *Chem. Rev.*, 1999, **99**, 2683–2710.

B. Sarkar, Treatment of Wilson and Menkes diseases, *Chem. Rev.*, 1999, **99**, 2683–2710.

P.S. Dobbin and R.C. Hider, Iron chelation therapy, *Chem. Britain*, (June), 1990, 565.

Cancer treatment and DNA interactions

E.R. Jamieson and S.J. Lippard, Structure, recognition, and processing of cisplatin-DNA adducts, *Chem. Rev.*, 1999, **99**, 2467–2498.

T.W. Hambley, The infuence of structure on the activity and toxicity of Pt anti-cancer drugs, *Coord. Chem. Rev.*, 1997, **166**, 181–223.

M.D. Hall and T.W. Hambley, Platinum(IV) antitumour compounds: their bioinorganic chemistry, *Coord. Chem. Rev.*, 2002, **232**, 49–67.

N.J. Wheate and J.G.Collins, Multi-nuclear platinum complexes as anti-cancer drugs, *Coord. Chem. Rev.*, 2002, **232**, 49–67.

T.W. Hambley, Platinum binding to DNA: structural controls and consequences, *J. Chem. Soc., Dalton Trans.*, 2001, 2711–2718.

E. Wong and C.M. Giandomenico, Current status of platinum antitumor drugs, *Chem. Rev.*, 1999, **99**, 2451–2466.

C.A. Claussen and E.C. Long, Nucleic acid recognition by metal complexes of bleomycin, *Chem. Rev.*, 1999, **99**, 2797–2816.

M.J. Clarke, F. Zhu and D.R. Frasca, Non-platinum chemotherapeutic metallopharmaceuticals, *Chem. Rev.*, 1999, **99**, 2511–2533.

K.E. Erikkila, D.T. Odom and J.K. Barton, Recognition and reaction of metallointercalators with DNA, *Chem. Rev.*, 1999, **99**, 2777–2795.

Gold drugs for rheumatoid arthritis

M.J. McKeage, L. Maharaja, S.J. Berners-Price, Mechanisms of cytotoxicity and antitumor activity of gold(I) phosphine complexes: the possible role of mitochondria, *Coord. Chem. Rev.*, 2002, **232**, 49–67.

C.F. Shaw, Gold-based therapeutic agents, *Chem. Rev.*, 1999, **99**, 2589–2600.

Diabetes

Y. Shechter, I. Goldwaser, M. Mironchik, M. Fridkin and D. Gefel, Historic perspective and recent developments on the insulin like actions of vanadium; towards developing vanadium based drugs for diabetes, *Coord. Chem. Rev.*, 2003, **237**, 3–11.

K.H. Thompson and C. Orvig, Coordination chemistry of vanadium in metallopharmaceutical candidate compounds, *Coord. Chem. Rev.*, 2001, **219–221**, 1033–1053.

K.H. Thompson, J.H. McNeil and C. Orvig, Vanadium compounds as insulin mimics, *Chem. Rev.*, 1999, **99**, 2561–2571.

Cardiovascular system

M.J. Clarke, Ruthenium metallopharmaceuticals, *Coord. Chem. Rev.*, 2002, **232**, 69–93.

D.P. Riley, Functional mimics of superoxide dismutase enzymes as therpeutic agents, *Chem. Rev.*, 1999, **99**, 2573–2587.

Radiotherapy

W.A. Volkert and T.J. Hoffman, Therapeutic radiopharmaceuticals, *Chem. Rev.*, 1999, **99**, 2269–2292.

Boron neutron capture therapy

A.H. Soloway, W. Tjarks, B.A. Barnum, F.-G. Rong, R.F. Barth, I.M. Codogni and J.G. Wilson, The chemistry of neutron capture therapy, *Chem. Rev.*, 1999, **99**, 1515–1562.

Subject Index